Arms Makers *of* Maryland

ARMS MAKERS *of* MARYLAND

by DANIEL D. HARTZLER

Longrifle Series

GEORGE SHUMWAY PUBLISHER
YORK • PENNSYLVANIA

Library of Congress Catalog Card No. 74–24434
ISBN 0 - 87387 - 054 - 9

This First Edition published 1977

Printed by W & M PRINTING, INC., Mechanicsburg, Pa.

GEORGE SHUMWAY PUBLISHER

R.D. 7 * YORK * PENNSYLVANIA * 17402

Dedication

This work is dedicated to my wonderful family. My mother, Thelma, and my dad, Byron, are avid large-game hunters and among their many trophies are twelve bear. In my youth they encouraged my interest in collecting guns. My lovely wife, Kit, accepted the view of the good financial investment in collecting arms and her assistance in this work, as in everything, was invaluable. If it were not for my son, Jeb, and my daughter, Sandra, this work could have been completed a year sooner.

Contents

Chapter		Page
	ILLUSTRATIONS	10
	PREFACE	11
	FORWORD	13
1	INTRODUCTION – THE SETTING FOR ARMS MAKING IN EARLY MARYLAND	17
2	MILITARY FLINTLOCK SHOULDER ARMS	21
3	LONGRIFLES	37
4	FOWLING PIECES, SMOOTHBORES & SHOTGUNS	57
5	PISTOLS	61
6	ARMS FOR THE CIVIL WAR	71
7	TARGET RIFLES & SHOOTING MATCHES	75
8	SWORDS	79
9	ARTILLERY	83
10	NAVAL SHIPBUILDING	87
11	GUNPOWDER	91
12	SHOT TOWERS	93
13	ACCOUTREMENTS & UNIFORMS	95
	* * * * * * * * * * *	
	BIOGRAPHICAL DATA – ARMS MAKERS OF MARYLAND & WASHINGTON, D.C.	115
	* * * * * * * * * * *	
	APPENDIX	279
	PATENTS	292
	INDEX	302

Illustrations

Figure

1 Map of Maryland, 1780.
2 Map of Maryland, 18??.
3 Conestoga freight wagons in 1827.
4 Musket signed "I KING" of Revolutionary period.
5 Lock of musket by I. King.
6 Side-plate and trigger-guard of I. King musket.
7 Musket of Revolutionary War period with fleur-de-lis proof mark.
8 French 1763 musket surcharged "US" & branded "MARYLAND"
9 Brown Bess musket branded "9B·ANNAPOLIS·M".
10 U.S. musket Model 1795 branded "MARYLAND".
11 U.S. musket Model 1798 marked "MARYLAND" "6MM".
12 Contract for muskets by White, Crabb, Metzger, Barnhizle.
13 J. J. Henry bill to Pikesville Armory mention-
14 Musket, Model 18?? marked "M" on the barrel.
15 Bayonet with "M" stamped on blade.
16 Cavalry saber with "M" stamped on blade edge.
17 Tool inventory, Frecerick gunlock factory.
18 Flintlock fusil with lock marked "Hodgson & Thompson".
19 Hodgson & Thompson fusil, right side.
20 Hodgson & Thompson fusil, left side.
21 Flintlock signed "Hodgson & Thompson".
22 Hodgson & Thompson fusil with bell-flower trigger guard finial.
23 Philip Creamer rifle, right side.
24 Toe plate of P. Creamer rifle.
25 Philip Creamer rifle, left side.
26 Mathias Shroyer's name in account book.
27 Philip Creamer and apprentices in account book.
28 John Armstrong's deposition giving his birth date.
29 Engraved patchbox on John Armstrong rifle.
30 Right side of John Armstrong rifle.
31 Left side of John Armstrong rifle.
32 "John Armstrong 1836" written on stock of percussion rifle.
33 Rifle signed "Wickham & Matthews".
34 N. Rowe percussion rifle.
35 Rifle by Jacob Metzger of Frederick.
36 Rifle by John Gonter of Frederick.
37 Rifle by Ben Troutman, right side.
38 Rifle by Ben Troutman, left side.
39 Rifle by D. & P. Marker, left side.
40 Rifle by D. & P. Marker, right side.
41 Rifle by Daniel Marker, right side.
42 Rifle by J. Marker, right side.
43 Rifle by George Kreps, left side.
44 Rifle by George Kreps, left side.
45 Rifle by C. & J. Hawken, left side.
46 Rifle by William Hawken, right side.
47 Rifle by John Stonesifer, right side.
48 Rifle by George W. Glaze, right side.
49 Rifle by George W. Glaze, left side.
50 Rifle by Martin Rizer, left side.
51 Smoothbore gun by John Armstrong.
52 Flintlock fowler by James Haslett.
53 Flintlock double shotgun by Gafford.
54 Trimble double-barreled shotgun.
55 McComas double-barreled shotgun.
56 Sneider breechloading double shotgun.
57 Flintlock pistol by Philip Creamer.
58 Flintlock pistol by George Kreps.
59 Flintlock pistol by George Kreps.
60 Flintlock matched pistols by Haslett.
61 Detail of Haslett pistol barrel.
62 Flintlock pocket pistol marker "Baltimore".
63 Haslett pistol, right side.
64 Haslett pistol, top view.
65 Haslett pistol, bottom view.
66 Haslett pistol, left side.
67 Detail of Haslet pistol breech and tang.
68 Flintlock on Haslett pistol.
69 Case with two Haslett pistols.
70 Detail of Haslett pistol grip and lock.
71 Double-barreled pistol by A. Escherich, top view.
72 A. Escherick pistol, bottom view.
73 Percussion pistol by A. McComas.
74 Sneider 11-shot rim fire pistol.
75 Merrill Carbine with action open.
76 Merrill navy carbine.
77 W. Pratt's breech-loading alteration.
78 Target Rifle by Jacob Pipino.
79 Target rifle by John C. J. Meyer.
80 Scheutzen rifle by F. Nestle.
81 Breechloading rifle by A. Escherlich.
82 Silver-hilt sword by Thomas Sparrow.
83 Silver-hilt sword by William Ball.
84 Silver-hilt sword by John Lynch.
85 Silver-hilt sword by Silliam Ball.
86 Cannon barrels from Mt. Etna Furnace site.
87 U.S.S. Constellation launched in Baltimore, 1797.
88 T. Ruckles painting "Assembly of the Troops "
89 Linen bags for shot produced in Baltimore.
90 T. Ruckles painting "The Battle of North Point."
91 Ross Winan pike, Civil War period.
92 Knife with blade marked "Reinhardt Bal."
93 Powderhorn "J^{no} W^{m} McComas Baltimore Town."
94 Maryland military buttons.
95 Silver shoulder-belt plate, "5thReg^{t} B U V G T D".
96 Maryland accoutrement plates, Civil War period.
97 Secession cockade or Maryland blue cockade.
98 Maryland union cockade.
99 Silver cross, C^{o} A, 1stMaryland Infantry.
100 Silver cavalry badge, Cole's Maryland Cavalry.
101 Advertising label of James Haslett.
102 Flintlock rifle by Thomas Crabb.
103 Double gun by A. Escherich, Baltimore.
104 Smith rifle marked "Poultry & Trimble Baltimore".
105 Flintlock detail, musket by J. J. Behr.
106 Flintlock musket by J. J. Behr, Baltimore.
107 Signature of Ben Trautman on rifle barrel.
108 Lock detail, 6-ga. shotgun by McComas.
109 Percussion shotgun by A. McComas, Baltimore.
110 Shotgun by C. E. Sneider, top view.
111 Shotgun by C. E. Sneider, bottom view.
112 Detail of shotgun by A. T. Baxter.
113 Shotgun by A. T. Baxter, Baltimore.
114 Shotgun by A. T. Baxter, top detail.
115 Shotgun by Baltimore Arms Co.
116 Touch-marks of some Baltimore silversmiths.
117 Silver-hilt sword by Geo. W. Riggs.
118 Flintlock pistol marked "Halbach & Son."
119 Portrait of Standish Barry, silversmith.
120 Letter from Wm. McKim re gunstock contract.
121 Elmer Wolf, gunsmith, photo.
122 Nathaniel Rowe, gunsmith, photo.
123 Nathaniel Rowe, gunsmith, photo.
124 Abel Browning, gunsmith, photo.
125 Abel Browning, gunsmith, photo.
126 Map of Emmitsburg, Md., 1808
127 Patent drawing No. 6,612 B. Chambers
128 Patent drawing No. 14,077 J. H. Merrill
129 Patent drawing No. 14,319 J. J. B. Latrobe
130 Patent drawing No. 20,954 J. H. Merrill
131 Patent drawing No. 32,032 J. H. Merrill
132 Patent drawing No. 34,703 C. E. Sneider
133 Patent drawing No. 40,884 J. H. Merrill
134 Patent drawing No. 46,054 C. E. Sneider
135 Patent drawing No. 46,516 W. L. Winans
136 Patnet drawing No. 46,612 C. E. Sneider
137 Patent drawing No. 171,442 C. E. & C. W. Sneider
138 Patent drawing No. 202,126 C. W. Sneider
139 Patent drawing No. 227,135 C. W. Sneider
140 Patent drawing No. 417,594 C. E. Sneider

Preface

This work is an attempt to solidly document those arms makers who, through their products, helped to mould the history of the Free State of Maryland. It is but a starting place for those who wish to carry on the research, specializing in certain individuals and rounding out the data located so far. I wanted to prove by documentation the activities of individuals working in the arms trade.

Information about old-time gunsmiths almost always is meager. Little scraps of information are to be found in old records and documents. I have gathered together as much as I could find of this kind of data and it is presented here. Occasionally I have added comments and conclusions, and have tried hard to separate facts from assumptions. The material on some of our forefather-arms-makers is meager and there is plenty of opportunity for interested students to carry on with the research. To my successors who are inflamed with the desire to further the knowledge about these old craftsmen I leave names such as Peter Walters, P. Lapold, Jacob Ernst, M. L. Rizer, John E. Dorsey, J. Grove, Peter Reichter and Daniel Wibb, which I have located in old records but have been unable to clearly establish occupations for.

Very little research has been done on the craftsmen of early Maryland, and because of this, the state has been overlooked to a certain extent by the antiques world. Artisans of great ability and associated mechanical geniuses have flourished and gone, forgotten and overlooked by subsequent generations. During the nineteenth century Baltimore had countless hardware stores that retailed the products of the arms trade, but it was not blessed with large gun factories that employed tradesmen in quantity. Probably the most outstanding arms makers of early Maryland were the gunsmiths who made longrifles during the late eighteenth and early nineteenth centuries in the northern and western parts of the state. These men and their work deserve detailed study and it is hoped that in time an entire book can be devoted to the subject.

I express my sincere thanks to all of those persons and institutions who in any way assisted me in this work. I am particularly indebted to the dean of Marylandia, Harry D. Berry, Jr., who is particularly knowledgeable about the guns and gunsmiths of early Baltimore, and who readily contributed information where it would be helpful. His large collection of arms and related materials was unselfishly put at my disposal.

Mrs. Mary D. Dudderar, the granddaughter of Captain Edward D. Danner, was an inspiration to me, and her knowledge of history was of great assistance. The artwork for the half-title page was prepared by Alban A. Abrams. Special recognition also extends to the following persons:

Alpheaus H. Albert
Hugh Benet, Jr.
William S. Bowers
Duane Brodersen
William L. Brown
Earl J. Coates
Rev. H. Austin Cooper
Erick Davis
William G. Gavin
Edward R. Flanagan
Frederick G. Gaede
Ernest L. Hoxter
Rev. Thomas C. Jensen
Joe Kindig, Jr.
Patsy Koontz
Dr. John K. Lattimer
P. Douglas Manger
Robert H. McCauley, Jr.
Maurice Miller
Steven C. Piepmeier
George C. Neuman
Harold L. Peterson
Dr. Raymond C. Phillips, Jr.
Dick Myers
William H. Reisner, Jr.
Dr. George Shumway
Albert M. Sullivan, Jr.
Albert M. Sullivan, Sr.

Foreword

This book is the result of a long and time-consuming study by the author to document the craftsmen who made sporting arms and military arms, and the manufactories of military equipment in early Maryland. It has taken a great amount of dedicated interest and devotion on the part of the author to compile the biographical data on the individual arms makers as well as the other material which comprises the work. The finished product is an important contribution to the history of the state.

In pursuing his interest in the antique arms of Maryland the author has amassed an impressive collection of arms and equipment made within the state. His collection includes many pieces fabricated within Maryland, but the majority, although manufactured elsewhere, were carried by Maryland soldiers during different historical periods. For example, he has the French light cavalry saber and the Leech and Rigdon revolver used by the famed Confederate cavalryman and raider, Harry Gilmor. Other arms from Revolutionary-period flintlocks through percussion arms and cartridge guns are included in the collection, which thus covers the entire period of Maryland arms history.

Each of the original thirteen colonies had a military history comparable to that of Maryland. If each of these colonies was represented by a detailed study of arms making such as this, it would provide the country's arms students with an invaluable asset in the study of U. S. military history and early arms manufacture. John Bivin's study, LONGRIFLES OF NORTH CAROLINA, and Dwight B. Demeritt Jr.'s study, MAINE MADE GUNS AND THEIR MAKERS, are two such works already completed and published. Others are on the way.

The history of the Indians of Maryland, prior to the arrival of the white man, purposely is not treated here. As is generally the case, little really is known about those early inhabitants except for the meager clues left behind by their simple artifacts. This book, of course, is a study of 18th and 19th century arms making, and it is not the place for a consideration of stone-age weapons and tools.

The first white men who came to Maryland possessed typical European military backgrounds and brought with them the standard military equipment of the period. In some instances they wore armor, and often were equipped with swords. Matchlock muskets, pikes and pole-arms undoubtedly were in use, as they were in the other colonies during this early period. Eventually these early weapons were replaced with the individual sword and flintlock musket produced in the colony of Maryland. Percussion arms came into widespread use after 1830. In this work the author treats both firearms and edged weapons, including shotguns, target rifles, and handguns of the era.

The author has made a particular effort to tell the story of the development of the weapons manufacturers during the period of the Revolutionary War. At the outbreak of this conflict, Maryland, like the twelve other colonies, found herself completely ill-equipped and unprepared to provide arms for carrying on a war. The author has done an excellent job of reading through Maryland records. The result is a fascinating and well-documented recital of the manner in which Maryland succeeded in establishing an arms industry. The data presented on the individual contracts with various specific arms makers have never been published previously and will be refreshingly new information for those with an interest in arms of the period. It is particularly inspiring to note the patriotic efforts which were made by Revolutionary Marylanders to achieve the objectives of providing adequate arms and equipment for the Maryland soldiers of the conflict. The Maryland Revolutionary story probably is typical of the other colonies, and we can only hope that arms making in this critical period will eventually be thoroughly researched for all of the other original colonies, and that the documentation will find its way into print.

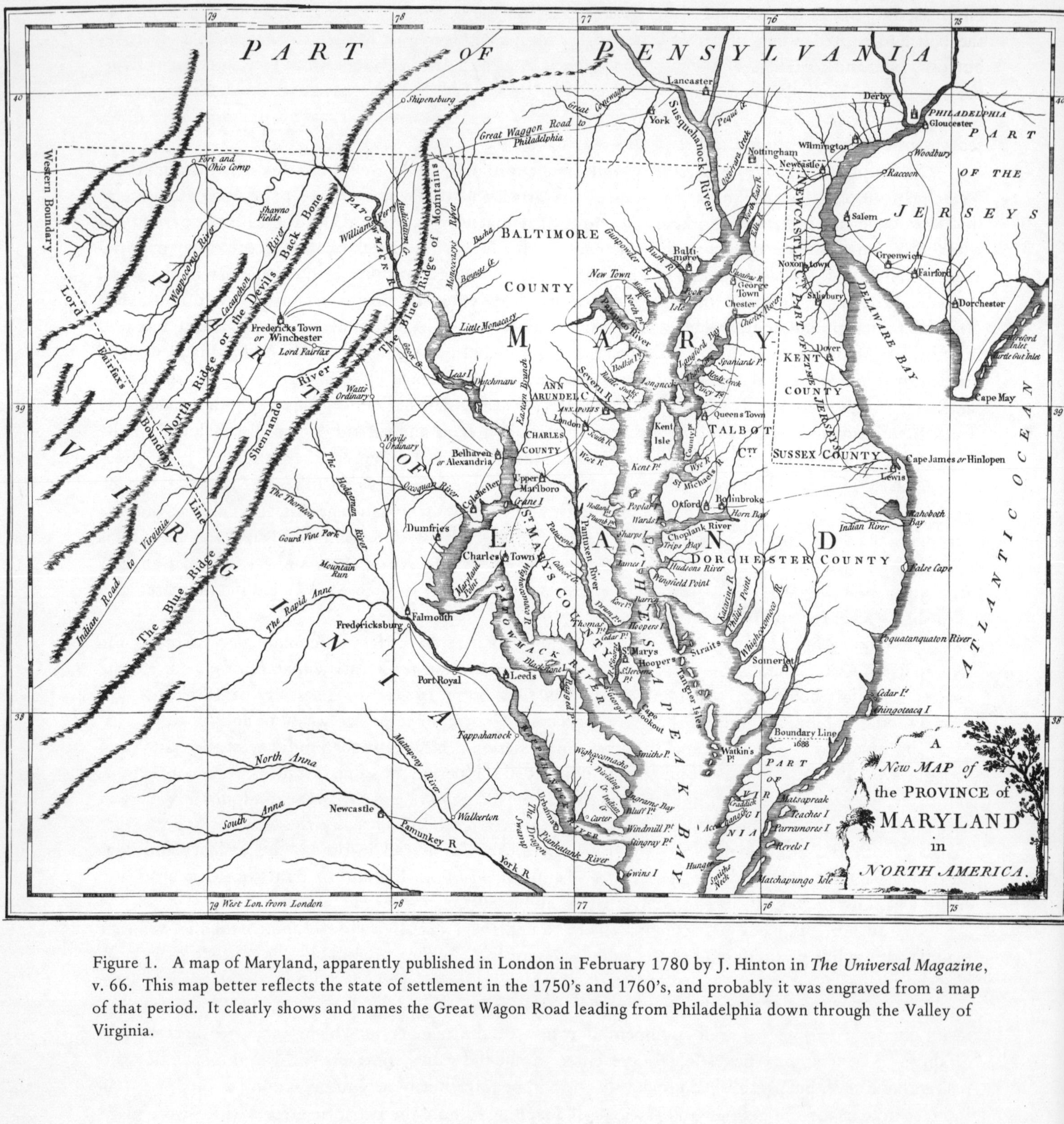

Figure 1. A map of Maryland, apparently published in London in February 1780 by J. Hinton in *The Universal Magazine*, v. 66. This map better reflects the state of settlement in the 1750's and 1760's, and probably it was engraved from a map of that period. It clearly shows and names the Great Wagon Road leading from Philadelphia down through the Valley of Virginia.

The author has thoughtfully included material on the uniforms, accoutrements, and insignia of the Maryland soldier. Too often this important category of military collecting is poorly covered or eliminated in works of this type. Data is presented on the buttons of the Maryland troops as well as the belt and cartridge box plates of state origin. Maryland troops are known to have worn distinctive buttons, belt and cartridge box plates during the Civil War. Although somewhat rare, specimens have been found on battle sites. It is important to note that troops of both Union and Confederate armies wore Maryland buttons and plates. Men of both sides were proud of their state origins and used state seal insignia whenever they could obtain it.

The historical roster of Maryland gunsmiths has been compiled over a number of years and after countless hours of research. The result is one of the best documented and complete listings for any state that is available to collectors. Essentially all data that could be found for each of the old-time gunsmiths is included here, including documented dates for work as a gunsmith or related tradesman, locations and addresses, etc. This list of arms makers has brought to light some new things. Instead of the dozen or so longrifle makers often thought of as being Maryland's share, the author's work shows more than 70. The Military History Department of the U. S. Army, the Naval Historical Foundation, and the Smithsonian Institution have found no records disclosing the manufacture of cannon within the state, but this publication reveals twelve different manufacturers and patentees. Most of the information presented in the list of arms makers has never before been published, having remained untouched in Maryland archives, trade directories, newspapers and other Maryland files.

William G. Gavin
Charles Town, West Virginia

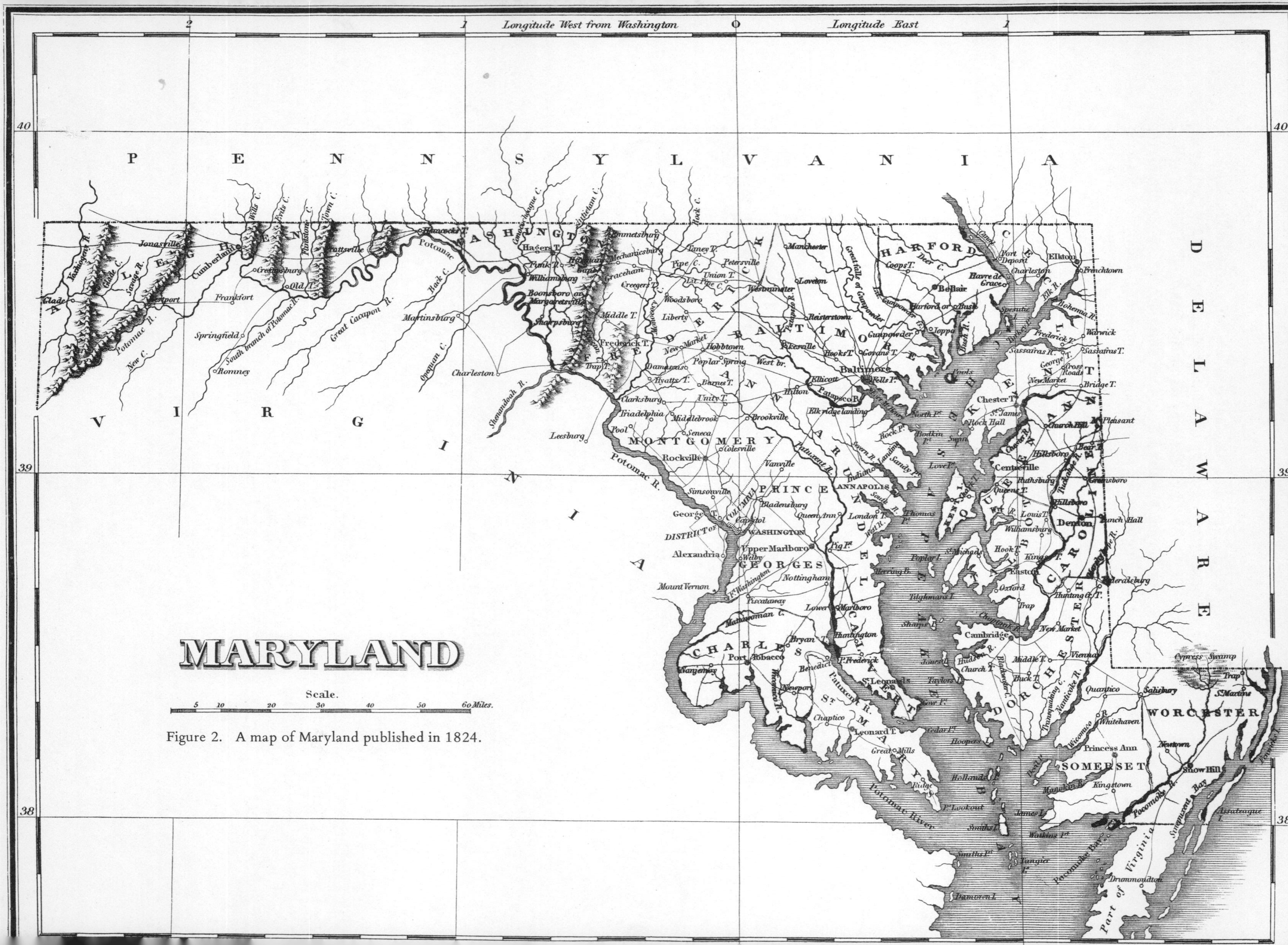

Figure 2. A map of Maryland published in 1824.

CHAPTER 1

Introduction
The Setting for Arms Making in Early Maryland

The manufacture of firearms in Maryland had its start a little over 200 years ago, in the decades preceding the American Revolution. The beginnings were small and our present knowledge about it is limited. In the 1760's and first half of the 1770's there were a few gunsmiths working in the sparcely-settled piedmont of western Maryland, turning out slender flintlock longrifles and smoothbore guns for settlers of the region and for others headed to the frontiers farther north and west. In Baltimore and perhaps some other towns of the Chesapeake Bay region there were a few other gunsmiths at work. When the Revolutionary War burst upon the scene in 1775 these gunsmiths suddenly became of considerable importance to the province.

Following the actions at Lexington and Concord, Maryland's Council of Safety was established to take responsibility for procuring arms for the province. During the long colonial period Maryland's needs for arms generally were provided by England, but when she suddenly had to provide muskets and other arms for her own troops, she was hard-pressed because adequate manufacturing facilities did not exist. The Council of Safety asked that a survey be made to list the places where muskets might be made. In August of 1776 it was reported that there were but 12 individual gunsmith shops operating in Maryland. Of these, seven were in the western piedmont and five in the lower lands. But these gunsmiths, with others trained to help them, provided the muskets needed by the Maryland troops.

During the decades of settlement and westward expansion that followed the Revolution there was a growing need for hunting and sporting arms, and a consequent growth in the number of manufacturing gunsmiths as well as in the number of gun dealers who imported arms from Europe. Throughout the nineteenth century Maryland was active in the sporting arms field both in manufacturing and in importing of arms from abroad. With the Industrial Revolution well advanced by mid-century, Maryland made arms and accoutrements for the Civil War.

Maryland is a two-part state, having its main eastern portion as coastal lowlands surrounding Chesapeake Bay, and a smaller western portion consisting of piedmont and mountains. In many respects these two portions of the state are separate units which grew up side-by-side during the eighteenth and nineteenth centuries, and in studying the early firearms of Maryland it is desirable to keep this separation in mind.

Eastern Maryland is an economic, cultural, and social unit centered about the Chesapeake Bay and the cities of Baltimore and Annapolis. During the colonial period this region was rather closely tied to England, and of course most of her people had English origins. Throughout the eighteenth century and the first two or three decades of the nineteenth century the firearms in use in this region generally were of English origin or were Baltimore-made in keeping with the latest English styles.

Western Maryland stands out in marked contrast. Her peoples were a mixture of English, Scotch-Irish, and Germans. The land, consisting of piedmont areas, broad valleys, and mountain ridges, is squeezed between similar lands of Pennsylvania to the north and Virginia and West Virginia to the south. During the eighteenth and early nineteenth centuries three important roads ran across western Maryland carrying the commerce of the times and providing avenues for immigration. In addition, the Potomac River is present along the southern boundary, and this provided an additional avenue for commerce and communication.

The first of the three important routes through western Maryland was the Monocacy Road which ran north from the Potomac River through the valley of Monocacy Creek to Frederick and then on to

Figure 3. Conestoga freight wagons and other traffic on the road north of Baltimore. This scene, painted in 1827 by Thomas C. Ruckles, is titled "The Fairview Inn." No other known painting so well depicts life on an important avenue of commerce in this period. The Fairview Inn was a "wagon stand," that is, an inn which catered to wagoners and which contained a large yard adjacent to it in which wagons and teams could stand overnight. Courtesy of The Maryland Historical Society, Baltimore.

the north and east to make connections to York, Pennsylvania, which in turn was tied to Lancaster and Philadelphia. This road was of special importance during the first three quarters of the eighteenth century, and Frederick became the principal town of western Maryland.

The second important road ran west-northwestward from Baltimore through Frederick and on to Hagerstown and Cumberland. This route anchored the western part of the state to the port and cultural center of Baltimore, but in addition, and far more important commercially, it gave Baltimore access to the trade of the Ohio Valley when it opened up early in the nineteenth century. Braddock's men had first cut the route from Cumberland westward in 1755, headed for Pittsburgh. Braddock's Road, as it came to be known, fell into disuse shortly after it was made. But when trade with the Ohio Valley became of ever-increasing importance in the decades after the Revolutionary War, the idea developed of building a National Road to connect Baltimore and the Potomac Valley with the Ohio Valley at Wheeling. Surveys began in 1806 and construction in 1811. The road was completed to Wheeling in 1818 by way of Uniontown, Brownsville, and Washington, Pennsylvania. A great trade was then carried on between Baltimore and the Ohio Valley by means of Conestoga wagons. The more important towns along this road were places where gunsmiths could carry on their business profitably, and thus Frederick, Hagerstown, Cumberland, and Uniontown were places where gunsmiths continued the manufacture of longrifles at least until mid-century.

The third great road that passed through western Maryland was the most important of all. This was the Great Wagon Road that tied together the colonies of Pennsylvania, Maryland, Virginia, and North Carolina during the late colonial period, and down which emigrants went by tens of thousands to reach the western promised lands of Kentucky. This road ran through the Valley of Virginia or Shenandoah Valley and was in active use as a wagon road in the 1750's. Before the white man took it over it had been an important Indian trail and was known as the "Indian Road to Virginia." But it soon came to be known as "The Great Wagon Road." At the southern end the road branched in a number of places, leading to the Wilderness Road that crossed Cumberland Gap into Kentucky, and to the settlements of the North Carolina piedmont such as Salem. Extensions of it continued along the piedmont to Atlanta. At the northern end the road ran from Winchester, Virginia, north to Williams Ferry on the Potomac, on to Hagerstown, and then over the Blue Mountain ridge to York, Pennsylvania, and on to Lancaster and Philadelphia. The Great Wagon Road was particularly important in the decades following the close of the Revolution when settlers streamed down it, headed for the lands of the Carolinas, Kentucky, and other places west.

At the time of the Revolution, and the decades preceding it, the Great Wagon Road helped considerably to give a feeling of unity to the people of the piedmont. Colonial people of the seaboard towns felt some of this unity through the abundant coastal shipping that went on, but in the back country it was the Great Wagon Road that stitched things together. Thus when the call to arms came in the summer of 1775, Daniel Morgan and his riflemen from the Valley of Virginia promptly marched up the road, as did others from Maryland and Pennsylvania.

Towns situated along the Great Wagon Road were centers for the production of longrifles. In Maryland both Hagerstown and Frederick were along the road, though they lie on opposite sides of the mountain ridge. From York south to Winchester there were various routes possible, one by way of Gettysburg and over the ridge to Hagerstown, another through Hanover and Emmitsburg and on down the Monocacy Road to Frederick. Small town though it always was and still is, Emmitsburg was the location of some outstanding longrifle making during the first half of the nineteenth century. To the east of Emmitsburg and somewhat off the beaten track is Taneytown where longrifle making as well as manufacture of military arms for the federal government was carried out.

The purpose of this brief introduction is to point out that there are reasons why the old-time gunsmiths of Maryland were located in certain places. As with most businesses, they were situated where there was a ready market for the particular products they made or dealt with. Longrifles were made in western Maryland because they were used by the people of the region for hunting and sporting purposes, and also because important trade routes passed through the region. In contrast, flintlock arms manufactured in Baltimore often were styled after English arms of the period, in keeping with the demands of the times.

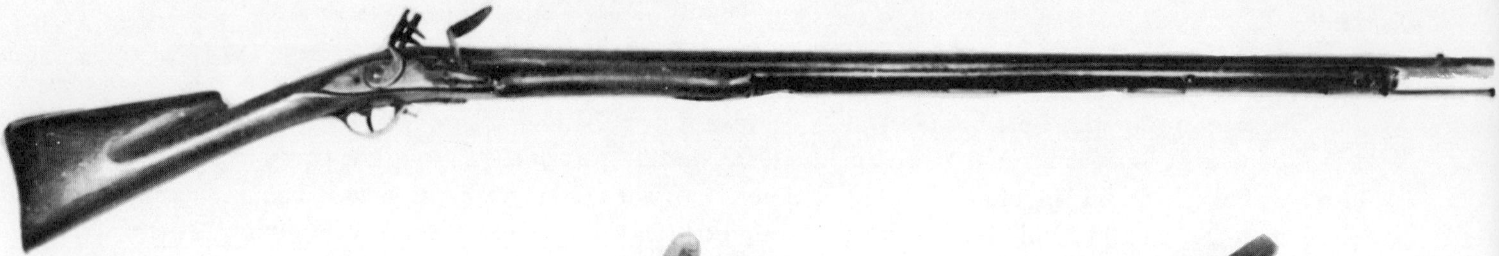

Figure 4 (top), Figure 5 (middle), Figure 6 (bottom). Eighteenth-century musket of the Revolutionary War period with barrel signed "I KING". Possibly this was made by a Maryland gunsmith, John King. The brass furniture is of English style patterned after furniture of Brown Bess muskets. The forestock holds four long ramrod pipes and a short nose cap. The number "33" is stamped on the walnut stock at the rear of the side-plate. Overall length of the piece is 57½ inches and the barrel is 41½ inches long. From the collection at Valley Forge Museum.

CHAPTER 2

Military Flintlock Shoulder Arms

Those who governed colonial Maryland and other colonies in America deemed it necessary to have shoulder arms available for the defense of the colony. In Maryland such arms were purchased by the colony, but there never were enough muskets to meet the needs, and Mother England constantly was petitioned for more. England encouraged her colonies to be self-sufficient, but not to the point of manufacturing her own arms. England's arms industry was well developed during the seventeenth and eighteenth centuries, due in part to the long-continuing needs of the fur trade in America, and also to needs of the colonies. Thus it was that during the long colonial period of about one and one-half centuries, Maryland relied upon England for her needs in military arms and in other arms as well. It was not until the French and Indian War and the build-up to the Revolutionary War that followed it that there was any real need for an arms industry to develop within Maryland.

An example of an early record concerned with the acquisition of arms, taken from the proceedings of Maryland's Council for 12 October 1681, is this appeal to Lord Baltimore III:

> We cannot but take notice thereof unto your Lordship praying and requiring you to cause the same if true to be speedily redressed, and that in the distribution of the Armes and Ammunition (which at the request of your Agent Nicholas Lowe Merchant, Wee have permitted to be transported for the defense of that Place) your Lordship do express your trust and confidence in His Majestie's Protestant Subjects by putting the said Armes into their hands.[1]

If the young colony of Maryland was to have arms in quantity then it was necessary to have gunsmiths present to keep them in repair. In 1631, when Bennett and Packson established a settlement on Kent Island, they personally supplied what arms were needed by the settlement, and may have been the first to set up an arms repair shop. Alexander Toulson, a gunsmith, was working at St. Mary's in 1663, presumably doing repair work. Maryland's Council proceedings for 27 October 1682 mention "An account exhibited by John Martin of Charles County Smith for scouring cleaning and fixing Arms amounting unto four thousand one hundred and eighty pounds of tobacco resd and thus andorsed Vist."[2]

Another early account of repair work being done to the public arms is the following, from the proceedings of the Council held at the "city of St. Maries at St. Johns, in 1688."

> Ordered that what publick Armes shall be or are Conveyed to Mattapany be put to William Haimes Gun Maker at Harvey Towne to be fixed in the way of his trade till those be done amended and finished And that afterwards the same be sent to the severall Counties againe according to the direccons of the proclamacon in that case issued And that authority be granted to presse the said Haimes to the service aforesaid.
>
> Ordered that the Armes at Coll Pyes be sent to St. Maries to Coll Digges to be fixed as afresaid And that the sheriff forewith receive them and send them downe accordingly and that his honr Coll Digges have the power to presse a smith for the fixing the said Armes.[3]

In Maryland's early colonial period there were squabbles among the colonists as well as Indian wars – Indians against Indians and Indians against Whites – but Maryland continued to attract people and the colony grew. The Puritans tried to do away with religious freedom, and after years of attempting to secure political control were successful, but religious freedom later was re-established. In 1694 the capital was changed from St. Mary's to Annapolis which formerly had been called Providence and Anne Arundel Town. Baltimore Town was founded in 1729 and soon became a very important port.

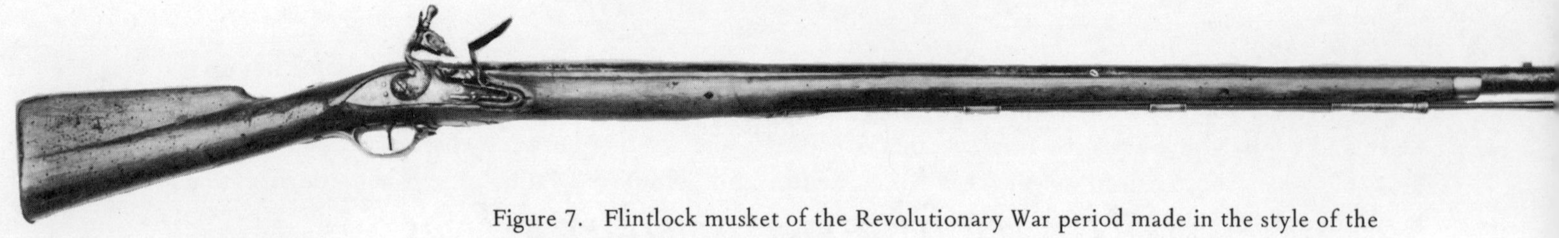

Figure 7. Flintlock musket of the Revolutionary War period made in the style of the English Brown Bess musket. The stock is of walnut. The barrel is 42 inches long and of 0.75 caliber. On the left side of the barrel 1 1/16 inches from the breech end is a fleur-de-lis type of proof stamp. A much enlarged version of this mark is shown at the left. This mark is believed to be the proof mark used during the Revolutionary War period by Captain Thomas Ewing and Major Gist for musket barrels made for Maryland troops. A number of muskets are known which bear this stamp, some with Maryland marks. Harold L. Peterson collection.

In the early 1750's jealousy developed between England and France over control of the Ohio Valley, a region that included western Maryland. French trading interests acting through their military support incited the Shawnee Indians against the frontiers of the English colonies, particularly in Pennsylvania, Maryland, and Virginia. The French and Indian War, which began with Washington's loss of Fort Necessity and Braddock's defeat while on his way to take Fort Pitt, brought numerous raids to western Maryland and adjacent Virginia. Fort Cumberland was built in 1754, but it was too far west to protect the whole frontier. Fort Frederick, built in 1756, provided additional protection.

The frontier forts of Pennsylvania and Maryland at this period almost always were equipped with one or more swivel guns. One such piece is known which is reported to have survived from Fort Cumberland. It has a heavy brass barrel, a massive stock, and iron flintlock, and a unique trigger to which is attached a large ring that extends outside of the trigger guard.

On 8 February 1756 Governor Horatio Sharpe wrote to the Lords of Trace, explaining the needed arms for the Province:

> I am to acquaint your ldps that there are in & belonging to this Province 19 Carriage Guns all Iron but very good 4 of them are 6 pounders and 15 four pounders that last not mounted. In the Magazine is 16 lb weight of 6 lb shot & 24 half Barrels of Powder; of Musquet Ball & Bar Lead we have a considerable quantity & there were last Spring about 500 stand of small Arms belonging to the Publick most of which are in the hands of the men that have been raised for the protection of the Frontiers & some were lost at the Action of the Monongahela. The Militia of the Colony are near 1,650, One third of whom at least are entirely destitute of Arms & many of the Guns that are the property of the Rest are very bad & scarcely fit for use. For want of a proper Militia Law (Which the assembly has been frequently in vain sollicited to make) the people are undisciplined as well as badly armed & cannot be compelled to serve in Defence of the Country. The white Inhabitants in Maryland are 107,963, the Black 46,225.[4]

On 9 June 1758 an order was issued for arms to be sent to the Maryland militia who were in the Virginia Service:

> . . . dispatched an Express to Annapolis and ordered the Person who had the Care of our Arms to pack them all up that they might be ready to be sent on the first notice to any Place that you should think proper to order them, a few Days afterwards most of them were brought to Frederick Town agreeable to Sir John St. Clairs Desire and 500 new Muskets, upwards of 50 old but the best Carbines that I could collect in the Province and 33 Pair of Pistols have been since sent from that place to Winchester to arm the Virginia Forces. I have likewise ordered 50 broad swords Scabbards and Belts 40 Pair of Holsters and as many Blanketts and Shoulder Belts with Springs as can be got in Annapolis to be sent thence to Winchester.[5]

During this war both Eastern and Western Shores of Maryland were much annoyed by French privateers, who made frequent landings, plundering the defenseless inhabitants. The Province at this time had no ships of war to defend her shores. Peace came with the Treaty of Paris in 1763, and these raids as well as the Indian troubles on the frontier came to an end.

During the eighteenth century Maryland's ability to produce supplies and materials for her own use as well as for export continued to grow, and economic independence slowly became a reality. Maryland's long squabble with Pennsylvania over the boundary line between them was finally settled in 1767 when the English surveyors, Charles Mason and Jeremiah Dixon, established the line that bears their name today.

In order to maintain a tight control over the growing American colonies, England passed and imposed on the colonies the Navigation Acts, the Trade Acts, the Stamp Act, and acts regulating manufacturing. This helped greatly to unite the colonies to thoughts and actions for gaining independence.

In 1774, when other colonists were only muttering about revolt, John Hanson chaired a meeting at which Frederick became the first colonial county to vote a blockade on all trade with England. In the following summer he played an important role in forming the Association of Freemen in Maryland, the first colony-wide group approving the use of force if necessary to repel British troops. When war broke out he was active in mobilizing men and supplies in Frederick County. In 1779 he was elected to Congress and was so popular that he became President of the United States of Congress Assembled, actually the first President of this new country.

The First Continental Congress met in September of 1775, helping to give sanction to the movement for independence then underway, and specifically advising the people to arm and to form their own militias. After blood was shed at Concord and Lexington, Maryland established her Council of Safety, in keeping with similar councils or committees of safety being formed in the other colonies. Each province was responsible for supplying its men with the arms of war.

A survey of Maryland's potential for producing muskets was given to the Council of Safety on 2 August 1776:

> There are three Gun Smith in Baltimore Town, one in George Town, four in Fredericktown, one near Fredericktown, two in Hagers Town, one in Jerusalem Town, in the whole amount to twelve. We are informed that each of these shops can in one month, complete twenty musquets.[6]

This succinct account shows that there was a surprisingly small number of gunsmiths in Maryland at the time. It also shows that eight of the twelve gunshops were located in the vicinity of Hagerstown and Frederick. Jerusalem Town was an early name for Funkstown, which is situated a few miles south of Hagerstown. These gunsmiths operated individually, from separate gunsmith shops, in general, assisted by apprentices when they were available. They were not set up for arms production on any large scale, and the sudden needs of the Revolutionary movement called for additional manufacturing facilities.

Charles Beatty, James Johnson, and John Hanson, Jr., were appointed to establish a gun lock factory at Frederick Town on 1 January 1776. The need for gun locks was especially critical in Maryland as in other colonies. The manufacture of locks was a specialty that had been carried on in England and continental Europe for a century or more, and at the time the Revolution erupted virtually all flintlocks used in America were supplied from across the sea. Labor costs in America were relatively high both in colonial times and in the post-Revolutionary decades and it was cheaper to manufacture locks in Europe. Furthermore, flintlocks require steel for springs and frizzen, and there was no steel produced in America until the 1830's.

John Yost corresponded with the Council of Safety 13 September 1776 concerning musket locks:

> I was told by the manager of the Gun Lock Manufactory at Frederick Town, that they forge Gun Locks much faster than they can finish them off; as that is the Case, I should be glad to furnish myself from thence with 300 ready forged Locks, provided the Terms are admissable. If this Proposal is agreeable, I hope I may shortly be furnished with proper Authority through you, or some other Department, to receive that Quantity, as it will greatly further my work, I have nine Hands employed at present and have engaged two more, who I expect will be at work for me in less than a fortnight.[7]

With the factory at Frederick supplying locks, the gunsmiths had one less item to produce and could complete muskets in shorter time. A great number of locks were supplied, but the Council continued to pressure the factories and gunsmiths for greater production. Unfortunately, the Council was consistently behind in its payments, as were other Committees of Safety in other states. On 10 December 1776 the Council received the following letter from Charles Beatty:

The Barer hereof Mr. Boon [overseer of the Gun Lock factory] waits on you in order to get some money to enable him to carry on the factory, for it cannot be supported any longer without. Mr. Hanson who keeps the books is not at home so that I cannot get to see them in order to draw for the money, the factory is entitled to, but I am clearly satisfied that it must be about 400 Dollars by the acct rendered, and the locks now in the factory 156 locks completely finished as will appear by the Inclosed certificate for nearly two months past I was compelled to take money on Interest to support the factory or let the hands go idle. It is not in my power to borrow any more for that purpose here – pray dont fail sending the money or I shall be compelled to sell locks to pay the money that I have taken up and let the servants shift for themselves, which will be a great loss to the publick.[8]

Men such as Henry Hollingsworth and John Yost erected horse mills for boring musket barrels, to speed production. Yost communicated with the Council of Safety on 13 September 1776:

I have to inform you that I have erected a Horse Mill for boring Gun Barrels; that I am now employed with all my Workmen I have in making Locks, screws, Mounting and forging Barrels ready for boring, but cannot proceed to that part of the Work before I receive the Materials which I purchased at Philadelphia sometime ago.[9]

Blacksmiths were persuaded to learn the gunsmith trade because of the great demand for arms. An added impetus may have been the fact that gunsmiths generally were exempted from the military service. On 23 May 1776 the Council was informed as follows:

Since my return home I have agreed with Robert Reed of this place blacksmith for 10 musketts for acct of the Province at 85/ each they are all finished and only wait your Determination to confirm my contract before they are delivered.[10]

Stephen West was a man of means who hired a number of hands to produce arms. He felt that the products of his factory were superior to most other muskets, and in the following letter addressed to Hall dated 4 May 1776 he supplied some valuable information concerning his operation and the costs of production:

I received your esteemed favor relative to my Muskett which I apprehend suffered from the rain, it is by no means equal to many that I have and shall have, yet I believe it is much better than what you generally get. I am warranted to say so by the People Officers and Soldiers from your City. I sent you one much better finished than that you have got and I desire that it may be compared with those you get. I defy all your People to produce such a Bayonet, Rammer, or a Barrel so well turned off, or such a Barrel of their make, like, Plain well poized and Proof, yet these Barrells are by no means so good as the rest that I have which are larger and better, if my opinion was worthy of notice I would deliver the lightest Arms to the Officers the largest and heaviest to the Privates for very obvious reasons but you cannot expect good Arms or the Art to be carried to perfection unless you give liberal Encouragement. I have many that will be worth 5 Guinnies at least, and those I send you really stand me 5..7..6 without a great Profit or allowance for my own time, Expense and Collecting Material. Your Rammers and Bayonets of your Countryside Arms I am told are rough Trash, and the Rammers fly out in Exercise, mine will not, they are kept by a fine spring in the lower pipe, people who are good Judges declare that my Arms on an Average are well worth fitted and finished as I do them 4 Guinnies, my own Carbine from London, not equal any way to the one I now send you-cost me 4 Guinnes out of John Baziers shop. We cannot make them so cheap as they do, the price I have asked to indemnifye me is 6 of our Paper not equal to 3½ guineas, here, and I affirm in the face of the sun they are well worth it; I prove all the locks, and if any way defective finish them with the best steel, many Guns that I have seen are fitt only to beat Homminy with, not for men to trust their lives to. pray have you summed up the price your Musketts will come to, from piece work from your favorite Harriss try them fairly and see what they will amount to and then when they are really finished as well as mine which I do not believe all your workmen together can do, please to sum up the amount.

first the barrell, to Harriss	1..15..
Loops for the Sling Swivels	..2..
A screw and nutt for the Butt	..1..
A trigger and guard	..1..
Fixing the loops and sight to the barrel, or if the price of 25/	..3..
A sett of brass mountings, if the cost only 7/6 might yet before it is put on properly filed & polished it will cost at least 15/	..15..
A Stock	..15..
A Lock	..15..

A Rammer	..5..
Screws and pins for the Butts, Guards & loops at least	..5..
A Spring for the Rammer	..1..
Polishing the Barrell	..5..
2 Sling Swivels and screws	..5..
A bayonet Scabbard hook & fixing suppose ..4	..6..
A good well finished Bayonet such as mine 16 or 17 inches in the blade, cannot be made compleat for less than	..12..6
To a workman for putting all the parts together fixing the Lock Brasses, loops swivels, &c in which a Joiner and Smith must both be employed, fitting in the lock to the stock &c will cost	..5..
	6..10..
A woollen case for the Gun, thread & making	..2..6
	6..12..6

This estimate my good friend you will find if the work be well done and the workmen paid will be pretty near the value of your Barrells of Mr. Harriss when delivered to you fit for use, there will not be much variation. there may possible be some saving from your cheap purchase of some of the articles or of their inferior Quality, or the business being ill done, or what is very often the case the undertaker getting nothing by the Job. Harriss asserted to the Committee that arms & accoutrements could be made for 3 or 3..10, but I presently demonstrated to him the impossibility & he would not undertake for 4.. or 4..10, the Pennsylvanians who have many advantages get from Virg 5..6..3 Wooden rammers, and their Guns are not equal any way to those of mine. I hope you will allow there is a great difference in the value of Arms. I think I can furnish from 70 to 100 of all sorts, they shall be good, and as I went to work with a sincere spirit to serve my Country I think I should have the value of my Labour. if you will appoint any Man of Candour to deal with me I shall treat him as a friend. I want Brasses & Locks. I wish you could spare me some out of your stock, it would expedite matters.[11]

Major Gist and Captain Ewing were appointed by the Council of Safety on 31 January 1776 to prove the muskets made by Maryland's gunsmiths. Ewing stamped the proven muskets with a proofing stamp believed to resemble a fleur-de-lis (Fig. 7). A few surviving muskets of the period are known which have a mark impressed onto their barrels. Definite proof that Ewing's stamp had this shape remains to be found, however. On 12 February 1776 Ewing wrote to the Council:[12]

Hon. Sirs. I now acquaint you that agreeable to your orders directed to Major Gist and myself, I have prov'd all the guns made by the Gunsmiths in Town, as below, in company with Mr Vanbibber, who has seen them several times try'd in England. Mr Keener after seeing the rest try'd refus'd to have his provd but upon my threatening him a good deal he comply'd. We charg'd all the guns with one ounce of powder and two Balls. I have a proof stamp with which I stamp'd all that prov'd good, which stamp I will not deliver to any without your orders.

Thos. Ewing

1776				
Feb 7	Peter Lydick	72 guns	64 good	8 bad
Feb 7	Sam Missersmith	27 guns	23 "	4 "
Feb 7	Sam Keener	32	13 "	19 "

Many muskets were not up to Council standards and could not be accepted. On 3 May 1776 the Council wrote to Henry Yost and John Unseld that they received 29 muskets very roughly made with pickers and wipers and one bullet mould, but no bayonets. After further explanation of the arms the following order was given on 14 June 1776:

Ordered that the Commissary of Stores deliver out of the Magazine at Annapolis to Henry Yost and John Unseld all the Musquets by them respectively made which were ordered by the late Convention returned to said Yost and Unseld as unfit for service.[13]

Men skilled in working with wood such as cabinet makers, joiners and carpenters worked along with gunsmiths in stocking the muskets. "Messrs John Shaw and Archibald Chisholm agreed with the Council of Safety to stock the Gun Barrels made by Mr Isaac Harris for the use of this Province at the price the same are done in Philadelphia."[14]

William Whetcroft, a goldsmith, entered into a contract with the Council of Safety to supply arms:

Figure 8 (top). French musket, St. Etienne Model of 1763, surcharged "US" on the lock and barrel, and with the word "MARYLAND" carved or branded on the forestock. Daniel D. Hartzler collection.

Figure 9 (second from top). English Brown Bess musket, 3rd model, stamped or branded "9B·ANNAPOLIS·M" on the buttstock, which probably stands for 9th Batallion Annapolis Militia. Daniel D. Hartzler collection.

Figure 10 (third from top). U. S. musket, Model 1795, dated 1800 and branded "MARYLAND" on the stock in front of the trigger guard. Daniel D. Hartzler collection.

Figure 11 (bottom). U. S. musket, Model 1798, made by Nippes & Company in Philadelphia and dated 1808. The stock holds the markings "Maryland", "6MM", and "T.P." and was carried by Thomas Pilkerton of the 6th Maryland Militia during the War of 1812. Daniel D. Hartzler collection.

> The Council promised . . . Mr. William Whetcroft that [in case he will next spring import] a sufficient number of Workmen, and [will deliver for the] use of this Province for every week for two [years] then next following, fifty good substantial proved Musquets 3½ feet long in the Barrell and of three Quarters of an Inch in the Bore, with good double Bridle Locks, black Walnut or maple Stocks, and plain strong Brass mounting Bayonets with Steel Blades 17 Inches long, steel Ramrods, double Screws, priming Wires and Brushes fitted thereto, with a pair of brass Moulds for every 80 Musquets, to cast 12 Bullets on one Side, and on the other Side to cast Shot of such Size as the Musquet will chamber three of them, wholly made within this Province, the Public will, provided the Differences continue between Great Britain and the Colonies, take the said 50 Stands per Week at the Rate of four pounds common money per stand . . . It is understood that the Importation of the said workmen shall be at the Expence & Risque of Mr. Whetcroft.[15]

Throughout most of the Revolutionary War period gunsmiths were encouraged by the Council to make arms. The Council also tried to purchase arms from other Provinces, but rarely could they be obtained for the demand generally exceeded the supply. This demand forced the Council of Safety to lower its standards, by accepting iron as well as brass mountings, and single-bridled as well as double-bridled locks. In 1777 General Washington ordered all muskets that were not privately owned to be surcharged with the mark "U S." At the close of the War the army was equipped with a heterogeneous assortment of arms – Council of Safety arms, French muskets, captured British Brown Bess muskets, other imported arms, and a variety of other arms once privately owned, some of which doubtless dated to the early part of the eighteenth century.

During the Revolution, the Maryland troops who used these arms became some of General Washington's favorite troops. Twice the American army was saved by the Maryland line. At the Battle of Long Island 400 Marylanders courageously charged thousands of British soldiers six times while the army retreated to a safer position. Over half of them were killed and only 20 escaped uninjured. At Monmouth, Washington found the whole army in danger and said to the Marylanders, "If you can stop the British for ten minutes I can save the army."[16] They were held for thrice that time.

Not all Marylanders were in favor of independence, however. The Tories or Loyalists in Somerset and Worcester Counties rose in arms against the Rebels in 1777, but the action was quickly put down.

Young America spent a few years catching her breath after the British surrendered at Yorktown, and then efforts began to establish and equip an army and navy. Likewise, militia units were established or in some cases, re-established, having been in existence before the Revolution. In 1794 the Whiskey Rebellion developed and President Washington called up militia units as well as federal troops to meet the threat. Maryland's militia was called to arms, and volunteers came out in numbers so that 10 companies were raised.

On 5 July 1798 Congress passed an act providing for the procurement of 30,000 stands of arms. The Charleville Pattern French musket Model of 1763 was chosen as a standard, and ultimately 40,200 stands were contracted for. On 10 September 1798 four Maryland gunsmiths, Nicholas White, Thomas Crabb, Jacob Metzger, and Christoph Barnhizle jointly contracted to produce 1000 stands of these muskets in Frederick Town (Fig. 12). By December 1801 they had produced 548 stands and the contract was canceled. Three other Maryland gunsmiths took contracts for these muskets: Eli Williams in Williamsport for 2000, Alexander Clagett in Hagerstown for 1000, and Mathias Shroyer in Taneytown for 1000.

Mathias Schroyer, a gunsmith of Taneytown, contracted for 1000 stands. Scharf's HISTORY OF WESTERN MARYLAND tells about this:

> In the latter part of the eighteenth century there stood on or near what is now the lime-kiln in Taneytown a long, low frame building, in which were manufactured, by a Mr. Sroyer, such implements as fireshovels, tongs, hoes, nails, and guns. The venerable Mrs. Elizabeth Thompson has in her possession a heavy pair of tongs made at this primitive factory, on which is inscribed the date of manufacture, 1796. The establishment was under the supervision of the government, or at least that portion which embraced the manufacture of firearms, and was annually visited and inspected by government officials. The machinery was very crude and simple. Instead of the belts, pulleys, emery-wheels, and ordinary appurtenances of a modern factory, regulated by steam, and by means of which a gun-barrel or other iron implements can be polished in a few moments, the only contrivance then known and used was a huge grindstone turned by an old horse. With these limited facilities, however, many guns were made for the government. The factory burned down

Articles of Agreement made on

the *Tenth* .. .. .. Day of *September* .. .. --One Thouſand Seven Hundred and Ninety-Eight, between OLIVER WOLCOTT, Secretary of the Treaſury of the United States, of the one part, and *Nicholas White, Thomas Crabb, Jacob Metzger and Christian Barnhizle* of *Frederick Town* .. .. .. in the State of *Maryland* .. .. .. of the other part.

FIRST. THE ſaid *Parties of the Second Part* — contracts and engages to manufacture within the United States, and deliver to ſuch perſon or perſons as ſhall be appointed by the Secretary of the Treaſury, or the Secretary of War, for the time being *One Thouſand* .. .. ſtands of Arms, or Muſkets, with Bayonets and Ramrods complete and fit for ſervice—One third of the whole number to be delivered within eight months—One other third within fourteen months, and the remaining third with eighteen months from the date of theſe preſents.

SECOND. THE ſaid Arms ſhall be delivered at *Frederick Town* —— in the State of *Maryland* .. .. and ſhall be made after the Charleville model. The Barrels ſhall be proved and the Muſkets Inſpected agreeably to the rules now practiſed and required by the United States; the Locks ſhall be duly hardened; the Ramrods and Bayonets ſhall be properly tempered; and the Mountings, Stocks, and every other particular ſhall be finiſhed in a workman-like manner, in all parts preciſely, or as near as poſſible, conformably to two patterns, which have been marked, and ſealed by the contracting parties to this Inſtrument; one of which patterns, the party of the ſecond part hereby acknowledges to have received.

THIRD. THE Barrels ſhall be proved, and the Muſkets inſpected at *Frederick Town aforesaid* .. .. by a perſon or perſons to be appointed by the Secretary of the Treaſury, or the Secretary of War. When due notice ſhall have been received by the party of the firſt part, that *One Hundred* — .. .. Barrels are ready for the teſt; the proof ſhall commence, on like notice that a further number of *Two Hundred* .. .. Barrels are ready, the ſame ſhall alſo be proved; after which the United States ſhall not be required to prove leſs than *Three Hundred & Fifty* .. .. Barrels at one time. The wages of the Inſpector; the expenſe of powder and ball; and all other expenſes attending the proving and inſpecting, ſhall be borne by the United States.

FOURTH. THE Stocks of the Muſkets ſhall be made of well ſeaſoned black walnut timber, when the ſame can be procured by the Contractor, otherwiſe, of well ſeaſoned, tough maple timber; or the United States ſhall cauſe to be delivered, at Philadelphia, to the party of the ſecond part, if the ſame ſhall be required, —— *One Thouſand* .. .. well ſeaſoned black walnut ſtocks in the rough, as they have been uſually received into the public ſtores, at the rate of twenty-five cents each.

FIFTH. THE party of the firſt part, contracts for and on behalf of the United States to pay or allow in ſettlement, Thirteen Dollars and Forty Cents, for each and every of the Stands of Arms which ſhall be manufactured according to the pattern, and delivered after proof and inſpection as aforeſaid.

IN witneſs whereof the ſaid Secretary of the Treaſury on behalf of United States, hath hereunto ſubſcribed his hand and affixed the Seal of the Treaſury; and the ſaid Nicholas ~~Crabb~~ White, Thomas Crabb, Jacob Metzger, and Christian Barnhizle - - - - - - - - hath hereto ſet their hands and ſeals the day and year firſt above written.

Oliv. Wolcott
Secy of the Treas.

Signed, Sealed and Delivered in the Preſence of

Jacob Shellman
John [illegible]

Nicholas White — (Seal)
Thomas Crabb — (Seal)
Jacob Metzger — (Seal)
Christof Barnhisel — (Seal)

The Contractors have delivered five hundred and forty eight stands of arms and the Contract is by consent of the parties this day cancelled

Department of War
Accountants Office
Decr 28. 1801

[illegible]
Acct Dept War

Figure 12. Articles of Agreement in which Nicholas White, Thomas Crabb, Jacob Metzger, and Christopher Barnhizle of Frederick, Maryland, agree to produce 1000 muskets for the United States. According to the document a total of 548 were completed. Reproduced from U. S. MILITARY FIREARMS 1776-1956 by Major James E. Hicks.

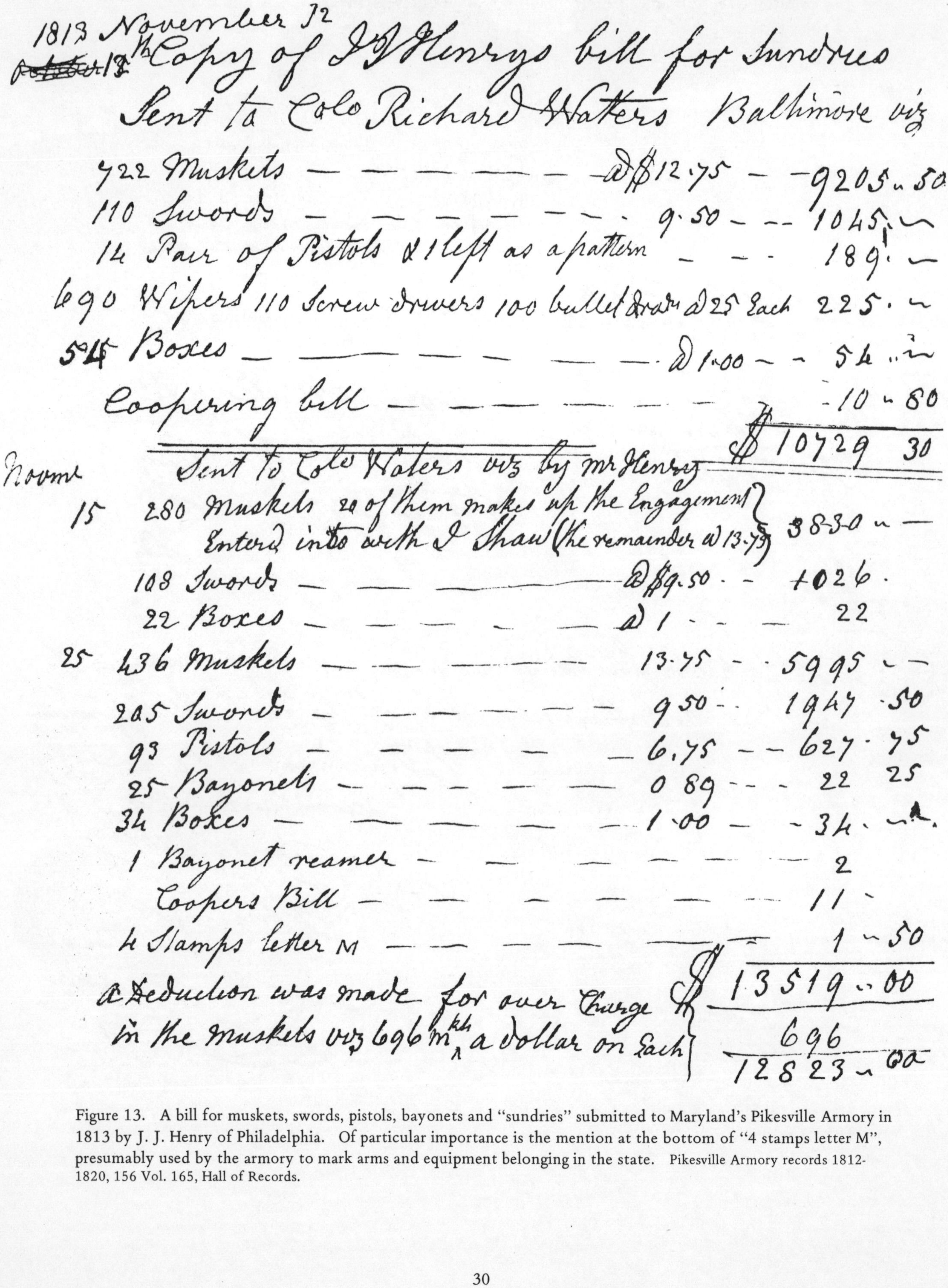

1813 November 12
~~October~~ 13th Copy of J J Henrys bill for Sundries
Sent to Colo Richard Waters Baltimore viz

	722 Muskets	@ $12.75	9205 .. 50
	110 Swords	9.50	1045 ..
	14 Pair of Pistols & 1 left as a pattern		189 ..
	690 Wipers 110 Screw drivers 100 bullet drawrs	@ 25 Each	225 ..
	54 Boxes	@ 1.00	54 ..
	Coopering bill		10 .. 80
			$ 10729 30
Novm	Sent to Colo Waters viz by mr Henry		
15	280 Muskets 20 of them makes up the Engagement Entered into with J Shaw (the remainder @ 13.75)		3830 ..
	108 Swords	@ $9.50	1026 .
	22 Boxes	@ 1	22
25	436 Muskets	13.75	5995 ..
	205 Swords	9.50	1947 .50
	93 Pistols	6.75	627 . 75
	25 Bayonets	0 89	22 25
	34 Boxes	1.00	34 ..
	1 Bayonet reamer		2
	Coopers Bill		11
	4 Stamps letter M		1 .. 50
			$ 13519 .. 00
	A Deduction was made for over Charge in the muskets viz 696 mkts a dollar on Each		696
			12823 .. 00

Figure 13. A bill for muskets, swords, pistols, bayonets and "sundries" submitted to Maryland's Pikesville Armory in 1813 by J. J. Henry of Philadelphia. Of particular importance is the mention at the bottom of "4 stamps letter M", presumably used by the armory to mark arms and equipment belonging in the state. Pikesville Armory records 1812-1820, 156 Vol. 165, Hall of Records.

Figure 14 (top). Flintlock musket, Model 1798, by Nippes & Company (see Figure 11), marked "M" on the barrel. This stamping, believed to designate ownership by the State of Maryland, though faint, can be seen about ½ inch forward of the proof mark "P". Daniel D. Hartzler collection.

Figure 15 (middle). Bayonet which accompanies and fits a Maryland-marked French musket, Model 1763 (see Figure 8) carries a deeply-stamped letter "M". Daniel D. Hartzler collection.

Figure 16 (bottom). Cavalry saber of the War of 1812 period and similar to the Starr saber carries the stamped letter "M" on the back edge of the blade, indicating Maryland ownership. Daniel D. Hartzler collection.

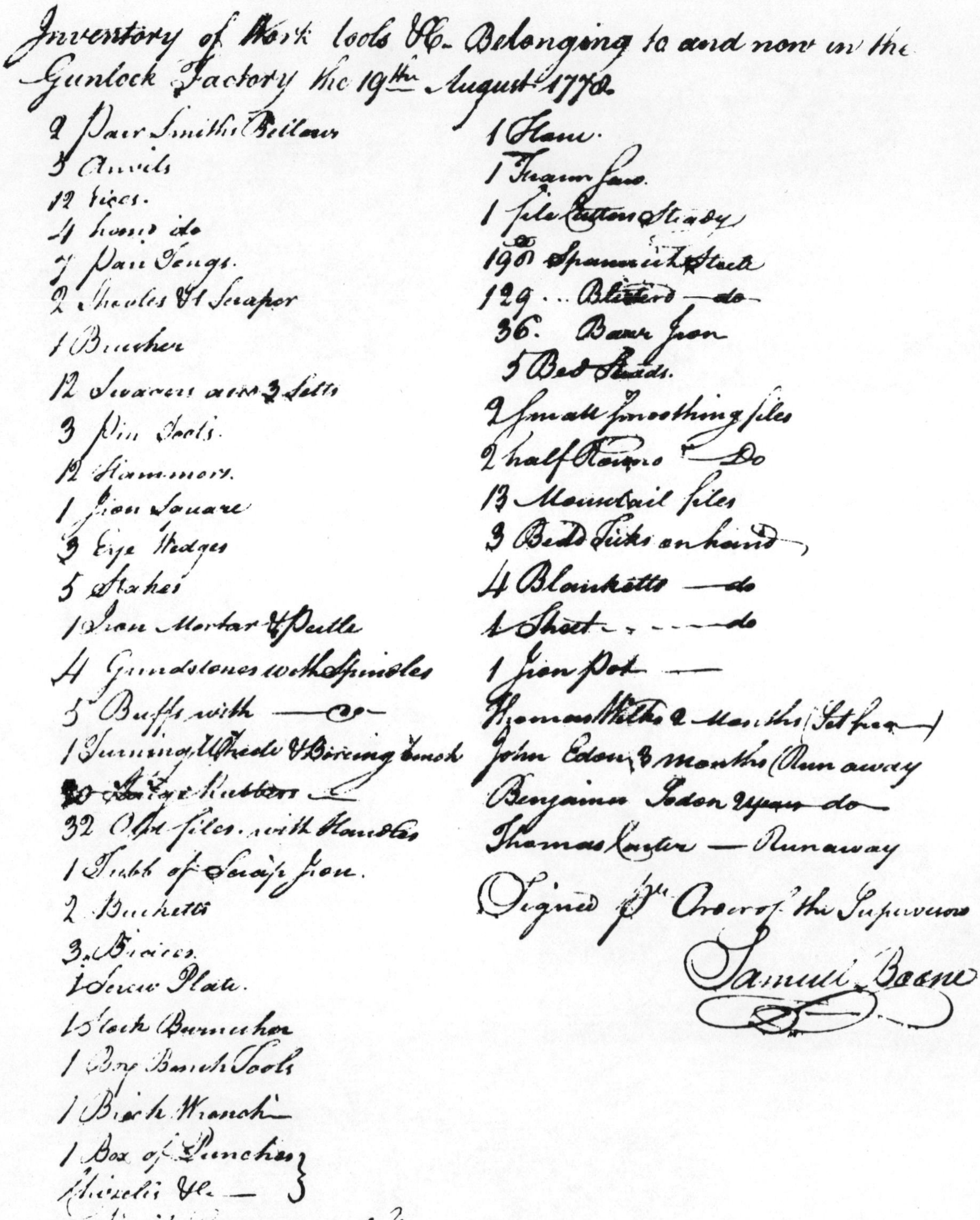

Inventory of Work tools &c. Belonging to and now in the Gunlock Factory the 19th August 1778.

2 Pair Smiths Bellows
3 Anvils
12 Vices.
4 hand do
7 Pair Tongs.
2 Shovels & 1 Scraper
1 Brusher
12 Swages are 3 Setts
3 Pin Tools.
12 Hammers.
1 Iron Square
3 Eye Wedges
5 Stakes
1 Iron Mortar & Pestle
4 Grindstones with Spindles
5 Buffs with — do
1 Turning Wheel & Boring Bench
20 [illegible] Rubbers —
32 Old files with Handles
1 Tubb of Scrap Iron.
2 Buckets
3 Braces.
1 Screw Plate.
1 Stock Burnisher
1 Boy Bench Tools
1 Brick Wrench
1 Box of Punches }
Chizels &c. — }
3 Knife files & 1 Watch file

1 Plane.
1 Frame Saw.
1 file Cutters Stady
190 lb Spanish Steel
129 ... Blistered — do
36. Barr Iron
5 Bed Steads.
2 Small Smoothing files
2 half Round — do
13 Mousetail files
3 Beds Ticks on hand
4 Blankets — do
1 Sheet ... do
1 Iron Pot —
Thomas Wilks 2 Months (Set free)
John Edon 3 months (Run away)
Benjamin Sedon 2 year do
Thomas Carter — Runaway

Signed per Order of the Supervisors
Samuel Boone

August 19th 1778 then Came Samuel Boone Overseer before me one of the Justices of the Peace for Frederick County, and made Oath on the Holy Evangels of Almighty God that the above List Contains a true Account of all the Effects belonging to the Frederick Town Gun Factory to the best of his Knowledge

Sworn Before John Haas

Figure 17. An inventory of the tools in the gunlock factory in Frederick on 19 August 1778. The list is of modest size and indicates that the factory was not a large establishment. Hall of Records, Red Book IX, p. 273.

early in the present century, and was never rebuilt, the government factory having been subsequently transferred to Harpers Ferry, Va.[17]

In 1803, on 3 June, James Haslett advertised, "Any state in the union desirous of contracting for muskets, may be supplied with any number (the Charleville pattern) at 10 dollars and 50 cents each, being 50 cents less than they have been manufactured for the United States."[18]

The Maryland legislature passed a controversial law in 1808 requiring militia men to provide themselves with arms. T. J. C. Williams gave the following account:

> In 1808 the whole country was stirred up over the Embargo and there was talk of foreign war. The legislature of Maryland enacted a militia law which required every man in the State military force to provide himself with a gun. This requirement caused great dissatisfaction as well it might. There were at the time in Frederick County five organized regiments of militia and one extra batallion containing on the rolls three thousand five hundred men. Each gun it was declared, would upon the average cost ten dollars. "The gun tax," said the Frederick Herald, "will create such a demand for arms, that we are confident that the price will be much higher than ten dollars a gun. But even rating them at that sum Frederick County will have to pay thirty five thousand dollars for this tax. All of this is to be paid in one year." Three of the delegates from Frederick County to the legislature, it was charged had voted to levy this tax upon their constituents and to exempt themselves from its payment. These delegates were Thomas Hawkins, Benjamin Biggs and Henry Kuhn, and all of them were candidates for reelection and the Herald urged that they be defeated. "A tax of thirty five thousand dollars," this paper declared, "would at any time be most oppressive on this County, but in these embargo times it is ruinous. It will perhaps be said that many men have guns. Admitting this to be so the tax will still be equally oppressive on those who have to buy: and as to those who have them the guns were heretofore worth their price to the owner – he could sell them for the money – they were his own property which he could dispose of as he pleased – and in such times as these many a man would be willing to sell his gun to raise money. But now he dares not sell it and if he is going to jail for ten dollars he is not allowed to sell his gun to raise it." It was said that General Nelson, at a meeting of Colonel Barach's regiment at Woodsbury had undertaken to defend this tax in the interest of Hawkins, Kuhn and Biggs and indicated that the law would be enforced unless men were sent to the legislature to repeal it. Another argument against the law was that there was to be no uniformity in the guns and in case the militia was called out they would be useless because of the impossibility of supplying ammunition. There were many people in the County who had conscientious scruples against bearing arms and these were to be exempted under the law upon payment of five dollars each and the Courts martial had the right to exempt from "the gun tax" all who were too poor to pay it.[19]

England and France were at war again, and ships with supplies from Baltimore set out to run the blockades, selling to both countries. English warships were constantly in need of sailors and began stopping American ships on the high seas to impress seamen from them. Some were Englishmen who had once served in the English navy, others were Americans who were high-handedly taken. British ships also landed on both of Maryland's shores and seized men. The War of 1812 soon erupted and the Free State again contributed its share and more to the effort. Because Baltimore sent out so many privateers, England called the city a "nest of pirates" and felt it essential to capture the town. The City of Washington was burned. The heroic deeds of the Marylanders on land and sea are well known, particularly for their efforts at Bladensburg, Caulk's Field, Fort McHenry, and North Point.

During the War of 1812 William Shaw's family and Notley Burnard provided timber to the government for use as gunstocks. Walnut logs which they had felled and dried were hauled to Westernport and then transported on flatboats down the Potomac to the armory at Harpers Ferry.

In the decades following the Revolutionary War full-time work at gunsmithing was not always available. Men with gunsmithing skills often spent part of their time working at other trades. Robert Reid, Thomas Crabb, Christian F. Behrens, and George Distelbart worked in blacksmithing as well as gunsmithing. Benjamin Meredith, Edward Evatt II, and William H. Lauderman advertised as gunsmiths and bellhangers. The demand for guns of all sorts fell off markedly during the recession that followed the War of 1812, beginning about 1816 and remaining relatively low during the 1820's, 1830's, and 1840's. James Haslett listed himself in the Baltimore directory for 1822 as a gunsmith and grocer;

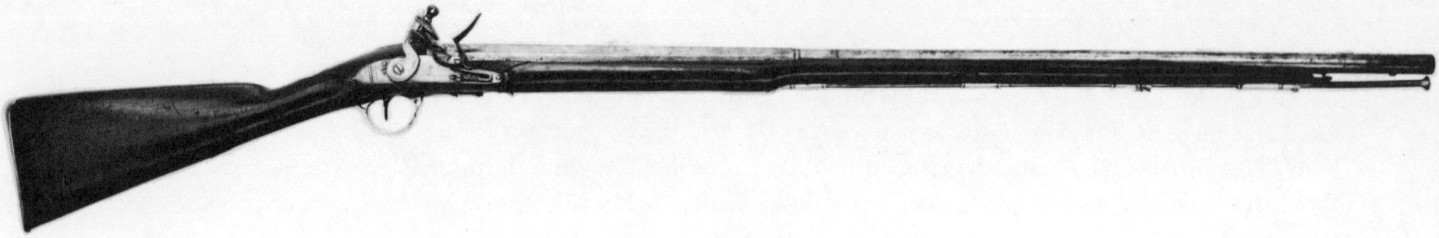

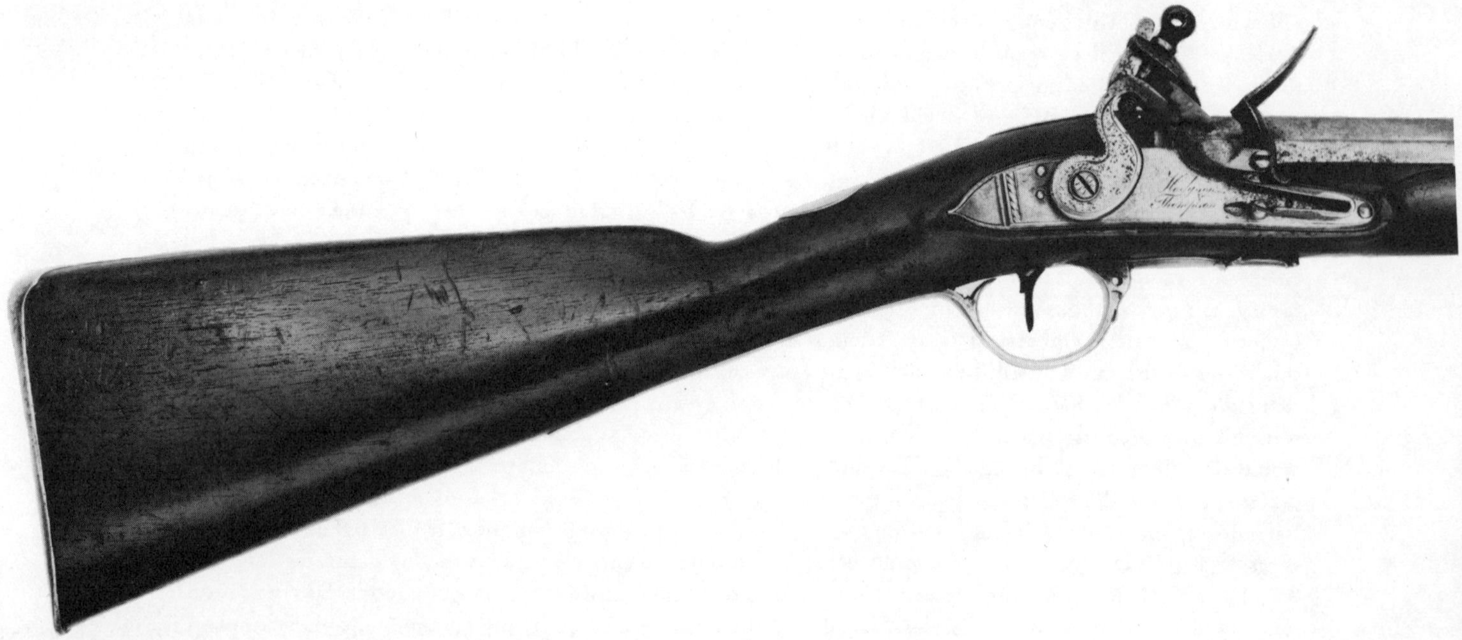

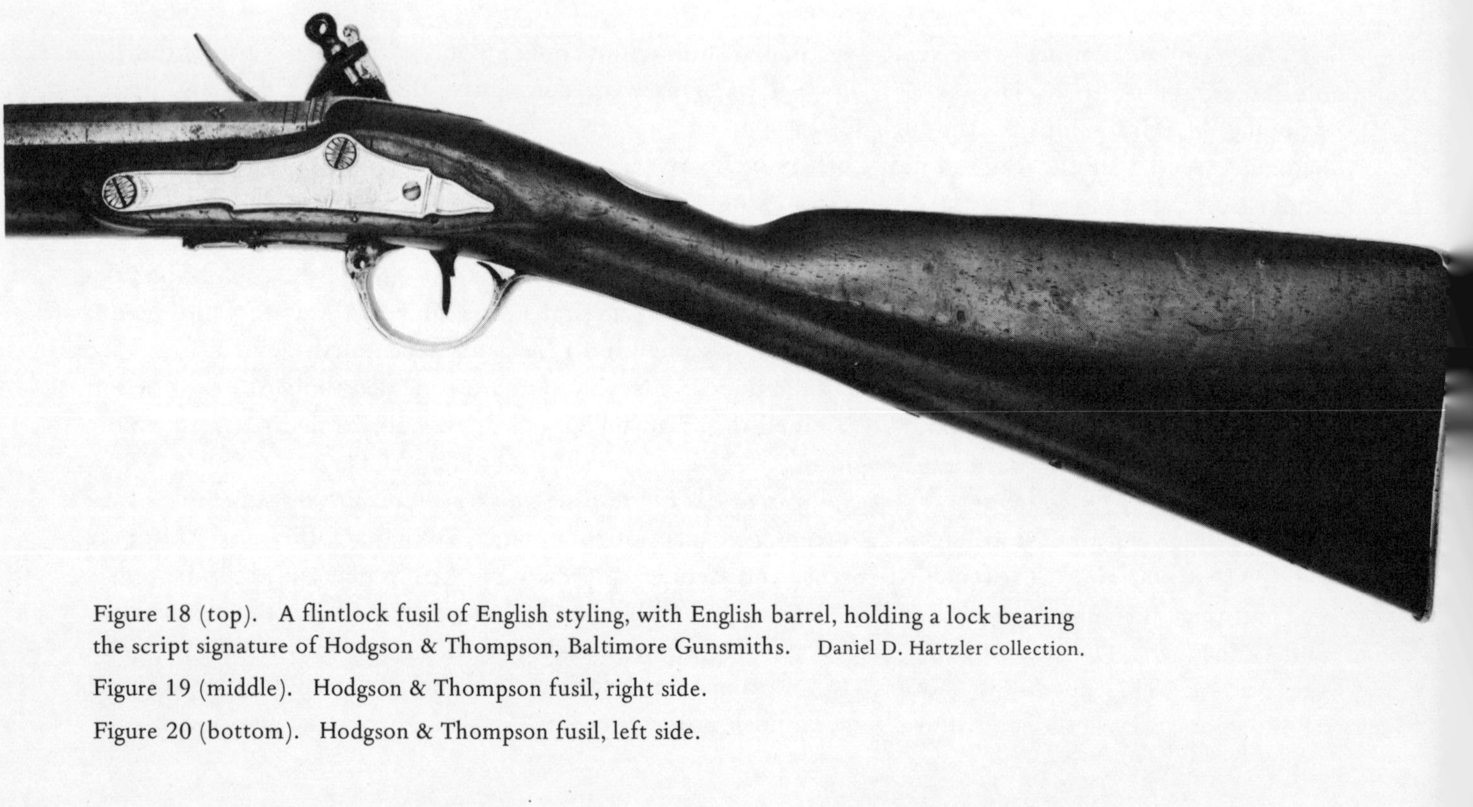

Figure 18 (top). A flintlock fusil of English styling, with English barrel, holding a lock bearing the script signature of Hodgson & Thompson, Baltimore Gunsmiths. Daniel D. Hartzler collection.

Figure 19 (middle). Hodgson & Thompson fusil, right side.

Figure 20 (bottom). Hodgson & Thompson fusil, left side.

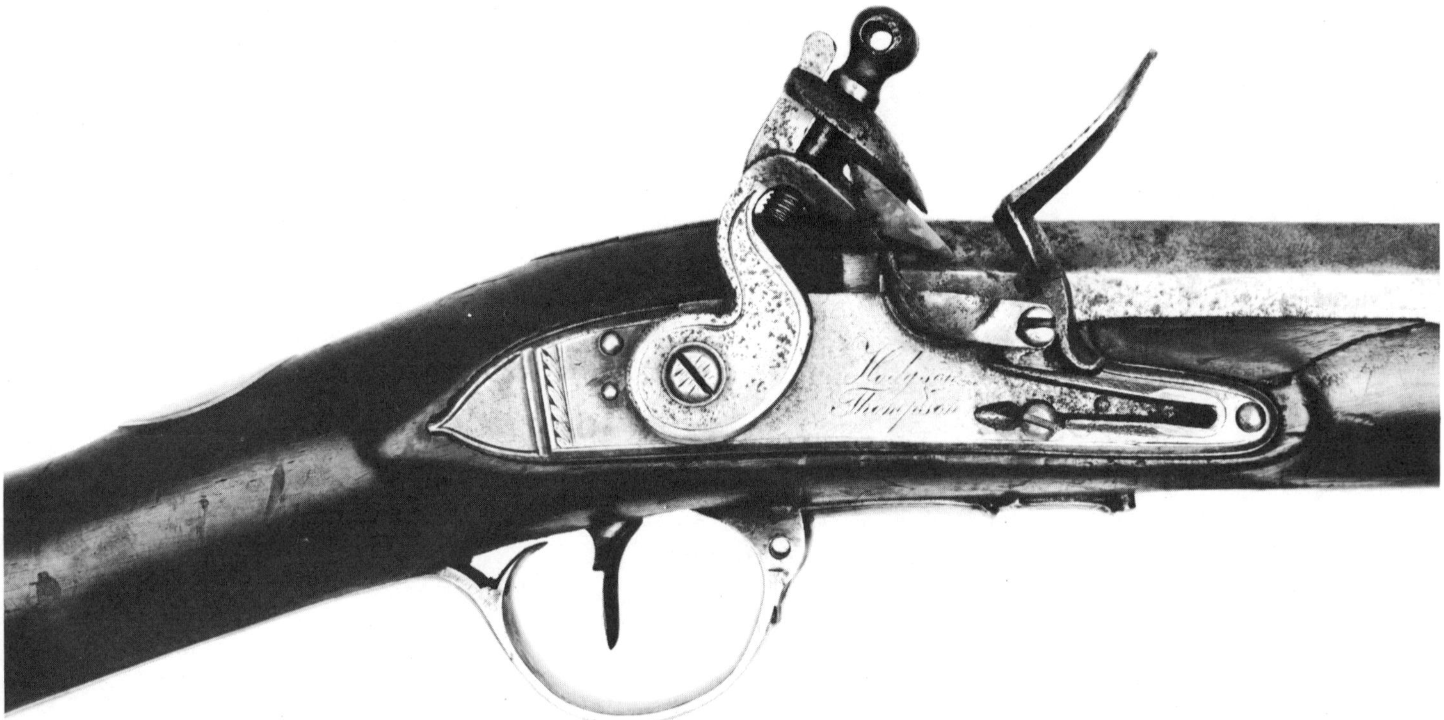

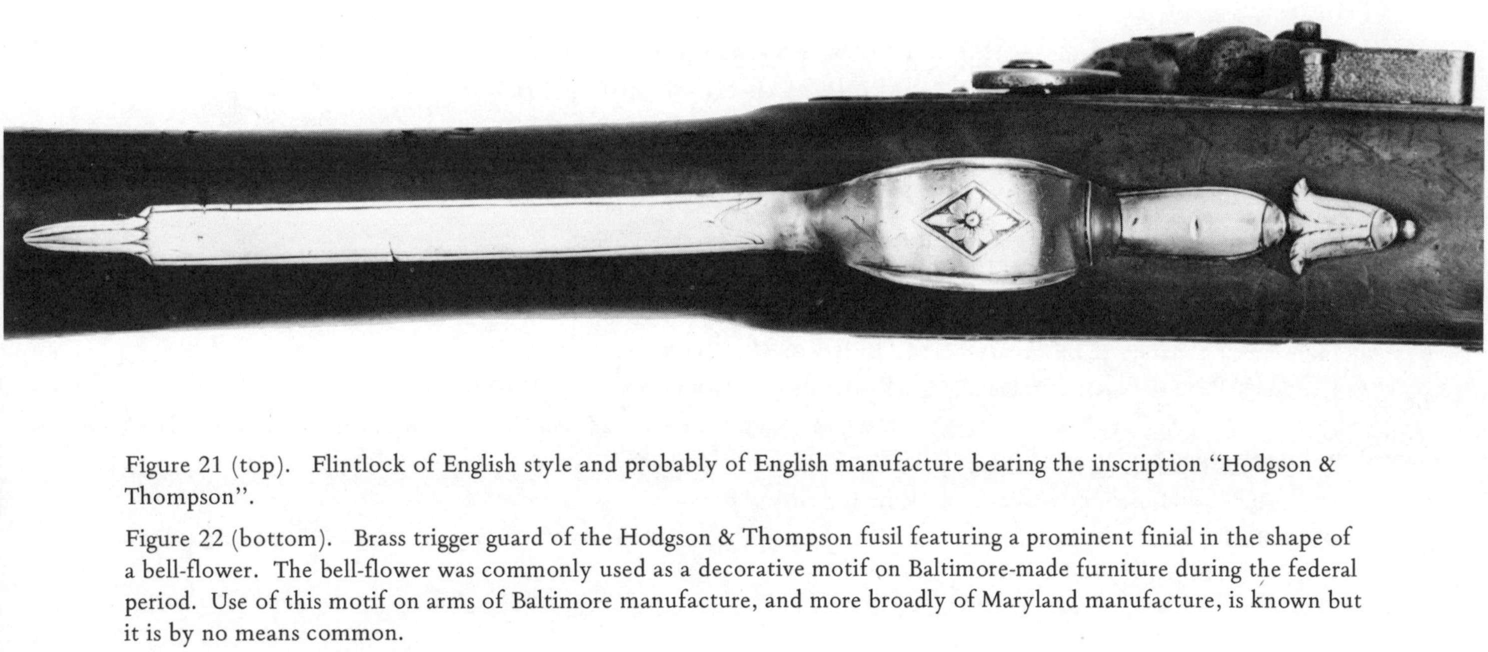

Figure 21 (top). Flintlock of English style and probably of English manufacture bearing the inscription "Hodgson & Thompson".

Figure 22 (bottom). Brass trigger guard of the Hodgson & Thompson fusil featuring a prominent finial in the shape of a bell-flower. The bell-flower was commonly used as a decorative motif on Baltimore-made furniture during the federal period. Use of this motif on arms of Baltimore manufacture, and more broadly of Maryland manufacture, is known but it is by no means common.

1823 was the last time he was listed as a gunsmith. In 1831 he opened a clothing store and one year later was appointed Judge of the 6th Election District. Anton Escherich operated a restaurant as a sideline near his shop on Baltimore Street near Green. Moritz & Kaidel, James Foy, Matthew Drake, and Levi Cromwell advertised as gunsmiths but occasionally added hardware at the bottom of their ads. James Marker farmed during the summer and concentrated on making rifles during the off season. John Armstrong and C. E. Stonesifer, like many gunsmiths, also engaged in whitesmithing.

Positive identification of Maryland Council of Safety muskets is difficult if not impossible. A number of surviving muskets of the period have the word "MARYLAND" branded or carved into the stock, and about half of these have a proof mark "M" on the barrel. No proofs such as this have been noted in Salem muskets or other Massachusetts pieces, so the probability is good that the mark refers to Maryland.

Proof that a metal stamp of the letter "M" was in use in a Maryland armory is found in a copy of a bill sent by J. J. Henry of Philadelphia to the Pikesville Armory in 1813 (Fig. 13). This shows that "4 stamps letter M" were delivered at a cost of $1.50. Henry's bill also is notable for the other items it contains – muskets, pistols, swords, and accoutrements.

REFERENCES

1. Council to Lord Baltimore (12 October 1681) ARCHIVES OF MARYLAND, pp. 300-301.
2. Assembly proceedings (27 October 1682) ARCHIVES OF MARYLAND, p. 336.
3. Proceedings of the Council (19 March 1688) ARCHIVES OF MARYLAND, p. 67.
4. Sharpe to Lord of Trace (8 February 1756) ARCHIVES OF MARYLAND, pp. 352-353.
5. Sharpe to Forbes (9 June 1758) ARCHIVES OF MARYLAND, p. 197.
6. Council of Safety (2 August 1776) ARCHIVES OF MARYLAND, p. 65.
7. Yost to Council (13 September 1776) ARCHIVES OF MARYLAND, p. 271.
8. Beatty to Council (10 December 1776) ARCHIVES OF MARYLAND, p. 519.
9. Yost to Council (13 September 1776) ARCHIVES OF MARYLAND, p. 271.
10. Smyth to Council (23 May 1776) ARCHIVES OF MARYLAND, p. 440.
11. West to Hall (4 May 1776) ARCHIVES OF MARYLAND, pp. 406-408.
12. Ewing to Council (12 February 1776) ARCHIVES OF MARYLAND, p. 155.
13. Council of Safety (14 June 1776) ARCHIVES OF MARYLAND, p. 489.
14. Council of Safety (23 February 1776) ARCHIVES OF MARYLAND, p. 180.
15. Council of Safety (21 October 1775) ARCHIVES OF MARYLAND, pp. 84-85.
16. Scharf, J. Thomas (1967) HISTORY OF MARYLAND, Vol. II: Tradition Press, Hatboro, Pennsylvania, p. 337.
17. *Ibid.*, p. 837.
18. (3 June 1803) *Baltimore Gazette & Daily Advertiser*.
19. Williams, T. J. C. (1910) HISTORY OF FREDERICK COUNTY, MARYLAND: L. R. Titsworth & Co. Publishers, Hagerstown, Maryland, p. 132.

CHAPTER 3

Longrifles

During the past 20 years the American longrifle rightly has come to be regarded as an important object of early American art. The subject was formally brought to public attention in 1960 with the publication of Joe Kindig's pioneering work, THOUGHTS ON THE KENTUCKY RIFLE IN ITS GOLDEN AGE.[1] On the basis of hundreds of illustrations of fine rifles from his collection and sound biographical data from primary sources, Kindig demonstrated that the artwork of these rifles, as shown by the wood carving, engraved metal inlays, and overall architecture, probably surpasses in originality, variety, and general excellence that of any other decorative art form produced in early America. And he showed that pronounced regional differences in decoration and construction details existed throughout the Appalachian piedmont.

Working chiefly with rifles from Pennsylvania, but including some from Maryland, Kindig showed that separate schools of gunsmithing existed in each of the important gunmaking towns. The term "school of gunsmithing" refers to a tradition, usually within a limited region, of shaping stocks in a certain way, of using characteristic art motifs in designing the decorative carving and the metal inlays, and in some instances of using particular construction details. In the 15 years that have passed since the publication of Kindig's work, much new information has come to light concerning the early gunsmiths who made the longrifles. Additionally, many more fine old rifles have come into the collector's realm where they can be seen, appreciated, and studied by those interested. The picture is clearer now than it was 15 years ago, and all the more interesting, for the new information has proven the soundness of the early synthesis that Kindig made. The concept of schools of gunsmithing brought out by Kindig, Hetrick,[2] Dyke,[3] and Kauffman[4] about 1960 now is known to be valid over a much wider range of territory than was envisioned at that time.

Much new information about the longrifle makers of Maryland was found in the course of doing research for this book, and it is presented in the biographical data section in the latter part of this volume. In addition, many fine old longrifles produced in Maryland will now be brought to the attention of collectors. A lot of these newly-discovered old rifles are worthy of being described and illustrated in detail in a publication, but it is beyond the scope of this present volume to give the subject the space and treatment it deserves. This present chapter can only serve as a brief introduction to the subject of the longrifles of Maryland. It will take a separate volume to do justice to the topic. Fortunately, it is the intention of the publisher of *The Longrifle Series* to bring out such a book within the next few years. There is no doubt that other beautiful old Maryland-made longrifles and other early arms exist in collections or as family heirlooms, unknown to the author and to the publisher. Hopefully these can be brought to the attention of the publisher for study or possible inclusion among the illustrated pieces of this future volume.

The first settlement in western Maryland was made in 1734 along the Monocacy Creek near the present site of Frederick. Within a few years the fertile Monocacy Valley was dotted with farms and small settlements. Many of these early settlers were of Germanic origin – Pennsylvania-Germans who had pushed to the southwest in search of land to farm. Others pushed further to the west and south, moving into the broad valley system beyond the Blue Mountain ridge, which became the Shenandoah Valley south of the Potomac River. And these people of the Shenandoah Valley and western Maryland, like their piedmont neighbors in Pennsylvania, took to the flintlock rifle.

The rifle-barreled gun, fired either by flintlock ignition or wheellock ignition, was the common

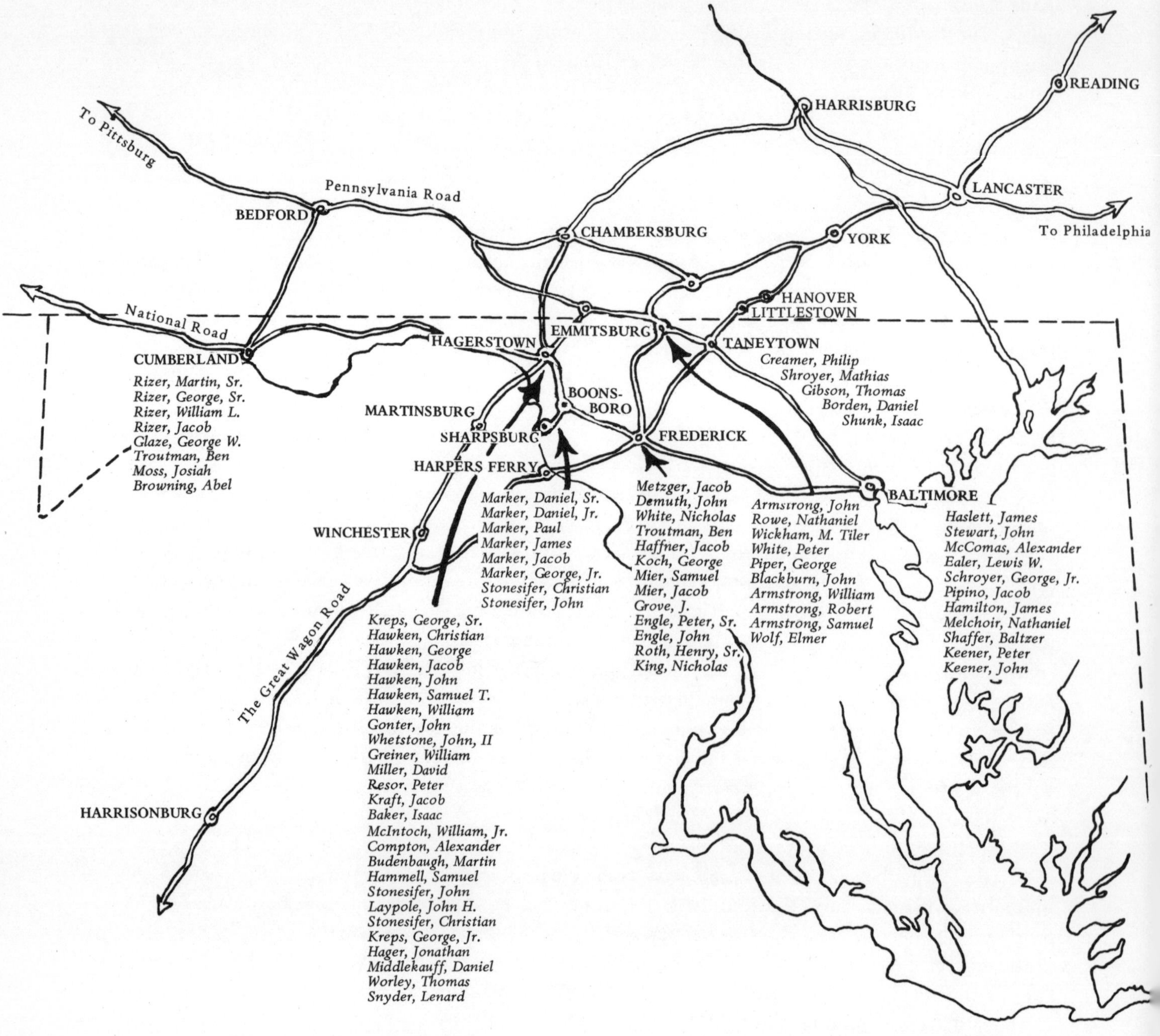

Longrifle-making Centers in Maryland

EMMITSBURG

CUMBERLAND HAGERSTOWN TANEYTOWN

SHARPSBURG -- BOONSBORO

FREDERICK

BALTIMORE

The principal centers are the seven mentioned and shown above. There were individuals making rifles in other locations also. The names listed are those of gunsmiths known to have made rifles and a few others suspected of having made them.

sporting arm of the Germanic peoples of continental Europe in the late 1600's and early 1700's. By the year 1700, before any rifle making is known to have been done in America, the rifled gun had been made in Europe in all possible barrel lengths and bore diameters and with peep sights as well as open sights. The Germanic settlers brought to Pennsylvania, Maryland, and Virginia their traditions of rifle usage, and doubtless some of them carried rifles with them from the old country. It is well to keep in mind, however, that many of these early settlers on the frontier were very poor peasants at the time of their departure from Europe. The rifle in Europe was not a necessity but rather a sporting arm for the well-to-do and middle classes. The poor peasant who came to America may not have had a rifle when he arrived, but he probably aspired to own one and when his personal economics made it possible, he bought one.

For the settlers of the piedmont in the early part of the eighteenth century the rifle became an important tool, chiefly as a means of procuring game. In Pennsylvania at this time the white settlers lived at peace with the Indians and serious thoughts of defense against Indian attack were all but unknown. In Virginia, on the other hand, difficulties with the Indians had taken place during the 1730's, 1740's, and early 1750's, and it probably was there that the first *organized* use of rifles in defense against Indian attack took place.

It is not now known, and probably never will be known, where the first rifles were produced in early America. For 50 years or more writers have claimed that the first rifles were made in Pennsylvania. It is entirely possible that a few rifle-barreled guns were made in the Philadelphia region, or in the Hudson Valley, or in New England before 1700, but statistically this is not important. What is important is that by the 1740's rifles were being used by, and doubtless made by, the peoples of the piedmont from eastern Pennsylvania across western Maryland and down into the Valley of Virginia.

At the beginning of the American Revolution two rifle companies from Maryland were asked to join Washington's army in Massachusetts. Frederick was the center of riflemaking and rifle usage in the western part of the province, and it was there that the companies were formed, one under the command of Captain Michael Cresap who was the son of Colonel Thomas Cresap, a famous early Maryland frontiersman, the other under Captain Thomas Price. A letter has survived, dated 1 August 1775, that describes these men shortly before they left for Cambridge on 18 July:

> Notwithstanding the urgency of my business, I have been detained three days in this place by an occurrence truly agreeable. I have had the happiness of seeing Captain Michael Cresap marching at the head of a formidable company of upwards of one hundred and thirty men from the mountains and back woods, painted like Indians, armed with tomahawks and rifles, dressed in hunting-shirts and moccasins . . . Yesterday the company were supplied with a small quantity of powder from the magazine, which wanted airing, and was not in good order for rifles; in the evening, however, they were drawn out to show the gentlemen of the town their dexterity at shooting. A clap-board, with a mark the size of a dollar, was put up; they began to fire offhand, and the bystanders were surprised, few shots being made that were not close to or in the paper. When they had shot for a time this way, some lay on their backs, some on their breast or side, others ran twenty or thirty steps, and firing, appeared to be equally certain of the mark. With this performance the company were more than satisfied, when a young man took up the board in his hand, not by the end, but by the side, and holding it up, his brother walked to the distance and very cooly shot into the white; laying down his rifle, he took up the board, and holding it as it was held before, the second brother shot as the former had done. By this service I was more astonished than pleased. But you will believe me, when I tell you, that one of the men took the board, and placing it between his legs, stood with his back to the tree while another drove the centre. What would a regular army of considerable strength in the forests of America do with one thousand of these men, who want nothing to preserve their health and courage but water from the spring, with a little parched corn, with what they can easily procure in hunting; and who, wrapped in their blankets, in the damp of night, would choose the shade of a tree for their covering, and the earth for their bed.[5]

Some additional observations have survived, written in a letter from a gentleman of Frederick Town to a friend in Baltimore Town, dated 19 July 1775:

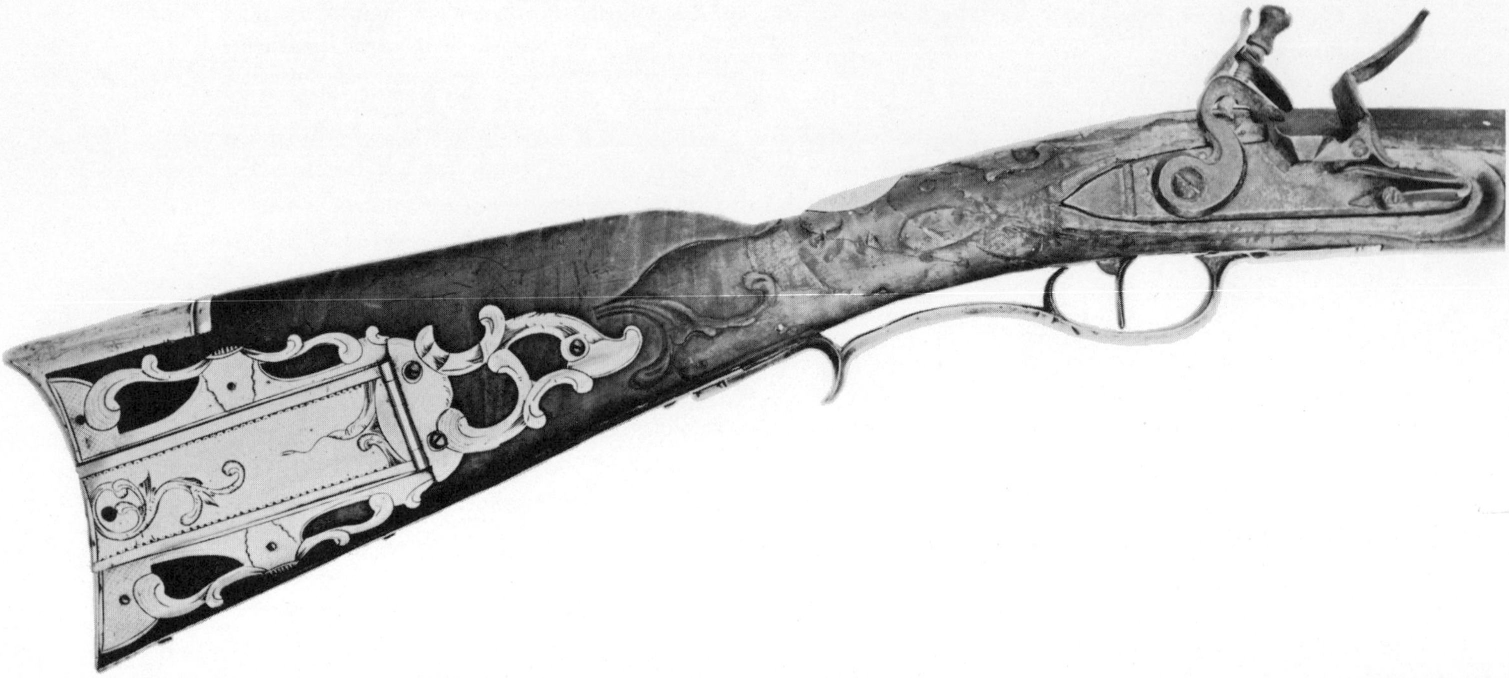

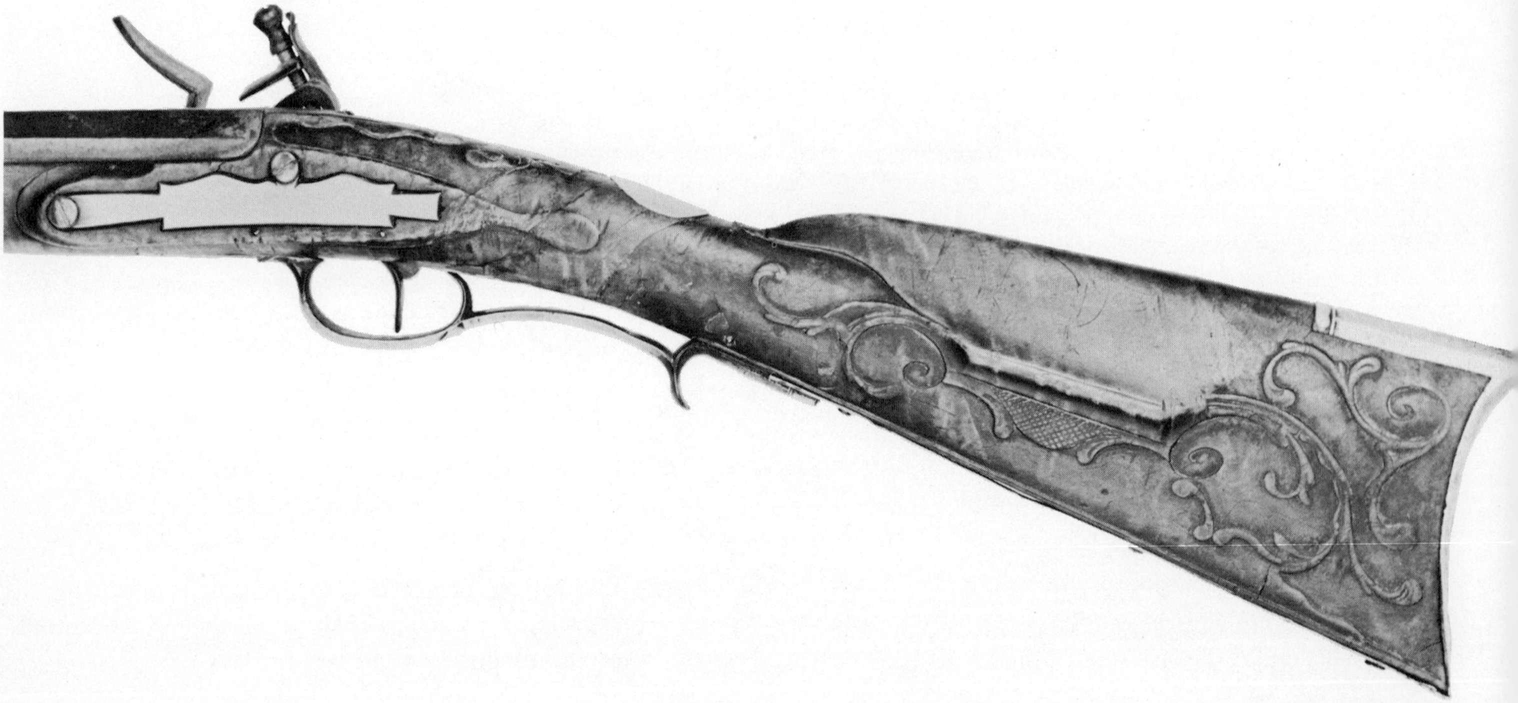

Figure 23 (top). Right side of a flintlock rifle signed "P. Creamer", probably made in Taneytown, Maryland, before 1805 when Creamer left the town. Courtesy of Daniel W. Shoemaker.

Figure 24 (middle). Full-length engraved brass toe plate on the rifle by Philip Creamer.

Figure 25 (bottom). Left side of Creamer rifle exhibiting particularly fine relief carving in the style of George Schroyer of nearby Hanover.

On Monday last, July 17th, Capt. Morgan, from Virginia, with his company of riflemen (all chosen), marched through this place on their way to Boston. Their appearance was truly martial, their spirits amazingly elated, breathing nothing but a desire to join the American army and to engage the enemies of American liberties. They were met a mile out of town by three companies, viz.: Capt. Price's company of riflemen, Capt. Grosh's and Capt. Beatty's companies of militia, and escorted a few miles out of town, amidst the acclamation of all the inhabitants who attended them. And yesterday Capt. Price with his company also marched, and surely never were two finer companies raised in any country more determined to conquer or die than those two companies are. Capt. Cresap also with his brave company have marched. I need not say anything of Capt. Cresap's undaunted courage. Not an American but knows him to be an intrepid warrior, and of course he knows his men, and has culled them from the many. We are also in hourly expectation of Capt. Stinson with his company in this town, on his way to Boston. *God* grant him a speedy and happy arrival there. So many offered to join the above companies that not one of them but might have had one hundred men at least.[6]

Cresap's and Price's companies of riflemen, together with Daniel Morgan's Virginia riflemen were the first soldiers from the southern colonies to reach Cambridge. They traveled 550 miles in 22 days, arriving on 9 August without any loss of men. Thatcher, writing in his military journal late in August from the camp in Roxbury stated:

They are remarkably stout and hardy men, many of them exceeding six feet in height. They are dressed in white frocks, or rifle shirts, and round hats. These men are remarkable for the accuracy of their aim, striking a mark with great certainty at two hundred yards' distance. At a review, a company of them, while on quick advance, fired their balls into objects of seven inches in diameter at the distance of two hundred and fifty yards. They are now stationed on our lines, and their shot have frequently proved fatal to British officers and soldiers who expose themselves to view, even at more than double the distance of common musket-shot.[7]

Thus it was that American longrifles, and the men who used them, gained fame and a romantic reputation that has survived to the present day, and which is destined to grow in the future as interest in muzzle-loading rifle shooting expands along with a general interest in the ways of early America.

In studying longrifles and longrifle makers of Maryland it is convenient to deal separately with each of the important localities where rifles were produced, taking them in a geographical order from east to west. This involves Taneytown, Emmitsburg, Frederick, Hagerstown, Cumberland, and Baltimore. Rifles were made in some other Maryland towns also, but the places mentioned are the principal ones. In the study of longrifles as objects of art it is the details of construction, of stock architecture, and particularly of artistic decoration that call for attention. These varied geographically, from region to region or town to town, but they also varied in time, so that rifles of the pre-Revolutionary and Revolutionary period differed from those of the 1790-1815 period. Further changes took place during the 1820's, 1830's, and 1840's.

Taneytown lies along the important old-time route, the Monocacy Road, that runs north from Frederick through Taneytown, Littlestown, Pennsylvania, Hanover, and on to York. In each of these towns there were gunsmiths who made beautiful rifles in the best traditions of the eighteenth century. A particularly important eighteenth century gunsmith by the name of George Schroyer (also spelled Schreyer, Schreyor) worked in Hanover, Pennsylvania, during the last quarter of the eighteenth century. He worked in a style of his own, producing rifles that were both beautiful and from the shooter's viewpoint, well-designed. Apparently he was rather influential with the younger gunsmiths who either apprenticed with him or who felt his influence through the soundness of his products.

Philip Creamer, a gunsmith of Taneytown, seems to have been such a person. Until now Creamer has been all but unknown as an early gunsmith of note, but judging from a number of the arms he produced that have been studied in the course of preparing this book, he can rightly be regarded as one of the finest gunsmiths of early Maryland.

A particularly beautiful relief-carved flintlock rifle with the script barrel signature "P. Creamer" has survived in a family from the southern part of York County, Pennsylvania, close to Taneytown. The carving on this piece resembles the work of George Schreyor very closely, as does the patchbox and the general architecture. It is probable that a good many other rifles of comparable quality were

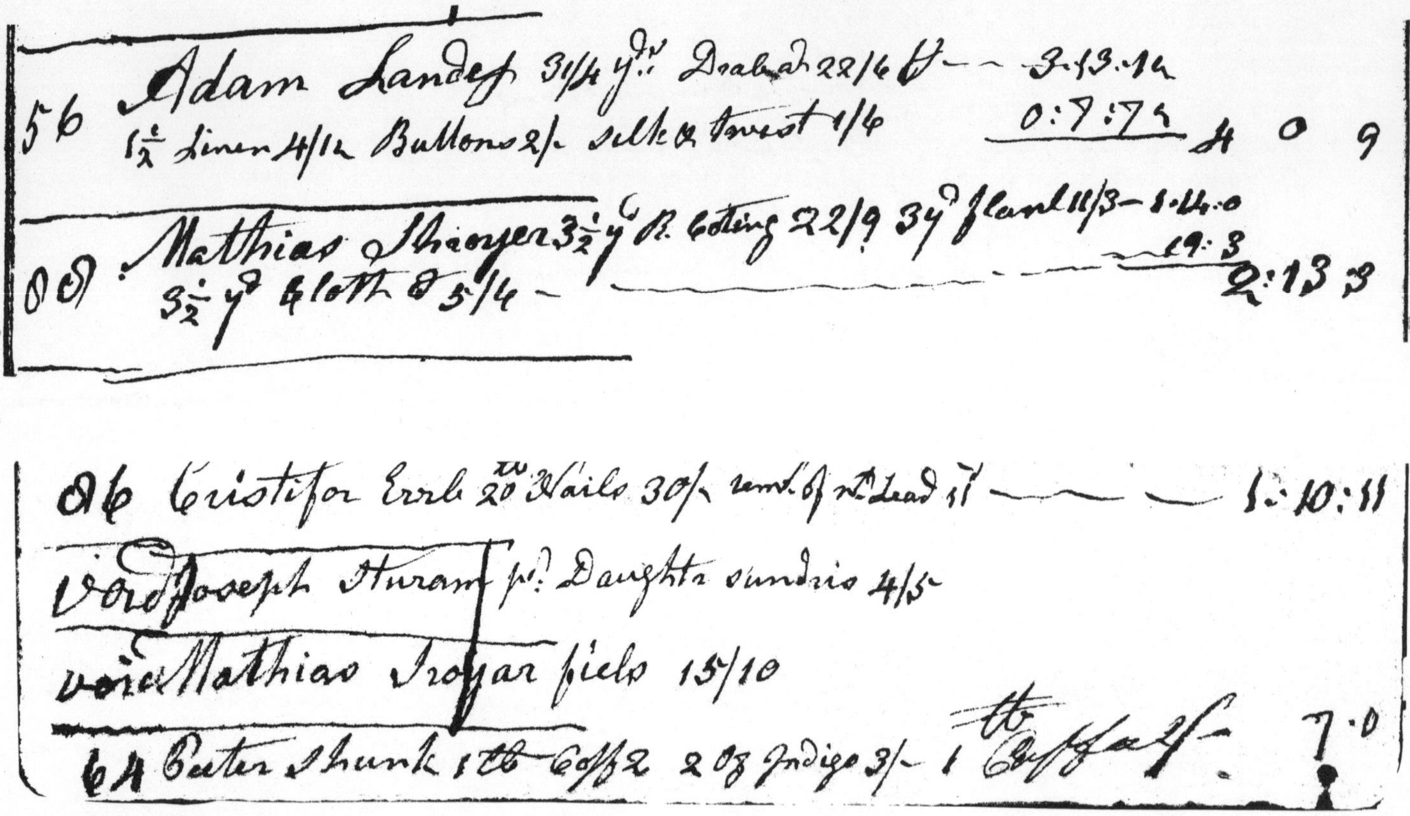

Figure 26. Mathias Shroyer, gunsmith of Taneytown, had an account at the general store of John M. Kaleb & Co. An account book has survived for the years 1801 through early 1809. Shroyer's accounts run from August 1802 through July 1803. The above two entries are for 29 November 1802 and 22 September 1802 respectively, showing his purchase of fabric and of files, doubtless for his gunsmith work. Courtesy of Edward R. and Helen Flanagan.

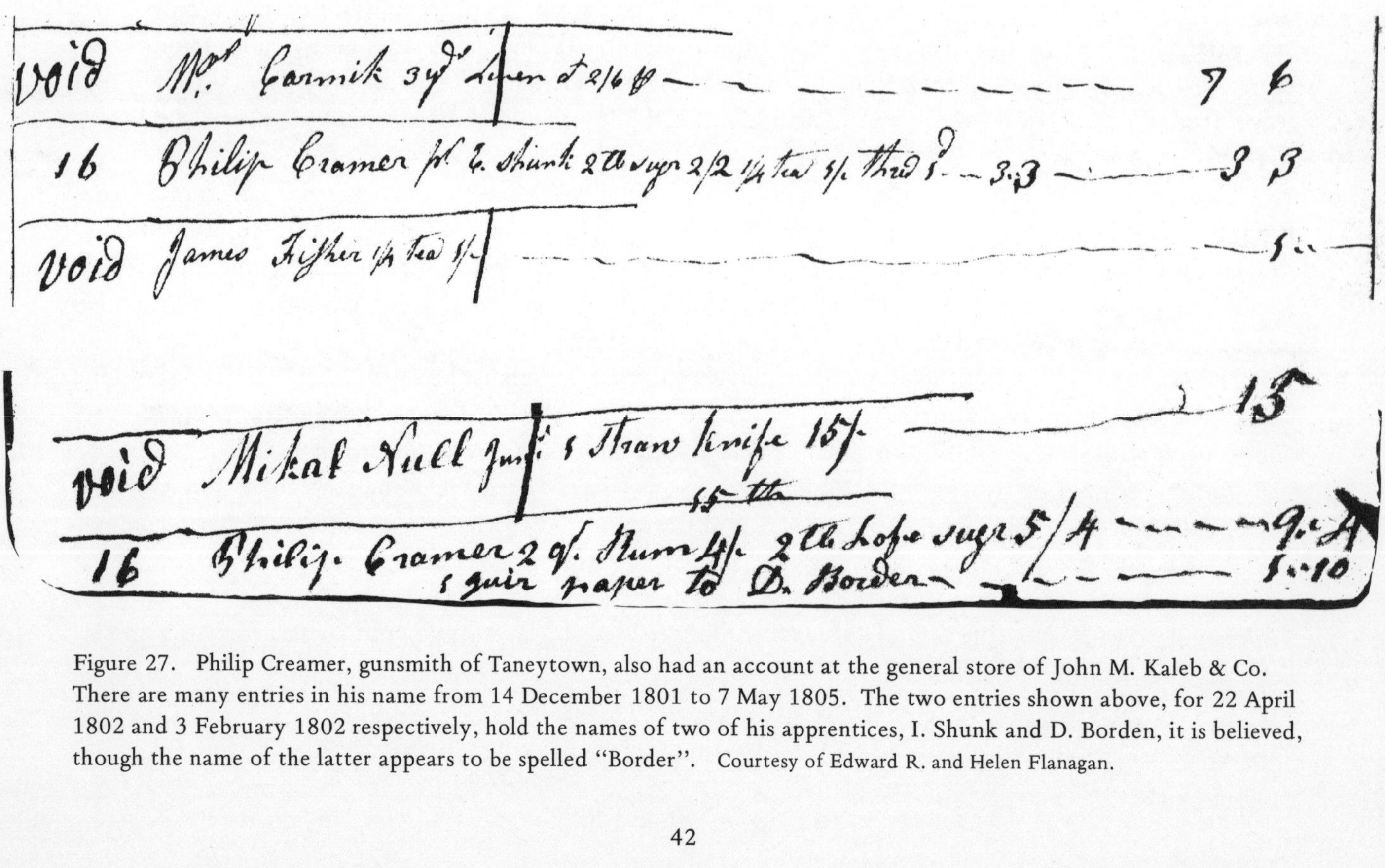

Figure 27. Philip Creamer, gunsmith of Taneytown, also had an account at the general store of John M. Kaleb & Co. There are many entries in his name from 14 December 1801 to 7 May 1805. The two entries shown above, for 22 April 1802 and 3 February 1802 respectively, hold the names of two of his apprentices, I. Shunk and D. Borden, it is believed, though the name of the latter appears to be spelled "Border". Courtesy of Edward R. and Helen Flanagan.

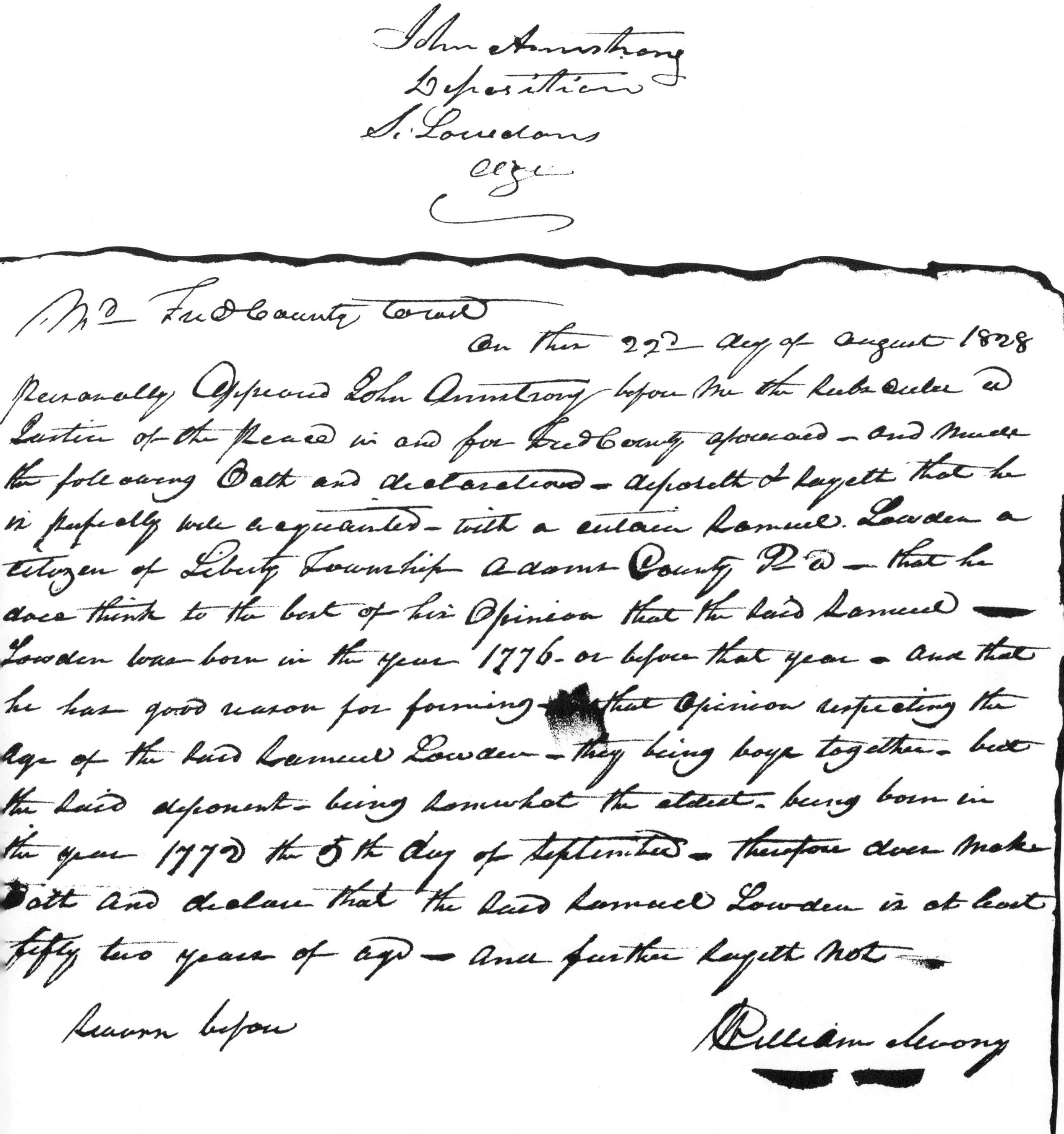

John Armstrong
Deposition
S. Lowdons
Age

Md Fred County towit

On this 22d day of August 1828
personally appeared John Armstrong before me the subscriber a Justice of the Peace in and for Fred County aforesaid — And made the following Oath and declaration — deposeth & sayeth that he is perfectly well acquainted — with a certain Samuel Lowden a citizen of Liberty Township Adams County Pa — that he does think to the best of his Opinion that the said Samuel — Lowden was born in the year 1776 — or before that year — and that he has good reason for forming that Opinion respecting the age of the said Samuel Lowden — they being boys together — but the said deponent — being somewhat the eldest — being born in the year 1772 the 5th day of September — therefore does make Oath and declare that the said Samuel Lowden is at least fifty two years of age — And further sayeth not

Sworn before

William [illegible]

Figure 28. John Armstrong's deposition in which he tells the date of his birth and the age of his boy-hood friend, Samuel Lowden, with whom he had property transactions late in life. This unusual document was purchased at an estate auction in Gettysburg in the summer of 1975. Courtesy of Albert M. Sullivan, Sr.

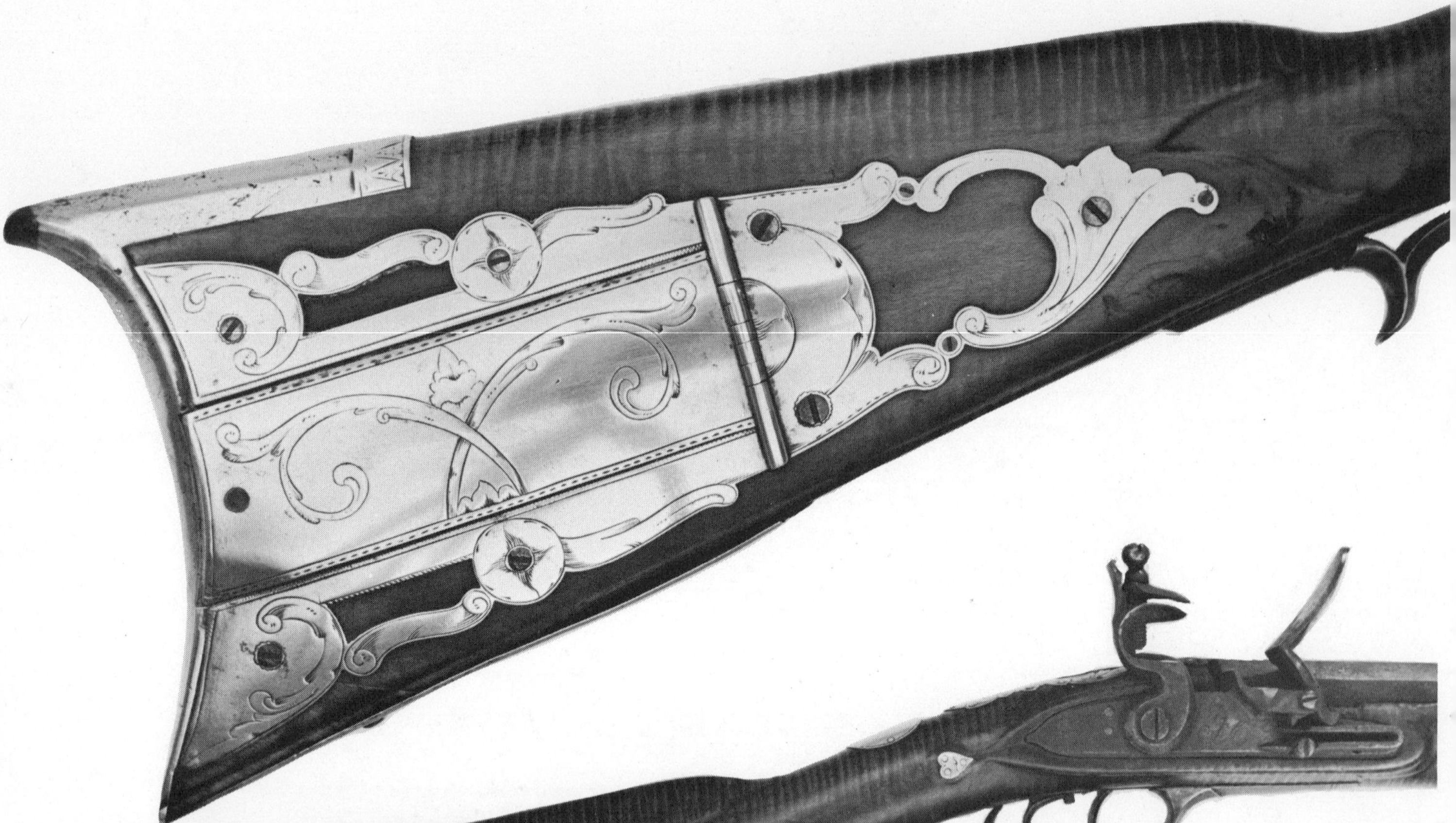

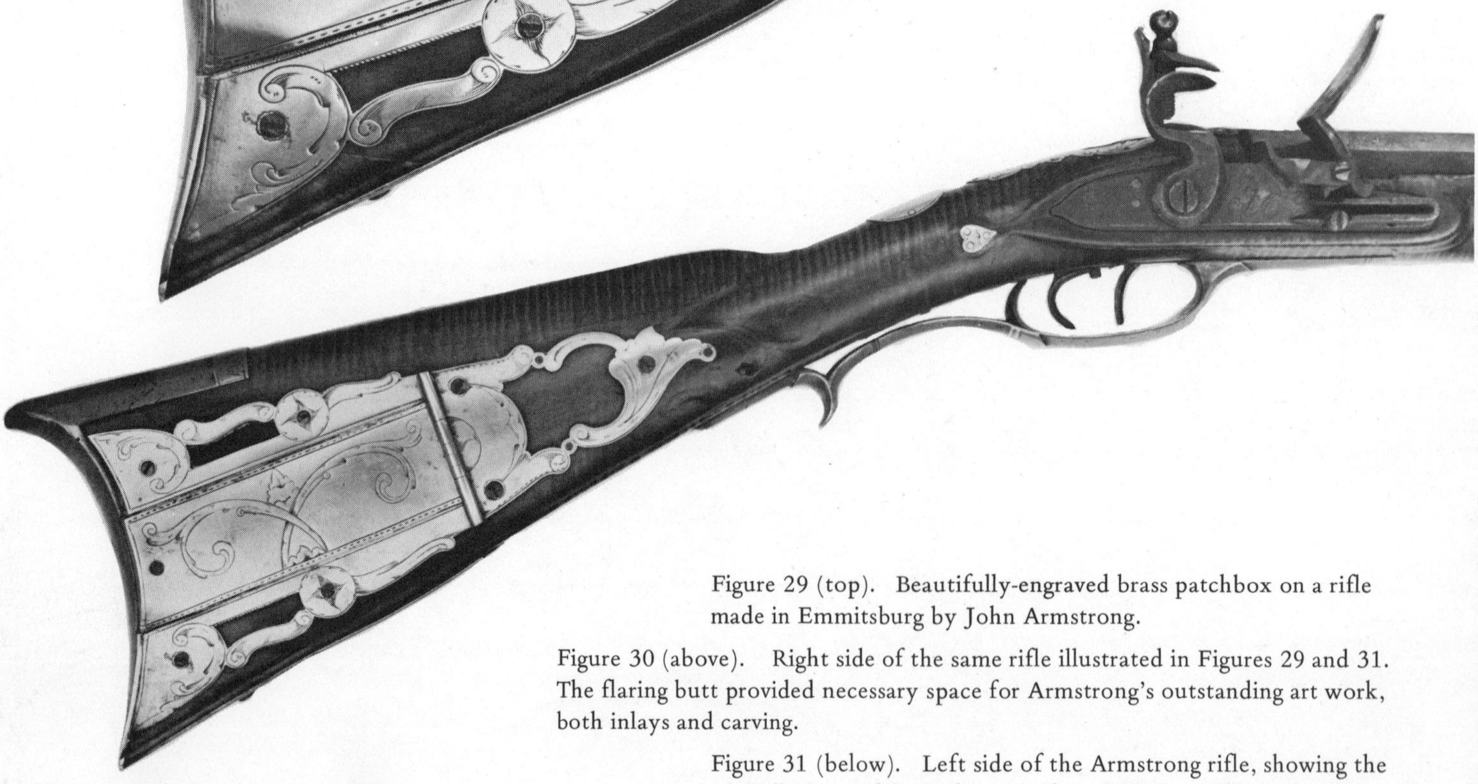

Figure 29 (top). Beautifully-engraved brass patchbox on a rifle made in Emmitsburg by John Armstrong.

Figure 30 (above). Right side of the same rifle illustrated in Figures 29 and 31. The flaring butt provided necessary space for Armstrong's outstanding art work, both inlays and carving.

Figure 31 (below). Left side of the Armstrong rifle, showing the typically beautiful relief carving for which the maker is famous.

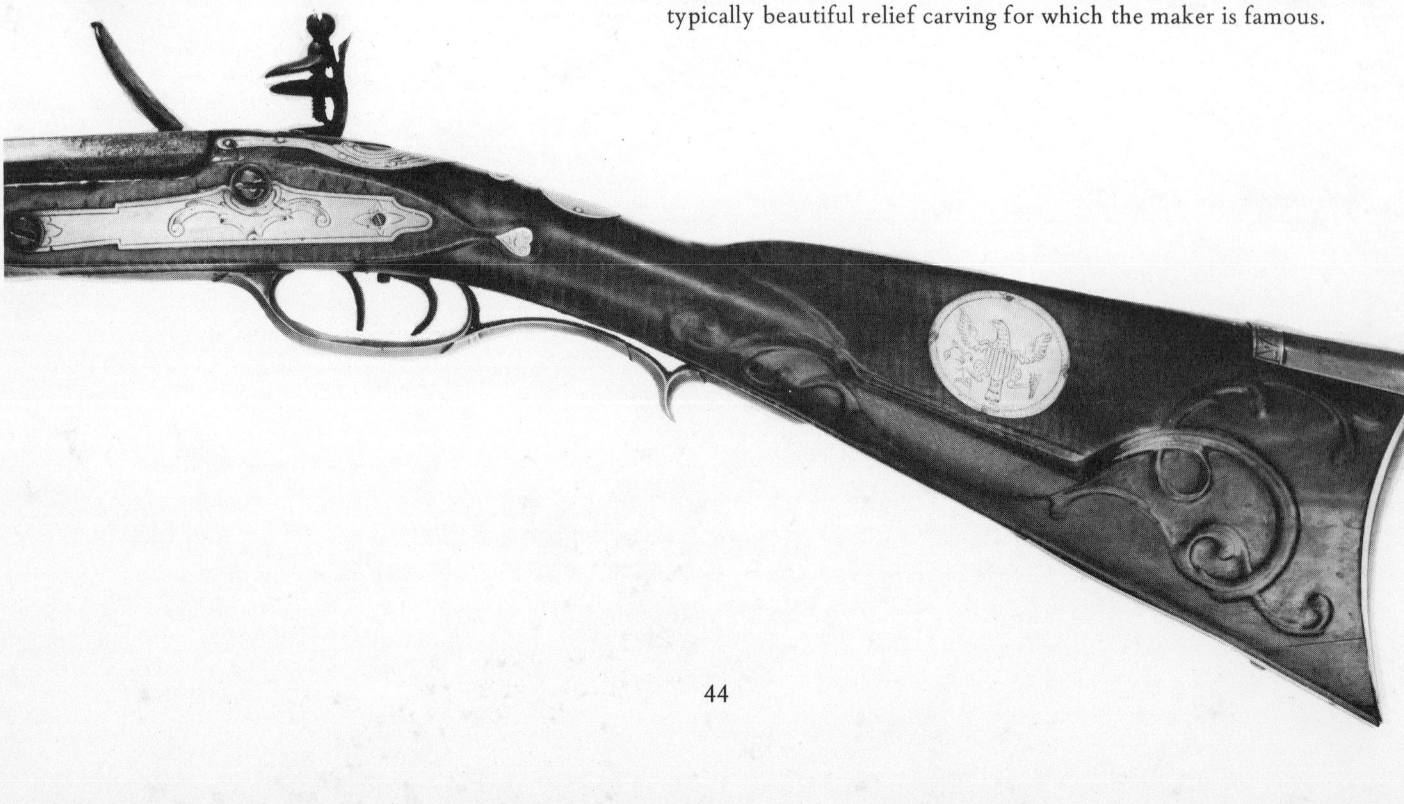

made by Philip Creamer, and hopefully a few of them have survived. If they do exist it is hoped that they can be brought to the attention of both the author and the publisher. One additional Creamer rifle has been reported, though it is of lesser note.

The early records show Creamer working as a gunsmith in Taneytown in 1799 and in 1801, and in each of these years he took an apprentice. His name is spelled "Cramer" in the records, which creates confusion because there apparently was a distiller of the same name in the same region at the same time. Along with a good many other gunsmiths of Pennsylvania and Maryland, Creamer moved west after his initial years in Taneytown. It is not known exactly when he left, but it probably was in the period between 1815 and 1820, after the War of 1812 concluded and the demand for rifles dropped off markedly. In 1820 he was working in Illinois, and his late years apparently were spent in St. Louis. Creamer made some very fine pistols, both in flintlock and percussion ignition, in the course of his working career, and a few of these have survived. A finely-made percussion pistol made by Creamer was owned by Andrew Jackson, and is on display in Jackson's house, the Hermitage, in Tennessee; it is unlikely that it was made in Maryland, however.

Mathias Schroyer was another gunsmith who worked in Taneytown. What his relationship was, if any, to George Schroyer, Sr., and George Schroyer, Jr., of nearby Hanover, Pennsylvania, is unknown, but it is probable that there was a family relationship. A longrifle signed by Mathias Schroyer has been reported by a collector but details about it are unknown. Mathias Schroyer had a shop in Taneytown in which he turned out a variety of iron implements along with guns, and in 1798 he took on a contract to produce 1000 muskets of Charleville pattern for the government.

Emmitsburg lies about 8 miles west of Taneytown. It was here that the famous longrifle-maker, John Armstrong, worked, and a number of others as well. Nathaniel Rowe, who apprenticed under Armstrong, carried on gunsmithing in Emmitsburg until about the end of the nineteenth century.

John Armstrong stands out as the most important gunsmith of Emmitsburg. His relief-carved flintlock rifles, many of which have survived, represent the art of longrifle building in its grandest form. The earliest information about Armstrong comes from Williams' HISTORY OF FREDERICK COUNTY, MARYLAND,[8] a secondary source to be sure, but probably correct. This mentions that the gunsmith's father, John, Sr., came from England and settled in the Cumberland Valley of Pennsylvania. John Armstrong, Jr., was born 5 September 1772. With whom he learned the trade of gunsmithing is not known, but there is a good possibility that it might have been with George Schroyer of Hanover, Pennsylvania, which is about 20 miles away.

Young John Armstrong apparently set up his gunsmithing shop in Emmitsburg shortly after finishing his apprenticeship, probably in the fall of 1793. Williams (1910)[9] comments that "The first gunsmith shop built by John Armstrong is standing at the present time [1910] and is in good state of preservation, over one hundred and twenty years." This suggests that the shop was built in the late 1780's, but late 1793 or 1794 would have been the likely time.

The names of four apprentices who bound themselves to John Armstrong to learn the gunsmithing trade are known:

	1799	Marine Tiler Wickham, age 19
	1801	George Piper, age 18
	1804	John Blackburn, age 17 years, 5 months
	c. 1837	Nathaniel Rowe, age 16

In the 33-year period after 1704 Armstrong must have had a number of other apprentices, but they remain unknown because shortly after 1805 Frederick County stopped the recording of most apprentice indentures.

Many old-time gunsmiths worked from time to time at other jobs than gunsmithing, and John Armstrong was no exception. Williams' account of the gunsmith mentions that he was a whitesmith and master mechanic.[10] The apprentice indenture of Tiler Wickham mentions Armstrong's occupations as "Gun smith and Lock smith."

According to Williams (1910), John, Jr., the gunsmith, married a Miss James of Emmitsburg. From a deed we learn that her name was Agnes. They had three daughters, Anna, Jane, and Elizabeth, and four sons, William, Robert, Samuel, and James, all of whom learned the gunsmithing trade.[11] William,

Figure 32. Buttstock of percussion rifle with butt-plate removed, showing signature of John Armstrong of Emmitsburg, and date of 1836. This is the only known rifle by Armstrong that bears a date, and it proves that he made rifles late in his life. Additional proof that Armstrong made rifles very late in his life comes from a newspaper interview that Armstrong's apprentice, Nathaniel Rowe, gave in 1908. He mentioned beginning his apprenticeship in gunsmithing in 1837 and told a few details about rifle-making. Albert M. Sullivan, Jr., collection.

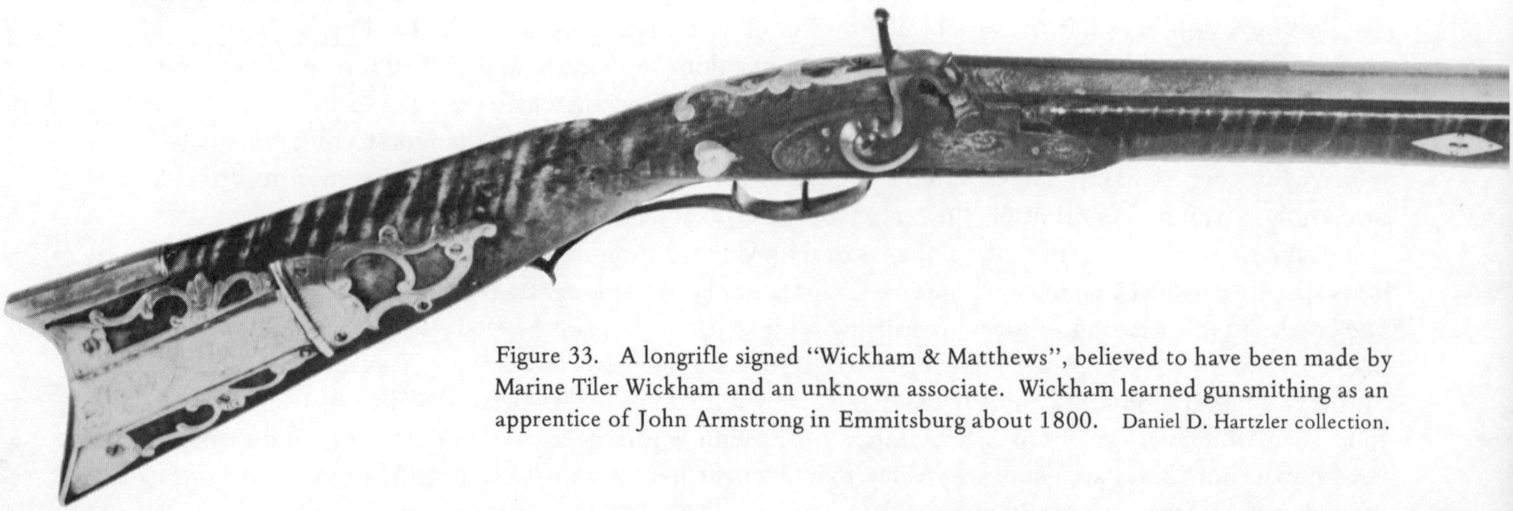

Figure 33. A longrifle signed "Wickham & Matthews", believed to have been made by Marine Tiler Wickham and an unknown associate. Wickham learned gunsmithing as an apprentice of John Armstrong in Emmitsburg about 1800. Daniel D. Hartzler collection.

Figure 34. Nathaniel Rowe, who apprenticed with John Armstrong late in the master's life, signed his name, "N. Rowe", in script on this percussion rifle. The barrel is 36¼ inches long and of .38 caliber. It is equipped with a peep sight as well as an open sight. Both are stationary and, depending on the target distance, either can be used without interfering with the other. Daniel D. Hartzler collection.

the only son who married, was a Master Armorer at the Marine Barracks in the Navy Yard at Washington, D. C., in 1822, and later may have been an inspector of arms at Harpers Ferry.

John Armstrong lived and worked in Emmitsburg for most of his adult life. He owned town lots #1 and #2, and in 1809 was living on lot #2. Hard times hit the gunsmithng trade during the recession that followed the close of the War of 1812, that is, during the decade 1815-1825, and many gunsmiths either went into other businesses or moved west. Armstrong may have been caught up in this recession, for in November 1822 he sold all of his gunsmithing tools to Jacob Harner for the sum of $28. He then may have left town for a period of five or six years. He got back into the gunsmithing business in Emmitsburg, however, after 1828, and apparently worked at it until his death which presumably occurred in the 1840's. A beautiful relief-carved percussion rifle in the collection of Albert M. Sullivan, Jr., made in typical Armstrong style, holds the signature of John Armstrong and the date November 1836, signed in ink on the wood underneath the buttplate. This proves, of course, that Armstrong did make percussion guns late in his life, and that he was working at that late a date. Apparently he continued to work during the late 1830's and early 1840's. Nathaniel Rowe, born in 1821, apprenticed with Armstrong. This would have been in the period 1838-1842, most likely. According to Helman (1906),[12] Rowe took over Armstrong's gunsmithing business when the old man died. Thus it seems likely that Armstrong died during the 1840's, probably in 1842, of about 70 years. Records relating to his death have not yet been located.

Nathaniel Rowe's early rifles resemble those of his master. A few surviving specimens exhibit relief carving, patchbox design, furniture, and stock architecture similar to the work of Armstrong. On signed specimens the last name is spelled "Row" though the preferred spelling apparently is "Rowe." As time went on, N. Rowe made rifles more in keeping with the styles of the mid-nineteenth century. The relief and incised carving was dropped, the butt stocks became thinner and smaller in dimension from toe to heel, and when patchboxes were used they were of simpler design. Rowe lived to be 94 years of age and died in Emmitsburg in 1915. Thus master and apprentice together lived and worked in three centuries.

Peter White was an outstanding craftsman who is best known for his rifle-building in Bedford County, Pennsylvania, and later in Uniontown. It is probable that he spent his early working years in Emmitsburg and perhaps learned the trade there also. Peter White was born in late 1777 or during the first eight months of 1778, according to information on his tombstone in Uniontown. The federal census for Frederick County lists a Peter White as the head of a household in Emmitsburg in the year 1800. The name appears on the same page as that of John Armstrong and there are only three or four other names separating the two. Peter White's age is given as being less than 26. Peter White, the gunsmith, had a son, John, who was a gunsmith and who was born in Maryland in 1800 according to the 1860 federal census concerning John.

Two outstanding rifles made by Peter White have survived which fit with the Emmitsburg school and suggest that the maker worked in Emmitsburg. One of these, illustrated by Hetrick (1959),[13] is dated 1794. The other piece was illustrated by Kindig (1960).[14] If the date of 1794 is correct, then this rifle was made by White when he was 16 years of age. Though an apprentice usually did not sign rifles while serving as an apprentice, it is not impossible that he made this piece at an early age. The apprenticeship system was not rigidly controlled by a guild, and a lenient master might very well allow an apprentice to sign a rifle, perhaps for his own personal use.

Peter White did not own any property while in Emmitsburg, if indeed he was there, nor did he own property later in Bedford or Uniontown. White arrived in Colerain Township, Bedford County, Pennsylvania, in 1806, and immediately went to work at gunsmithing. It is reasonable to suppose that he worked in Emmitsburg until moving to Bedford.

Little is known about two of John Armstrong's apprentices, John Blackburn and George Piper. A beautiful relief-carved longrifle is known bearing the signature of Jacob Sell, the younger, of Littlestown, Pennsylvania, on the barrel, and the name "G. Piper" on the lockplate. The work is done in a style similar to that of Armstrong and does not appear to be by the hand of Sell. What happened to Blackburn after he finished his apprenticeship is unknown.

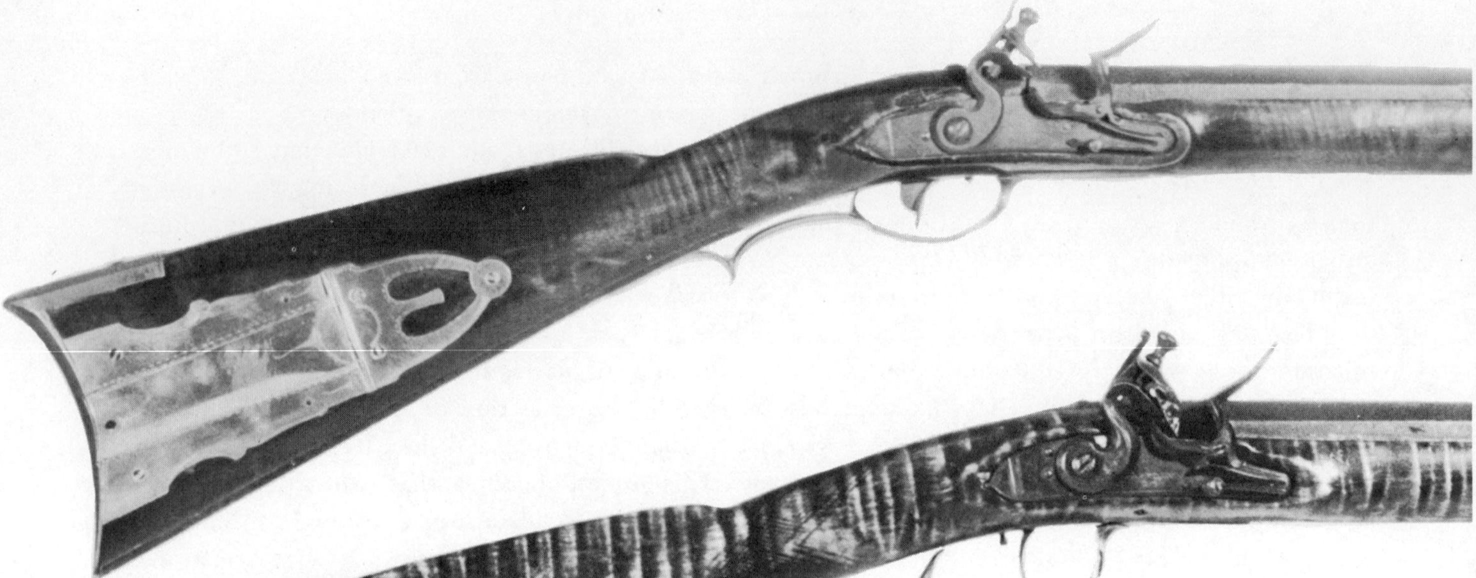

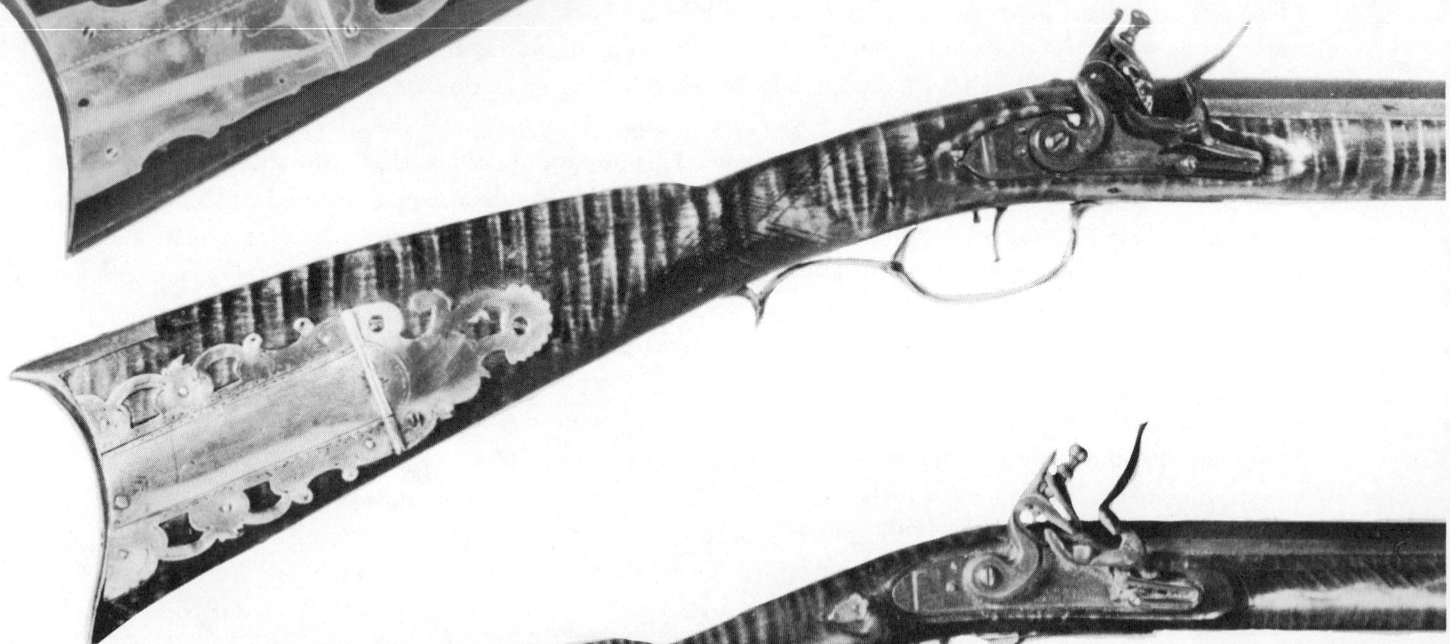

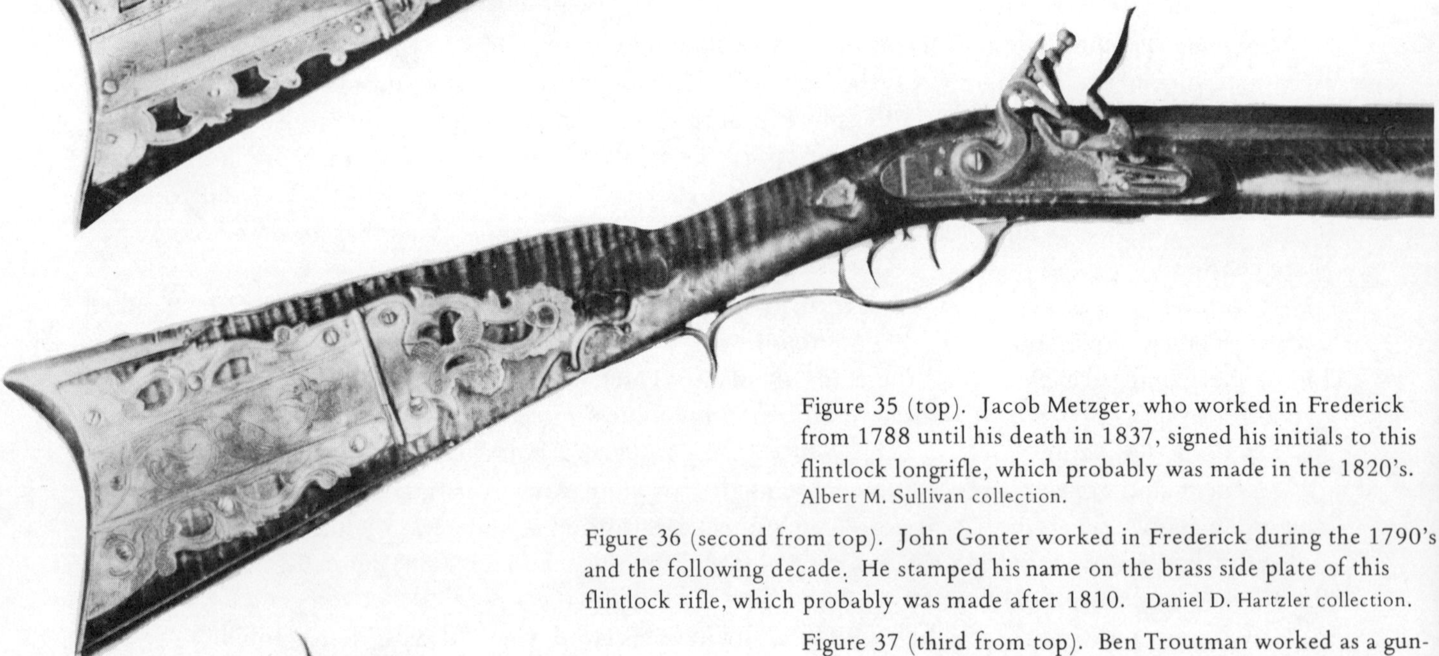

Figure 35 (top). Jacob Metzger, who worked in Frederick from 1788 until his death in 1837, signed his initials to this flintlock longrifle, which probably was made in the 1820's. Albert M. Sullivan collection.

Figure 36 (second from top). John Gonter worked in Frederick during the 1790's and the following decade. He stamped his name on the brass side plate of this flintlock rifle, which probably was made after 1810. Daniel D. Hartzler collection.

Figure 37 (third from top). Ben Troutman worked as a gunsmith in Frederick after learning his trade in Cumberland. This particularly ornate rifle is the finest known piece by his hand. Joe Kindig, Jr., collection.

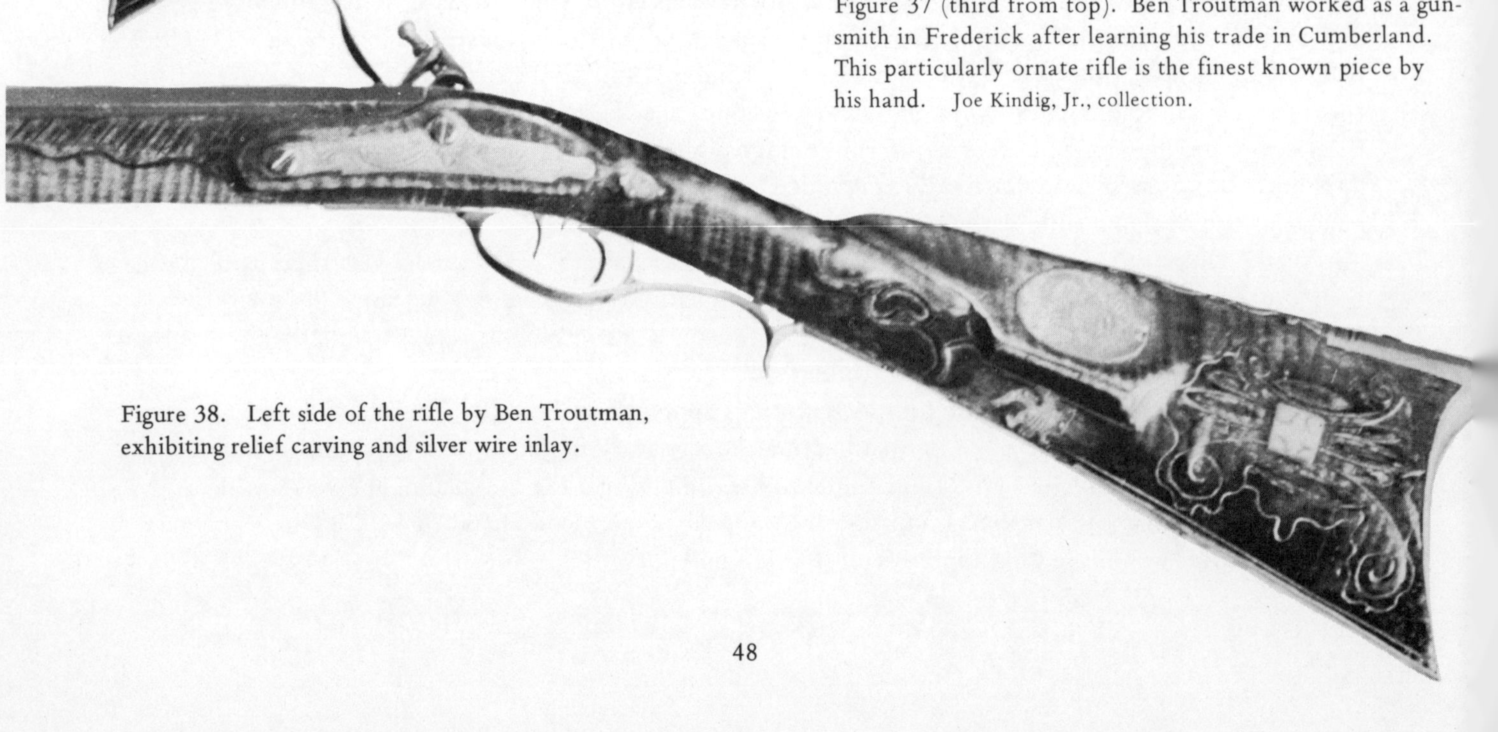

Figure 38. Left side of the rifle by Ben Troutman, exhibiting relief carving and silver wire inlay.

Tiler Wickham, John Armstrong's first known apprentice, had a long and distinguished career in the military arms field. After completing his apprenticeship and reaching the age of 21 in 1802, he bought a small town lot in Emmitsburg and apparently made rifles there for a few years. Then he moved to Jefferson County (now West Virginia), presumably to work at Harpers Ferry. By 1808 he was at work as an arms inspector, and at some time during the next few years he moved to Philadelphia where he lived and worked until his death in 1833. He worked as Inspector of Arms at Schuylkill Arsenal near Philadelphia and was the designer of the U. S. Model 1812 Musket and the Model 1813 Pistol. By 1814 he was Chief Armourer there. He went into business for himself after 1817, as a gunsmith, hardware merchant, and military hardware dealer. His full name was Marine Tiler Wickham, and generally he went under the name of M. T. Wickham. Beginning in 1829 he manufactured military arms under the name of M. T. Wickham & Co.

One additional comment about rifle-making at Emmitsburg deserves to be made here. The great majority of gunsmiths who made longrifles in America used flintlocks, and later percussion locks, that were made in Europe. In the eighteenth century these locks often came from the Germanic regions of the continent, but after the Revolutionary War was settled, flintlocks typically came from Birmingham, England, often with the name of an American hardware dealer stamped on the lockplate. John Armstrong stands out as a gunsmith who usually made and signed his own flintlocks, contrary to the usual practice. His apprentices learned to make gun locks as well as the other parts of the rifle and the apprentice indentures make specific mention of the lock making instruction. It is noteworthy that Peter White, who probably worked with Armstrong and who possibly apprenticed under him, also made most of his own flintlocks until his later years. Furthermore, there is a similarity of design to the locks made by Armstrong and White.

One of the two most important centers for longrifle making in old Maryland was Frederick. In some respects Frederick was a jumping-off place for people heading west down the Great Wagon Road. Emigrants and other travelers always could count on finding an abundance of rifles for sale there. It already has been noted that the town was a center for rifle activity in the Revolutionary War period, and that a gun lock manufactory was in operation there during the war. But in spite of the importance of Frederick as a rifle center, the overall picture of rifle-making there during the flintlock period is none too clear. Research on the gunsmiths who worked there is seriously hampered by the theft of the early tax assessment records from the court house during the 1960's.

The earliest-known gunsmith in Frederick County was Gabriel Friend, who made out a curious will in 1759 leaving money and his gunsmith tools to his brothers Charles and Jacob. He died shortly afterward, in early 1760. There are no known surviving examples of his work.

It was noted earlier that at the outbreak of the Revolution there were four operating gunshops in Frederick, and a fifth nearby, which implies that there were five master gunsmiths, possibly some apprentices, and perhaps a few journeymen gunsmiths. We cannot be certain who these five master gunsmiths were, but it is likely that one of them was Nicholas White who was an important maker of muskets there during the early years of the war. White was a gunsmith who it would be desirable to know more about. He was established as a gunsmith there in 1775 and doubtless made rifles there, but unfortunately we do not know of any rifles made by his hand that have survived. The possibility exists that Nicholas White was the father of the gunsmith Peter White. There is no documentary evidence proving this hypothesis or even suggesting it, but the facts that are known do not rule it out. In the fall of 1775 Nicholas White, gunsmith, purchased lot No. 179 in Fredericktown. For the next two years he was actively engaged manufacturing muskets for the Council of Safety, and there are a number of documents relating to this activity. He apparently continued working in Frederick as a gunsmith until 1788 when he removed to Berkeley County, Virginia. How long he stayed there is not known, but he was back in Frederick nine years later. In May 1797 Nicholas White, gunsmith, took as an apprentice an orphan boy named Nicholas King, age 16 years. The following year White joined together with Frederick gunsmiths Jacob Metzger, Thomas Crabb, and Christoph Barnhizle to take a government contract for 1000 stands of muskets. White's activities as a gunsmith are unknown after 1800. In 1807 he purchased lot No. 27 in the town of Berlin (Brunswick) on the Potomac River south of Frederick. His name ap-

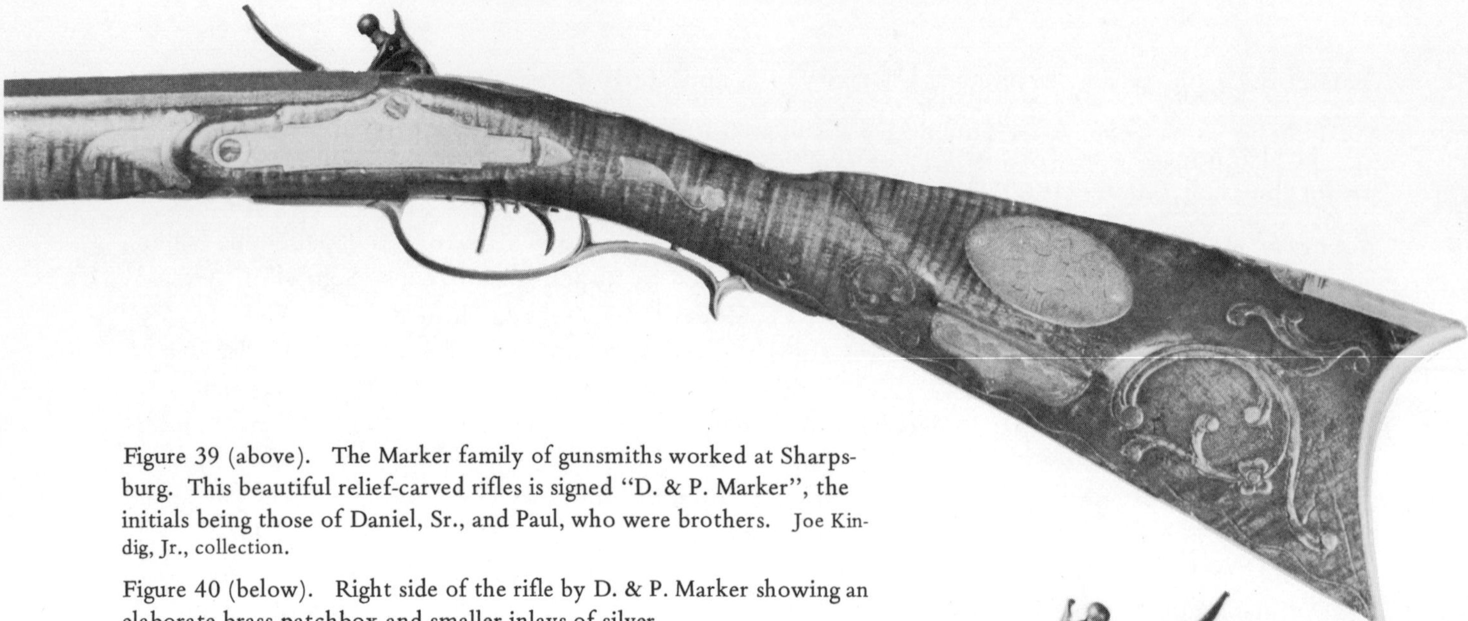

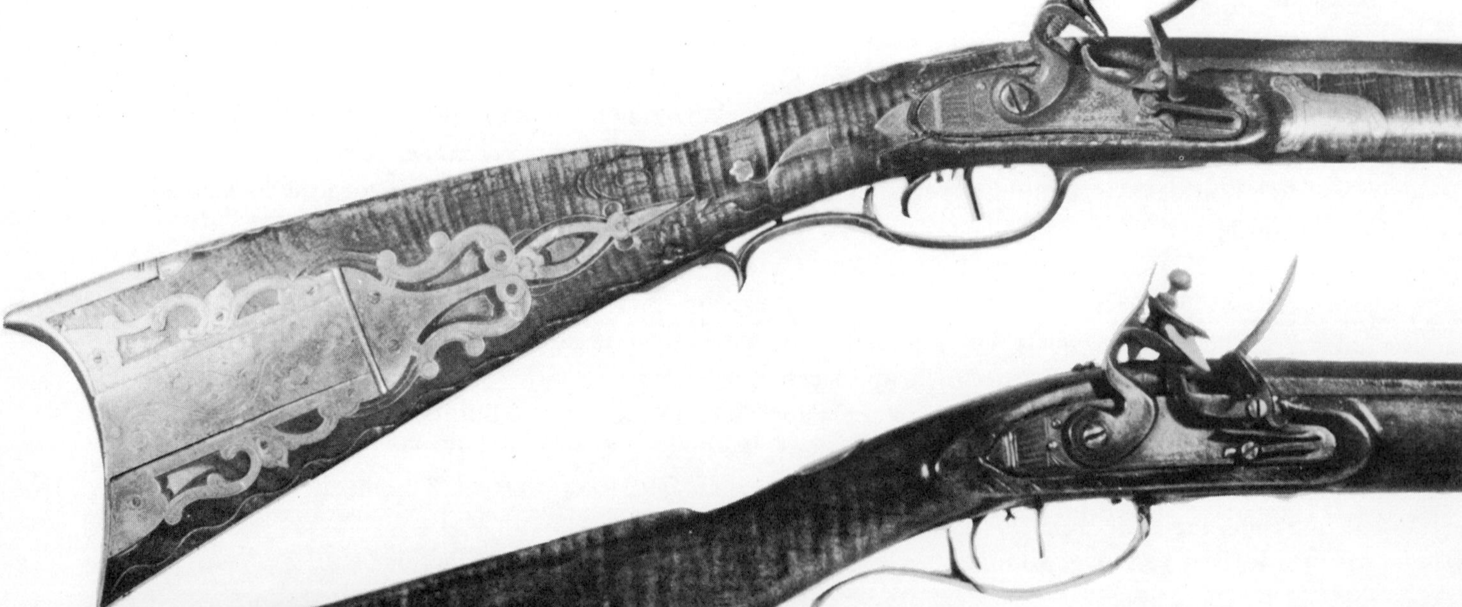

Figure 39 (above). The Marker family of gunsmiths worked at Sharpsburg. This beautiful relief-carved rifles is signed "D. & P. Marker", the initials being those of Daniel, Sr., and Paul, who were brothers. Joe Kindig, Jr., collection.

Figure 40 (below). Right side of the rifle by D. & P. Marker showing an elaborate brass patchbox and smaller inlays of silver.

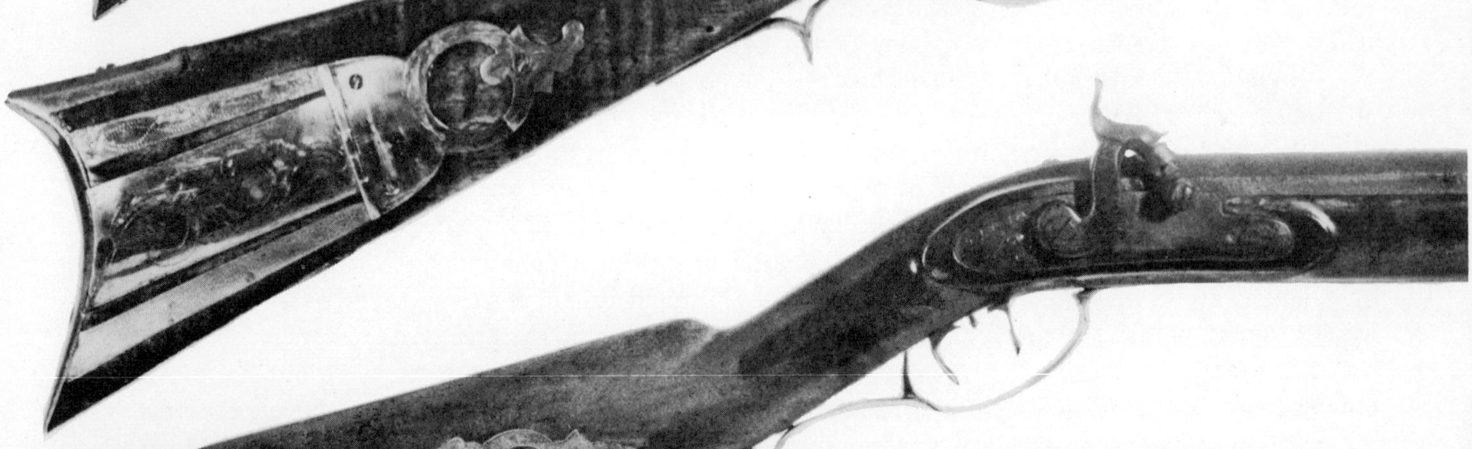

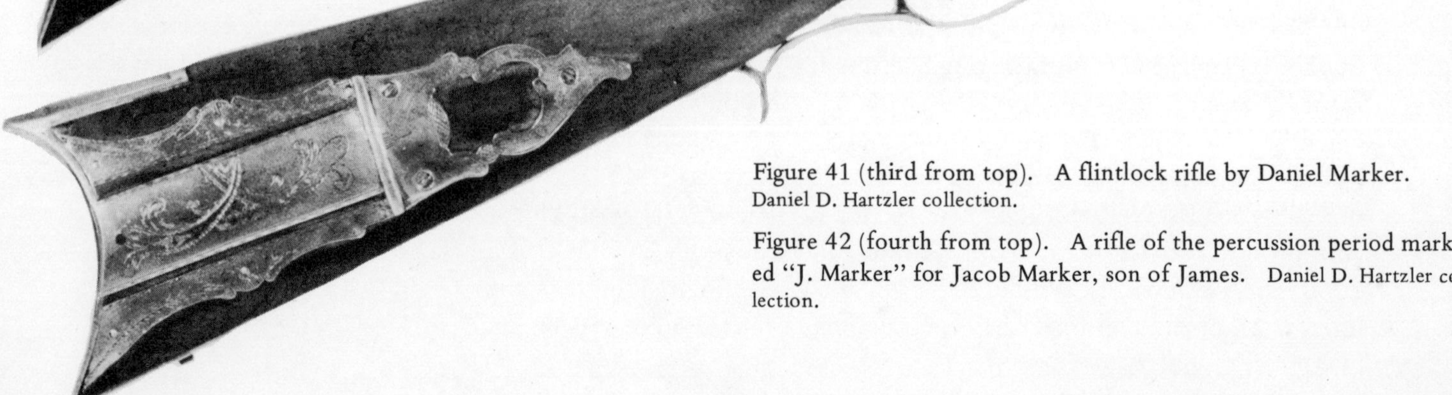

Figure 41 (third from top). A flintlock rifle by Daniel Marker. Daniel D. Hartzler collection.

Figure 42 (fourth from top). A rifle of the percussion period marked "J. Marker" for Jacob Marker, son of James. Daniel D. Hartzler collection.

pears in the 1800 census for Frederick County, and apparently he was over 45 years of age then. His name again appears in the census for 1810.

Jacob Metzger is one of the better-known gunsmiths of Frederick, but signed examples of his work are not abundant. Kindig (1960)[15] discussed him briefly and illustrated three of his rifles, all classic examples of the Maryland longrifle in its golden age. Metzger produced beautiful flintlock rifles decorated with engraved brass and silver inlays, and with relief and incised carving.

Metzger was born between 1760 and 1765 according to census information. In 1777 his name appears with the designation of gunsmith as a "Non-Associator" in the Borough of Lancaster, Pennsylvania. Presumably he was an apprentice or trainee between 15 and 17 years of age. At some time during the next ten or eleven years he settled in Frederick and in January 1788 he purchased a town lot there. Deeds of 1788 and 1793 mention his occupation as gunsmith, and the 1788 deed mentions that his wife was named "Christona," doubtless a misspelling for Christina. It already has been mentioned that Metzger was involved in the 1798 musket contract with White, Crabb, and Barnhizle. It may be presumed that Metzger continued active in gunsmithing for 20 years or more after 1800, but little is known about his later years. He lived a long life and died in 1836 or 1837 at an age of 75 plus or minus one or two years. Jacob Metzger, Sr., had a son, Jacob, Jr., of unknown occupation, and at least one other child.

Disturbances between the patriots of Frederick County and various religious denominations, known as the dissenters, resulted in large migrations out of the county. The majority of the denominations harassed were German; the members said they owed no allegiance to England or the Colony and refused as they often had previously in Europe to serve in the military, pay taxes or fines. In this way the Frederick area lost at least eight gunsmiths. Henry Roth, Sr., a member of the Church of the Brethren, removed from the county sometime after 1776 to the Carlisle, Pa., area. Jacob Mier and his son, Samuel, migrated from Frederick County to Salisbury, Pa., shortly before the Revolutionary War. Peter Engle and his family of 13, of whom 4 were gunsmiths, removed from the county in the middle of the 1790's to what is now Garrett County, Md.; in 1800 he returned from Pennsylvania to dispose of his remaining property in Western Maryland. Their membership had been in the Frederick German Reformed Church.

A number of other gunsmiths worked in Frederick during the flintlock period, but they cannot all be treated here in detail, both because space is limited and because information about them is limited. For example, Ben Troutman advertised in a Frederick newspaper for September 1810 that he "carries on the gunsmith business on Market Street at the sign of the gun," but that is all we know about his work in Frederick. The name of Benjamin Troutman, gunsmith, also appears on tax lists in Somerset County, Pennsylvania, for the years 1810, 1811, and 1817.[16] Presumably this is the same man.

John Demuth worked in Lancaster, Pennsylvania, during the late 18th century and his family operated a tobacco shop there.[17] In 1794 Demuth was one of the group of gunsmiths who supplied rifles to General Edward Hand for government purchase. But in 1796 John Demuth, gunsmith, of Frederick County, Maryland, took as an apprentice a youth of 16 years named Henry Koch.

Daniel Marker, George Marker, Jr., and Paul Marker were three brothers who worked during the flintlock period as gunsmiths in the region of Sharpsburg and Myersville, about halfway between Frederick and Hagerstown, and later in life emigrated to western Ohio where they settled in the region of Darke, Preble and Montgomery Counties. Daniel made beautiful relief-carved rifles, some of which were ornamented with attractive engraved inlays of brass and silver. The work of George, Jr., is less notable. A few rifle by Paul Marker have survived showing very fine work in both carving and engraving. Daniel's son, James, was a gunsmith also in Sharpsburg, and he carried on the work there after his father died.

Another family of gunsmiths working midway between Frederick and Hagerstown were the Stonesifers, who were located at Boonesboro. John Stonesifer, born in 1813, was working there as a gunsmith in 1850. His son, Emory C., born in 1838, advertised himself as a gunsmith and whitesmith in Boonesboro in 1860. It is possible that John was actually John, Jr., for the 1790 federal census lists one John Stonesifer in Washington County, with an age between 16 and 26.

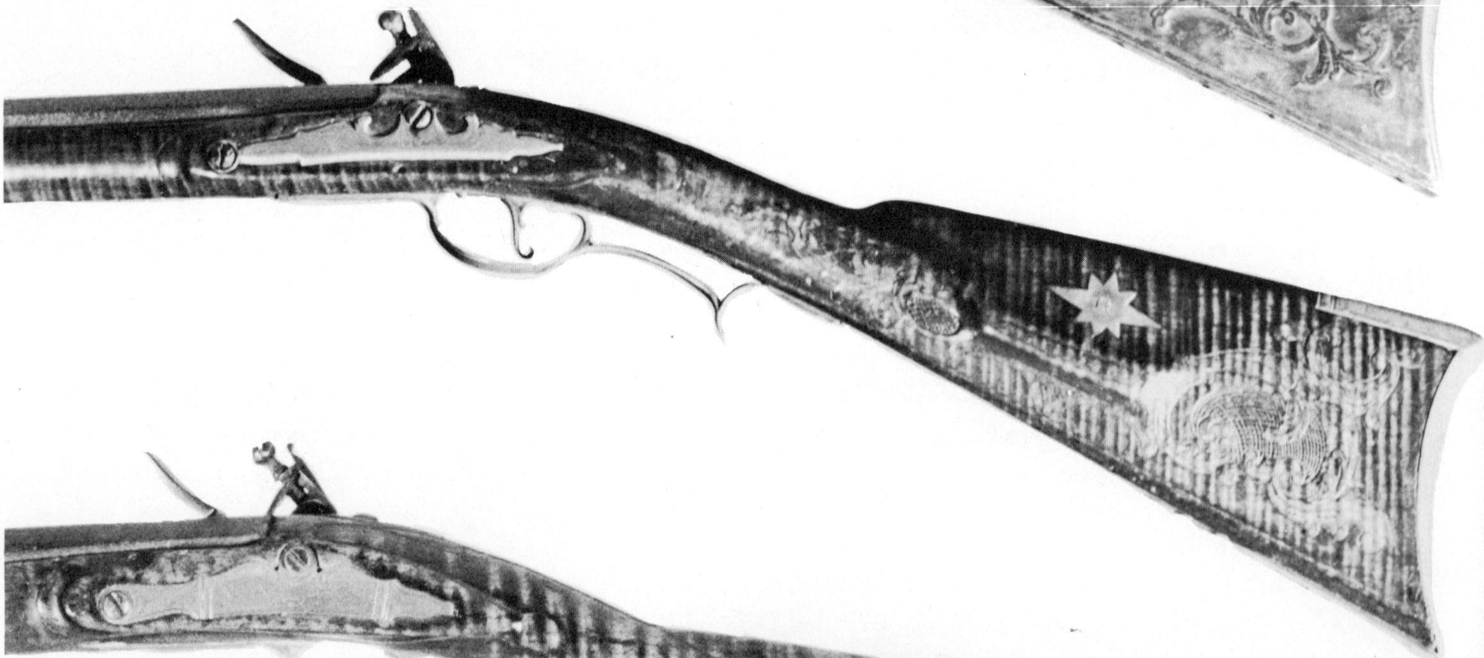

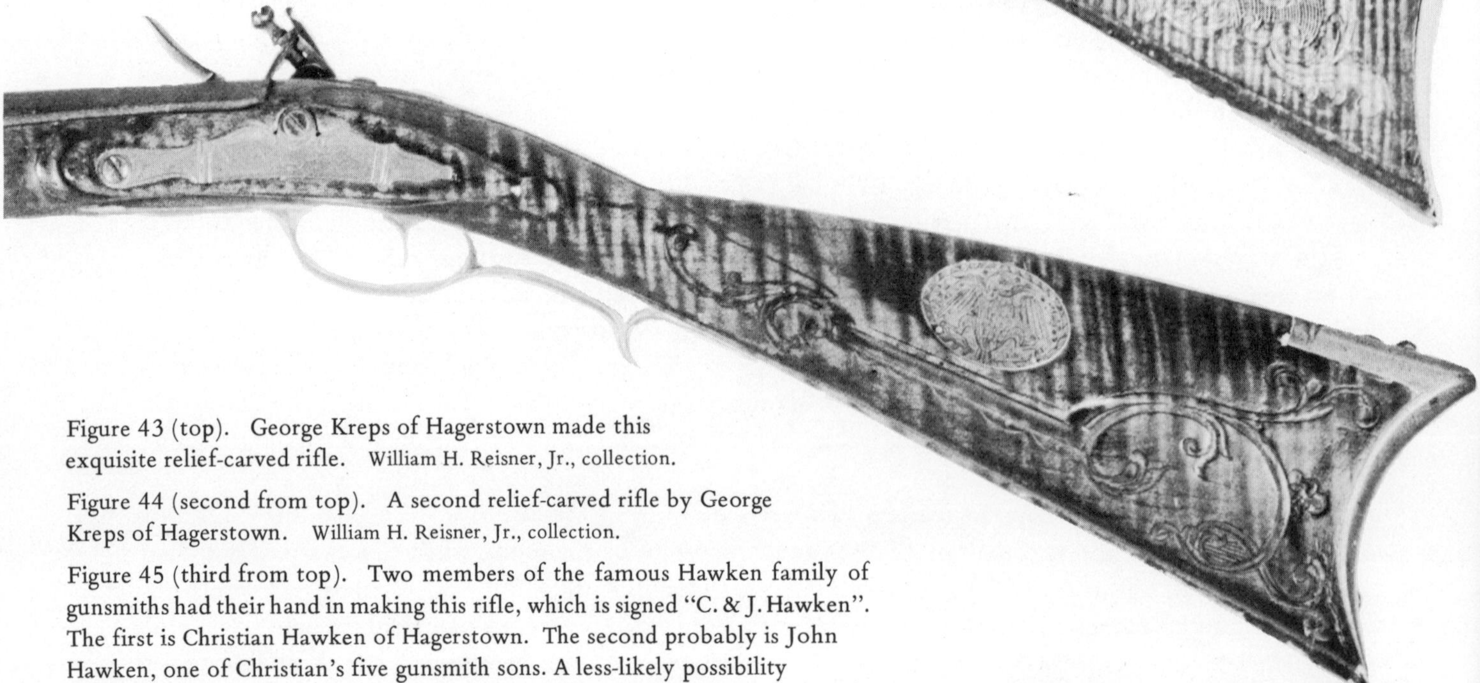

Figure 43 (top). George Kreps of Hagerstown made this exquisite relief-carved rifle. William H. Reisner, Jr., collection.

Figure 44 (second from top). A second relief-carved rifle by George Kreps of Hagerstown. William H. Reisner, Jr., collection.

Figure 45 (third from top). Two members of the famous Hawken family of gunsmiths had their hand in making this rifle, which is signed "C. & J. Hawken". The first is Christian Hawken of Hagerstown. The second probably is John Hawken, one of Christian's five gunsmith sons. A less-likely possibility would be another of Christian's sons, Jacob, but he left Hagerstown upon reaching 21 years of age in 1807, and moved to St. Louis where he gained fame in later years for his plains rifles. Harry D. Berry, Jr., collection.

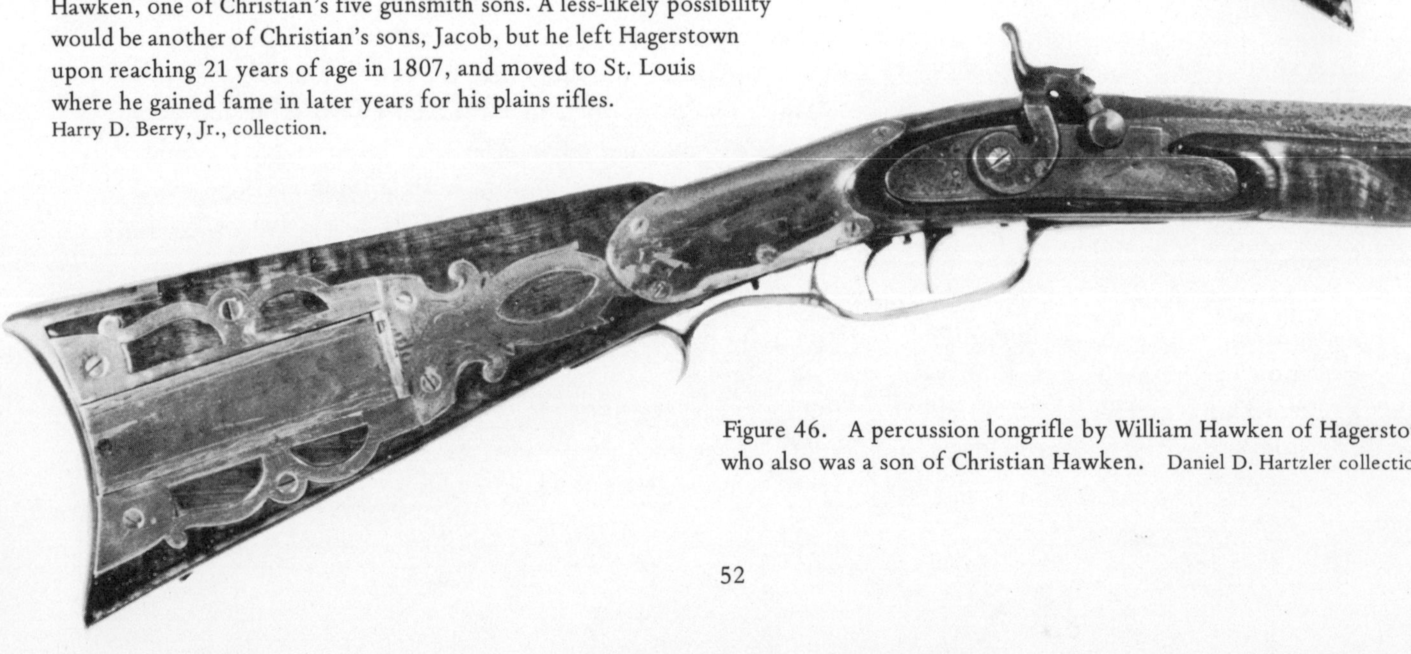

Figure 46. A percussion longrifle by William Hawken of Hagerstown, who also was a son of Christian Hawken. Daniel D. Hartzler collection.

Hagerstown, which lies west of Frederick and is separated from it by a mountain ridge, was another important rifle making center in the late 18th and early 19th centuries. The names of more than 20 gunsmiths who worked at Hagerstown are known. Again, it is not feasible to cover all of these men here, and very little is known about a number of them.

George Kreps of Hagerstown made particularly beautiful relief-carved rifles in the late 18th and early 19th century. The earliest record of Kreps in Maryland is a deed of 1795 by means of which "George Kreps late of Frederick County in the Commonwealth of Virginia" purchased a tract of land in Elizabethtown, Washington County. However, he may have been in Maryland for a number of years before this, for he had a son, George, Jr., born about 1789 or 1790 and apparently in Maryland. In January 1798 Kreps took as an apprentice to the gunsmithing trade a youth of 16 years named Martin Budenbaugh. Another youth, William Criner, was apprenticed to Kreps to learn "The art, trade and mystery of a Gun smith and Rifle smith" in 1802. At some time after 1810 George Kreps apparently turned to cabinet making, or combined it with gunsmithing, for he took an apprentice to this trade in 1813, another in 1814, and two in 1816.

George Kreps, Jr., clearly was no match for his father. In 1807 he was apprenticed into the cordwainer or shoemaking trade, but he got into gunsmithing also. He attempted to make rifles for the Office of the Indian Trade in 1811 but his work was so poorly done that it was rejected. He continued with the profession of gunsmithing but then joined the U.S. Army. In 1819 he was being sought as a deserter from the army.

The famous Hawken family of gunsmiths had their beginnings in Hagerstown. Christian Hawken, who worked throughout his long life as a gunsmith in Hagerstown, was the patriarch of the family. He made flintlock rifles ranging from beautiful relief-carved pieces to the more common type. Christian was born in 1756, probably in Switzerland or Germany, and apparently came to Hagerstown in the Revolutionary War period. By 1784 he was established there as a gunsmith and bought property. He died in 1821 at 65 years of age, leaving an estate inventory that lists many gunsmithing tools. He had three daughters and six sons. All but one of the sons learned the gunsmithing trade and worked at it. William remained in Hagerstown and continued in his father's shop after 1821. Jacob moved to St. Louis at some time before 1818. Samuel Hawken probably worked as a gunsmith in Hagerstown for a short period, but he came of age in 1813 while the War of 1812 was underway. He served in the War, and after it was over he married and moved to Xenia, Ohio, where he worked as a gunsmith until 1821; in 1822 he moved to St. Louis to join his brother Jacob. Jake and Sam Hawken were the well-known gunsmiths of St. Louis who made the famous Hawken plains rifles, so highly prized by the mountain men of the Rockies. George Hawken apparently worked as a gunsmith until his death in the 1820's. Of the two remaining sons, Christian, Jr., and John, one was a gunsmith and one was not. John, who probably was the gunsmith, died before 1821, and Christian, Jr., in 1822.

Additional members of the Hawken family worked in Hagerstown as gunsmiths. Nicholas Hawken, who was a bit older than Christian Hawken, Sr., was at work as a gunsmith in Hagerstown from the late 1770's until after 1800. His relationship to Christian, Sr., is uncertain, but it is probable that he was a brother. John C. Hawken, possibly a son of George, took over William Hawken's gunsmith stand for a short time in 1840.

A gunsmith named Henry Hawken may have had his beginnings in Hagerstown, possibly as a son of Nicholas Hawken. Between 1796 and 1815 he worked in Greensburg, Pennsylvania, and his later years were spent at Columbus and Springfield, Ohio. He was the father of the Ohio gunsmiths John Hawken and Daniel Turney Hawken.

John Gonter was another early gunsmith of Hagerstown. In 1794 he took as an apprentice to the gunsmithing trade Alexander Compton. About two years later Gonter took another apprentice, David Whetstone. David's brother, John Whetstone, apprenticed with Gonter also. William McIntosh became an apprentice of Gonter's in 1799, and Samuel Hammell became an apprentice in 1800. Gonter left Hagerstown by 1810, when he was working in Columbia, Pennsylvania, and then he moved on to Lancaster. In 1824 he was established in the gunsmithing business in Reading, Pennsylvania.

Proceeding westward from Hagerstown, the last important center for rifle-making was Cumberland

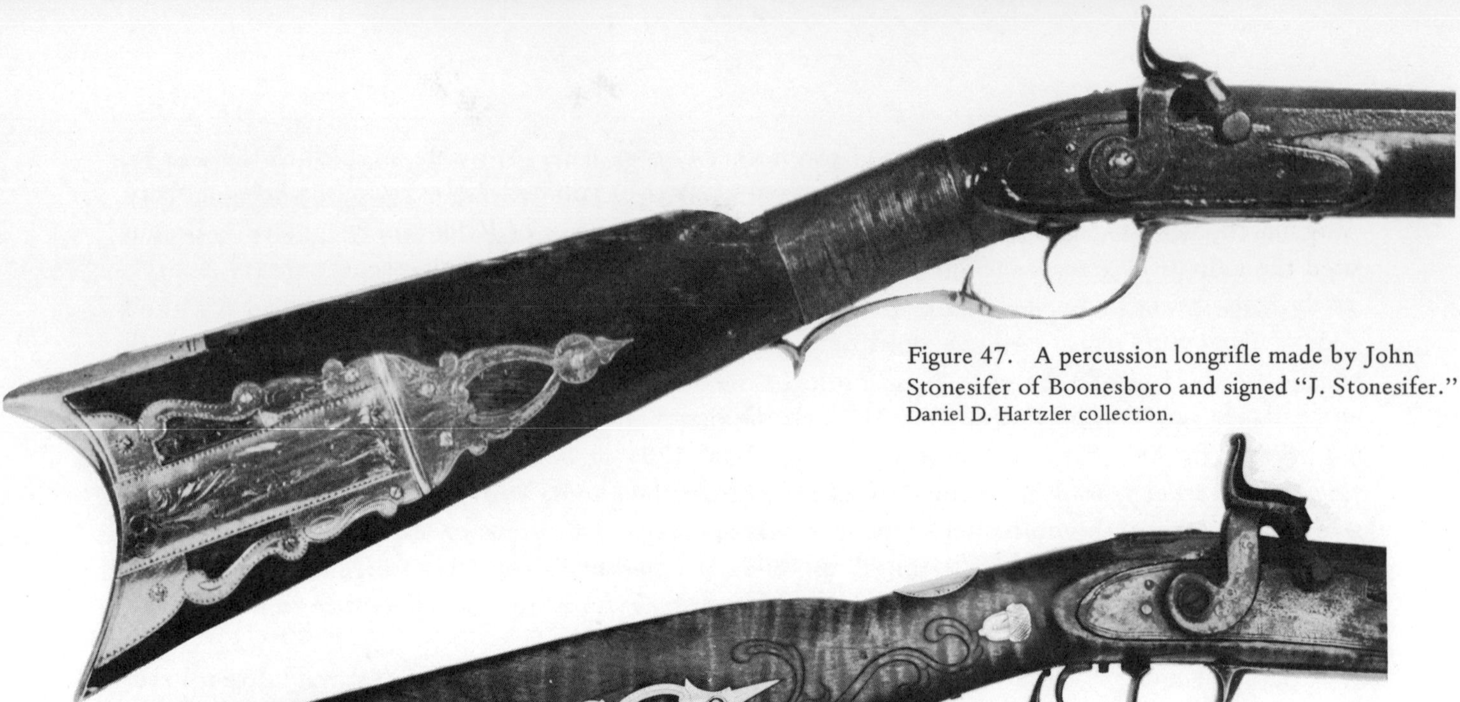

Figure 47. A percussion longrifle made by John Stonesifer of Boonesboro and signed "J. Stonesifer." Daniel D. Hartzler collection.

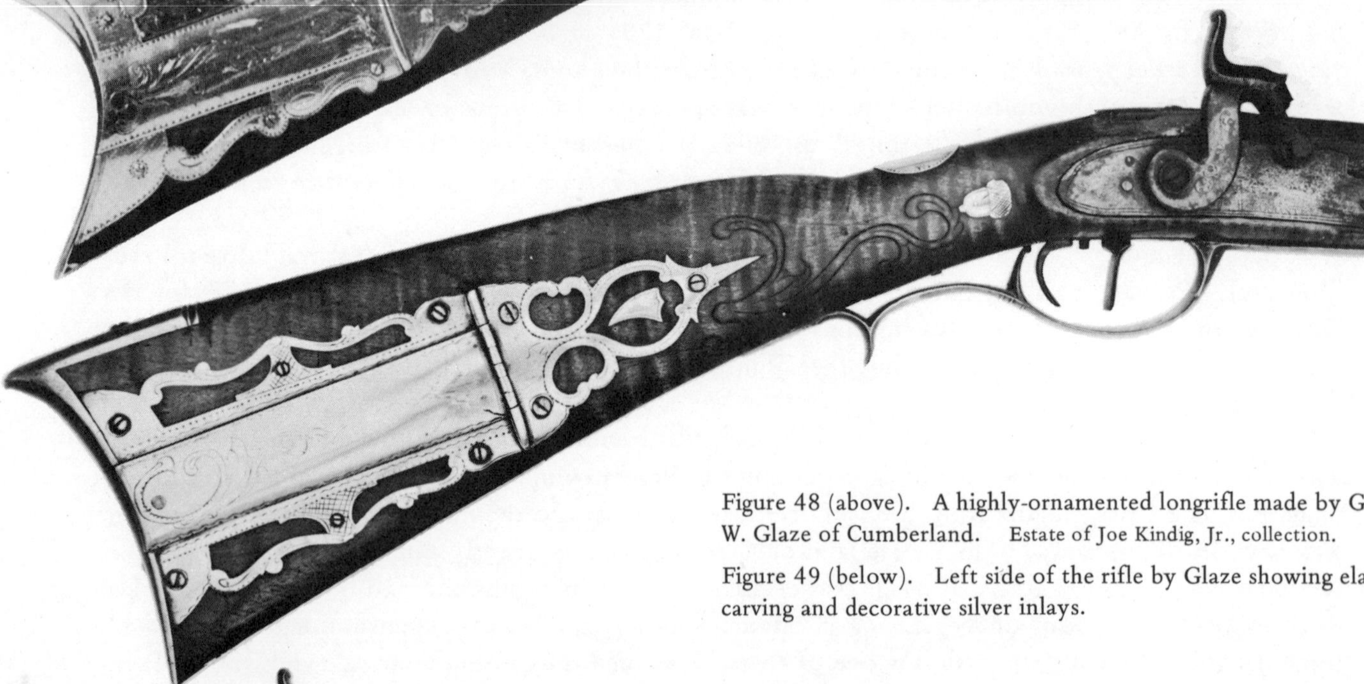

Figure 48 (above). A highly-ornamented longrifle made by George W. Glaze of Cumberland. Estate of Joe Kindig, Jr., collection.

Figure 49 (below). Left side of the rifle by Glaze showing elaborate carving and decorative silver inlays.

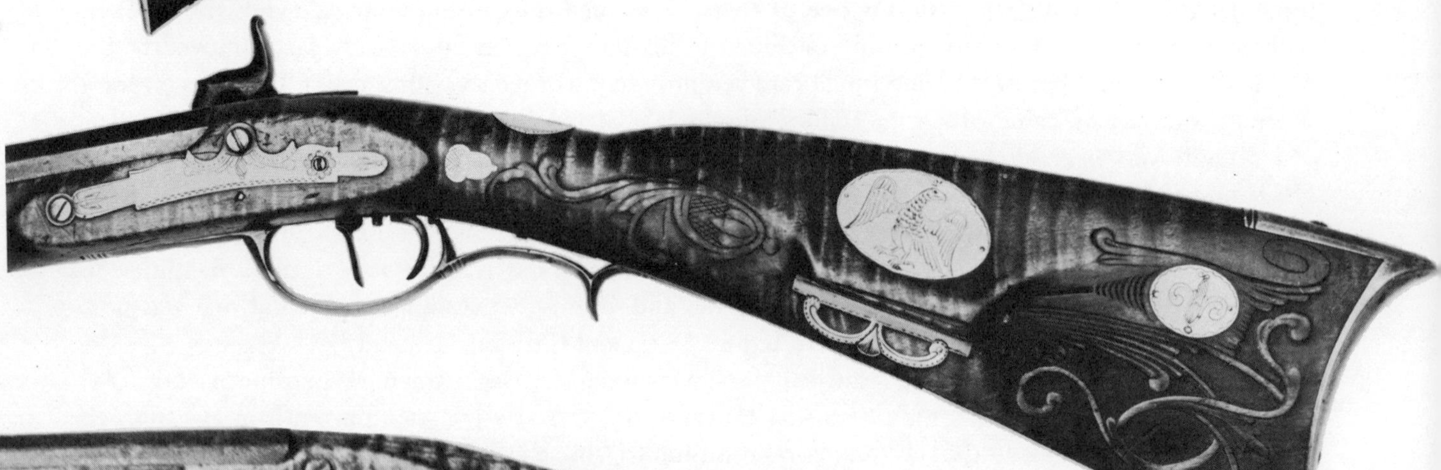

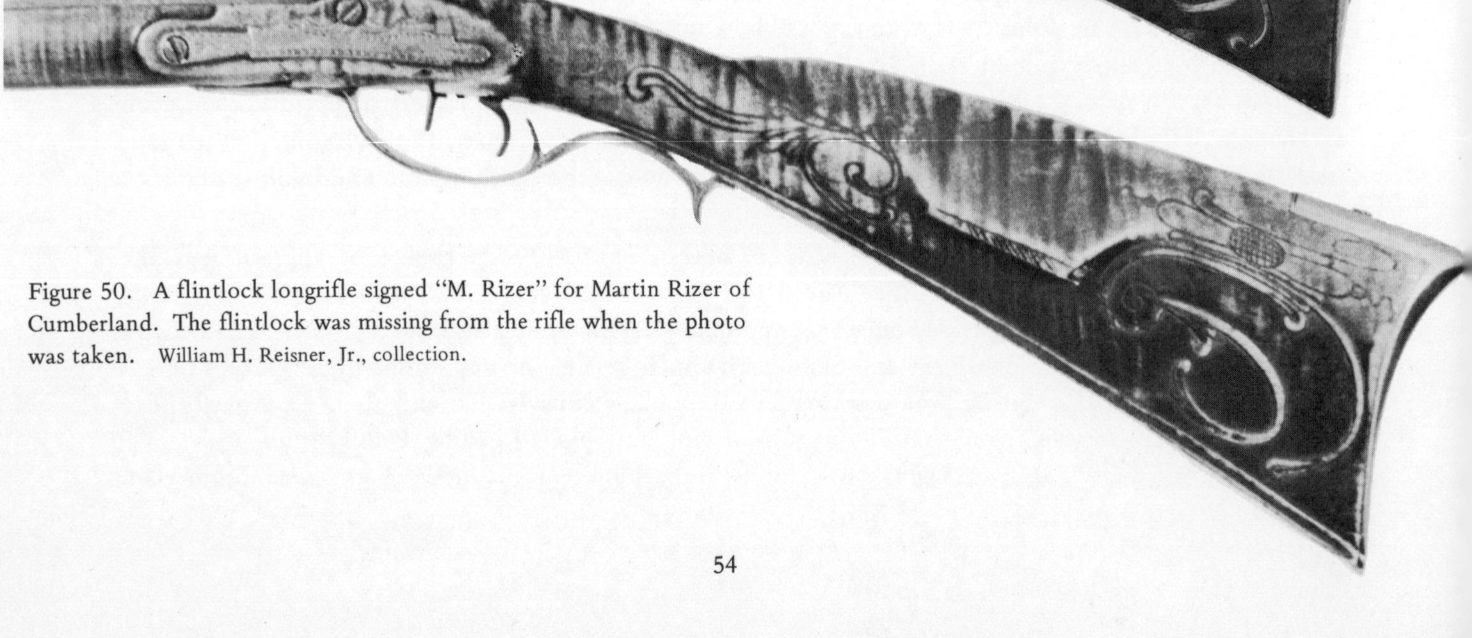

Figure 50. A flintlock longrifle signed "M. Rizer" for Martin Rizer of Cumberland. The flintlock was missing from the rifle when the photo was taken. William H. Reisner, Jr., collection.

in Washington County. Martin Rizer and his brother George Rizer, Sr., were established there in the 1790's having spent their early working years and likely their apprenticeships too in Berkeley County, Virginia, probably at the town of Martinsburg (now West Virginia). The Rizer family apparently dominated the gunsmithing scene in Cumberland from the 1790's until the mid-nineteenth century, but relatively little can be found in the old records to document it. In 1806 the two brothers inquired of the federal Purveyor's office in Philadelphia about making rifles for the government. In November 1808 the Office of Indian Trade agreed to take 1000 rifles from "Mr. Riser" the following March, but it is unknown if this contract was carried out. Longrifles signed by the Rizers are scarce, but at least two are known bearing Martin's signature, one of which is relief-carved, and both of which hold well-designed and engraved patchboxes. Strohfeldt (1974) reported that in 1800 George Rizer took as an apprentice to the gunsmithing trade one Jacob Rizer, born in Berkeley County, Virginia, in 1785. By 1806 Jacob and his brother Phillip moved to Bardstown, Kentucky, and Jacob was active as a gunsmith there until his death in 1859. William L. Rizer was at work as a gunsmith in Cumberland in 1850, but his relationship to the families of George and Martin Rizer is unknown.

The last area to consider for longrifle making is Baltimore. It already has been mentioned that European influence on arms making was strong in Baltimore. During the flintlock period it was fine English arms that set the standards for the arms business there. But during the 1830's, 1840's and 1850's an influx of German gunsmiths and German emigrants as well to both Baltimore and Washington, D.C., showed its effect on percussion rifles.

Longrifle-making in Baltimore during the flintlock period seems to have been done on a limited basis, judging from the relatively few specimens that have survived and the general absence of knowledge about such work. One or more typically fine silver-mounted longrifles have survived that were made by James Haslett. Some rifles were made by John Stewart while he worked in Baltimore between 1810 and 1819. George Schryer worked in Baltimore between 1810 and 1815. This probably was George Schroyer, Jr., son the the famous and highly skilled gunsmith of Hanover, Pennsylvania. George, Jr., was no match for his father and his work at best was average and more typically it was below average if not poor for the period. George, Jr., made longrifles and a number of them have survived, but it is impossible to determine if any of them were made in Baltimore rather than in Hanover where he also worked. During the percussion period, Charles C. C. O'Brien made rifles and other guns in the period between 1833 and his death in 1841. His apprentice, Alexander McComas, carried on with the manufacture of fine sporting rifles from 1841 until late in the century.

Fortunately the manufacture of longrifles in Maryland did not stop with the invention of the breech-loading rifle. There were a few who carried on the old traditional work into the 20th century. One of these was Abel Browning who lived north of Oakland at the extreme western end of the state. Browning was born in 1854, and he was still active in his trade during the 1930's. Elmer P. Wolf, born in 1868, was another. He worked near Thurmont in Frederick County until about 1940.

To complete the picture and bring it down to the present time, mention must be made of those who in the present day and over the past five or ten or twenty years have made longrifles in Maryland. A revival of interest in the American longrifle began in the 1920's and 1930's and grew slowly until the 1960's when it mushroomed. This activity involved shooting the muzzle-loading rifle, making rifles, and studying the art of the rifle. Foremost among the modern makers and widely regarded as the "dean" of modern longrifle makers is Carl Pippert of Bladensburg. Carl has built rifles for more than 30 years, and still works on them. Carl's ability to design beautiful rifles and to execute the work necessary to produce them equals that of the finest makers of the past. In recent years George Stanford has become a most accomplished maker also. Rudolph P. Bahr, a cabinetmaker of College Park, is highly skilled in the manufacture of beautiful rifles. Robert C. Orr makes a fine rifle as does Ronnie Griffie, and there are a good many others at work also. The future will see many additional longrifle makers in Maryland as interest grows in them.

REFERENCES

1. Kindig, Joe, Jr., (1960, 1976) THOUGHTS ON THE KENTUCKY RIFLE IN ITS GOLDEN AGE: George Shumway Publisher, York, Pennsylvania.
2. Hetrick, Calvin (1959, 1975) THE BEDFORD COUNTY RIFLE AND ITS MAKERS: George Shumway Publisher, York, Pennsylvania.
3. Dyke, Samuel E. (1959, 1974) THOUGHTS ON THE AMERICAN FLINTLOCK PISTOL: George Shumway Publisher, York, Pennsylvania.
4. Kauffman, Henry J. (1960) THE PENNSYLVANIA-KENTUCKY RIFLE: Stackpole Press, Harrisburg, Pennsylvania.
5. Scharf, J. Thomas (1967) HISTORY OF WESTERN MARYLAND, Vol. II: Tradition Press, Hatboro, Pennsylvania, pp. 182, 183.
6. Ibid.
7. Ibid.
8. Williams, T. J. C. (1910) HISTORY OF FREDERICK COUNTY, MARYLAND: L. R. Titsworth & Co., Publishers, Hagerstown, Maryland, p. 805.
9. Ibid.
10. Ibid.
11. Ibid.
12. Hellman, James A. (1906) HISTORY OF EMMITSBURG, MARYLAND: Citizens Press, Frederick Maryland.
13. Hetrick, Calvin, op. cit., p. 4.
14. Kindig, Joe, Jr., op. cit., p. 541.
15. Kindig, Joe, Jr., op. cit., pp. 465-469.
16. Kauffman, Henry J., op. cit., p. 352.
17. Kauffman, Henry J., op. cit., pp. 206, 207.

CHAPTER 4

Fowling Pieces, Smoothbores and Shotguns

In early America the most widely-used and versatile gun of all was the long-barreled smoothbore gun. To the settler of the eighteenth or early nineteenth century such a piece would serve a dual purpose, or even a triple purpose. It could be loaded with a round ball, or with buckshot, or with both, and used for deer hunting. It could be loaded with smaller shot for small game and birds, which was its more usual service. And when its owner went for militia duty, which was widespread and more or less mandatory, it was accepted as the musket or gun which he was obliged to provide for himself.

Smoothbores suitable for the above-mentioned purposes were imported from Europe in quantity during the flintlock era, but they also were locally made. The longrifle gunsmiths of northern and western Maryland produced many of them. Some were made in the usual manner for smoothbore guns, with full stocks of curly maple, barrels part octagon and part round, a low button for a front sight and no rear sight, and with a simple brass trigger guard. The smoothbore barrels typically were between 44 and 48 inches in length, and bores typically between .50 and .60 caliber.

A second type of smoothbore rifle made by the same piedmont gunsmiths was the "two-sighted smoothbore" or "smooth rifle." Such pieces usually had barrels part octagon and part round, but otherwise were almost exactly like rifles in external appearance. The barrels again typically were 44 to 48 inches in length and the bores typically between .50 and .60 caliber. But the barrels were equipped with two sights, just like rifles. The full-length curly maple stocks often held patchboxes, brass and silver inlays, and incised and relief carving. And the trigger guard was of the kind used on rifles, with a grip rail extending to the rear of the trigger bow. Functionally these pieces were no different from the more usual smoothbore gun except for what advantage the rifle-type sights might have provided.

There is a third type of gun made by the longrifle gunsmiths that fits halfway between smoothbores and rifles that deserves mention here. This is the rifle with grooves inside the barrel that run straight down the bore instead of spiraling down. Such pieces are called straight-rifles. They could be loaded with a patched round ball and shot like a rifle with reasonable accuracy, though not as good as rifle accuracy. But also they could be loaded with buckshot or smaller shot and thus be usable for almost any kind of hunting. More old rifles than one might imagine had this straight rifling.

Maryland's most famous maker of longrifles, John Armstrong of Emmitsburg, produced a small number of smoothbores along with his rifles. Nathaniel Rowe, his apprentice, and a later gunsmith of Emmitsburg, also made them. And they were made at Hagerstown and at Frederick, and elsewhere too. The estate inventory of Christian Hawken, gunsmith of Hagerstown, mentions "1 single barrel shotgun, 1 double barrel shotgun, . . . "

But much of central and southern Maryland lies near to the waterways of Chesapeake Bay and the rivers that lead into them. Waterfowl abound today as they did in the past, though doubtless a large human population and the polluting effects that go with it have altered the scene from what it was in earlier times. In the eighteenth century extra-large fowling pieces sometimes were made to shoot water birds with more efficiency. Many were made in England and imported to America, where they sometimes were called "shallopmen's guns," because they were often carried by men who used rowboats or shallops in the rivers and bays. And of course a variety of other shotguns and fowling pieces were used on the abundant waterfowl.

In the flintlock period double-barreled shotguns were in use in Maryland, but few of them were made in Maryland or anywhere else in America. One of the rarest of all American-made flintlock arms

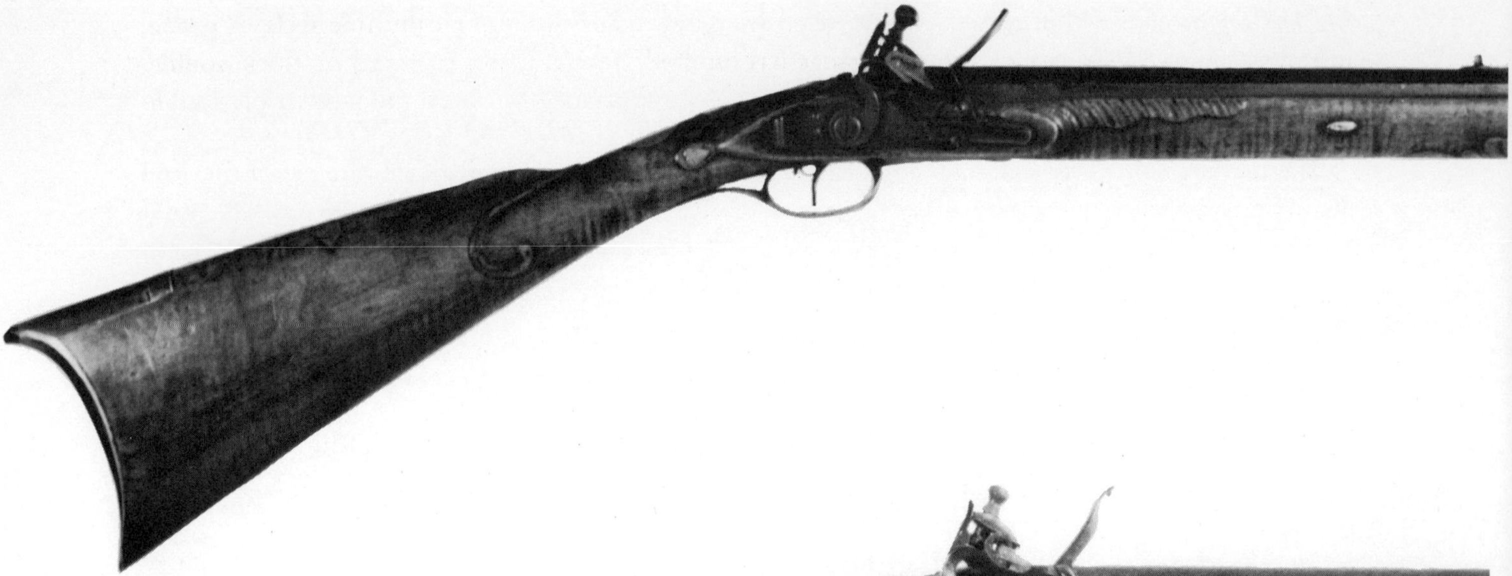

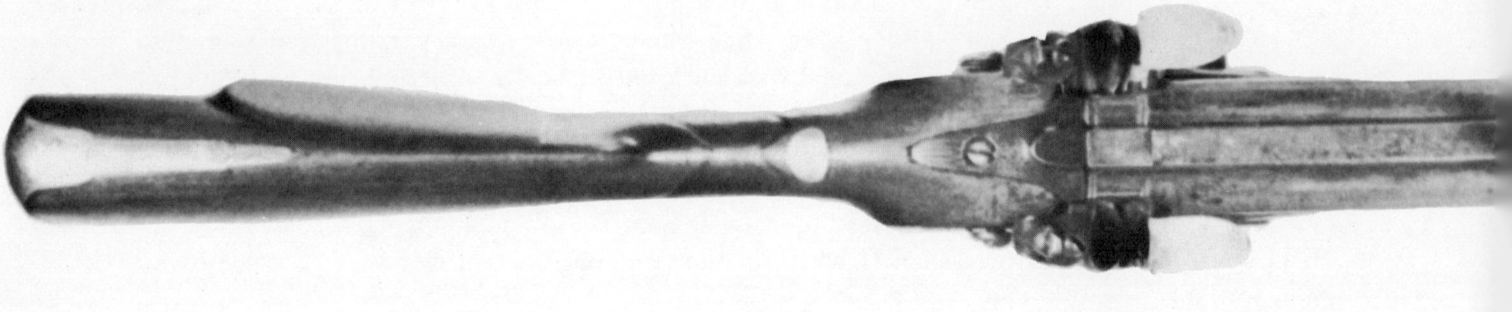

Figure 51 (top). John Armstrong, the famous longrifle maker of Emmitsburg, made this smoothbore flintlock gun. As was typical for smoothbore guns built with rifle lines, this piece has the kind of trigger guard used with smoothbore guns, and it lacks the usual Armstrong patchbox. The gun is equipped with a rear sight and front sight, and was used with a patched ball for deer, with shot for smaller game, or with buckshot and a ball for deer. Albert M. Sullivan, Jr., collection.

Figure 52 (middle). A flintlock fowling piece by James Haslett of Baltimore in 16 gauge. Rev. Thomas C. Jensen collection.

Figure 53 (bottom). A double-barreled flintlock shotgun in 16 gauge, mounted in iron, with a script signature on the barrel rib, "J. Gafford Baltimore". Double-barreled flintlock shotguns of American manufacture are very rare arms. Harry D. Berry, Jr., collection.

is the double-barreled shotgun. A surviving specimen made in Philadelphia is known. In Baltimore, James Haslett produced flintlock double-barreled shotguns, made in his typically fine style. A petite, beautifully-engraved specimen of about 20-gauge has survived. Several finely-engraved flintlock double-guns made by Joseph Gafford have survived. John Keener advertised flintlock and percussion double and single barreled guns.

The market for fowling pieces was considerable in Maryland and many hardware merchants and dealers entered the competition with imported arms. John Walraven of Baltimore advertised on 24 November 1803: ". . . from London and Liverpool – fowling pieces, double and single barreled in cases complete from 50 to 90 dollars. Single ditto common and good from 5 to 55 dollars highly finished."

In the late eighteenth century the Damascus barrel grew in popularity to the point where it was widely accepted for shotgun barrels. These barrels were made by wrapping a ribbon of iron spirally around a mandrill and welding it together in the forge as it was wrapped. Many guns were imported from European gunsmiths and sold by Maryland gunsmiths such as T. A. Calderwood of Baltimore who advertised in 1874 in the Baltimore City Directory: "Guns, Pistol and Rifle guns manufactured and imported, guns repaired, restocked, and altered to percussion in superior manner, moderate terms – gun barrels rebored on improved principles, guaranteed not to recoil when discharged." Damascus barrels retained a good reputation in the black powder period, but when higher-powered smokeless powder came into use, the Damascus barrels were not able to take the increased stresses and many accidents occurred.

During the percussion period and early breech-loading period many shotguns were produced by Marylanders. Gunsmiths such as Arthur Baxter, Levi Cromwell, and William Harris were among these makers. J. H. Merrill produced a 10-gauge double-barreled shotgun with 39-inch barrels and back-action locks. Peter A. Favier advertised in the City Directory: "Gun, Rifle, Pistol Maker, having constantly on hand a large & general assortment of double and single barrel bird and ducking guns."

E. Remington and Sons opened a gunshop in Baltimore about 1876, as did W. W. Greener in 1888. Shotguns were made for gun clubs, as a surviving example illustrates. It is an 8-gauge Damascus double-barrel piece marked "Made for the Chesapeake Gun Club." The gunsmith Green of Cheltenham, England, made 6-gauge shotguns for A. M. McComas, who ran Baltimore's largest gun store.

Charles E. Sneider had several patents issued to him during the Civil War concerned with the conversion of common muzzle-loading double and single barrel shotguns to breechloaders. Sneider's patent of 22 December 1869, #85,252, for a rotary-bolt shotgun, along with his invention of the solid grip and double grip had particular value. This new locking of the breech, the retracting bolt, relief check, anti-friction lock, and actuation of the expended cartridge, brought success to the firm of Clark and Sneider. Charles W. Sneider, also of Clark and Sneider, met with much success with his several patents during the 1880's dealing with hammerless shotguns. His cocking mechanism, safety catch, and free air vent enabled this firm to dominate the Maryland area in the early period of hammerless shotguns. William C. Dodge's experimentation with various breech loading shotgun action systems led him to apply for ten different patents from 1864 to 1872.

During the nineteenth century certain people along the shores of the Chesapeake hunted ducks and geese on a commercial basis. The shooting was done by means of a gigantic fowling piece called a punt gun which was used from a small boat or punt. The gun usually was a crude affair with a barrel 6 or 7 feet in length and a bore measuring about two inches in diameter. The stocks were large and bulky, and were painted for protection from the elements. The percussion locks were rough and massive, sometimes as long as 9 inches.

The Baltimore Arms Company dominated the shotgun manufacturing field in the late 19th century. They held patents for a new type of locking bolt, and advertised barrels of Damascus or monumental steel,and of *Ortin* steel (nitro spelled backwards). Their "Baltimore hammerless guns were produced in various gauges and in five grades. After seven years of production the factory was destroyed in the great Baltimore fire of 1904. They advertised in 1905 and then ceased to exist. Two former employees were successful later. Frank A. Hollenbeck moved to Wheeling and produced three-barreled drilling guns. Ansley H. Fox became a shotgun manufacturer in Philadelphia.

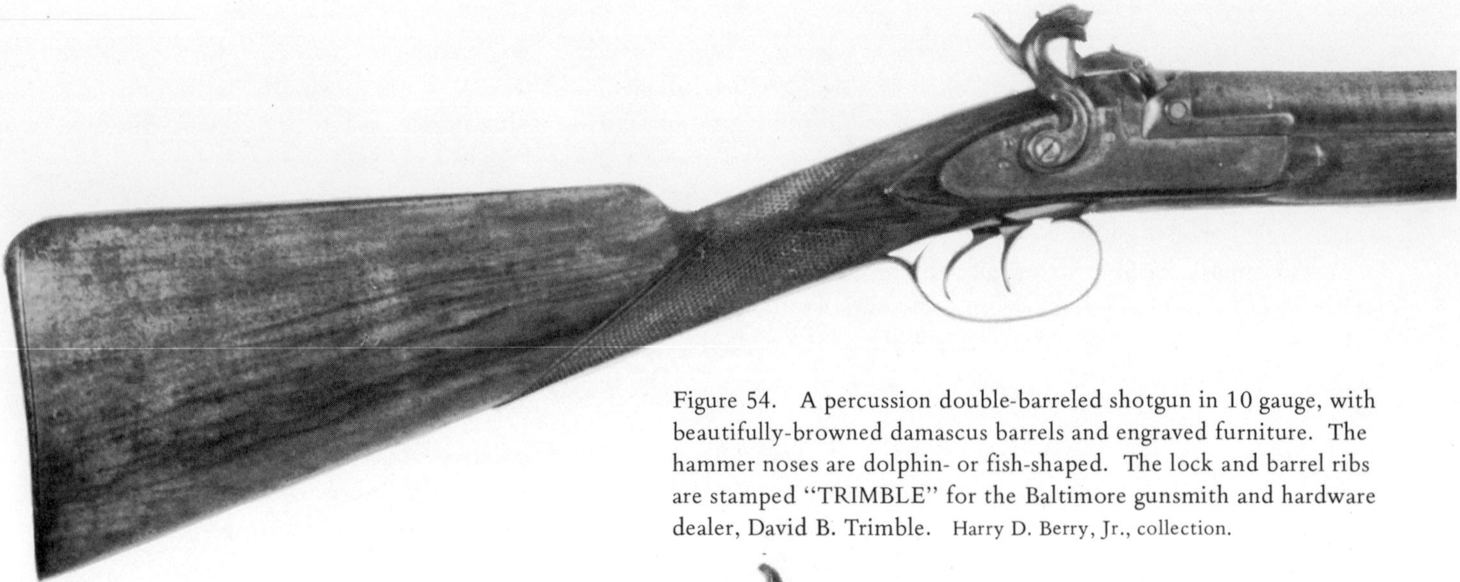

Figure 54. A percussion double-barreled shotgun in 10 gauge, with beautifully-browned damascus barrels and engraved furniture. The hammer noses are dolphin- or fish-shaped. The lock and barrel ribs are stamped "TRIMBLE" for the Baltimore gunsmith and hardware dealer, David B. Trimble. Harry D. Berry, Jr., collection.

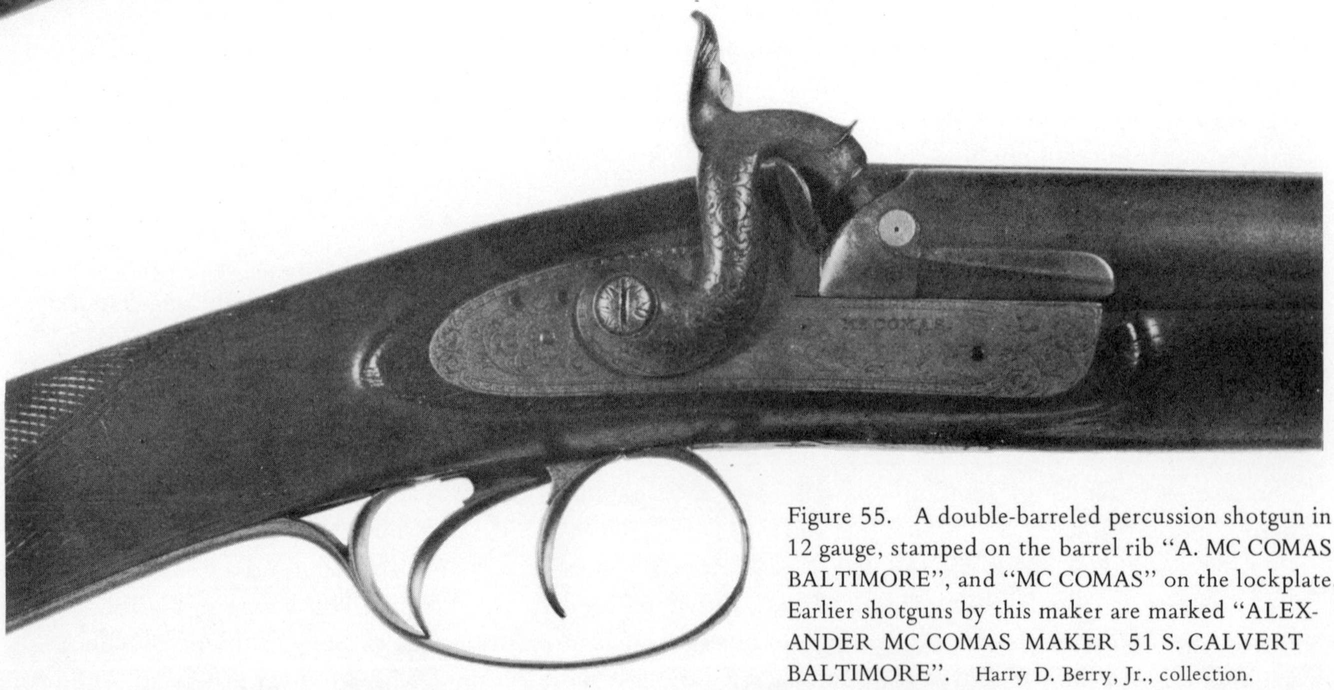

Figure 55. A double-barreled percussion shotgun in 12 gauge, stamped on the barrel rib "A. MC COMAS BALTIMORE", and "MC COMAS" on the lockplate. Earlier shotguns by this maker are marked "ALEXANDER MC COMAS MAKER 51 S. CALVERT BALTIMORE". Harry D. Berry, Jr., collection.

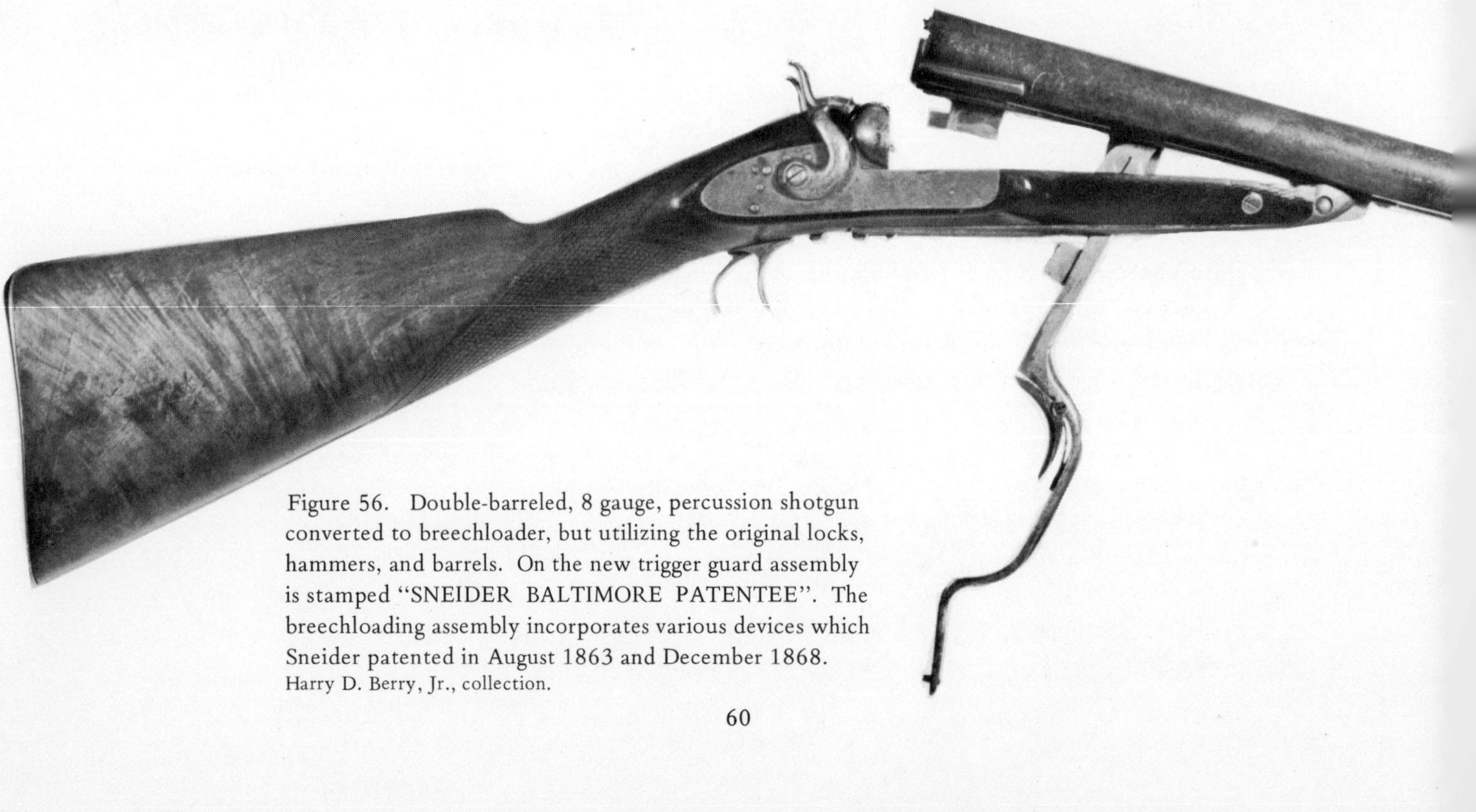

Figure 56. Double-barreled, 8 gauge, percussion shotgun converted to breechloader, but utilizing the original locks, hammers, and barrels. On the new trigger guard assembly is stamped "SNEIDER BALTIMORE PATENTEE". The breechloading assembly incorporates various devices which Sneider patented in August 1863 and December 1868. Harry D. Berry, Jr., collection.

CHAPTER 5

Pistols

In Maryland as in the neighboring gun-producing states of Pennsylvania and Virginia, the use as well as manufacture and commerce in rifles was concentrated in the piedmont and mountain areas. In contrast, commerce in pistols was concentrated in the larger coastal towns and cities, particularly in Philadelphia and Baltimore. The reasons for this are fairly obvious. The rifle was a hunting arm of particular importance to the people of the piedmont and mountains and to others traveling through these regions to more remote lands farther inland. The pistol is basically a weapon intended for personal defense, and in the old days it found its demand more in the cities and towns than in the hinterland. In addition, there was a certain demand for pistols from sea captains, seamen, and others in trades related to the sea. Furthermore, many pistols of European manufacture entered the American trade through Philadelphia, Baltimore, and other important seaports. Thus, in Maryland, Baltimore was the center for pistol commerce in the late eighteenth century and throughout the nineteenth century.

A variety of pistol sizes, shapes, and styles was used in the old days, ranging from beautifully made cased sets to small and rudimentary pocket pistols. In flintlock days cavalrymen and military officers often carried a pair of large pistols in saddle holsters. Landholders in the coastal regions were likely to have pistols in their houses as a means of protection. Sea captains needed pistols as an ultimate means of insuring authority. And travelers carried them while on the road. Colonel Harry Gilmor, in his book FOUR YEARS IN THE SADDLE [1], tells how effective the pistol was in the War Between the States when placed in the hands of an excellent shot.

Baltimore's most famous and important pistol maker of the flintlock period was James Haslett who learned the gunmaking trade in Europe from one of the leading craftsmen. He established himself in Baltimore in 1803 at 64 North Gay Street and carried on his business until about 1824. Not only did he make many pistols, but he also was most active as an importer of them from Europe. On 12 November 1806 he advertised in the Federal Gazette & Baltimore Daily Advertiser that he offered for sale "an elegant assortment of . . . duelling pistols, both of his own and London manufacture." Haslett produced pistols of many varieties and styles – octagonal and round barreled duellers, small pocket pieces, and English-style saw-handled pistols have survived from his shop. All of his pistols carry a characteristic ribbon of gold let into the top of the barrels and stamped "Haslett Baltimore", and an oval of gold let into the lock plate stamped "Haslett". The quality of his work attracted prominent people of the day, and his pistols were owned by Edward Lloyd, the Governor of Maryland, Senator Robert Wright, and the Governor of Virginia.

The firm of Halbach & Sons of Baltimore sold many pistols marked with their name on the lock. The barrels of these pistols sometimes carry Spanish proof marks and their styling is different enough to suggest manufacture abroad, probably in Spain. The author has one such piece in his collection, with a barrel having cannon-like turnings near the muzzle, Spanish proofs, and the script marking "Halbach & Sons" on the lock. No proof has yet been found that Halbach & Sons did any manufacturing in Baltimore.

Several pistols have survived with locks marked "McK & Brothers Baltimore," and other locks are known marked "McKim & Brother Baltimore." In the late flintlock period between 1824 and 1833 William D. McKim and Michael M. Maslin were in business together as wholesale hardware dealers. In this capacity they imported many locks, presumably from Birmingham, stamped with their names. McKim & Brother are listed in the Baltimore directories between 1835 and 1841.

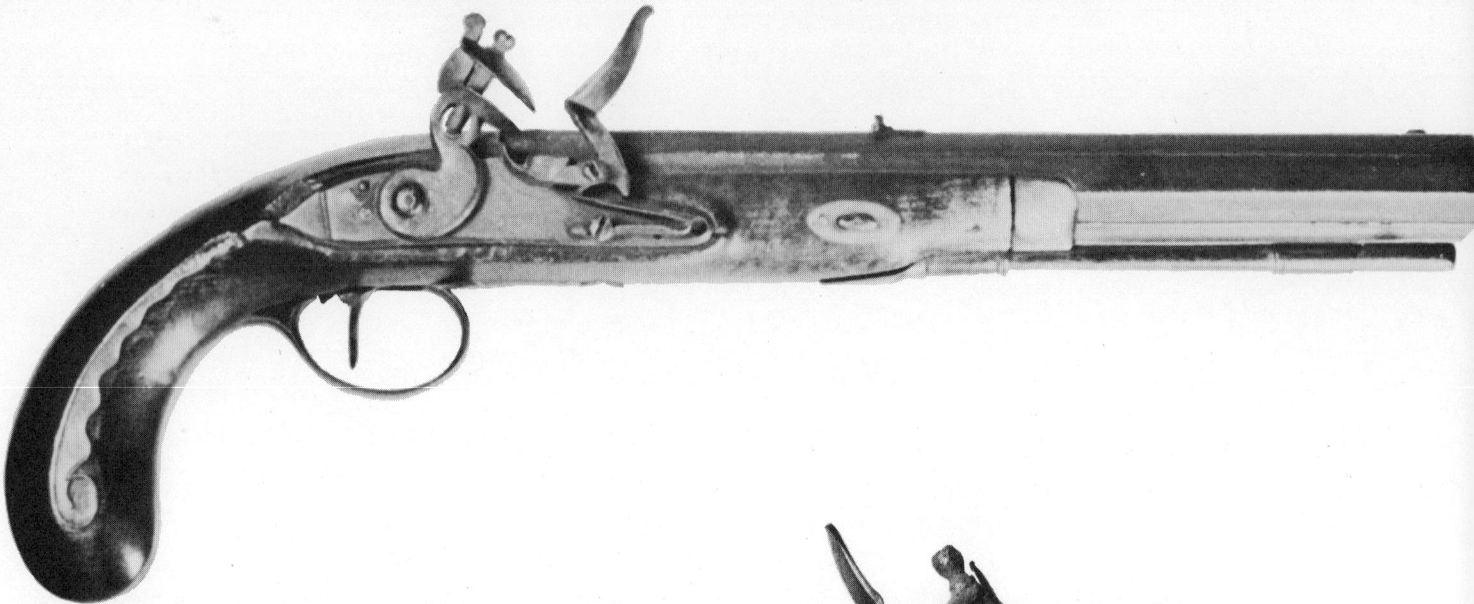

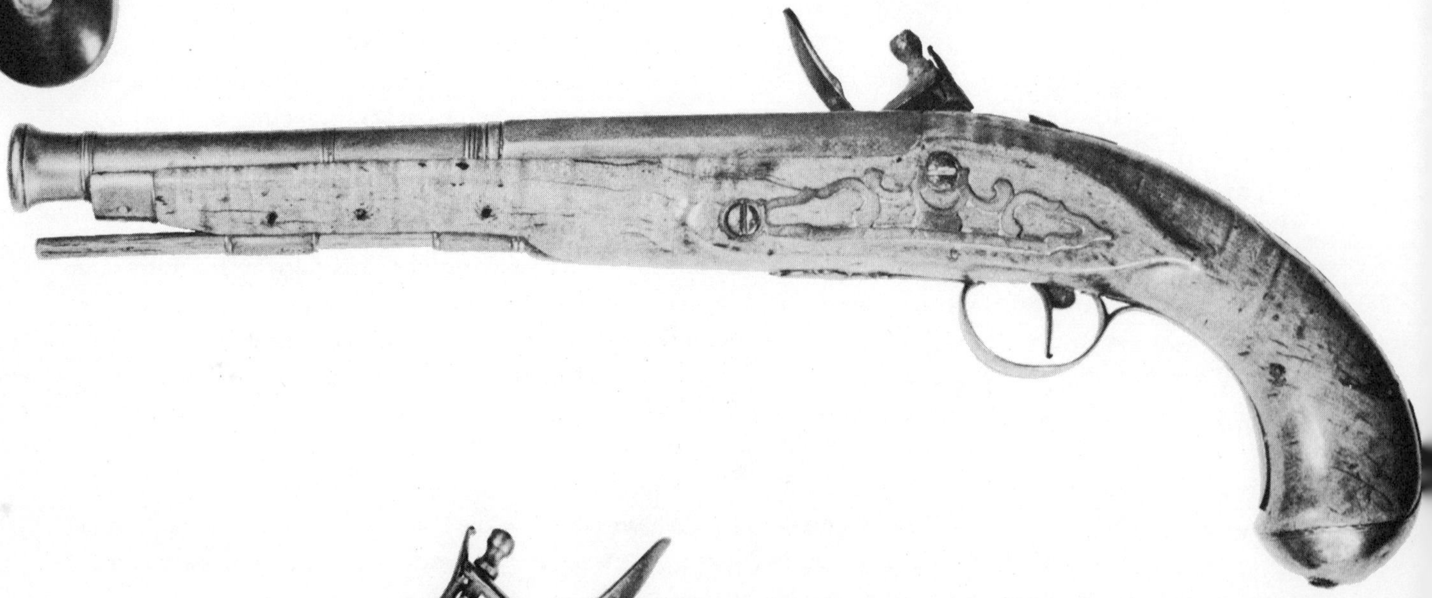

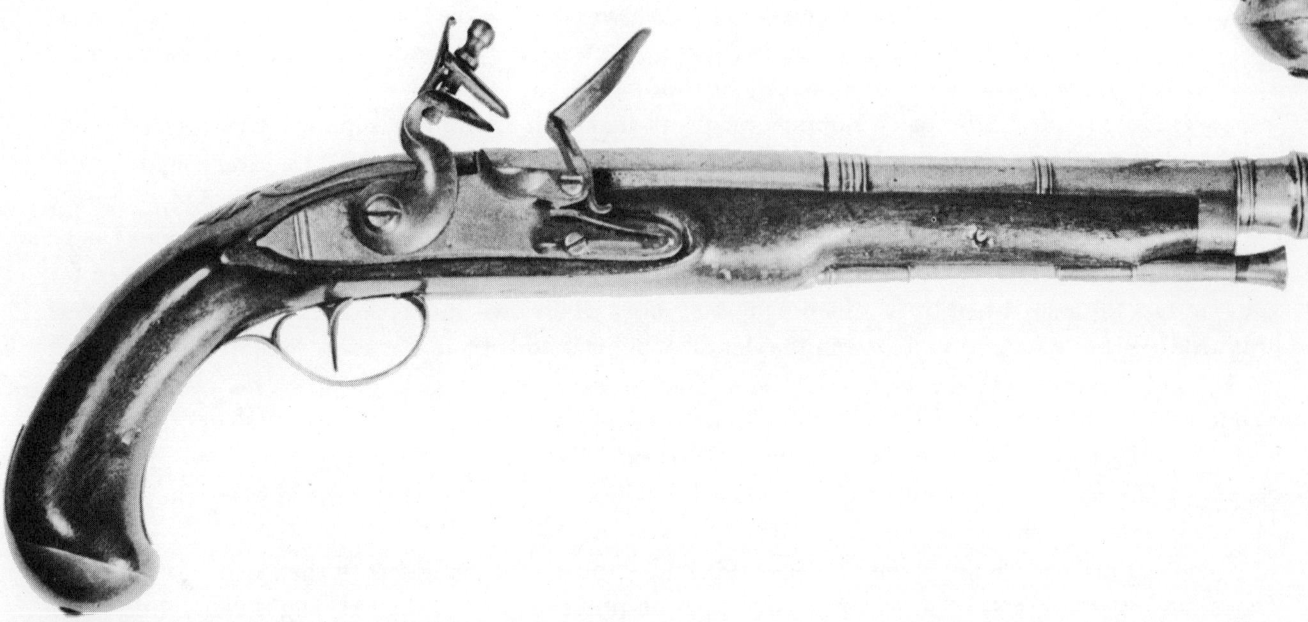

Figure 57 (top). Flintlock pistol with a silver plate let into the barrel holding the script signature of "P. Creamer" who worked in Taneytown. The grips are inlaid with silver wire, and the escutcheons and muzzle cap are of silver. The flint hammer is an inappropriate replacement. Joe Kindig, Jr., collection.

Figure 58 (middle). Brass-barreled flintlock pistol stocked in maple and mounted in silver by George Kreps of Hagerstown; length overall 15¼ inches, barrel length 9¾ inches.

Figure 59 (bottom). One of a pair of pistols belonging to Capt. Charles McCaulley, this piece also is signed "Geo. Kreps" in script on the brass barrel. It is brass mounted and stocked in walnut, holds an oval silver inlay engraved "C. M^{C}"; length overall 14 inches, barrel length 8¾ inches, 56 caliber. His will, probated 7 May 1817 in Washington County, Maryland, states, "I give and bequeath to my son Charles my Black Stud Horse, a Saddle and bridle, a silver watch, a pair of pistols and a sword and no more." Joe Kindig, Jr., collection.

Anton Escherich learned the gunsmithing trade on the European continent in the mid-nineteenth century and established himself in Baltimore just prior to the Civil War. His work in metal was excellent and he is noted for his engraving and low-relief chasing. An elegant double-barreled percussion pistol bearing his name has survived.

The company of E. Hutchins & Company, also spelled E. Hutchings & Company, listed themselves in a Baltimore directory for 1836 as "agents for the mfrs. of the long & picket rifles & dealers in guns, rifles, & pistols of every description." Surviving specimens of underhammer or bootleg pistols are known bearing the date 1836 and the barrel stamping "E. Hutchings & Co. Agents Baltimore." The author owns three such pieces, with barrels ranging from 6 inches to 12 inches in length. All have the Hutchings stamp on the barrel, and on two of them it is stamped on top of the maker's stamp. Some of the pieces handled by Hutchings were manufactured by A. Ruggles, by the Quinnebaug Rifle Company and Gibbs Tiffany and Company.

A few flintlock military-style pistols are known which could be classed as secondary martials. These have only the name "Baltimore" on the barrels, and the makers remain unknown.

The opening remarks of this section indicated that relatively few pistols were made and used in the piedmont and mountainous regions of western Maryland. Some pistols were made there, however, by the gunsmiths who made longrifles. John Armstrong of Emmitsburg produced a few pistols. One fine specimen that has survived is fully stocked in curly maple, bears the maker's name in script on both barrel and lock, and has a beautiful silver inlay in the region of the barrel tang. Other attractive flintlock pistols are known that were made by George Kreps, Christian Hawken, and Philip Creamer.

Philip Creamer was an outstanding gunsmith of the flintlock period who has remained all but unknown down to the present. He worked in Taneytown, a few miles south of Maryland's northern border. A few pistols that he made have survived and these leave little doubt about his skill. A percussion pistol that he made was owned by Andrew Jackson and currently it is located at Jackson's home, The Hermitage, near Nashville, Tennessee.

With the coming of the Industrial Revolution and the general use of the percussion system, in the 1830's, a new field for pistol invention, development and manufacture opened up. Maryland played her part in this. For example, Samuel Colt, in developing his famed Colt pistols, took out patents in Europe as well as the United States. Some of the perfecting as well as actual manufacturing of the arms was done in Baltimore by Hiram Parsons, a gunsmith employed by Colt as early as 1832.

The 1847 model Walker Colt was named for Samuel H. Walker, a famed Indian fighter from Prince George County. He was a veteran of the Creek and Seminole Wars. In 1839 Walker was impressed with the Patterson Colt revolver. Complimenting Col. Colt, he suggested that a heavier model be made for cavalry use. On 4 January 1847, while recruiting in Maryland for the Mexican border trouble, he contracted on behalf of the federal government for 1000 heavy revolvers made in keeping with his suggestion and now known as Walker Colts.

Edward N. Spiller, a Baltimorean, went to Richmond in 1861, and together with David J. Burr and James H. Burton formed the partnership of Spiller and Burr, manufacturing revolvers for the Confederacy.

Charles Edward Sneider, on 18 March 1862, patented a single-action, fourteen-shot, rimfire, cartridge revolver with two cylinders mounted on the same base pin. The two cylinders were separated by a disc. After the front cylinder was discharged, the hinged barrel was dropped; the loaded and discharged cylinders then were reversed. The forward cylinder always was discharged first by the hammer which had a long hinged head reaching over the rear cylinder. The invention was unsuccessful and probably few were manufactured. The several that have survived are of .22 and .32 caliber, and are engraved "C. A. Sneider's Pat. Baltimore, Md." on the rib of the octagonal barrel. The brass frames and iron barrels are nicely engraved. Under patent No. 46,612 (28 February 1865) for another revolver, he used pins within the rear ends of the chamber and when the pins were struck by the hammer they were propelled forward against the primer of the cartridge to produce the explosion. The patent was assigned to himself and Thomas Poultney.

Under patent No. 27,518 of March 20, 1860 William H. Bell registered his development of a system by which the cylinder, containing a priming wafer that continuously capped the nipples, revolved on its axis. Although it could be loaded rapidly, and the mechanism of the wafer functioned well under favorable conditions, in inclement weather it did not possess the reliability of the metallic percussion cap.

REFERENCES

1. Gilmor, Col. Harry (1866) FOUR YEARS IN THE SADDLE ,Harper & Brothers, New York.

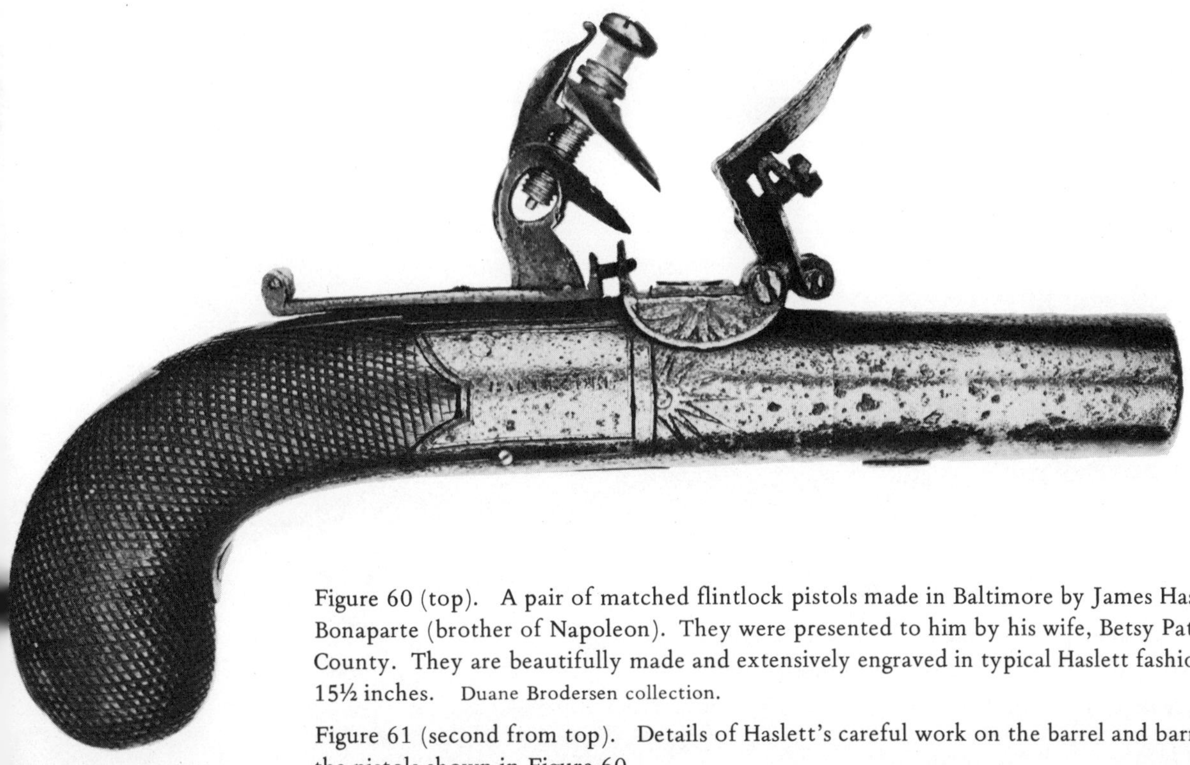

Figure 60 (top). A pair of matched flintlock pistols made in Baltimore by James Haslett for Jerome Bonaparte (brother of Napoleon). They were presented to him by his wife, Betsy Patterson of Carroll County. They are beautifully made and extensively engraved in typical Haslett fashion; length overall 15½ inches. Duane Brodersen collection.

Figure 61 (second from top). Details of Haslett's careful work on the barrel and barrel tang of one of the pistols shown in Figure 60.

Figure 62 (third from top). A flintlock pocket pistol marked "Baltimore" and with the name of Haslett. Harry D. Berry, Jr., collection.

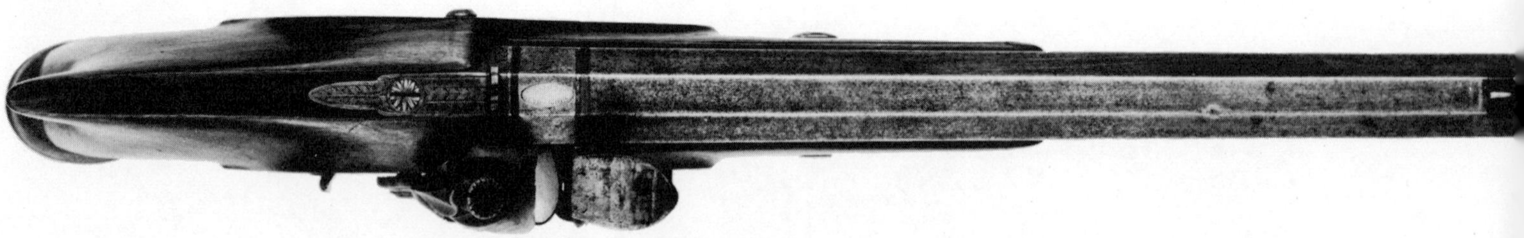

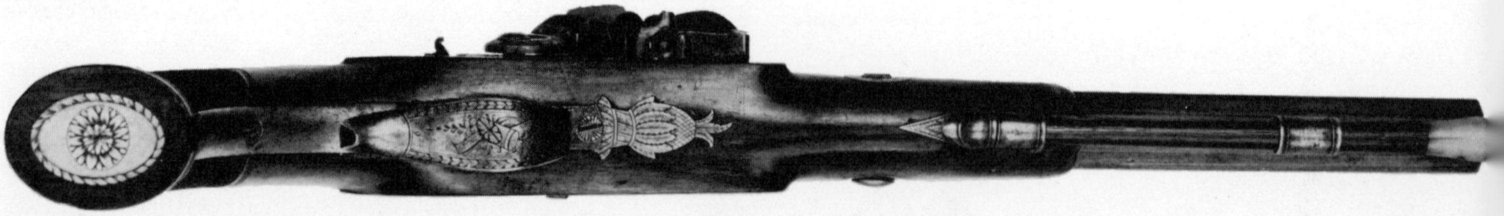

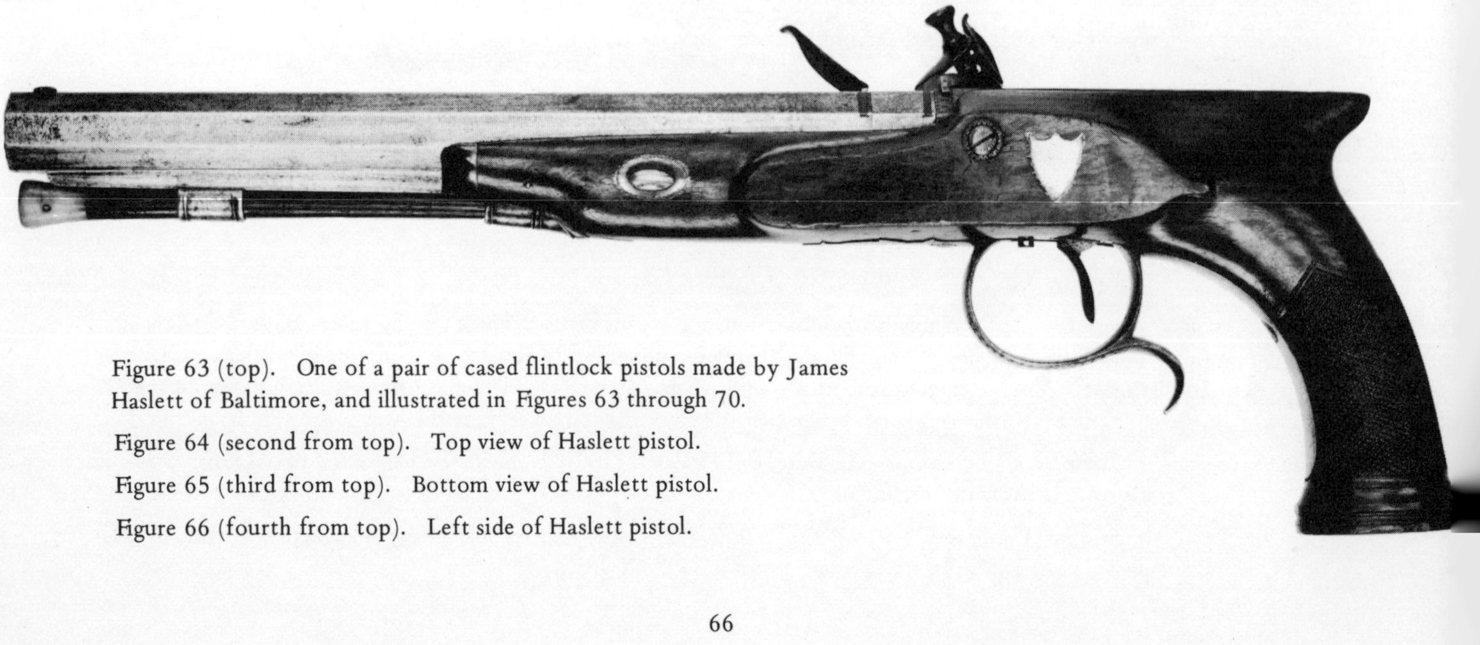

Figure 63 (top). One of a pair of cased flintlock pistols made by James Haslett of Baltimore, and illustrated in Figures 63 through 70.

Figure 64 (second from top). Top view of Haslett pistol.

Figure 65 (third from top). Bottom view of Haslett pistol.

Figure 66 (fourth from top). Left side of Haslett pistol.

Figure 67. Detail of barrel breech and tang on Haslett pistol.

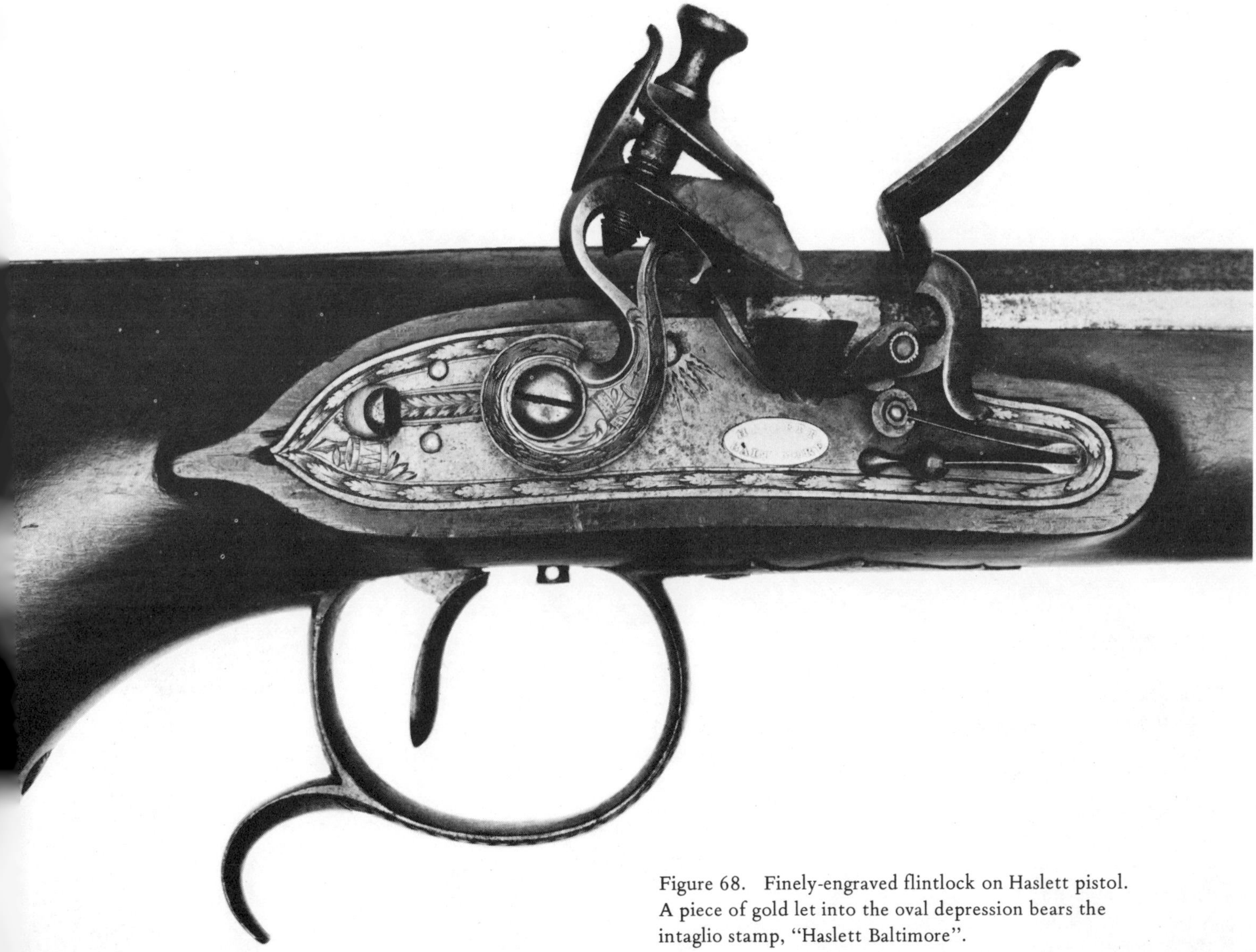

Figure 68. Finely-engraved flintlock on Haslett pistol. A piece of gold let into the oval depression bears the intaglio stamp, "Haslett Baltimore".

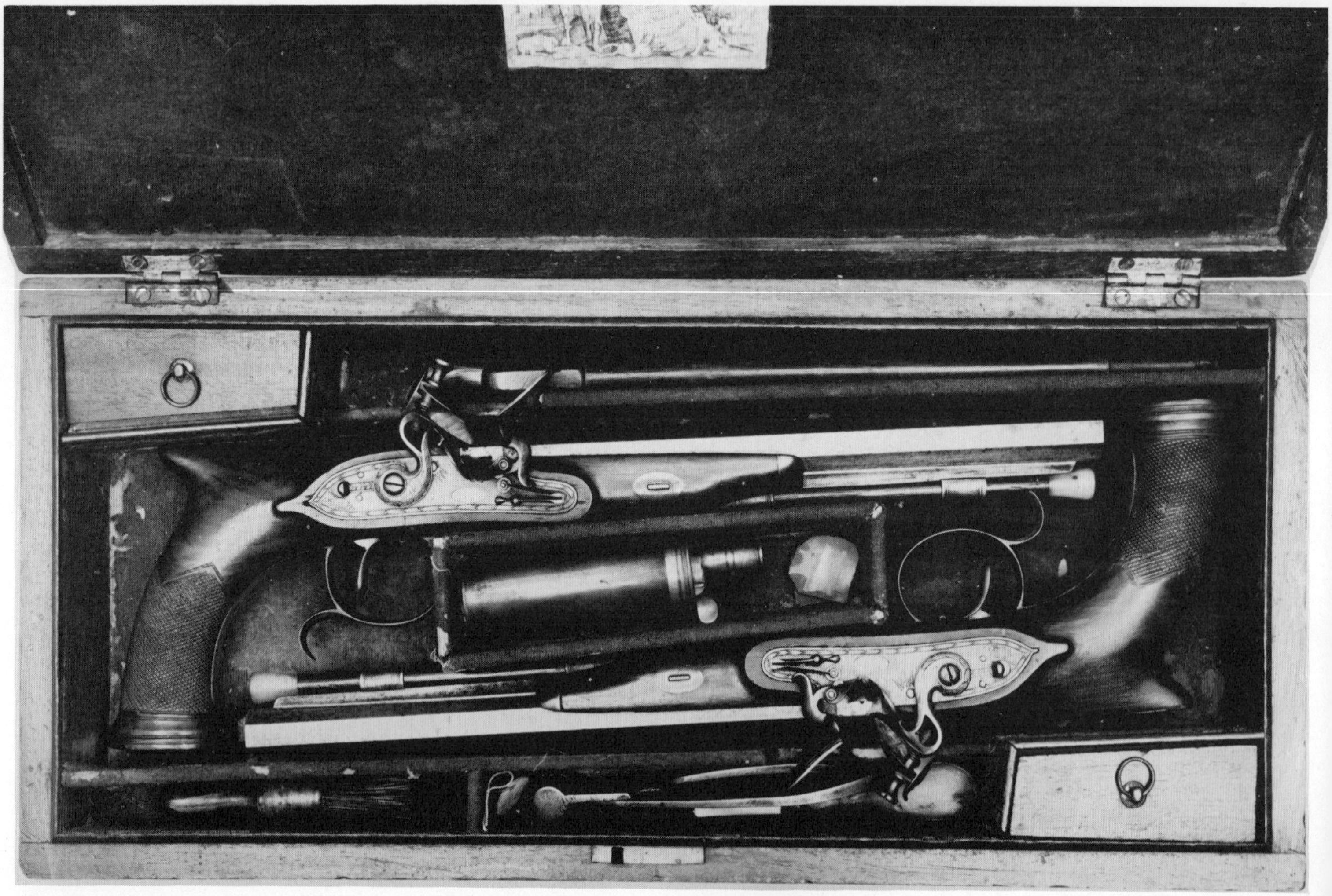

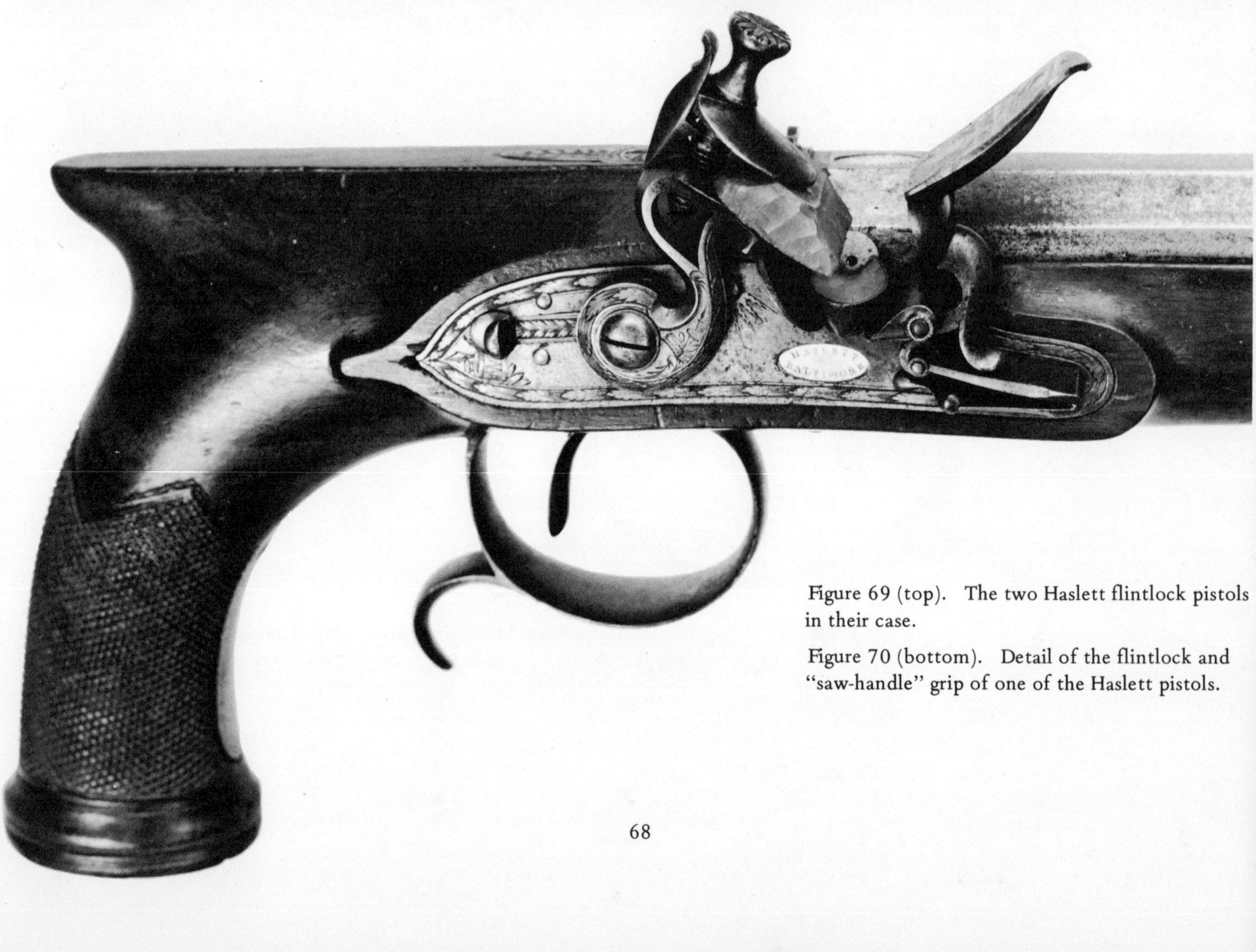

Figure 69 (top). The two Haslett flintlock pistols in their case.

Figure 70 (bottom). Detail of the flintlock and "saw-handle" grip of one of the Haslett pistols.

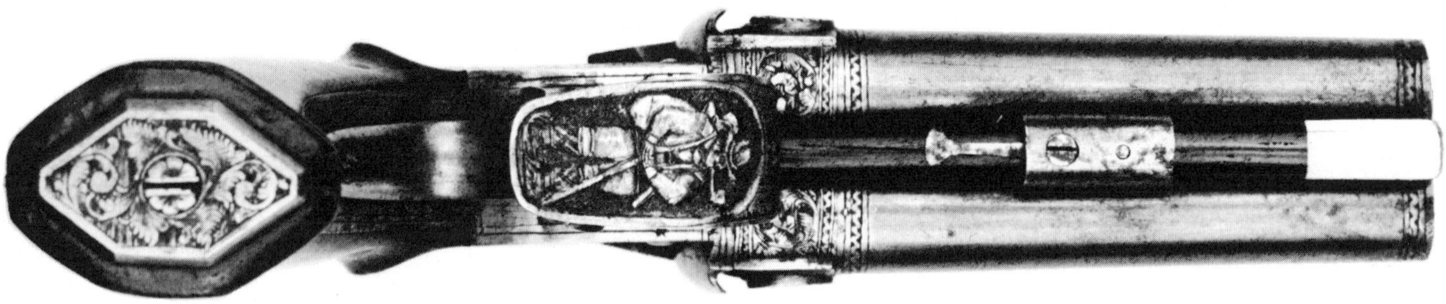

Figure 71a (top). Double-barreled percussion pistol, elaborately engraved and chased in a Victorian Germanic style. The rib between the barrels is engraved "A. Escherich Baltimore".

Figure 71b (middle). Bottom view of the percussion pistol by A. Escherich, exhibiting chased designs on the trigger guard and butt cap.

Figure 72 (bottom). Right side of the percussion pistol by A. Escherich. Daniel D. Hartzler collection.

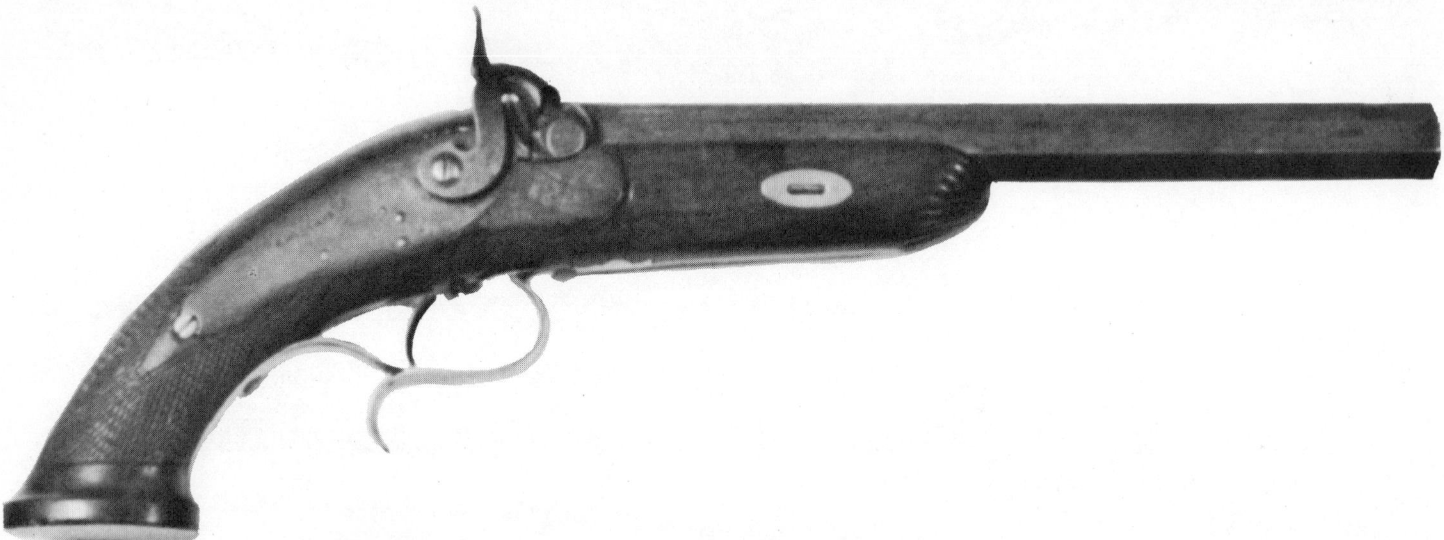

Figure 73 (top). A silver-mounted smoothbore percussion pistol with barrel stamped "A. McComas Baltimore" and lock stamped "A. McComas". Harry D. Berry, Jr., collection.

Figure 74 (bottom). An 11-shot, .32 caliber, breechloading rim-fire two cylinder revolver marked on the barrel "C. E. Sneider Pat. Baltimore Md." The mechanism of this piece is far superior to Sneider's pistol patent designs of March 1862 and February 1865. After the front cylinder is fired an ingenious central disc with a tubular arbor enables the rear cylinder to be discharged. The bullet passes through the one empty chamber of the front cylinder and out the barrel.

CHAPTER 6

Arms for the Civil War

During the period of relative peace and quiet between the War of 1812 and the Civil War, Maryland's arms industry concerned itself almost exclusively with sporting arms. The production of military arms was carried on by government arsenals at nearby Harpers Ferry, at Springfield, and by private contractors in industrialized New England. However, by 1855, with the Industrial Revolution well advanced, there was a widespread interest among arms people and inventors to devise breech-loading systems and other improvements in firearms, and Maryland was involved along with most of the other eastern states.

Thomas Poultney and D. B. Trimble of Baltimore became agents for the Smith carbine. This mechanism was patented by Gilbert Smith in 1856 but he was not involved in its development. Charles E. Sneider was responsible for the development of it and for making further improvements. His patent of 20 March 1860, #27,600, concerned a change in the awkward motion of unlocking the breech as well as a number of other modifications. The patent rights were assigned to Thomas Poultney who thereby gained exclusive rights to produce the arm. Manufacturing of the Smith Carbine was done by three companies, American Machine Works, American Arms Company, and Massachusetts Arms Company, all of Massachusetts. Between 1860 and 1865 the federal government purchased 30,362 Smith carbines from Poultney and Trimble. On 16 May 1865 Sneider applied for a patent to enable this carbine to take a metallic cartridge.

James H. Merrill was a gun, rifle, and pistol manufacturer who had experience working in Europe before he arrived in Baltimore in the early 1840's. Between 1855 and 1859 he and two associates, Ferdinand C. Latrobe and Phillip E. Thomas operated a small armory in Baltimore under the name of Merrill, Latrobe and Thomas. Merrill had developed a breechloading mechanism for use with a paper cartridge, and on 26 July 1855 the firm delivered to the Ordnance Department 170 Merrill, Latrobe and Thomas carbines utilizing this system. Trials showed the arms unfit for military service and apparently no more were produced. Patent #14,077 for this arm was issued on 8 January 1856.

James H. Merrill and associates also were at work developing a breechloading system based on the old Jenks system, with improvements so that a paper cartridge could be used. A number of patents were issued to him in the period between 1858 and 1861 concerning this. The firm altered a small number of regulation military muzzle-loading arms to the Merrill system, and it went to work on the problem of altering obsolete Jenks carbines to the improved Merrill system. Eventually it produced 300 altered Jenks carbines and in January 1861 the Navy Department accepted 240 of these after further modification. Beginning in 1859 the firm was known as Merrill, Thomas and Company. Using the Merrill system, a carbine and a rifle suitable for military service were eventually produced. The federal government purchased about 14,500 Merrill carbines, and 770 Merrill rifles for naval use. In 1864 the firm was reorganized under the name of Merrill Patent Fire Arms Manufacturing Company.

William Pratt of Baltimore, a dental surgeon with engineering talents, devised a breechloading system operated by a large lever. His patent model has survived, marked "W Pratt Baltimore 1853 Patent Model," but apparently nothing further developed. On 27 January 1852 Charles V. Nickerson patented a breechloading gun following after the original patentee, Tourrette of Paris.

William S. Polk of Baltimore, offering his services to the Confederacy, was appointed a captain and for nine months was superintendent for altering old flintlock muskets to percussion. As rapidly as possible the flintlocks were altered for immediate use in the field, with the intention of replacing them with better arms as soon as it was possible to do so.

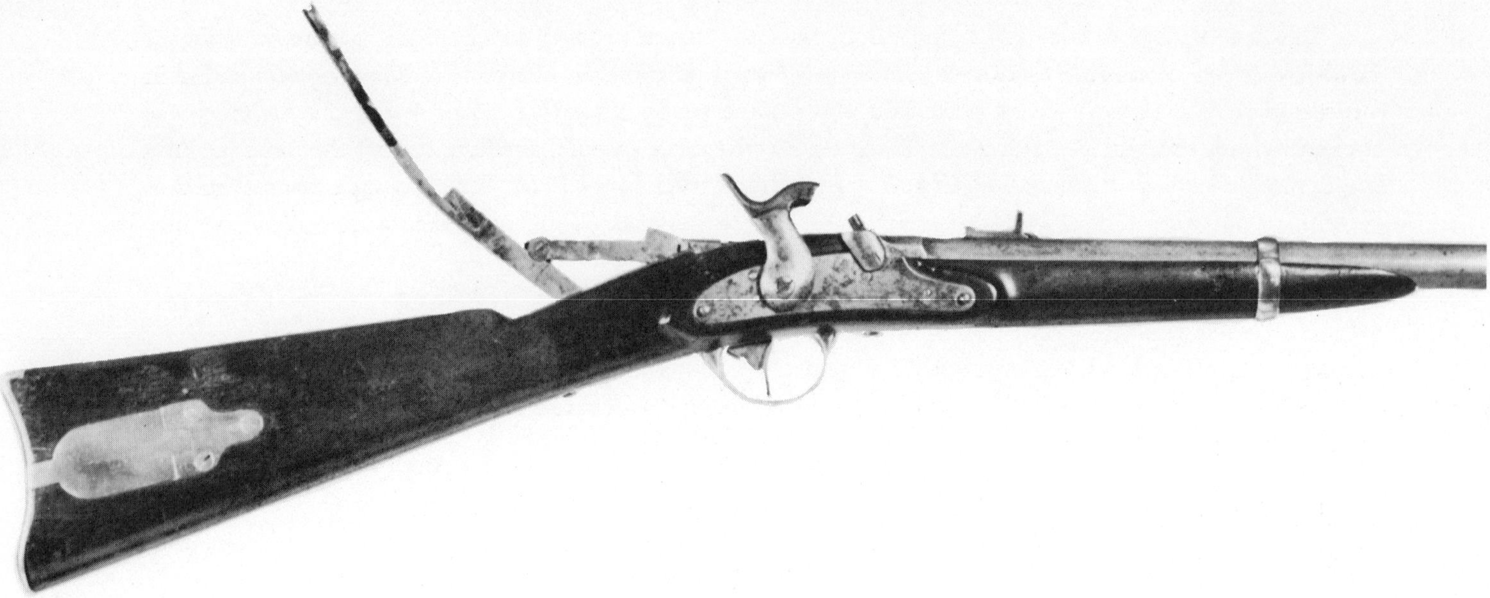

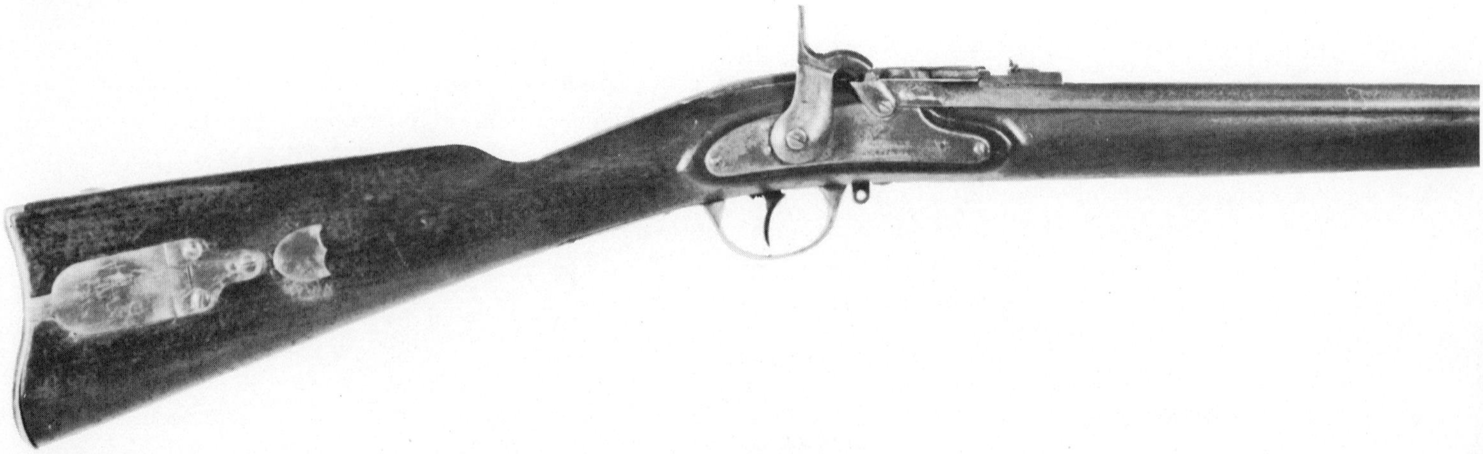

Figure 75 (top). The breechloading Merrill carbine produced in Baltimore during the Civil War period. About 14,500 of these were for the Federal government. Daniel D. Hartzler collection.

Figure 76 (second from top). The breechloading Merrill navy rifle. About 770 of these were manufactured in Baltimore for the Federal government. Daniel D. Hartzler collection.

Figure 77 (third from top). This alteration of a Wickham percussion musket to a breechloader is marked "W Pratt Baltimore 1853 Patent Model". The breech opens and closes by means of a large lever. Apparently this model never was patented. The overall length of the arm is 33 inches. Harry D. Berry, Jr., collection.

Edward Maynard of Washington, D.C. was the inventor and patentee of a narrow strip of varnished paper of double thickness having deposits of fulminating compound. The strip was coiled in a recessed magazine in the lock plate and was pushed up by a toothed wheel when the hammer was cocked, which brought the fulminating cap up on the cone (nipple) while it cut off the paper behind the exploded cap. In 1845 the government purchased the rights to the Maynard tape primer ignition system, permitting unreserved use for $50,000. The Maynard tape primer system was applied not only to the manufacture of new arms, but to the modification of old as well.

On November 15, 1859, W. H. Arnold designed and patented a side hammer, clumsy top lever breech rifle for the insertion of a nitre-treated paper cartridge. Since Arnold's invention has remained unknown, apparently it was not successful. William H. Bell and Francis H. Bell submitted separate patents for self-priming percussion rifles with the magazine located in the hammer of the lock. Although they were not officially adopted William H. Bell applied to the government and was compensated $2,000 on January 25, 1859 to alter some rifles to his ignition system. With the refinement of metallic cartridges, gunsmiths throughout the war were not slow to devise new systems. Josiah V. Meigs and Samuel Strong both had several patents in traverse-and-gate hinged-breechloading rifles.

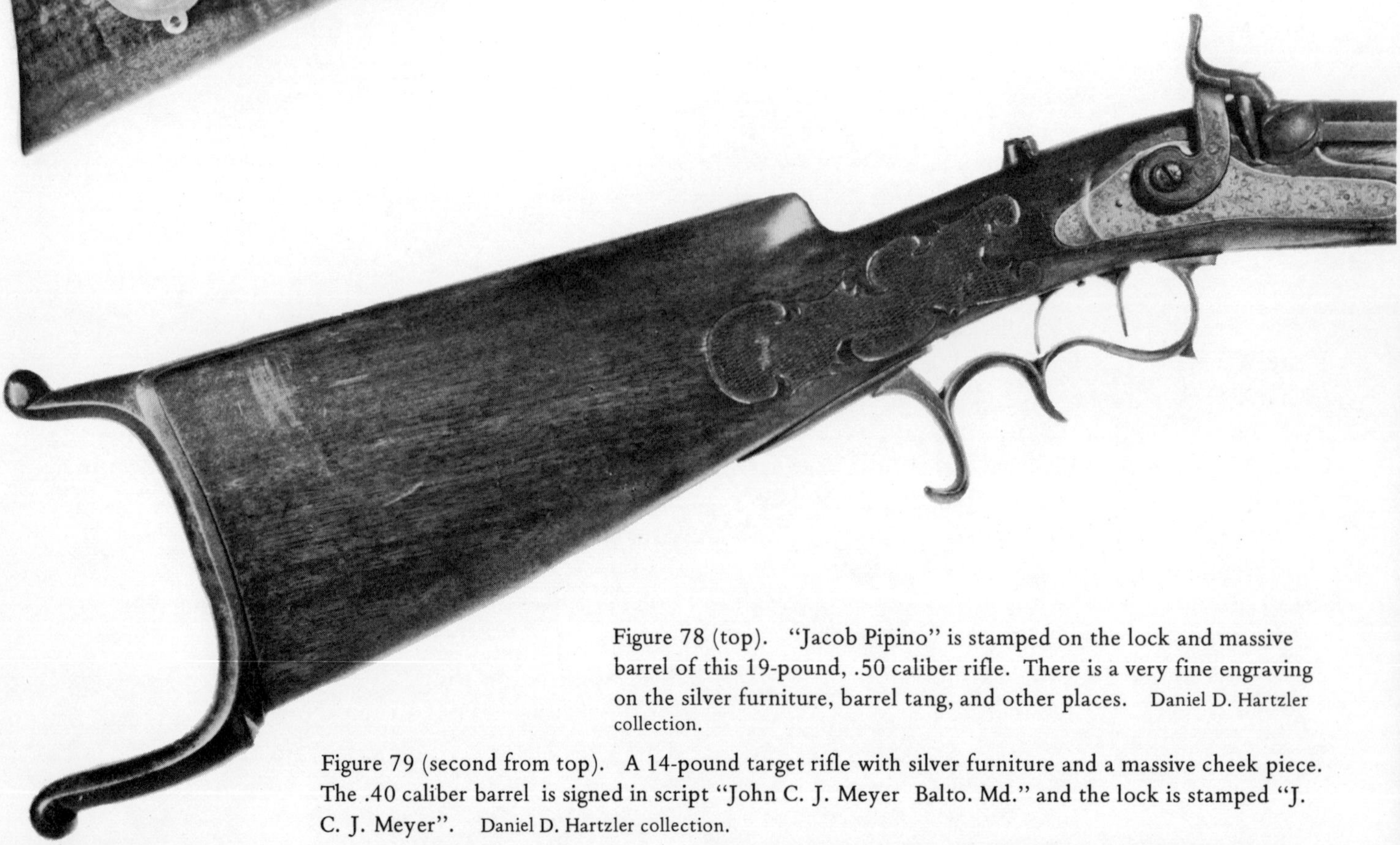

Figure 78 (top). "Jacob Pipino" is stamped on the lock and massive barrel of this 19-pound, .50 caliber rifle. There is a very fine engraving on the silver furniture, barrel tang, and other places. Daniel D. Hartzler collection.

Figure 79 (second from top). A 14-pound target rifle with silver furniture and a massive cheek piece. The .40 caliber barrel is signed in script "John C. J. Meyer Balto. Md." and the lock is stamped "J. C. J. Meyer". Daniel D. Hartzler collection.

Figure 80 (third from top). A percussion scheutzen target rifle in .40 caliber equipped with bullet starter, forehand rest, and socket for ball rest. The barrel is stamped "F. Nestle Baltimore Md.". There is fine engraving on the piece, and the barrel also is marked "Engraved by Escherich and Son, Baltimore, Md." Nestle apparently was not an engraver for a number of other rifles by his hand are known and they hold no engraving. Harry D. Berry, Jr., collection.

CHAPTER 7

Target Rifles and Shooting Matches

The shooting match was a common sport in early America, wherever and whenever rifles were found in abundance. The tradition of the match or tournament of skill was widespread in continental Europe from the days of crossbows and before. In eighteenth century America, rifles became popular in the piedmont lands of Pennsylvania, Maryland, Virginia, and the Carolinas during the second half of the century, and along with them came the test of skill in their use. Hunting, fishing, shooting, and cock-fighting were recreations for men in the early settlements. Meshach Browning, in writing of his courtship in the backwoods in the late 1700's, mentioned: "I continued with great glee to partake of all pastimes, attending the dances, shooting matches, &c."[1]

Unfortunately, very little has been recorded about these early matches except for brief mentions and descriptions of them in newspaper advertisements, and some of the comments from early travelers. It was usual for the contestants to vie for the "five quarters" of a beef animal, the hide and tallow being considered as the fifth quarter and being valued the highest. Almost any item of value might be put up as a prize for a match, including prize horses. A highly unusual shooting match was announced in a Maryland newspaper in 1810, and reprinted as a news item by a Harrisburg, Pennsylvania, newspaper, the *Oracle of Dauphin,* on 23 June 1810:

> RIFLEMEN ATTENTION
>
> A MAN to be shot at for the benefit of his wife and children – 1 dollar a shot – 100 yards distance, with rifles, on Wednesday the 13th inst. at Govan's town, at 3 P.M. The above mentioned man is in a very low state of health, and wishes to leave his family snug.

The rifle match really came into its own in the decades of relative peace and quiet that followed the War of 1812, that is, during the 1820's, 1830's, 1840's, and 1850's. Matches became so popular that heavy-barreled rifles intended chiefly for shooting at the mark were made by longrifle gunsmiths of the piedmont in Pennsylvania, Maryland, and Virginia. Manufacture of these match rifles began in the 1820's in the late flintlock period, and continued throughout the percussion era. The heavy rifles were full-stocked, weighed between 15 and 25 pounds, and were equipped either with open sights or with adjustable peep-sights. Though intended primarily for shooting at a mark, doubtless some of these rifles saw use in deer hunting when it was possible for the hunter to wait at a stand. These heavy round-ball rifles would be used at target ranges of 50, 60 and 100 yards primarily, but if used as a hunting rifle from a stand could be effective at 200 and even 250 yards.

A much more typical shooting match advertisement appeared in the *Maryland Herald & Hagers-Town Weekly Advertiser for* 23 Sept. 1823, though even a live bear was not a typical prize:

> Notice to Marksmen! ON Thursday the 2nd of October next, will be shot for, at Williamsport, a fine large Bear – and on the 3rd there will be a bear beat. Those who are disposed for sport will have a fine opportunity to gratify themselves.

Meshach Browning tells of two cub bears that he caught, one named Cuffee. "My sons took him to a Mr. Peters, who was a good rifleman. He kept Cuffee until he was two years old, when he put him up at shooting matches, and won him several times himself, until he made forty dollars off him; after which, poor Cuffee was butchered and divided among the sportsmen." [2]

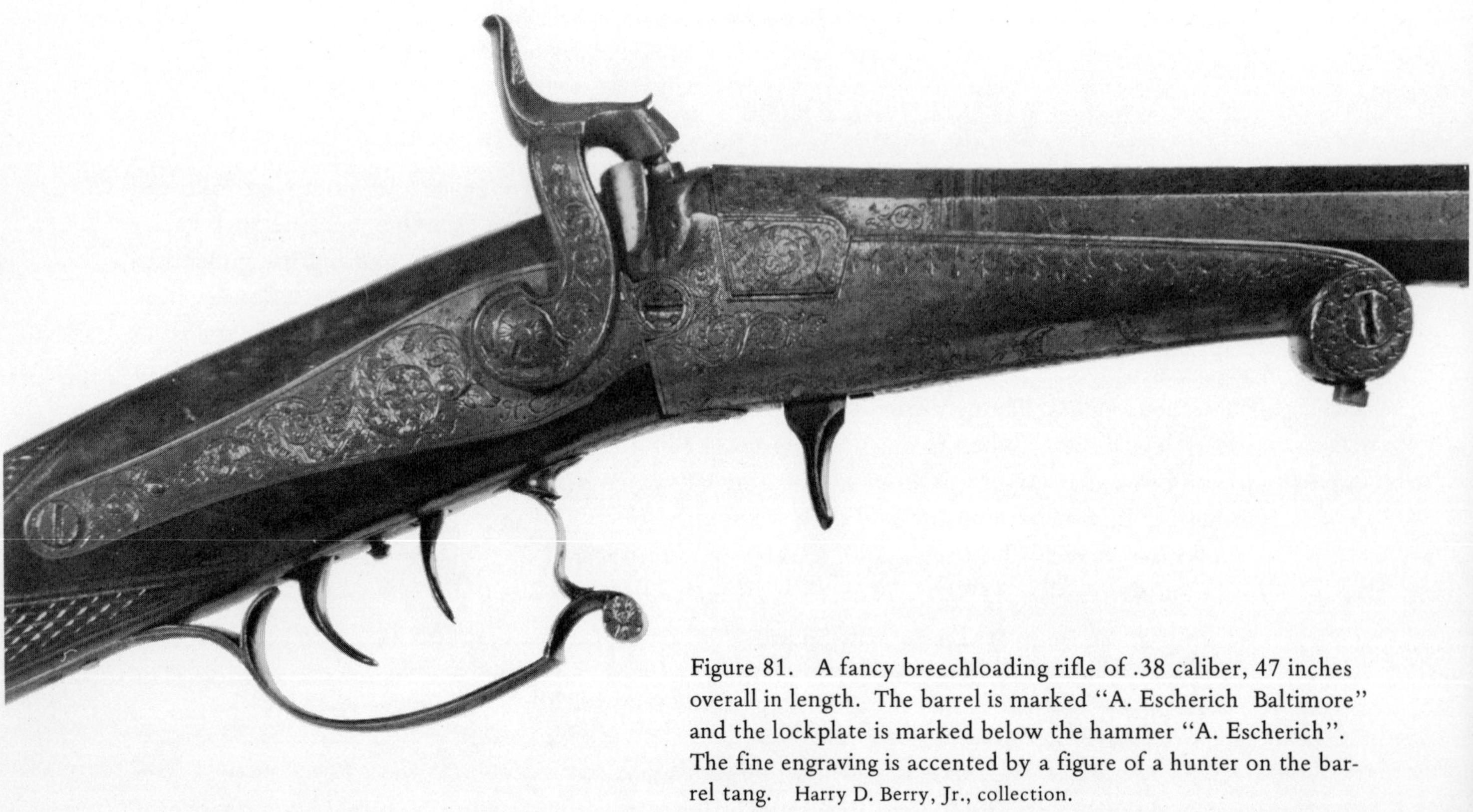

Figure 81. A fancy breechloading rifle of .38 caliber, 47 inches overall in length. The barrel is marked "A. Escherich Baltimore" and the lockplate is marked below the hammer "A. Escherich". The fine engraving is accented by a figure of a hunter on the barrel tang. Harry D. Berry, Jr., collection.

During the 1840's target shooting took on a new aspect as percussion arms shooting elongated bullets came into use. Target ranges were greatly increased, with 40 rods (220 yards) being common and even longer ranges in use also. The rifles used for shooting on these ranges were heavy octagon-barreled pieces ranging in caliber from about .28 to .50. Muzzle rests of metal or wood sometimes were attached to the barrel near the muzzle. Shooting benches or bench rest blocks became essential pieces of equipment.

The use of the false-muzzle began in the 1840's as a means of helping to get an over-size conical or elongated bullet wrapped in a paper patch started down the barrel correctly. The grooves and lands in the false-muzzle coincide exactly with the rifling in the bore, so the bullet and patch, once started down the false-muzzle, continue on into the rifle bore without interruption. Telescopic sights came into use during this period also.

From the 1840's through to the end of the century the long-range matches were held with the very specialized rifles made for them. Improvements in accuracy were inevitable as marksmen and gunsmiths strived for perfection. With the evolution and development of breechloading cartridge rifles during this period, a new dimension was added. During the 1870's and 1880's Sharps, Ballard, and Remington single-shot rifles became popular for long-range shooting, and controversy developed as to whether muzzle-loading or breechloading rifles were superior. Great enthusiasm was aroused as a result of the 1876 International Rifle Match when the U. S. team using breechloaders, mainly Sharps rifles, defeated the invincible Irish team using muzzle-loaders.

Of the early Maryland gunsmiths, James Marker is known for having made a number of heavy-barreled bench-rest rifles with full-stocks. Alexander McComas produced a large number of small, light-weight, half-stock percussion target rifles with walnut stocks and oval patchboxes or capboxes. J. H. Merrill made a finely-rifled breechloading target rifles with rear peep sight and front globe-type sight. The Merrill chamber with improved patent allowed no gas to escape, and the rifle could be fired 500 times without cleaning.

Frederick Nestle produced large, walnut-stocked, German-type scheutzen rifles with massive trigger guards and large-pronged butt plates. Such pieces were intended for offhand target shooting. The .38 caliber barrels of octagon shape had false-muzzles, and large, folding, disc-shaped tang peep sights. With the large number of German families in Maryland, shooting muzzle-loaders remained popular after the turn of the century, and scheutzen-fests were common.

Charles W. Sneider patented a one-piece, falling-breech, cartridge rifle on 9 April 1878. The breech was operated by means of a second trigger forward of the trigger guard. According to specifications, it was intended for light arms of small caliber.

REFERENCES

1. Browning, Meshach, (1965) FORTY-FOUR YEARS IN THE LIFE OF A HUNTER; Winston Printing Co., Winston-Salem, N. C., p. 79.
2. Ibid., p. 301.

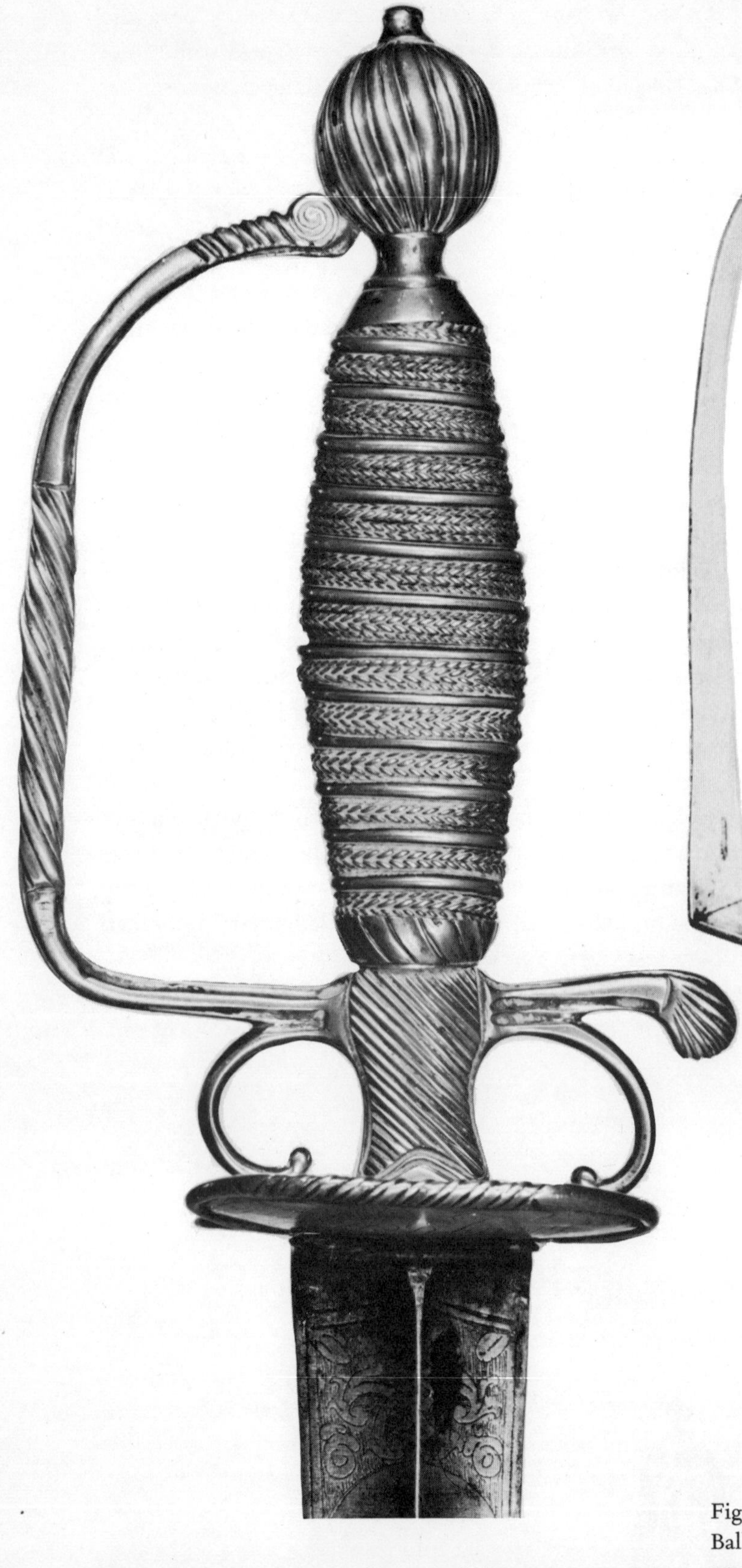

Figure 83. Silver-hilted sword with the mark of William Ball inside the knuckle bow. Ball was a well-known Baltimore silversmith who produced many silver-hilts of this general type, featuring an eagle's head or bird's head pommel of the characteristic design shown here. Collection of Harry D. Berry.

Figure 82. Silver-hilted colechemarde bearing the mark of Thomas Sparrow on the obverse side of the knuckle bow. The blade is of European origin, and bears the engraved inscriptions "Liberty & Property", "W R" and "Sept. 1 1775". Collection of Harold L. Peterson.

CHAPTER 8

Swords

Swords have had their special appeal to men for many centuries, serving both as practical arms for warfare and as defensive arms for civilians. A second and very real function of the sword has been to serve as a place for decorative arts to flourish, as true for the samurai sword of Japan as for the seventeenth century rapier and eighteenth century small sword of western Europe. In the military field swords carried by the lowest ranking foot soldier were not eliminated from his armament until more than three centuries had passed since the development of gunpowder. Then the bayonet took the place of the sword though its uses were somewhat more limited. In the social field the decorated, non-military sword became an important object of dress for the man of real or pretended social standing during the seventeenth and eighteenth centuries, and during the eighteenth century it was the silver-hilted sword that carried special prestige. The special aura of the silver-hilt sword carried on into the early decades of the nineteenth century.

The making of a sword calls for two separate specialists, the blade maker and the hilt maker. Steel is most necessary for the manufacture of sword blades, and steel was not produced in America until the 1830's. During the eighteenth and nineteenth centuries there were highly developed blade manufactories in Germany, with blades from Solingen being well known, in Spain, with blades from Toledo being famous, and in England.

Many imported blades came to America during the eighteenth century, with or without hilts, but there were times when imported blades were not to be had and American blacksmiths were called upon to make blades from imported steel. During the French and Indian War Philadelphia's leading cutler and mechanic, Stephen Paschall, was called upon to make a pair of sword blades for two colonial officers, but a letter written from Philadelphia reported that "Stephen Paschall is the only man in Town who makes Sword Blades, & he is so busy making Scythes & Sickles for Harvest that he says he cannot make any Blades for a month or two; However I have since met with two English Blades & have employed young Phil Sung to make the Handles, they are to be fluted, not chased, the latter being quite out of fashion . . ." [1]

During the Revolutionary War the demand for swords in Maryland increased and some blades apparently were made. The following portion of a letter written in 1781 provides a suggestion of this: "It will be Impossible to Equip. Coll. Nicholsons Troop of Horse in less than one Month. I have sent Ellicott to work on the swords &c He can make about 10 p week." [2] Of course almost any blacksmith could turn out a blade of iron and case-harden the lower portion to yield a piece that would do service in time of need, and doubtless a good many were turned out in the Revolutionary period. During the war years there already existed in America a considerable quantity of swords brought from Europe during preceding decades. Many hunting swords and decorative small swords intended for civilian dress were pressed into service and worn off to war. Maryland made the bayonet mandatory for infantry use, but swords were in demand by officers, non-commissioned officers, navy officers, and cavalrymen.

The silversmith's art of making silver hilts for swords was finely developed in eighteenth century Europe and along with almost everything else in the arts field spilled over into America. Early American silversmiths produced hilts in keeping with the fashions of England beginning in the early decades of the eighteenth century.

Thomas Sparrow, silversmith, was one such colonial craftsman. In the *Maryland Gazette* for 21 March 1765 he announced that he had opened shop in Annapolis, having previously been in Philadelphia, and noted that he made all sorts of gold and silver work, including plain sword hilts.

Charles Oliver Bruff, a silversmith as was his father, was born in Talbot County and may have worked there briefly. By 1775 he resided in New York and advertised on 19 June 1775 that he was "making swords for those Gentlemen who are forming themselves into companies in defence of their liberties."[3] Bruff was a Tory, however, and left New York with the British evacuation. In his testimony before the Loyalist Commissioners he said that "he was by inclination a friend to Gt. Britain but did duty in the City Watch of New York with the Rebels. But as he was a good sword cutler and worked for them he was excused turning out in the militia."[4] His loyalty was rewarded by a grant of fifty acres of land in Nova Scotia. He moved there to continue his trade, accompanied by his family of eight persons and by fifteen servants.

William Ball, Jr., was an outstanding silversmith who worked in Baltimore in the decades following the Revolution. On 1 November 1794 he advertised in the *Baltimore Daily Intelligence:* "Offering silver mounted swords and epaulets."[5] American-made silver-hilted swords with the maker's touch are rare, but Ball must have produced a great many, for more than a dozen have survived. Often Ball used a short-beaked eagle head for a pommel, a form that was used by others also and which is considered to be a mark of Baltimore silversmiths. The knuckle-bow that he used often was the simple but popular hammered sheet silver piece cut out to resemble a divided guard.

John Lynch, another well-known Baltimore silversmith, used the Philadelphia-style eagle-head pommel as well as the characteristic Baltimore style, and the pierced knuckle-bow of the form popular in Baltimore.

During the 1820's the silver-hilt sword finally dropped out of social usage and seldom was worn after then. A certain demand for silver-hilt swords continued, however, for a number of decades for it became popular for cities and states to present silver- and gold-hilted swords to heroes of the Revolutionary War, Tripolitan War, War of 1812, Mexican War, and Civil War. These swords usually had etched blades and ornate rococo hilts. The City of Baltimore as well as the State of Maryland presented many of them.

The War Between the States was the last major conflict in which swords, usually in the form of sabers, were used for hand-to-hand combat. Canfield Brothers of Baltimore were military outfitters, importers, and manufacturers. They carried military goods of every description, selling both to the North and South. In 1864 they advertised "Presentation swords on hand & made to order at short notice."[6] There exists a silver-gripped sword presented to Major General H. E. Paine dated September 1861. The blade is marked "Tiffany N. Y." with a solid gold ribband-shaped plaque marked "Canfield Bro. Baltimore." In the author's collection there are several infantry line officers' swords marked "Oehm & Co." and "Barnard & McAinn" for two other Baltimore dealers in military supplies.

Naval cutlasses with the following characteristics are found in the Baltimore area and are believed to have been manufactured there in the early 1800's. The guard, which forms a figure-eight, is cut from sheet iron, with "U. S." stamped on the knuckle-bow. The grips are cast corrugations. The blade is slightly curved and flat with a single edge, 28 inches long to the hilt and 1½ inches wide at the hilt, bearing a proof mark "P."

A militia cavalry saber similar to the 1808 Starr saber also is thought to have been made in the Free State. The grips are wood covered with leather and are wire-wrapped. The hilt is of iron and the knuckle-bow has a double curve and is pierced near the pommel for a sword knot. The blade, approximately 33½ inches in length, is curved, single-edged, and has a fuller running to within 7½ inches of the tip. The letter "M" is stamped on the blade near the hilt (Fig. 16). The scabbard is iron, with two small iron rings. Two of these sabers have been documented, one stamped with the "M" and one without. Both were carried in the War of 1812 by Marylanders in the 1st Regimental Cavalry District from Frederick County.

REFERENCES

1. Shippen, Joseph, Jr., letter of 16 June 1757, PENNSYLVANIA ARCHIVES.
2. Smith to Lee, letter, 24 June 1781, ARCHIVES OF MARYLAND, p. 311.
3. NEW YORK MECURY, 19 June 1775.
4. Loyalist Commission Manuscripts, 14 March 1787, New York Public Library
5. Baltimore DAILY INTELLIGENCE, 1 Nov. 1794.
6. BALTIMORE CITY DIRECTORY, 1864.

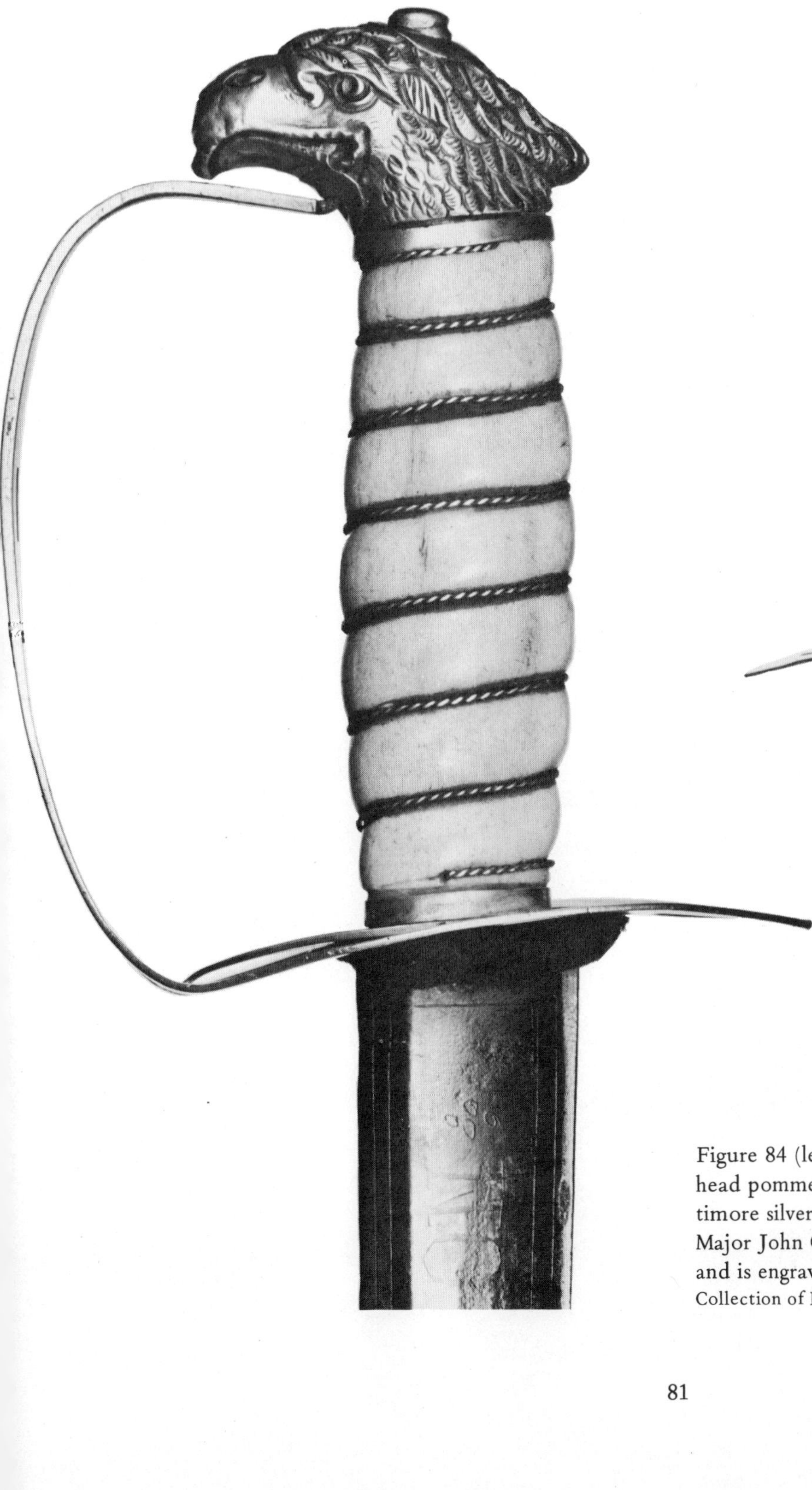

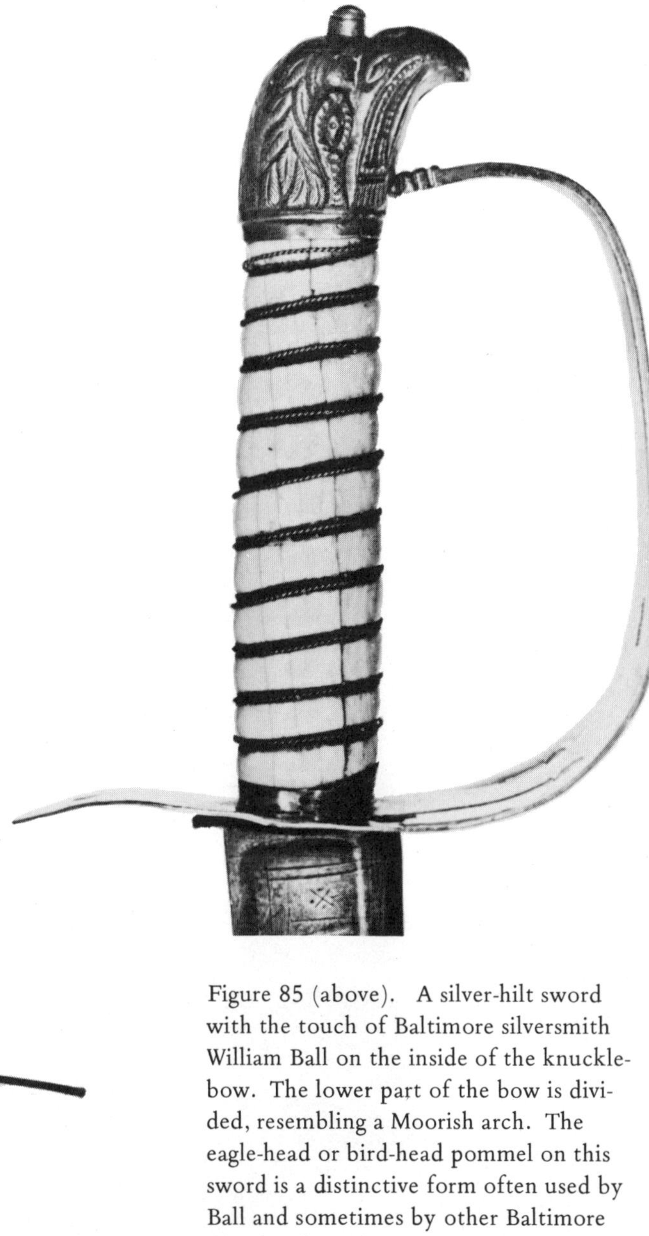

Figure 85 (above). A silver-hilt sword with the touch of Baltimore silversmith William Ball on the inside of the knuckle-bow. The lower part of the bow is divided, resembling a Moorish arch. The eagle-head or bird-head pommel on this sword is a distinctive form often used by Ball and sometimes by other Baltimore silversmiths. The curved blade is 31¼ inches long. John K. Lattimer collection.

Figure 84 (left). Silver-hilted sword with Philadelphia-type eagle-head pommel and a pierced knuckle-bow where the touch of Baltimore silversmith John Lynch is found. This sword belonged to Major John Griffith Johnson who was at Fort McHenry in 1814, and is engraved with his initials. The Spanish blade is 30¼ in. long. Collection of Dr. John K. Lattimer.

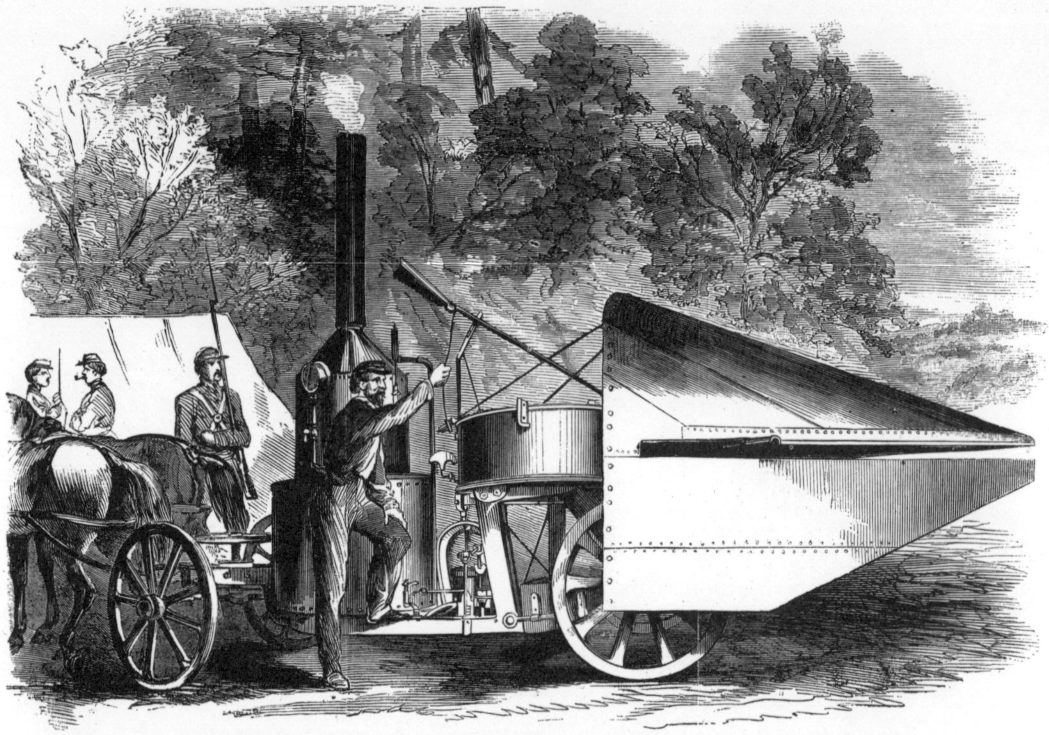

THE WINANS STEAM-GUN, RECENTLY CAPTURED NEAR THE RELAY HOUSE, BY MAJOR-GENERAL BUTLER'S COMMAND.

Figure 86a (above). The steam cannon patented in 1859 by Charles S. Dickinson and constructed at Winan's locomotive works the following year. This illustration of it appeared in *Leslie's Illustrated Weekly* in 1861. Courtesy of Erick Davis.

Figure 86b (below). Three of the five 18th century cannon barrels unearthed on the site of the Mt. Etna Furnace in Washington County. The large barrel painted black is a six-pound seige gun 7 feet 6 inches in length with a circumference of 19 inches at the breech and 16 inches at the moulding near the muzzle. The other two are twenty-four-pound field guns in unfinished condition, 5 feet 8 inches in length, with a circumference of 14 inches at the vent and 10½ inches at the muzzle moulding. Each has a sprue of about 10 inches length attached to the muzzle. All three are vented and have cascabel, trunions, reinforcing rings and muzzle mouldings.

CHAPTER 9

Artillery

During the colonial period England of course did not encourage the manufacture of cannon in the colonies, and of course there was no need for colonial manufacture inasmuch as the military needs of the day were met with pieces from abroad. But this picture changed drastically when the need for colonial arms became so apparent just prior to the outbreak of the Revolution, and Maryland soon became a colony that manufactured iron cannon.

According to Charles W. Miller, "a foundry was established on the Potomac, about Georgetown, where the first cannon were made in this country."[1] The first known cannon contracts let by the Council of Safety were made with Daniel and Samuel Hughes of upper Frederick County, George Matthews of Baltimore, and John Yost of Georgetown. Miller probably was writing about Yost who is documented as having a number of muskets ready for the Council in January 1776. In the fall of 1776 he erected a horse-powered mill. Yost had many workmen helping him and possibly made cannon in 1775 before the other contractors started. In 1781 he was repairing arms and informed the Council that he could no longer maintain his family without regular payment.

On 14 October 1775 George Matthews responded to Purviance, an official of the Maryland Council of Safety, about the casting of guns:

> I could not say anything with any degree of certainty till now as I have fully considered the madder I find that it will require about 200 worth of Apparatus to go into business of making guns Wch. will be of little value for any other purpose, the expense of Making Boring drilling and finishing compleat will be about 40 Currency p ton and any size or bore from 9 Pounds downwards maybe cast at this Country Furnaces.[2]

Henry Howard advertised in the *Pennsylvania Gazette* on 14 August 1776:

> Any person who understands the casting and finishing of cannon by applying to Henry Howard, at Northampton Furnace about ten miles from Baltimoretown, in the Province of Maryland, will meet with great encouragement.[3]

Cannon were proved, just as muskets were, for the protection of those using them. Proving inspectors were appointed by the Council of Safety to certify guns for both sea and land service. On 22 February 1776 the Council was informed that 20 nine-pound cannon were ready to be proven at the Hughes foundry. A charge of 180 wt. was used for the nine-pounders, and 900 wt. for the 50 eighteen-pounders that were being tested at the same time. On March 10 the Council was informed that four out of the first five cannon burst. Matthews, one of the proving inspectors, was killed. Daniel and Samuel Hughes apparently improved their quality of manufacturing, for they furnished many pieces for the Continental Army. Sprigg wrote to the Council on 17 June 1781, "Agreeable to Your directions I provd. 20 four pound Guns at Mr. Hughes works . . . they stood the first proof so well it inducd. me to give them something more than the usual proof in the second Charge, I believe them to be Extry Guns."[4]

Carl W. Shafer of Mt. Etna owns a lot that is part of the ground formerly owned by the Hughes brothers. While digging a foundation for a house a few years ago he unearthed an iron cannon barrel. With the help of a metal locator four other complete barrels have been found in his garden. In addition, a number of iron cannon balls ranging in diameter from 1 inch to 8 inches have been found on the site. These are important discoveries. The barrels must have been stored or stock-piled. It is amazing that they were not re-melted. Original correspondance shows that many cannon were manufactured at the Mt. Etna furnace, ranging from four-pounders to eighteen-pounders.

Toward the end of the Revolution, Maryland artillery companies were adequately equipped and the demand for cannon, at least in Maryland, was reduced. William Russell of Northampton Furnace advertised in the *Maryland Journal & Baltimore Advertiser* for 31 July 1781:

> To be sold at Northampton Furnace, in Baltimore County, about 10 miles from Baltimore Town, 12 miles from the navigation of Gunpowder River, 20 miles from Otterpoint on Bush River, and 30 miles from the Susquehanna lower ferry, Cannon of any size from two to twelve pounders (eighteen may be had if applied for sometime before they are wanted). Also swives, shot, salt pans, and plates, pots, kettles, dutch ovens, tea kettles, waggon-boxes, flat-irons, etc. Guns made at this Furnace, have been allowed by the best judges, to be equal in quality to any yet made on the continent. [5]

In completing the defenses of Baltimore during the Revolution over 250 negroes were employed in providing timber, building batteries, and mounting guns. Cannon carriages and cassions were built and mounted by individuals contracting with the Council of Safety. For example, "Ordered That the Treasure of the Western Shore pay to James Taylor five Pounds Currency in Part Payment for making Gun Carriages." [6]

The lull in ordnance production in Maryland following the Revolution came to an end in 1794 when a government contract was signed on 28 June with Samuel Hughes for 50 thirty-two pound cannon, 50 twenty-four pounders, and 90 naval twenty-four pounders at $106.66 per ton. On 25 October 1796 a second contract was let with Hughes for 40 twelve-pounders at $133.25 per ton.

When the War of 1812 developed, Maryland and other states again had need of cannon. James Haslett, in writing to the Governor of Virginia for an arms contract on 29 December 1810, remarked as follows:

> The Foundry and Boring Mill will receive my particular attention, but not having at present an accurate knowledge of the expence of manufacturing Cannon, cannot state with precision what I can make them for. However, I will engage that part of the work on as moderate terms as they can be manufactured in Richmond. [7]

During the War of 1812 the British fleet sailed up Chesapeake Bay, burning supplies and plundering towns. On 3 May 1813, after the burning of Havre de Grace, they ascended on Samuel Hughes' cannon foundry about eight miles above the town and destroyed a large number of finished as well as unfinished cannon. The entire factory with machinery for boring cannon, valued at twenty thousand dollars was destroyed. Hughes immediately rebuilt his facilities and on 16 December 1813 he again contracted for making 40 twenty-four-pound cannon, but he was in a state of extreme financial stress and his business never returned to the height he once enjoyed. He was so heavily mortgaged that his Cecil industry eventually had to be sold. At the time of the sale it consisted of "one blast furnace, two air furnaces, a boring mill with the most perfect machinery to bore five Cannon at a time, a stamping mill, a Stone grist mill, mangers and workmen's houses, Coal House, Stables & Smith's shop, with Iron flasks, tools & machinery for making Cannon."

John Mason, who operated the Columbia Foundry in Georgetown, contracted for and delivered to the government 167 twenty-four-pound cannon, 40 thirty-two-pound cannon, and a considerable quantity of cannon balls in the period between 1827 and 1830. During this period Mathew McLaughlan, Underhill Dann, and T. S. Savin contracted to supply cannon carriages.

In 1835, R. Beale, a Marylander, received two patents for cannon ignition by the use of a type of percussion lock inset in the top of the cannon. Benjamin Chambers invented and patented a lighter weight, moderately thick, wrought iron cannon construction by using staves of iron and hoops in alternating layers. He also invented further improvements in a moveable breech, covered under patent No. 6612 of 31 July 1849. On August 18, 1863, Daniel E. Somes received patent No. 39,592 for a shorter, thicker cannon, but of less weight than normal, that would resist much greater breech pressure.

At the start of the War Between the States, Charles Dickinson invented a unique "bullet-proof" steam cannon. It was manufactured at Winan's locomotive works, Parkin and Hollins Streets, Baltimore. The device was mounted on four wheels for mobility. It could project missles of any size from a musket ball to a one-hundred pound ball for a distance of two miles, and was said to spit them out at rates up to 200 per minute. The boiler was similar to that of the steam fire engine, and with one barrel the ammunition was fed through the hopper. Winan was shipping the gun to the Confederacy disguised as farm machinery when it was captured by the 6th Massachusetts at the town of Relay near Baltimore on the road to Harpers Ferry on 11 May 1861. The cannon was used by the Federals to guard the viaduct over the Patuxent River on the Washington branch of the Baltimore and Ohio Railroad.

In 1858 and 1859 James H. Merrill patented two breech-loading cannon. Other Marylanders secured cannon patents after the Civil War. Julius L. Bohrer invented a cannon with an additional chamber in the rear of the bore in the summer of 1868. When the charge was fired this chamber filled with air, eliminating a vacuum and reducing the recoil. Also in 1868, Frederick W. Alexander patented a Caliber-Diminisher, which when placed in the bore of a loaded cannon prevented or reduced the windage of the projectile upon discharge.

REFERENCES

1. Miller, Charles W., GENERAL DIRECTORY OF FREDERICK CITY & FREDERICK COUNTY, p. 127.
2. Matthews to Purviance, 14 October 1775, ARCHIVES OF MARYLAND, p. 82.
3. *Pennsylvania Gazette*, 14 August 1776.
4. Sprigg to Council, 17 June 1781, ARCHIVES OF MARYLAND, p. 298.
5. *Maryland Journal & Baltimore Advertiser*, 31 July 1781.
6. Council of Safety, 1 May 1776, ARCHIVES OF MARYLAND, p. 395.
7. Flournoy, H. W., Haslett to Governor of Virginia, 29 December 1810, CALENDAR OF VIRGINIA STATE PAPERS, Vol. X.

Figure 87. The U. S. S. *Constellation*, a frigate of 36 guns, was authorized on 27 March 1794 to be built. She was constructed in Baltimore by David Stodest and launched 7 Sept. 1797. She was armed with 28 long twenty-four-pounders, and 12 twelve-pounders, and carried a crew of 12 officers and 300 men. This photograph, in the Daniel D. Hartzler collection, was taken in 1914. The ship may be visited today in downtown Baltimore.

CHAPTER 10

Naval Shipbuilding

Shipbuilding in Maryland originated near the town of St. Michaels on the Eastern Shore in 1670 and proved to be a most profitable business. Before the Revolution most new vessels were sent to the West Indies laden with products, then on to Great Britain where they were sold. During the French and Indian War, when French privateers raided the shores of Maryland, the province was defenseless because she had no warships. This was to change, and Maryland ports have gone down in history as homes of celebrated clippers as well as some of the world's finest frigates and fightingest sailors.

The first two ships in the American Navy, the cruiser *Wasp* with eight guns and the cruiser *Hornet* with ten guns, were outfitted in Baltimore just prior to the Revolution. During the war Marylanders altered merchant ships into fighting ships and outfitted the great majority of privateers. These privateers were privately-owned armed vessels authorized by the United States to capture ships belonging to nationals of Great Britain. They were allowed to sell whatever ships and cargoes they could take.

The Maryland Council of Safety contracted for some building, outfitting, and equipping of ships. On 27 April 1780 Col. Hollingsworth wrote to the Council concerning Samuel and Daniel Hughes, who manufactured cannon at the Antietam Furnace: "I am ordered by the Navy Board to have a quantity (24 pieces) of Cannon haul'd from Mr. Hughes Furness to Baltimore to see them forwarded to Philadelphia."[1]

Cannon were not plentiful and had to be shared between the army and navy, as the following letter shows:

> May it Please Your Excellency I Received in the Course of Last Week Eighteen Pieces of Cannon (from Mount Etna Furnace) Said to be Four Pounders but are not all of a Boare some of them will Carry a Six Pound Shot. The Guns are very Fine to Judge from Appearance I think them Equal to any Ever Cast on the Continent But are By no means Proportioned or Proper for Field Peices as they are much too short & Heavy. They are only of the Common Length for Ship Guns, as I was not Informed wether Those are the Guns Intended for the Travelling Carriages I was Ordered to have made.[2]

One of America's first naval heroes, Joshua Barney, was only 23 years of age, when as captain of a converted merchant ship, he captured a 20-gun British man-of-war, the *General Monk*, during the Revolution. During the war he was captured three times, was exchanged twice and escaped a third time by disguising himself in a British uniform and walking out of prison. Returning to merchant service he was recalled by the navy during the War of 1812 when Barney's schooner *Rossie* captured ships and cargo valued over one and a half million dollars. Being ordered to command a flotilla of armed river barges, Barney led his 500 sailors and marines to Bladensburg where they worked five naval cannon until they were over run.

In 1789 the U. S. Navy Department was established and Benjamin Stoddert of Maryland was appointed the first Secretary of the Navy. On 27 March 1794 the Navy Department authorized the construction of the U. S. S. *Constellation.* She was constructed in Baltimore by David Stodest and launched on 7 September 1797. The 36-gun frigate had 28 long twenty-four pounders, 12 twelve-pounders, and carried 12 officers and a crew of 300 seamen.

In 1799 two sloops of war were made in Baltimore, the *Maryland* and the *Chesapeake.* Merchants also outfitted three armed privateers of 20 guns each at this time. Some well-known early Maryland ships included *Hyder Ally, Defence, Virginia, Friendship,* and *Amelia.*

In Maryland the War of 1812 was popular from the start. As soon as war was declared, Maryland privateers swarmed the seaways, attacking not only the commerce of Great Britain but her armed vessels as well. Baltimore outfitted more privateers than any other American port. A total of 126 ships left the Chesapeake Bay, capturing or sinking 556 British ships. More officers of the navy came from the Free State than any other – 46 out of 240. The privateers and letters-of-marque ships were a great thorn in the side of the British, who called Baltimore a "nest of pirates." Some of the most successful Maryland ships included the *Falcon, Globe, Nonesuch, Dolphin,* and *Chasseur* or *Pride of Baltimore*.

During the war with Tripoli, Stephen Decatur on the night of 16 February 1804 led a party of 80 men on board the frigate *Philadelphia*, an American ship recently captured and anchored in Tripoli harbor. After twenty minutes of hand-to-hand combat with cutlasses and pistols, Decatur seized control of the vessel and destroyed it. He was promoted and was presented a beautiful sword from Congress. In the War of 1812 in command of the warship *United States*, Decatur engaged the faster British frigate *Macedonian* and by feats of maneuver and accurate long-range gunnery, captured the foe. For this capture he received from Congress a gold medal.

During the War with Mexico, Maryland regiments embarked from Fort McHenry on transport steamships in 1846 and 1847 for Vera Cruz.

In the War Between the States, Maryland supplied 5,636 sailors in defense of the Union. Rear Admiral Louis M. Goldsborough had been appointed a midshipman at the age of seven, and was a veteran of the Seminole War and the Mexican War. During the Civil War he commanded the Atlantic blockade.

At the onset of hostilities the majority of officers from this state in the armed services resigned their commissions and 28 naval officers received commissions in the Confederate navy. Considering Maryland's seafaring history, there must have been many men who crossed the Potomac to join the Southern navy.

A lively incident was the capture of the bay steamboat *St. Nicholas* by 25 members of Zarvona's Zouaves, a Confederate company raised in Maryland. On 28 June 1861 the *St. Nicholas* left Baltimore on her regular run to Georgetown with the Zouaves aboard in disguise. Richard Thomas, leader of the Confederates, who went by the name of Colonel Zarvona, was masqueraded as a French lady, with weapons and uniforms hidden in "her" trunks. Shortly after midnight the Zouaves overpowered the crew and Commander (later Commodore) George H. Hollins of the Confederate Navy, a Baltimorean, assumed command. The boat took on Zouave reinforcements and steamed for Fredericksburg, Va., capturing three other U. S. vessels en route. Many of the passengers aboard the *St. Nicholas* were secretly on their way south to join Confederate forces.

Admiral Franklin Buchanan, who had been the first Superintendent of the U. S. Naval Academy at Annapolis, resigned his post as head of the Washington Naval Yard in 1861 and returned to his Maryland home in hope that the state would secede. When this did not come about he offered his services to the South and was put in command of the iron-clad *Virginia* (*Merrimac*). He destroyed the U. S. S. *Congress* on which his brother, McKean Buchanan, was an officer. Rear Admiral Raphael Semmes, a midshipman at Annapolis who had served in the Mexican War, resigned as a commander in 1861 and accepted the same rank in the Confederate Navy. He was given command of the C. S. S. *Sumter* which was the first Confederate ship of war, and took many prizes. In August of 1862 he was given the *Alabama,* and in the next 22 months made history by sinking, capturing, burning, or ransacking 82 Union vessels. *Alabama* was the most formidable of the Confederate privateers, and it is estimated that she destroyed U. S. shipping to the value of between 8 and 10 million dollars.

A patent was granted in 1865 to William Louis Winans for a new type of mounting for cannon on ships and gunboats. This invention made possible more mobility of the guns by raising the piece from the bulwarks of the vessel for firing, and lowering it to a place of relative security below deck for reloading. This enabled gun crews to work in a non-exposed position and did away with ports or embrasures in the sides of vessels.

REFERENCE

1. Hollingsworth to Council, 29 April 1780, ARCHIVES OF MARYLAND, p. 483.
2. Deeport to Lee, 25 June 1781, ARCHIVES OF MARYLAND, p. 315.

Figure 88. Thomas Ruckles of Baltimore painted this War-of-1812 scene during the war years. It depicts Maryland troops gathering prior to battle and is titled "Assembly of the Troops." A variety of uniforms can be seen in the large painting, but are difficult to distinguish in this small reproduction rendered in black and white. Courtesy of The Maryland Historical Society, Baltimore.

CHAPTER 11

Gunpowder

Throughout the colonial era Maryland's needs for gunpowder were readily supplied by Europe, and there was no need for a domestic industry to develop. The situation changed greatly when the Revolutionary War got underway and ties with England were cut. In August 1775 a crude powder mill was built near Baltimore and in the following year saltpeter plants were in operation in Cecil and Harford Counties. From this small beginning Maryland rose to lead the young nation in gunpowder production.

At the outbreak of the war the colony was in a precarious position for lack of powder, so the Council of Safety offered liberal proposals for the erection of powder mills. George Lindenberger and John McClellan constructed a mill near Baltimore. A second mill was built by John and Walter Hanson in Charles County in 1776.

In post-war years other mills were erected, spotted throughout the colony in part, but the majority were concentrated near Baltimore. In the Federal Census of 1810 the first summary of gunpowder production shows Maryland ranking first, manufacturing over one-fifth of the country's output. At this time gunpowder mills in the vicinity of Baltimore produced twice as much as the total output from the DuPont mills. Maryland was the first important gunpowder center in the country. It was only in later years, after the DuPont Company had become firmly established, that the Baltimore mills were seriously rivaled. The Bellona Mills were established in 1800 on Jones Falls about seven miles from Baltimore and became the leading Maryland producer and gained a national reputation.

Gunpowder is produced from a mixture of charcoal, sulfur and saltpeter, which is potassium nitrate. The most difficult of the three ingredients to secure was the saltpeter which at that time came from the bacterial fixing of nitrogenous wastes from animals and plants. This often was found as thin deposits in houses, cellars, stables, barns and caves, and was gathered by sweeping the dust into containers. The nitrate dissolved in water and was purified by repeated crystallization. Not all powder makers made their own saltpeter, but usually they refined it because the quality of the gunpowder depended on the purity of this primary ingredient. The Levering mill in 1811 was selling its product at such low rates that DuPont was forced to reduce their prices to compete.

Time after time gunpowder mills were demolished by explosions resulting from accidental ignition by sparks, lightning, spontaneous combustion and carelessness. Thomas Ewell established a mill near Bladensburg in 1811, and on 7 December 1813 he received a patent for three improvements that diminished the risks, the waste and the expense of manufacture. W. H. Richardson, a millwright of Baltimore, patented a new process for granulating powder on 3 August 1822.

In the early decades of the nineteenth century Maryland remained in the lead among the producing states, turning out 669,125 pounds during the year 1840. The output declined in the middle of the century to about fifth place, and continued downhill so that by 1860 only the Bellona mill remained active.

A study of the manufacture of black powder in early Maryland was made by Arlan K. Gilbert and published in 1957 with the title "Gunpowder Production in Post-Revolutionary Maryland" (Maryland Historical Magazine, v. 52, p. 187-201.

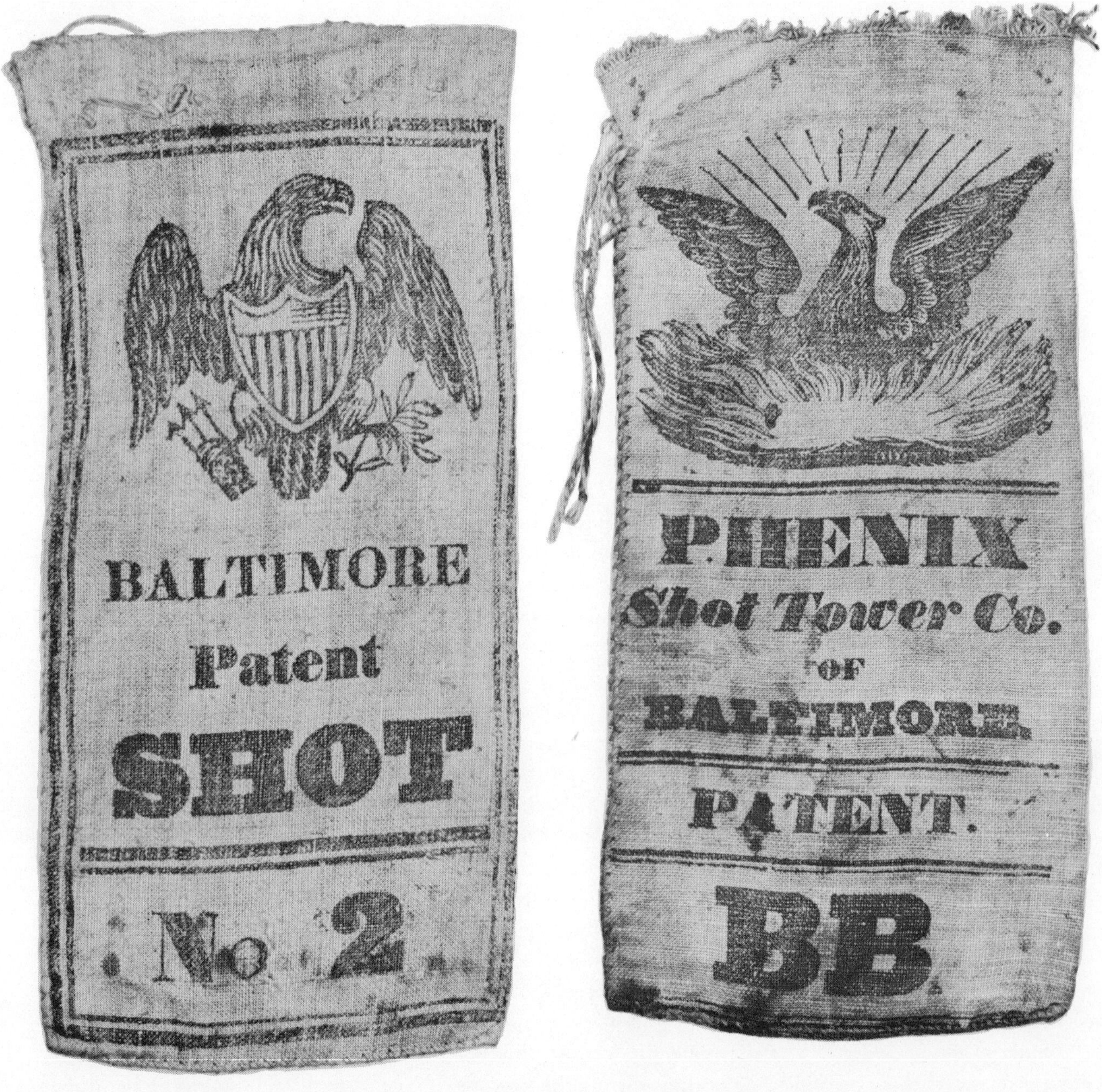

Figure 89. Linen tow shot bags for shot produced in Baltimore. When mounted in a picture-frame behind glass they become an attractive antique display. Collection of Harry D. Berry., Jr.

CHAPTER 12

Shot Towers

Lead shot for shotguns and fowling pieces has been produced by various means over the centuries, but it was not until shot towers came into use early in the nineteenth century that nearly perfect shot was made in large quantities. Baltimore became a center for shot production, and the high quality of the product made there created a large market. Basically, the shot tower was a tall platform from which molten lead could be poured through a sieve. Surface tension pulled the molten lead into small spheres of various sizes as they fell into the water below.

The manufacture of shot by dropping from a tower began in Baltimore in 1822 with the erection of a 160-foot high tower on North Gay Street by Col. Joseph Jamison and Company. This tower was destroyed in 1845.

The Merchants Shot Tower Company built what then was the world's largest shot tower on Eutaw Street in 1828 and the structure still stands. It was 50 feet wide at the foundation and 25 feet in diameter at the top, with a height of 250 feet. They advertised shot from "ooo/85 to 3/340 to the pound. Together with an assortment of well finished Rifle and Musket Balls, Bar Lead, &c."[1] The capacity of this tower was 500,000 bags annually.

Gist's Compressed Shot Works were manufacturers of musket, rifle, pistol ball and buck-shot. They advertised "the superiority of this Shot consists of its being pressed in iron rollers, consequently is much heavier than the drop or tower shot. Also, Drop Shot of all sizes, from No. 1 to 10."[2] The demand for dropped shot was so great that they produced it also. William L. Waits in 1835 was employed by Gists as a shot maker. By 1847 he had turned to gunsmithing and was working on his own.

The Baltimore Lead Works tower was erected in 1877 on Howard and Montgomery Streets. This tower was relatively low, being but 110 feet in height, but good shot was produced there because a current of cool air was forced upward through the structure, allowing the shot to harden in a shorter fall.

By 1885 a gun wad manufacturer was advertising the famed Baltimore shot with wads by Merino Elastic Gun Wad Company at 106 South Charles Street. In the 1930's G. W. Dobbin, J. Willar and A. Duval, all of Baltimore, had patents for various types of machinery used in the manufacture of shot.

REFERENCES

1. BALTIMORE CITY DIRECTORY, 1895.
2. *American & Commerical Daily Advertiser*, Sept. 2, 1834.

Figure 90. Thomas Ruckles of Baltimore depicted a War of 1812 scene in the large painting reproduced here and titled "The Battle of North Point." Many uniforms of the Maryland militia can be seen in the original painting. Courtesy Maryland Historical Society, Baltimore.

CHAPTER 13

Accoutrements and Uniforms

Military uniforms and accoutrements have helped to lure impetuous youths to the fortunes and misfortunes of soldiering throughout the ages. In the early decades of our country's history, people tended to live with a minimum of luxuries. The soldier was regarded with honor and esteem. His colored uniform with its bright buttons, buckles and other trappings helped create the aura that went with the soldier.

On 14 June 1775 the Congress called for two rifle companies to be formed in Maryland. Frederick County, Maryland's center of longrifle activity, supplied both the men and officers. These two companies, comprising about 130 men, armed with rifles and tomahawks, painted like Indians, and dressed in linen hunting shirts, buckskin leggings and moccasins, became part of the First Continental Regiment.

The province called for the organization of a battalion of nine companies on 1 January 1776. Three days later Smallwood's Battalion of nine companies of regular troops was raised in Baltimore and Annapolis. By 5 January seven independent companies from the counties were so inflamed with the cause that they accompanied Colonel Smallwood to New England as volunteers.

A company of marines also was raised, which saw sea duty on Maryland's vessels in the Chesapeake Bay protecting the shores and conveying troops and supplies. For the most part they served on coastal vessels fitted out for war-time duty. In 1777 seven galleys were built, and in 1778, eleven barges. A crew of about 65 men typically manned each of these galleys, including the marines who manned the two moderately large cannon that each carried. The largest of the craft was the *Defense,* which carried 22 six-pounders and cruised to the West Indies.

Six companies of Matrosses were raised to serve Maryland as artillerists. On 3 June 1776 the Continental Congress resolved that additional troops were needed and flying camps should be established. Maryland's quota was 3400 men. Four battalions, with a total of 3405 men were raised throughout the province and sent to New York.

The Maryland Line, composed of seven regiments, was organized from Smallwood's Battalion, the Independent Companies, the Flying Camp and additional recruits during the course of the retreat across New Jersey. The army was reformed in the latter part of 1780. The remaining 1400 Marylanders were reorganized into five regiments and they saw service in the southern campaigns.

Many Marylanders served in additional battalions that were connected with other colonies. Two companies of Germans were recruited in Baltimore County and two others were raised in Frederick County; they became part of the German Continental Regiment. Young men from Maryland were found in abundance in the Continental Dragoons and in the Continental Artillery.

During the American Revolution a great variety of uniforms and accoutrements were in use. Generally they were contracted for by individuals or groups of individuals within the colony. The variety of uniforms was greatest during the first two years of war, and when those uniforms wore ragged, necessity often kept them in further use, or they were exchanged for almost any usable clothing.

Buttons of brass and pewter were common, and when pewter was scarce, lead was substituted. Some buttons were produced marked for individual military units. Companies raised and equipped by the colony for specific regiments wore marked buttons, but when they reached the army they sometimes were assigned to other commands. Thus buttons were likely to be in the wrong places. The following button markings are mentioned in newspaper accounts of the Revolutionary War period.

1st Maryland Battalion	1 MB	1st Maryland Matrosses	1 MM	3rd Maryland Regiment	3MR
2nd Maryland Battalion	2 MB	3rd Maryland Matrosses	3 MM	4th Maryland Regiment	4MR
3rd Maryland Battalion	3 MB	5th Maryland Matrosses	5 MM		
		2nd Maryland Matrosses	2MM		

A few buttons with these markings have survived the years, and of course they are highly prized by collectors. They include the 1st, 2nd and 3rd Maryland Battalions, and the 5th Maryland Matrosses.

Two goldsmiths, William Whetcroft of Annapolis and Christopher Hughes of Baltimore, are known to have set gold into elaborate buttons.

The following descriptions of uniforms and buttons worn by men from various Maryland units during the Revolution are found in newspapers of the period:

* RIFLEMEN *

1st COMPANY CAPTAIN MICHAEL CRESAP FREDERICK COUNTY

2nd COMPANY CAPTAIN THOMAS PRICE FREDERICK COUNTY

Painted like Indians, armed with tomahawks and rifles, dressed in hunting shirts and moccasins.
They are dressed in white frocks, or rifle shirts, and round hats.
Hunting shirts, drawers, green leggings.
A suit of nankeen clothes, leggings and hunting shirt. [*Constitutional Gazette,* 12 June 1776]

* * SMALLWOOD'S CONTINENTALS * *

* COLONEL WILLIAM SMALLWOOD *

5th COMPANY CAPTAIN NATHANIEL RAMSEY

Green coat turned up with white, hunting trousers, calico jacket, a pair of pumps with carved silver buckles. [*Pennsylvania Evening Post,* 1 August 1776]

8th COMPANY CAPTAIN SAMUEL SMITH

White coat and waistcoat, mixed bearskin breeches, brown coat and waistcoat lined with oznaburg and made short. [*Maryland Journal,* 10 April 1776]
Castor hat, blue hunting shirt, osnaburg trousers with five buttons at the ankle, a short light colored double-breasted waistcoat.
Felt hat, brown coat, red waistcoat, leather breeches.
Brown coat and waistcoat with brass buttons, buckskin breeches, osnaburg trousers, and undyed hunting shirt, felt hat.
Buckskin breeches, dark hunting shirt, buff casimere waistcoat.
[*Pennsylvania Evening Post,* 10 August 1776]

* INDEPENDENT TROOPS *

3rd COMPANY CAPTAIN JOHN WATKINS WORCESTER COUNTY

Fourteen deserters, all dressed in black uniforms.

4th COMPANY CAPTAIN JAMES HINDMAN TALBOT COUNTY

Purple hunting shirts with red cuffs and cape, buckskin breeches.
Purple hunting shirts with red cuffs and cape, brown coat, swanskin jacket, leather breeches.
[*Pennsylvania Journal,* 14 August 1776]

MARINES

That the uniform of the land forces and marines to be hunting shirts (linen).
The hunting shirts of the marines to be blue and those of the land forces to be other colors.
[Proceedings of the Conventions, 14 January 1776]

* * FLYING CAMP * *

* COLONEL CHARLES G. GRIFFITH'S 1st BATTALION *

CAPTAIN VALLENTINE CREAGER'S COMPANY FREDERICK COUNTY

Felt hat, gray cloth, brown waistcoat, leather breeches.
Felt hat, gray surtout coat, leather breeches.
Short light colored jacket, long trousers, felt hat.
Blue shalloon breeches, light colored coat.
Yellow hunting shirt, long trousers.
White hunting shirt, leather breeches, felt hat.
White hunting shirt and trousers, felt hat. [*Pennsylvania Packet,* 29 October 1776]

Light colored coat, long skirted, scarlet breeches, felt hat.
Fustian frock, light colored waistcoat, light blue breeches, felt hat.

[*Pennsylvania Packet,* 5 November 1776]

CAPTAIN DANIEL CLAPSADDLE'S COMPANY WASHINGTON COUNTY

Black hunting shirt, brown jacket with sleeves, leather breeches, fur hat.
Black hunting shirt, grey double breasted jacket with sleeves, linsey-woolsey jacket without sleeves, leather breeches, light blue stockings, long white leggings with red spots, wool hat.

[*Pennsylvania Packet,* 29 October 1776]

* COLONEL JOSIAS C. HALL'S 2nd BATTALION *

CAPTAIN AQUILLA PACA'S COMPANY HARFORD COUNTY

Sky blue coat turned up with green, a brown jacket with silver basket buttons, buckskin breeches.

[*Pennsylvania Evening Post,* 12 October 1776]

CAPTAIN FRANCIS HOLLAND'S COMPANY HARFORD COUNTY

Brown coat, light colored broadcloth waistcoat, check shirt, castor hat, leather breeches, black stockings.

[*Pennsylvania Journal,* 13 November 1776]

* COLONEL THOMAS EWING'S 3rd BATTALION *

CAPTAIN EDWARD NORWOOD'S COMPANY ANNE ARUNDEL COUNTY

Scarlet cloth jacket without sleeves, new buckskin breeches.
Yellow striped waistcoat without sleeves. [*Maryland Journal,* 16 October 1776]

* COLONEL WILLIAM RICHARDSON'S 4th BATTALION *

CAPTAIN WALTER ALEXANDER'S COMPANY CECIL COUNTY

Short green jacket made sailor fashion, hunting shirt. [*Pennsylvania Gazette,* 23 October 1776]

CAPTAIN JAMES MAXWELL'S COMPANY CECIL COUNTY

Hunting shirt, brown colored coat, cat skin hunting cap.
Light blue coat, Scotch flannel jacket. [*Pennsylvania Gazette,* 30 October 1776]

CAPTAIN THOMAS SMYTH'S COMPANY KENT COUNTY

County linen hunting shirt. [*Pennsylvania Journal,* 11 September 1776]

* * MARYLAND LINE * *

* COLONEL JOHN H. STONE'S 1st REGIMENT *

CAPTAIN NATHANIEL EWING'S COMPANY

Blue coat with red facing, a white cotton jacket, light colored thickset breeches, white yarn stockings.

[*Maryland Journal,* 9 December 1776]

CAPTAIN THOMAS HARWOOD'S COMPANY

Dark brown coat with lapels of the same, a grey waistcoat, buckskin breeches, blue yarn stockings, a round hat. [*Maryland Journal,* 9 December 1777]

CAPTAIN ALEXANDER MURRAY'S COMPANY

Light colored coat, faced and lined with scarlet, with pewter buttons marked 10, light colored waistcoat, buckskin breeches.
Green sagathy coat and waistcoat, black knit breeches.
Brown cloth coat, lined with part of a tent cloth, red waistcoat, leather breeches, dark yarn stockings, a cocked hat. [*Maryland Journal,* 9 December 1777]

CAPTAIN LEVIN WINDER'S COMPANY

A light infantry cap, with silver lace around the edge, and B.L.I. (Baltimore Light Infantry) in cypher in front, short blue coat lined with scarlet and pewter buttons marked U.S.A., one red and one white waistcoat, buckskin breeches, white overalls, white hunting shirt, cotton stockings, large silver buckles.
Blue regimental coat lined and faced with scarlet and pewter buttons marked 3MR, a brown cloth waistcoat, buckskin breeches, woolen stockings, a fine hat with a set hat buckle, shoes with silver buckles, linen shirts. [*Maryland Journal,* 9 December 1777]

* COLONEL THOMAS PRICE'S 2nd REGIMENT *

Brown coat, white jacket, leather breeches, a Carolina hat with a gold button.
Whitish colored short coat, osnaburg shirt. [*Maryland Gazette,* 13 February 1777]
Blue coat faced with scarlet, blue waistcoat and breeches. [*Maryland Gazette,* 8 May 1777]
Brown coat, not turned up, and a flopped hat. [*Pennsylvania Journal,* 14 May 1777]
Brown coat faced with red, brown cloth waistcoat, country linen breeches, castor hat.
White linen coat and waistcoat, white linen breeches, country made shirt, thread stockings, a castor hat.

[*Dunlap's Maryland Gazette,* 24 June 1777]

Brown coat faced with buff.
Blue coat faced with white. [*Maryland Gazette,* 24 July 1777]

* COLONEL MORDECAI GIST'S 3rd REGIMENT *

CAPTAIN JOSEPH MARBURY'S COMPANY

Short blue coat, jacket and breeches, cocked hat. [*Maryland Gazette,* 1 April 1777]
Light colored coat, osnaburg trousers. [*Maryland Gazette,* 24 April 1777]
Brown coat and waistcoat, blue great coat, buckskin breeches, white leggings, castor hat. [*Dunlap's Maryland Gazette,* 27 May 1777]
Wilton coat and jacket, checked shirt, old hat, country made shoes with plated buckles. [*Maryland Journal,* 24 June 1777]

CAPTAIN HENRY RIDGLEY'S COMPANY

County made coat, of a mixed blue grey color, lined with mixed grey linsey, the back parts lined with yellow, a mixed blue grey jacket lined with brown stuff, buckskin breeches, a small round hat, a blue surtout bound with blue tape, a matchcoat blanket. [*Maryland Journal,* 15 April 1777]

CAPTAIN SAMUEL GRIFFITH'S COMPANY

Felt hat, white country linen shirt, short grey country cloth coat, striped linsey jacket, leather breeches, grey yarn stockings, new shoes. [*Maryland Journal,* 10 June 1777]

CAPTAIN EDWARD HINDMAN'S COMPANY

Blue coat, leather breeches, new dowlas shirt, blue worsted stockings, white linen leggings, new shoes, a new hat bound with white. [*Maryland Gazette,* 10 June 1777]
Fashionable cat hat, grey surtout, white linen shirt. [*Maryland Gazette,* 22 July 1777]

CAPTAIN JACOB BRICE'S COMPANY

Blue coat, light colored jacket with gold vellum holes, buckskin breeches worked on the front, white cotton stockings, pumps and silver buckles.
Short brown jacket with metal buttons, buckskin breeches, white yarn stockings, long quartered pumps and silver buckles, cocked hat. [*Maryland Gazette,* 24 July 1777]
Striped silk jacket, fur hat, silver shoe buckles.
Lead-color country cloth coat, tow trousers.
Osnaburg shirt, drilling breeches, blue jacket, a soldier's hat bound with white. [*Maryland Journal,* 9 September 1777]
Sailor's dress. [*Maryland Journal,* 18 February 1777]

LIEUTENANT JOHN DEAVER'S COMPANY

A blue upper jacket with sleeves, a blue under jacket, buckskin breeches, light yarn stockings. [*Maryland Journal,* 25 February 1777]
Blue regimental coat, lined and faced with scarlet, pewter buttons marked 3MR. [*Maryland Journal,* 9 December 1777]

* COLONEL JOSIAS C. HALL'S 4th REGIMENT *

A new brown coat faced with red, a brown coat and vest, two shirts, one white, the other brown, new buckskin breeches, old sheepskin breeches, woolen stockings.
A new brown coat faced with red, two new shirts, one white, the other brown, new buckskin breeches, drab leggings, a black stock, a white hunting shirt and a pair of overalls, neatly fringed. [*Maryland Journal,* 1 July 1777]
A brown coat faced with red, the buttons lead, new buckskin breeches, drilling leggings with black buttons, a black stock, new shoes, one brown French made shirt and overall neatly fringed.
A white hunting shirt and overall neatly fringed. [*Pennsylvania Journal,* 2 July 1777]

CAPTAIN JONATHAN SELLMAN'S COMPANY

Green plush coat turned up with white, light cloth waistcoat, leather breeches, osnaburg shirt, yarn stockings.
Blue sailor's jacket and trousers, osnaburg shirt, yarn stockings.
Blue cloth coat lapelled with the same, with white metal buttons, a cloth waistcoat, osnabrug shirt, and yarn stockings.
Short light cloth jacket, leather breeches, osnaburg shirt, yarn stockings. [*Maryland Gazette,* 17 April 1777]

CAPTAIN EDWARD OLDHAM'S COMPANY

A blue regimental coat with white lining, edged with yellow, plain metal buttons, a black leather belt, with a brass buckle, the end tipped with brass. [*Maryland Journal,* 9 May 1780]

CAPTAIN DANIEL DORSEY'S COMPANY

Blue regimentals, lapeled with scarlet. [*Dunlap's Maryland Gazette,* 3 June 1777]

CAPTAIN THOMAS YATE'S COMPANY

Blue hunting shirt trimmed with white fringe, a dark blue coat, with buttons of the same color, a short blue napped jacket, with a narrow belt and lead buttons, blue and white cotton trousers, white thread or cotton stockings, a pair of round silver plated buckles, a beaver hat.
A dark blue coat with buttons of the same, light colored cloth breeches, large square silver shoe buckles.
Brown double-breasted coat, light country cloth breeches, a new regimental hat.
Double-breasted light colored coat, very long waisted, with buttons of the same color, buckskin breeches, brown yarn stockings, a round hat, new shoes and large brass buckles.
[*Maryland Journal,* 22 January 1777]

Flowered linsey jacket, a brown coat, a brown greatcoat, buckskin breeches.
Blue regimental coat and jacket faced with red, with lead buttons marked $\frac{4}{M}$.
[*Maryland Journal,* 11 February 1777]

* COLONEL WILLIAM RICHARDSON'S 5th REGIMENT *
CAPTAIN JOHN LYNCH'S COMPANY

Whitish country made cloth short coat, breeches of the same, a snuff colored under-jacket without sleeves.
[*Maryland Journal,* 18 March 1777]

Light blue coat, a whitish jacket, a round hat.
Whitish cloth coat, buckskin breeches. [*Maryland Journal,* 18 February 1777]
Whitish coat. [*Maryland Journal,* 16 January 1777]

CAPTAIN JOHN DEAN'S COMPANY

A country made cotton coat, almost white, with white capped metal buttons, cloth waistcoat, osnaburg shirt, buckskin breeches, yarn stockings, and country made shoes. [*Maryland Gazette,* 3 April 1777]

CAPTAIN JOSIAH JOHNSON'S COMPANY

Uniform coat of beaver coating faced with green.
Reddish colored coat, buckskin breeches.
Brown coat and breeches. [*Pennsylvania Journal,* 23 April 1777]
Brown sagathy coat, white jacket, blue breeches. [*Pennsylvania Journal,* 11 June 1777]

CAPTAIN JAMES HENRY'S COMPANY

Bound hat, short white hunting shirt, white waistcoat, blue cloth breeches, mottled stockings.
Snuff colored cloth coat, scarlet cloth waistcoat, buckskin breeches.
Light colored seaman's jacket, felt hat, white trousers. [*Pennsylvania Journal,* 7 May 1777]
Light colored German serge coat, white linen jacket, black cloth breeches, cocked beaver hat.
[*Pennsylvania Journal,* 7 May 1777]

Blue nap coat, with white cuffs and cape, with flat white buttons, blue jacket, leather breeches, a large cocked hat bound with white tape, with a small white button.
Blue cloth coat, with scarlet facings and yellow buttons, white kersey jacket, buckskin breeches.
[*Pennsylvania Journal,* 28 May 1777]

CAPTAIN LEVIN HANDY'S COMPANY

Brown coating coat faced with red, spotted swanskin vest with brass oval buttons, brown broadcloth breeches, white worsted stockings, continental shoes. [*Pennsylvania Journal,* 10 September 1777]

* COLONEL OTHO H. WILLIAMS' 6th REGIMENT *
CAPTAIN ANDREW HYNE'S COMPANY

Blue regimental coats with white cuffs and collar, white jacket, drilling breeches, Germantown stockings.
[*Pennsylvania Evening Post,* 26 April 1777]

CAPTAIN HENRY DOBSON'S COMPANY

Brown coat faced with green. [*Dunlap's Maryland Gazette,* 2 September 1777]
New brown freize coat turned up with green or blue, a light colored jacket, osnaburg trousers, a felt hat.
[*Dunlap's Maryland Gazette,* 9 September 1777]

CAPTAIN ROBERT HARRIS' COMPANY

A regimental coat and waistcoat of grey cloth faced with green, a hat bound with white ferret, tow trousers, new grey cloth breeches. [*Pennsylvania Packet,* 12 August 1777]
New hat, grey regimental coat and jacket, the coat turned up with green.
Yellow hunting shirt, linen drawers, white woolen stockings, a fur cap bound with white and yellow tapes.
Blue coat, tow trousers. [*Maryland Journal,* 2 September 1777]

* COLONEL JOHN GUNBY'S 7th REGIMENT *

CAPTAIN FREDERICK DEAM'S COMPANY

Light colored sailor's jacket, leather breeches, yarn stockings, round hat, osnaburg shirt and trousers.
A sailor's green jacket, leather breeches, round hat, osnaburg shirt.
A regimental suit of blue turned up with green, leather breeches, yarn stockings, osnaburg shirt.
Light colored Wilton coat, blue waistcoat, buckskin breeches, yarn stockings, silver shoe buckles.
Regimentals of blue turned up with green, leather breeches, yarn stockings, silver shoe buckles.
Light coat, blue waistcoat, leather breeches, yarn stockings.
A sailor's brown jacket, blue greatcoat, sailor's trousers, yarn stockings, silver shoe buckles.
[*Maryland Journal,* 25 February 1777]

Short green lapelled coat with small plain metal buttons, trousers made of country linen, grey woolen stockings, good shoes and buckles, felt hat.
Blue coat, red waistcoat, country linen trousers. [*Maryland Journal,* 16 October 1776]
Hunting shirt and trousers, fringed with green.
Country made cloth blue and white coat, country linen trousers, felt hat.
[*Maryland Journal,* 13 November 1776]

CAPTAIN ELISHA O. WILLIAMS' COMPANY

Grey turned up with red, and small white metal buttons. [*Maryland Journal,* 2 September 1777]

* MATROSSES *

1st COMPANY CAPTAIN NATHANIEL SMITH ANNAPOLIS

Blue coat, red waistcoat, buckskin breeches, grey yarn stockings, round brass shoe buckles.
New felt hat with white loops and a pewter button marked $\overset{M}{\underset{I}{M}}$, a light colored short coat, a grey waistcoat with pewter buttons marked $\overset{M}{\underset{I}{M}}$, a white shirt, black handkerchief, new buckskin breeches, light blue yarn stockings, good shoes.
New felt hat with white loops, and a pewter button marked $\overset{M}{\underset{I}{M}}$, brown coat and waistcoat, buckskin breeches, blue yarn stockings, white shirt, Dresden tape stock, good shoes with open steel buckles.
[*Maryland Journal,* 17 April 1776]

Narrow brimmed hat with white binding round it, blue jacket, buckskin breeches, blue yarn stockings.
Blue outside jacket, light grey cloth under jacket, buckskin breeches, spotted cotton and blue yarn stockings, canvas trousers.
Regimentals of blue faced with white, buckskin breeches.
Light mixed broadcloth coat, light grey regimental jacket, leather breeches, blue yarn stockings, new shoes, plated buckles, a castor hat.
Spotted flannel jacket with sleeves, red waistcoat, buckskin breeches, osnaburg trousers, thread and yarn stockings, blue cloth coat, short in the waist. [*Dunlap's Maryland Gazette,* 22 April 1777]
Regimental clothes of blue turned up with light colored grey, waistcoat with pewter buttons marked $\overset{M}{\underset{I}{M}}$, new hats with white loops. [*Maryland Journal,* 1 May 1776]
Light blue coat, light grey waistcoat, white Russian drab breeches, thread stockings.
Buff coat, green cloth waistcoat, leather breeches, thread stockings.
Regimental clothes of blue faced and lined with white, light grey waistcoat, pewter buttons marked $\overset{M}{\underset{I}{M}}$, leather breeches, country linen trousers, blue yarn stockings. [*Maryland Journal,* 14 August 1776]

2nd COMPANY CAPTAIN JOHN FULFORD ANNAPOLIS

Hunting shirt, buckskin breeches, a country linen shirt. [*Maryland Journal,* 10 April 1776]
Blue regimental coats, faced with red and trimmed with yellow ferret, buckskin breeches.
[*Maryland Gazette,* 19 September 1776]
Light colored cloth jacket, regimental made, with buttons marked $\overset{2}{\underset{M}{M}}$, country made shoes with a remarkable small pair of brass buckles in them, an old blue coat with short skirts, felt hat.
[*Maryland Gazette,* 5 June 1777]

3rd COMPANY CAPTAIN ALEXANDER FURNIVAL BALTIMORE

Mixed blue and white jacket, with sleeves, a light colored waistcoat.
Blue regimental coat faced with red, the uniform buttons marked MM3, the buttonholes made with white bobbin. [*Maryland Journal,* 22 January 1777]

* INDEPENDENT ARTILLERY *

CAPTAIN RICHARD DORSEY'S COMPANY

Blue coat, country made coat and jacket, buckskin breeches, black stockings, felt hat, shoes with brass buckles. [*Maryland Gazette,* 2 December 1777]

In the decades following the Revolution, Maryland, along with the other states, built up her militia. A variety of uniforms came into use, each unit adopting what suited its taste and whim, as was more or less the standard of the times. By 1811, with clouds of the War of 1812 already over the horizon, an act was passed by the Maryland legislature to bring a certain amount of uniformity to the state troops. This long act, titled "An Act To Regulate & Discipline The Militia Of This State," contains the following information about uniforms:

> AND WHEREAS, In addition to the many other advantages which would result, an immense sum of money would be saved to the citizens by adopting a uniform for the state.
>
> 53. BE IT ENACTED, That the uniform for the infantry of this state shall be, for general officers, their aids-de-camp and brigade inspectors, long dark blue coat faced with buff, buff collar and cuffs, yellow buttons and buff under clothes, long boots and chapeaux-de-bras.
>
> For field, company and staff officers, long dark blue coat edged with red, red collar and cuffs, white buttons, dark blue pantaloons edged with red for winter, and white under clothes for summer, black gaiters and shoes or half boots.
>
> For non-commissioned officers and privates, short dark blue coat, red cuffs and collar, white buttons, blue pantaloons, black gaiters and shoes or half boots.
>
> For riflemen, dark green frock or hunting shirt and trowsers.
>
> For cavalry, short coats and pantaloons of dark blue cloth edged with red, long half boots, a helmet or cap of leather covered with bear skin.
>
> For artillery, long dark blue coats faced with red, red collar and cuffs, yellow buttons, blue pantaloons black gaiters or half boots and chapeaux-de-bras.
>
> 54. AND BE IT ENACTED, That all companies, troops or corps, which shall be raised after the passage of this act who shall uniform themselves, shall conform to the uniform of the state and those companies, troops or corps who have at present a different uniform, shall adopt that of the state within three years after the passage of this act.

This act was generous in allowing a three-year period for the necessary changes to be made, and when the call came for militia units to assemble for battle in 1812, the men of course used the uniforms and accoutrements on hand, whether or not they complied with the terms of the act.

When the War of 1812 action came to Maryland, 3200 Maryland militia men marched out to meet the invading British at Patapsco Neck. They fought a delaying action, inflicting heavy casualties before retiring. Thomas C. Ruckles, a militia participant in the war, made a fine painting of the Battle of North Point in which the uniforms of the Independent Light Dragoons, Fells Point Dragoons, Maryland Chasseurs and Baltimore Hussars are distinguishable (Fig. 90). Another superb painting by the same veteran depicts the defense of Baltimore and illustrates the dress of the Mechanic Volunteers, Independent Blues, First Batallion Light Infantry, Grey Yagers, Union Yagers, Fells Point Riflemen, Baltimore Union Volunteers, Baltimore Patriots and Baltimore United Volunteers (Fig. 88).

A haversack which belonged to Private William Gwynn is painted red, white and gold and holds the inscription: "4th COMP: 1st BATT'N 39th REG." A plume socket cockade medallion which belonged to an officer was dug up at North Point. It is marked "Fells Point Light Dragoons," and features a horseman with raised sword. Another fine silver oval hat plume is marked "Baltimore Volunteer Guards 39th Maryland Regt."

The Fells Point Light Dragoons wore a two-piece, plated, five-pointed star button with rays on their coats. The First Baltimore Battalion was organized in 1792 by Major Lowry and was comprised of a company of grenadiers, two companies of hatmen and one of light infantry, Their uniforms were lavish, made in the French style. The two-piece button was gilded, with an eagle holding a ribbon in its beak and bearing the inscription "First Baltimore 1810."

Jacob Seeger was a silver-plater working on Fayette Street in Baltimore who advertised buttons in 1837. He made several types of fire department buttons, which very much resemble the above two buttons. George Mattocks operated a brass foundry in Baltimore and produced one-piece military buttons of white metal for the government in 1811.

The state militia in the 1830's, so deep in tradition, was still active and flourishing, as evidenced by military patriotic marches and the large number of song sheet covers depicting the militia. The

Figure 91. Civil War pike of the type manufactured in Baltimore and known as the Ross Winan pike, although sometimes referred to as the Marshall Kane pike. The blade is 9½ inches long, with a double edge. This is mounted on a wooden staff of 8 feet length and secured with the help of an iron collar 1½ inches long. Daniel D. Hartzler collection.

Figure 92. Knife with checkered ivory hilt, silver guard, and blade marked "Reinhardt Bal." Length overall 7¼". Daniel D. Hartzler collection.

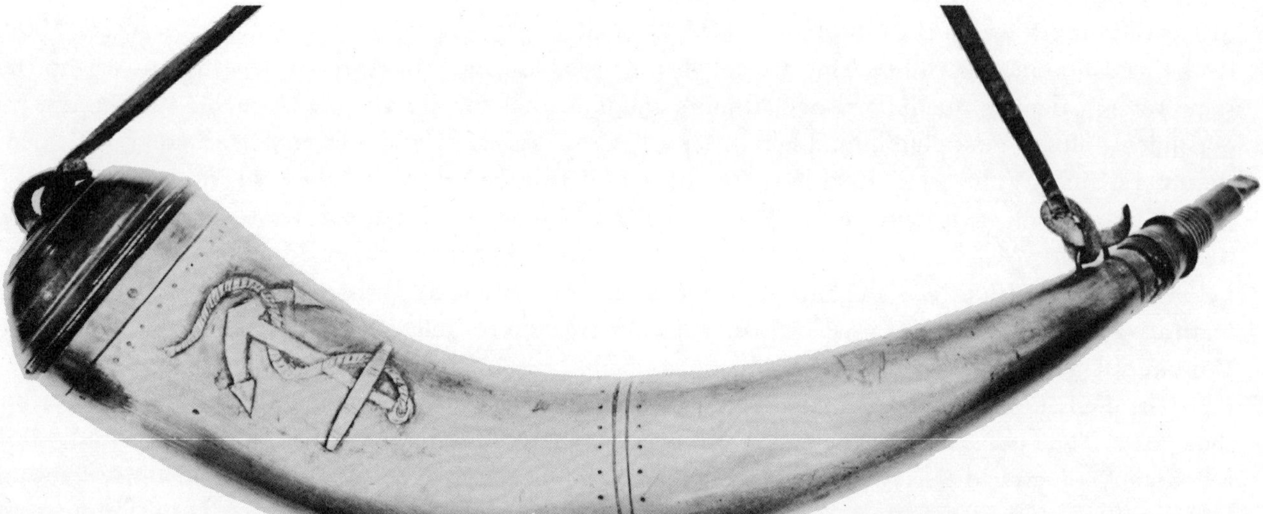

Figure 93. Powderhorn with neatly turned wooden end and a screw-on tip, engraved "Jno Wm McComas Baltimore Town". Daniel D. Hartzler collection.

dress of the Eutaw Infantry, Maryland Cadets, Marion Grays, Independent Grays, and National Guard, was so elaborate that they must have been fashioned for wives and sweethearts. The uniform of the 1st Battalion Invincibles and the Baltimore City Guard continued to change, becoming ever more extravagant.

At Camp Baltimore in Frederick, 43 volunteer corps of Maryland and neighboring states were assembled in 1842, with great pomp and ceremony, to be reviewed by President Tyler and General Scott. The officers of the white-plumed Independent Grays wore brass-plated shakos with a large "G" while the rank and file had eagles on theirs.

As the war clouds slowly gathered during the 1850's, the state militia units began preparing themselves for something more than military reviews. Inspiration for new companies being formed was provided by the exotic Zouaves of the Maryland Guard, fitted out in their uniforms of French-Algerian influence.

Knives, always one of Man's most personal weapons and tools, gained overwhelming popularity during the Mexican War. No doubt many were made in Maryland by local blacksmiths and others and remained unmarked. Many knives were marketed in Baltimore throughout the nineteenth century, some completely made in the city, some probably made in part there, and some imported from Europe. When considering the manufacture of knives in nineteenth-century America it is well to bear in mind that steel suitable for knife manufacture was not produced in the United States until the 1840's, and even after that much of the steel used for knife manufacture was imported from Germany, England and other European countries. In addition, it was common practice for certain specialized products to be manufactured in Europe but to be marked with the names and addresses of American dealers or manufacturers. This is well-known for the case of gunlocks manufactured in Birmingham. The knife manufacturers of Sheffield, England, made many a knife for the American trade, usually stamped with the name of some Sheffield manufactory but at times with an American name instead. In Baltimore in the early nineteenth century Halbach marketed crude wooden-handled single-edged knives marked with his name. Halbach is believed to have been an importer of flintlock pistols marked with his name, which suggests that the knives also may have been imported. However good quality knives were produced in Baltimore, such as those of Samuel Jackson and Charles Reinhardt. A large brass-handled Bowie knife is known marked with the name of Schmidet. Earlier, petite silver-mounted dirks were hilted by silversmiths in Baltimore. Willick, a Philadelphia silversmith, produced silver-mounted dirks that were sold in Maryland, as is proven by Baltimore assay marks on them.

During the War of 1812 Nathaniel Lock supplied fifes and drums to military units. Later, in Civil War time, Maryland units on both sides marched to the beat of drums made by Christopher H. Eisenbrandt.

Some leather goods of the 1860's bear the name of James H. Merrill. On the inside flap of certain cartridge boxes can be found the following in five lines: "Merrill Patent F. A. Co. – Pat. Applied For – Baltimore."

Among the most sought-after accoutrements in the eyes of many collectors are buttons and buckles, particularly of the Civil War period. The Baltimore City Guards, formed in 1830, were active until the War, when many of their members joined Confederate units. Their button was initialed in Old English "B C G" with the letters enclosed in a wreath of laurel and oak, with thirteen stars above. The name "Jacob Gminder Baltimore" can be found on the back of two-piece eagle staff buttons (1789 style) encircled with "GEORGETOWN COLLEGE CADETS", made for Georgetown University. St. John's College at Annapolis used a flat one-piece button initialed S. J. C. on the bottom with a structure on top of a hill and a man starting to ascend. St. Timothy's Boys School at Catonville was known as the Maryland Military Institute. They used two-piece buttons encircled with "MD. MIL. INS. *VIM ANIMUMQUE CONCITAMUS*" around a quill and sword crossed over a Bible, stamped on the reverse Canfield Bro & Co. Baltimore.

Maryland, because of her geographical location south of the Mason-Dixon Line and her association with southern states through the Chesapeake Bay, has traditionally been classed with the southern states. But when the concept of northern states and southern states became well-established during the

1850's, Maryland found herself on the fence between, with sympathies and feelings and loyalties in both directions, with divided feelings on states' rights, secession, slavery, tariffs and other issues of the day. Marylanders joined up to serve in the war on both sides and a host of military units emerged, each eager to have an identity and show it with distinctive accoutrements.

After the secession of South Carolina, President Lincoln called for volunteers to quell the hostile action. Authorities ordered volunteer troops to Washington to protect the Capitol. Intense excitement ran through Maryland in anticipation that Northern troops would cross her territory to strike her sister states. On 19 April, 1861, the 6th Massachusetts, while proceeding from President Street Station to Camden Street Station in Baltimore, was harassed by citizens. Stones began to fly and a shot rang out. Whether fired deliberately or by accident, by soldier or civilian, is unknown. The troops fired by order of their untrained officers and the first blood of the war was shed. Militia units were put on alert and the entire state was in turmoil. Gun shops were broken into, and eight firms placed claims against the city for loss of firearms. Poultney and Trimble presented the largest claim, $2,583.25, for loss of guns and pistols taken from their store by the mob.

City officials appropriated $5,000,000 for defense to resist the "Northern invasion." Isaac R. Trimble, later a Confederate major general, enrolled over 15,000 volunteers on 20 April, the day after the incident, for the defense of Baltimore.

Recruiting offices had been opened by the Federal government in various parts of the state. The state of South Carolina in February sent two men to recruit Marylanders for the Confederacy. In the border state of Maryland loyalties were mixed even within families. But those who remained loyal to the Union were in the minority.

The issues of the war had been present for many years prior to 1861, and the pre-war militia companies held many men who were partisan in their feelings one way or the other. In Westminster there were two companies of state soldiers before the war. The Smallwood Infantry was almost to a man Southern in sympathy, while the Carroll Guards were decidedly Northern. Other units that remained loyal to the Union with nearly 90 per cent of their men included the Union Riflemen, the Huntington Riflemen, the Orangeville Horse Guards, Sharpsburg Riflemen, Cecil Guards, and the Union Guards. The nucleus of the "Maryland Line, C. S. A." drew heavily from the Garrison Forrest Rangers, Independent Grays, Maryland Guard, Howard County Mounted Dragoons, Baltimore City Guard, Baltimore County Horse Guard, Patapsco Light Dragoons, Johnson's Frederick Volunteers, Linganore Mounted Guard, Liberty Riflemen, Vansville Rangers, and the Law Grays.

The geographic position of Washington, D. C., surrounded on three sides by Maryland and facing Virginia across the Potomac, made it imperative that Lincoln not allow Maryland to secede. General Butler was ordered to put down any uprisings in Maryland, and in the midst of a violent thunderstorm on the night of 13 May 1861 his troops took Federal Hill from which his cannon could control the city and harbor. Soon the entire state was under military control and remained so until after the war. All pro-Southern legislators, the Baltimore Commissioners, Police Marshall George Kane, Mayor Brown and other officeholders of Southern inclination were imprisoned.

After General Butler took control of the city he seized the following arms from a warehouse at the corner of Gay and Second Streets: [1]

58 Boxes, marked "Va. muskets"	20 in each box	1160
15 Boxes, marked "Va. armory"	20 in each box	300
40 Boxes, Hall's Carbines	20 in each box, 141 loose	941
13 Boxes, Hall's Rifles	20 in each box, 74 loose	334
46 Minie Muskets	percussion locks	46
38 Old Muskets, Flint locks		38
23 Old Muskets, Percussion locks		23
27 New Muskets, Flint locks		27
50 Yager Rifles		50
		2919

1 Box Accoutrements

115 Boxes Pikes

* * * * * * * * * *

. . . statement of arms at present to be found at Fort McHenry.
60 Boxes of Pikes, containing 60 each
11 Boxes of Carbines
300 Carbines
5 Guns, 7 Carriages, 13 Rammers

Ross Winan, a member of the Maryland legislature, had been ordered by city officials to make pikes or spears to arm the citizens. He is not credited with any pikes under the city ordinance and probably served to get the contracting done by firms who supplied them: [2]

3285 iron spears from A. & W. Denmead & Son, who had an iron foundry
302 pikes from Hayward, Bartlett & Co., who had stove foundries
414 pikes from George Page & Co., who were machinists and manufactured and advertised naval stores in 1864

The iron pikes have a double edge, a head shaped like a peach leaf, and a tang securely held in the staff by a thick collar. They bear the name of Winan pikes. Only a few of the pikes that have survived are still on the original eight-foot staffs; most have been sawed off about eight inches below the collars. This probably was done by Butler's troops to make them useless, and only those that got in the hands of civilians survived uncut.

General Butler, in writing his autobiography, commented on Winan as follows: "A Baltimorean by the name of Ross Winan, a gray-haired old man of more than threescore and ten, a bitter rebel and reputed to be worth $15,000,000 had made 5,000 pikes of the John Brown pattern to be used by the rebels of Baltimore to oppose the march of the United States troops. I thought that if such a man worth $15,000,000 were hanged for treason it would convince the people of Maryland at least that the expedition we were on was no picnic. I knew that he was going to Frederick to make a secession speech and I believed if we captured him he would be a very proper specimen traitor to be hanged." [53]

Thus Maryland was not able to decide her own fate. The Federal government controlled elections and to ensure the Union Party a majority of votes, furloughs were given the 1st, 2nd and 3rd U. S. Maryland Infantry Regiments. Any citizen thought to be of Southern conviction or engaged in helping the South was required to swear an oath of allegiance before voting. Soldiers from other states who were stationed in Maryland also were permitted to vote.

During the war, Maryland furnished 12 generals or flag rank men to the Union, and 17 to the Confederacy. She also furnished 50,316 white troops and 8,718 black troops to the Federal government. Though Confederate records have been destroyed, Maryland is credited with supplying about 25,000 men to the Southern effort. The actual figure may be higher because many Marylanders enlisted in units other than the well-known Maryland state Confederate regiments to be with friends or relatives. There were many companies consisting entirely of Marylanders which served under other states. Among these were Lanier Guards, Company G of the 13th Virginia Infantry, Zarvona's Zouaves, Company H of the 47th Virginia Infantry, Maryland Guerilla Zouaves, Company C of Velligan's Louisiana Infantry, Washington Volunteers, Company E of 1st Virginia Infantry, Company G of the 1st South Carolina Infantry, Company A of Johnson's Independent Prison Command, Baltimore Heavy Artillery, Company B of the 9th Virginia Infantry, Lucas' Company C of the South Carolina Artillery, Rhett's 1st South Carolina Artillery Regulars, Deas Maryland Artillery, Company C of the 19th Virginia Artillery, Breathed's Battalion of Stewart's Horse Artillery, Comanches Company B of the 35th Virginia Cavalry, Company G of the 7th Virginia Cavalry, Mounted Riflemen, Company I of the 2nd Virginia Cavalry, Howard County Dragoons, Company K of the 1st Virginia Cavalry, Davis' Cavalry Battalion. Also, there were many scattered companies in the Confederate army in which Marylanders comprised 80 per cent or more of the compliment. Among these were Smallwood's Rifles, Tilghman's Rifles, Danton Guards, Maryland American Rifles, Snovell's Local Defence, Centreville Home Guards, Halbrook's Artillery Battery, Sotheron's Artillery, and Lee's Maryland Cavalry.

After Maryland was under Federal control, which tightened as the war went on, it took a lot of courage for a young man to cross the Potomac and offer his services. He risked loss of all his possessions in Maryland, and his family faced threats and harassment, or even imprisonment if they were found aiding him. Many young men were Southern sympathizers but were not willing to make the great sacrifice.

Generals George H. Steuart and Bradley T. Johnson tried to organize all Maryland soldiers into the Maryland Line, a carry-over from the famous unit of the Revolutionary War. This was difficult because men of Maryland descent were scattered throughout the Confederacy in nearly every regiment of the army. In 1863 they were able to establish six of the eight officially-recognized Maryland state commands into the Maryland Line. There were several types of Maltese crosses that were worn by men of the Maryland Line on their uniforms, including several designs made of silver.

Maryland Confederate privates, as well as generals, sported Maryland buttons on their jackets. From the butternut coat of Private Henry Holiday to the coat of General Robert E. Lee, made by ladies of Carroll and Frederick Counties, they were worn with pride. Many of the Maryland buttons came from militia uniforms. McHenry Howard and other members of the Maryland Guard took their dress uniforms as well as their fatigue uniforms south. Marylanders serving with other states also wore Maryland state seal buttons, such as Captain John Donnell Smith of Battery A, Alexander's Batallion, so that his origin could be easily seen.

Military agents in Baltimore at that time were Canfield Brothers, Benjamin Greenwalt, Jacob Gminder, Sisco Brothers and A. Stowell Jr. "In the winter of 1862-1863 Company F, 1st Maryland Cavalry was recruited by three rich young Baltimoreans . . . They furnished uniforms, horses, accoutrements and arms for their company at an immense expense, for everything except the horses had to be smuggled through the blockade from Baltimore." [4] In the spring of 1864 two members of the 2nd Maryland Infantry ran the blockade and went to Baltimore to raise money so that this regiment might be mounted. They soon secured $25,000 for this purpose, but the man carrying the money was captured on his return. Ex-Police Marshall George Kane was shocked at the condition of the rags that clad the 2nd Maryland Infantry in the winter of 1865. He promised all of them new uniforms and a change of underclothing when he returned, and they were received on 4 March.

After the Civil War was over many of the militia and National Guard uniforms used Maryland state seal buttons. The pre-war and war-time buttons had the eagle at the top, whereas the later post-war ones usually had the crown. The 5th Maryland Regiment, National Guard, in Baltimore was of large size, and it used a button marked "5th M N G". This regiment was formed on 8 May, 1867, largely from Confederate veterans. They adopted a cadet gray uniform trimmed in black, the coat being patterned after the Austrian coter. They also wore a black pike-type hat with their own state seal medallion, as did the 2nd and 4th National Guard regiments who wore a similar seal medallion.

Accoutrement plates, or belt buckles, many of which bore the Maryland state seal, were worn by some Maryland troops in the Civil War. The pre-war Maryland Guards as well as the post-war Company C Linganore Guards, the Dandy Fifth, and other units had their own special plates. Relatively few of these Maryland plates have survived. Among the plates bearing the state seal (Fig. 96), the rectangular pieces are less scarce than the oval ones, perhaps by a ratio of two to one. The two-piece plate of round configuration is very rare, only a few being known today. The shoka-type buckle illustrated here is the only known specimen.

MANUFACTURERS & RETAILERS OF MARYLAND MILITARY BUTTONS

The following list holds names of manufacturers as well as retailers. The names found on the back of old buttons are not necessarily those of manufacturers. In this large list only Jacob Gminder, M. C. Lilley, the Scoville Manufacturing Co., Steele & Johnson and the Waterbury Button Co. are known to have been manufacturers. All of these manufacturers made buttons depicting the Maryland state seal.

	Canfield Bro. & Co.	Baltimore	[inscription on button B, Fig. 94]
	Canfield Bro. & Co.	Baltimore	[inscription on button C, Fig. 94]
1835 thru 1840	Canfield & Brother	Baltimore & Charles Sts.	jewellers
1842	Canfield & Brother	Baltimore & Charles Sts.	jewelry & fancy military store
1845	Canfield & Brother	227 W. Baltimore St.	
1850	Canfield & Brother	227 W. Baltimore St.	military goods, a full assortment, military Co. furnished complete
1855 thru 1859	Canfield & Brother Ira C. & W. B. Canfield	229 W. Baltimore St.	
1860	Canfield Brother & Co.	I. C. & W. B. Canfield, Jos. H. Meredith	
1864	Canfield Brother & Co.		military goods, of every description as swords, belts, sashes, spurs, buttons, lances, embroideries, haversacks, pistols, field & marine glasses, etc. Presentation swords, on hand & made to order at short notice, also jewellry.
1864 thru 1881	Canfield Brother & Co.		[Baltimore city directories]
	Devlin & Co. N. Y.		[inscription on button H, Fig. 94]
	Devlin & Co. **N. Y.**		[inscription on button M, Fig. 94]
1846	Devlin, Daniel	31 John St. (N.Y.C.)	clothier
1847	Devlin, Daniel	31 John St.	clothier
1848 thru 1889	Devlin, Daniel & Jeremiah	33 & 35 John St.	[New York City directories]
	Extra * Quality *		[inscription on button C, Fig. 94]
	Extra * Quality *		[inscription on button I, Fig. 94]
1860	Gminder, Jacob	10½ S. Calvert St.	silver plater & manufacturer of coffin trimmings, military ornaments, etc.
1864	Gminder, Jacob		
1865 thru 1867	Gminder, Jacob	10 S. Calvert St. & 117 W. Baltimore St.	silver plater & manufacturer & dealer in military equipment of all description, embroidered & metal hat & cap ornaments, large letters, corps badges, shoulder straps, etc., fine swords of elegant design for presentation furnished at short notice.
1868 thru 1884	Gminder, Jacob	10 S. Calvert St.	
1885 thru 1889	Gminder, Jacob	14 S. Calvert St.	military goods [Baltimore city directories]

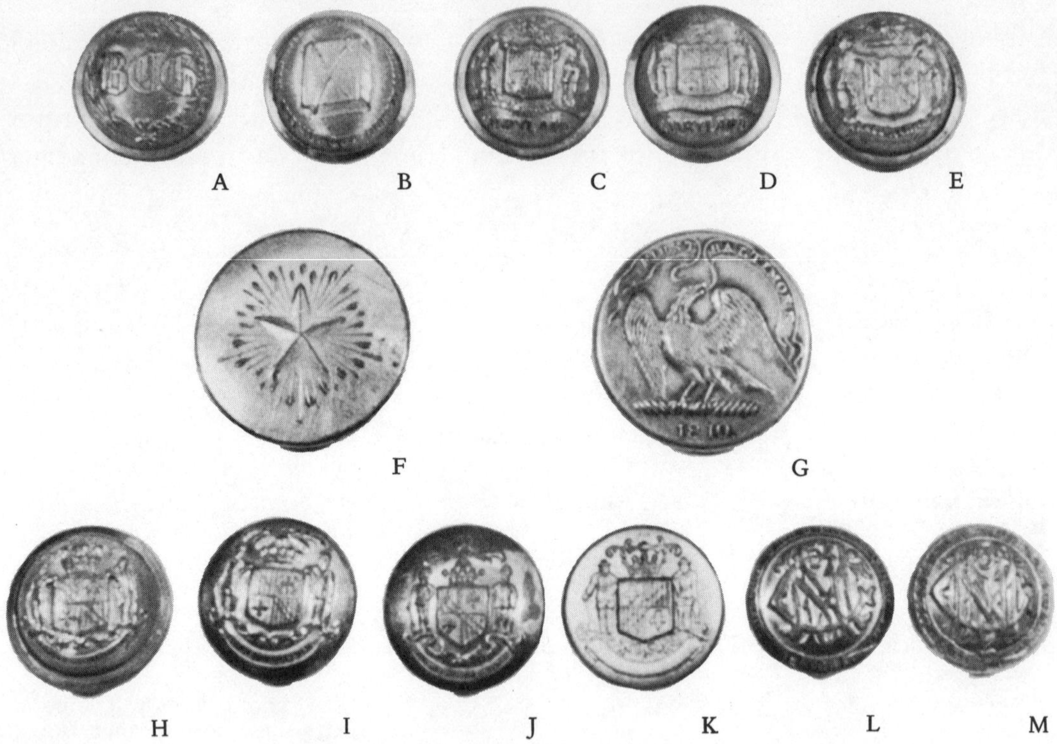

Figure 94. A selection of Maryland military buttons of the 19th century, from the collection of the author.

A – Baltimore City Guard – 13 stars above, initials encircled with laurel and oak wreath.
B – Maryland Military Institute – book, quill and sword; marked "Md. Mil. Inst. Vim Animumque Concitamus."
C – State seal – eagle above, "Maryland" below ribbon, fisherman's right arm at side of shield.
D – State seal – eagle above, "Maryland" below ribbon, both men with hand on top of shield.
E – State seal – eagle above, concave shield, fisherman's right arm at side of shield.
F – Fells Point Light Dragoons – plated, five-pointed star with rays.
G – First Baltimore Regiment – plated, eagle's ribbon "First Baltimore," below "1810".
H – State seal – crown and helmet above, both men with arms on top of shield.
I – State seal – crown and helmet above, both men with arms on top of shield, no outside ring.
J – State seal – crown and helmet above, both men with arms not touching shield, no outside ring, plated, flatter face.
K – State seal – crown and helmet above, both men with arms touching shield, crosses in shield in reverse position, no outside ring.
L – Fifth Regiment National Guard – "5" above with superimposed initials, flat face, four openings on motto belt.
M – Fifth Regiment National Guard – "5" above with superimposed initials, three openings on motto belt.

Figure 95. A silver shoulderbelt plate engraved "5th Regt B U V G T D". This was worn by George Towers Dunbar of the Baltimore United Volunteers, 5th Maryland Regiment, who rushed to the defense of Washington in August 1814, and and who also engaged in the Battle of North Point in defense of Baltimore in September 1814. Silver medallions such as this have long been associated with officers of the regular military only. This exquisite piece is unique in that it was worn by a private of a militia company. From the collection of the 1st Bn. 175th Infantry, Maryland A. R. N. G. through Frederick G. Gaede.

	Goodwin's Pat. July 27, 1875		[inscription on button D, Fig. 94]
	Goodwin's Pat. July 27, 1875		[inscription on button H, Fig. 94]

	J. A. Griffeth & Co. Baltimore. Md.		[inscription on button C, Fig. 94]
1853	Griffith, John A.	29 Fayette St.	merchant tailer
1854	Griffith, John A.	29 Fayette St.	merchant tailer
1855	Griffith, John A.	29 Fayette St.	merchant tailer
1856	Griffith, John A.	187 W. Baltimore St.	merchant tailer
1857	Griffith, John A.	187 W. Baltimore St.	merchant tailer
1858	Griffith, John A.	187 W. Baltimore St.	merchant tailer
1860	Griffith & Griffith		John A. Griffith
1865 thru 1880	Griffith, John A. & Co.	Baltimore & Charles Sts.	importers & dealers in merchant tailers & clothiers trimmings. J. A. Griffith, Jo Neil, Jos Maguire
1881	Griffith, John A. & Co.	247 W. Baltimore St.	Edward A. & Henry G. Griffith, John J. Griffith [Baltimore city directories]
1882	Griffith, John A. & Co.	247 W. Baltimore St.	[Baltimore city directory]
1883	Griffith, John A. & Co.	2 Hanover St.	[Baltimore city directory]
1884 thru 1888	Griffith, John A. & Co.	2 & 4 Hanover St.	[Baltimore city directories]
1889	Griffith, John A. & Co.	8 Hanover St.	Edward A. Griffith [Baltimore city directory]

	J. H. F. Hahn · Baltimore		[inscription on button H, Fig. 94]
	J. H. F. Hahn		[inscription on button I, Fig. 94]

	Heiberger Washington		[inscription on button M, Fig. 94]

	Horstmann · Philadelphia		[inscription on button C, Fig. 94]
	Horstmann Bros. & Co. Phil.		[inscription on button D, Fig. 94]
	Horstmann Bros & Co. Phil.		[inscription on button H, Fig. 94]
	Horstmann Bros & Co. Phil.		[inscription on button L, Fig. 94]
	Horstmann Bros & Co. Phil.		[inscription on button M, Fig. 94]
	Horstmann Philadelphia		[inscription on button M, Fig. 94]
1816	Horstmann, William H.	59 N. Third St.	

William H. Horstmann established a military equipment store in Philadelphia in 1816, and by 1828 he was a manufacturer of military equipment. By 1843 his sons William J. and Sigmund H. Horstmann were in business with him and the firm was known as William H. Horstmann & Sons. The firm prospered during the 1850's and 1860's.

1851	Horstman, William H. & Sons	51 N. Third St.	military store & fringe factory
1852	Horstman, William H. & Sons	51 N. Third St.	military store & fringe factory
1853	Horstman, William H. & Sons	51 N. Third St.	military store & dress trimmings
1854	Horstman, Wm. H. & Sons	51 N. Third St.	military store & dress trimmings
1855	Horstman, Wm. H. & Sons	51 N. Third St.	military store & dress trimmings
1856	Horstman, Wm. H. & Sons	51 N. Third St.	military store & dress trimmings
1857	Horstman, Wm. H. & Sons	223 Chestnut St.	military store & dress trimmings
1858	Horstman, Wm. H. & Sons	723 Chestnut St.	military store & dress trimmings
		5th & Cherry	factory
1859	Horstman, Wm. H. & Sons	NE 5th & Cherry	manuf. importers & Furnishers of military goods
		723 Chestnut	store
1860	Horstman, Wm. H. & Sons	Cherry bel Fifth	ladies' dress trimmings
1861	Horstman, Wm. H. & Sons	NE 5th & Cherry	ladies' dress trimmings
1862	Horstman, Wm. H. & Sons	NE 5th & Cherry	manufs. dress trimmings, carriage laces, &c
1863	Horstman, Wm. H. & Sons	NE 5th & Cherry	manufs of military goods & ladies' dress trimmings
1864	Horstman, Wm. H. & Sons	Fifth & Cherry	manufs of military goods & dress trimmings
1865	Horstman, Wm. H. & Sons	Fifth & Cherry	manufs of military goods & dress trimmings

1866	Horstman, Wm. H. & Sons	Fifth & Cherry	manufs of military goods & dress trimmings
1867	Horstman, Wm. H. & Sons	Fifth & Cherry	ladies' dress trimmings
1868	Horstman, Wm. H. & Son	N Fifth & Cherry	trimmings
1870	Horstman, Wm. H. & Sons	N Fifth & Cherry	trimmings
1871	Horstman, Wm. H. & Sons	N Fifth & Cherry	dress trimmings
1871 thru 1889	Horstmann Brothers & Co	Fifth & Cherry	
			[Philadelphia city directories]
	A. Jacobs & Sons Baltimore Md. (center marked "Made in England")		[inscription on button J, Fig. 94]
	M. C. Lilley & Co. Columbus O.		[inscription on button H, Fig. 94]
1865 thru 1869	Lilley, M. C. & Co.	26 North High St.	Odd Fellows regalia manufacturers M. C. Lilley & John Siebert
1870 thru 1889	Lilley, M. C. & Co.	253 South High St.	regalia manufacturers M. C. Lilley, John Siebert, Henry Lindenburg, C. H. Linderburg [M. C. Lilley & Co. catalog, Columbus, Ohio]
	J. A. Lunerick Baltimore Md.		[inscription on button I, Fig. 94]
	Made in England		[inscription on button J, Fig. 94]
	N. Y. Clothing House, Balto. Md.		[inscription on button H, Fig. 94]
	N. Y. Clothing House, Balto. Md.		[inscription on button I, Fig. 94]
	N. Y. Clothing House, Balto. Md.		[inscription on button M, Fig. 94]
1878 thru 1884	N. Y. Clothing House	184 West Balto. St.	clothier I Becker manager
1885 thru 1889	N. Y. Clothing House	102 East Balto. St.	clothier
			[Baltimore city directories]
	Oehms Acme Hall · Baltimore		[inscription on button H, Fig. 94]
1860 thru 1864	Oehm, Charles N.	150 Light St.	merchant tailor
1867 thru 1884	Oehm, Charles N. & Brother	150 Light & 230 Pratt Sts.	merchant tailor & clothier Charles N. Oehm & Franz F. Oehm
1885	Oehm & Co. – Acme Hall Clothing House	209 West Balto. St.	Alfred Haupt manager
1887 thru 1889	Oehm's Acme Hall Clothing House	5 & 7 West Balto. St.	Charles H. Oehm & Charles W. Oehm proprietors
			[Baltimore city directories]
	Pettibone Mf'g Co. Cin.		[inscription on button C, Fig. 94]
1872 thru 1877	Pettibone, James	92 West Fourth St.	fancy goods & toys, military, theatrical & regalia trimmings

1878 thru 1884	Pettibone Manufacturing Co.	99 West Fifth St.	
1885 thru 1889	Pettibone Manufacturing Co.	165 to 169 Elm St.	contractors for military & band goods
			[Cincinnati city directories]
	Purvis N. Y.		[inscription on button M, Fig. 94]
	Ridabock & Co. N. Y.		[inscription on button M, Fig. 94]
1883 thru 1889	Ridabock, James H.	141 Grant St.	military goods
1887 thru 1889	Ridabock, Henry C.	141 Grant St.	military goods
1864 thru 1887	McKennsy, James H.		military outfitters (purchased by Ridabock & Co. in 1888)
			[New York City directories]
Rosenfeld Bros. Baltimore			[inscription on button I, Fig. 94]
1867	Rosenfeld & Co.	23 & 25 West Balto. St.	clothiers M. Rosenfeld, G. Rosenfeld
1868 thru 1872	Rosenfeld Brothers	23 & 25 West Balto. St.	
1873 thru 1884	Rosenfeld Brothers	Balto. & Centre Market	Michael & Goody Rosenfeld
1885	Rosenfeld Brothers	617 East Balto. St.	J. Goody & I. Rosenfeld [Baltimore city directories]
	Scovill MFG Co. * Waterbury		[inscription on button A, Fig. 94]
	Scovill MFG Co. Waterbury		[inscription on button C, Fig. 94]
	Scovill MF'G Co. Waterbury		[inscription on button J, Fig. 94]
	Scovill MF'G Co. Waterbury		[inscription on button L, Fig. 94]

The Scovill Manufacturing Company of Waterbury, Connecticut, had its beginnings at Abel Porter & Co. from 1802 through 1810. From 1811 through 1813 it became known as Leavenworth & Scovill. From 1814 through 1826 it was Leavenworth, Hayden & Scovill. From 1827 through 1835 it was J. M. L. & W. H. Scovill.

1802 thru 1810	Abel Porter & Co.	Waterbury, Conn.	
1811 thru 1813	Leavenworth & Scovill	Waterbury, Conn.	
1814 thru 1826	Leavenworth, Hayden & Scovill	Waterbury, Conn.	
1827 thru 1835	J. M. L. & W. H. Scovill	Waterbury, Conn.	
1836 thru 1839	Scovill & Buckingham	Waterbury, Conn.	
1840 thru 1849	Scovill & Co.	Waterbury, Conn.	
1850 thru 1889	Scovill Manufacturing Co.		[Scovill Mfg. Co. records]

Date	Name	Location	Source
	Steele & Johnson		[inscription on button E, Fig. 94]
1852 thru 1855	Steele & Johnson Manufacturing Co.	Waterbury, Conn.	
1856	Waterbury Jewelry Co.		
1857	Waterbury Jewelry Co.		
1858 thru 1874	Steele & Johnson Button Co.	Waterbury, Conn.	
1875 thru 1889	Steele & Johnson Manufacturing Co.	Waterbury, Conn.	
			[Steele & Johnson Mfg. Co. records]
	Superior Quality		[inscription on button I, Fig. 94]
	Superior Quality		[inscription on button K, Fig. 94]
	Ten Stars		[inscription on button C, Fig. 94]
	Waterbury Button Co.		[inscription on button D, Fig. 94]
	Waterbury Button Co.		[inscription on button H, Fig. 94]
	Waterbury Button Co.		[inscription on button M, Fig. 94]
	Waterbury Button Co. * Extra *		[inscription on button M, Fig. 94]
1823 thru 1828	Benedict, A.	Waterbury, Conn.	
1829 thru 1833	Benedict & Coe	Waterbury, Conn.	
1834 thru 1842	Benedict & Burnham	Waterbury, Conn.	
1843 thru 1849	Benedict & Burnham Manufacturing Co.		
1850 thru 1889	Waterbury Button Co.		[Waterbury Button Co. records]

REFERENCES

1. Baltimore City Correspondance Manuscripts, 1862, No. 695, p. 8.
2. Ibid.
3. Butler, Benjamin F., AUTOBIOGRAPHY OF BENJAMIN F. BUTLER, Peabody Library
4. Johnson, Bradley T. (1899) CONFEDERATE MILITARY HISTORY, Confederate Publishing Co., Atlanta, Ga., v. II, p. 49.

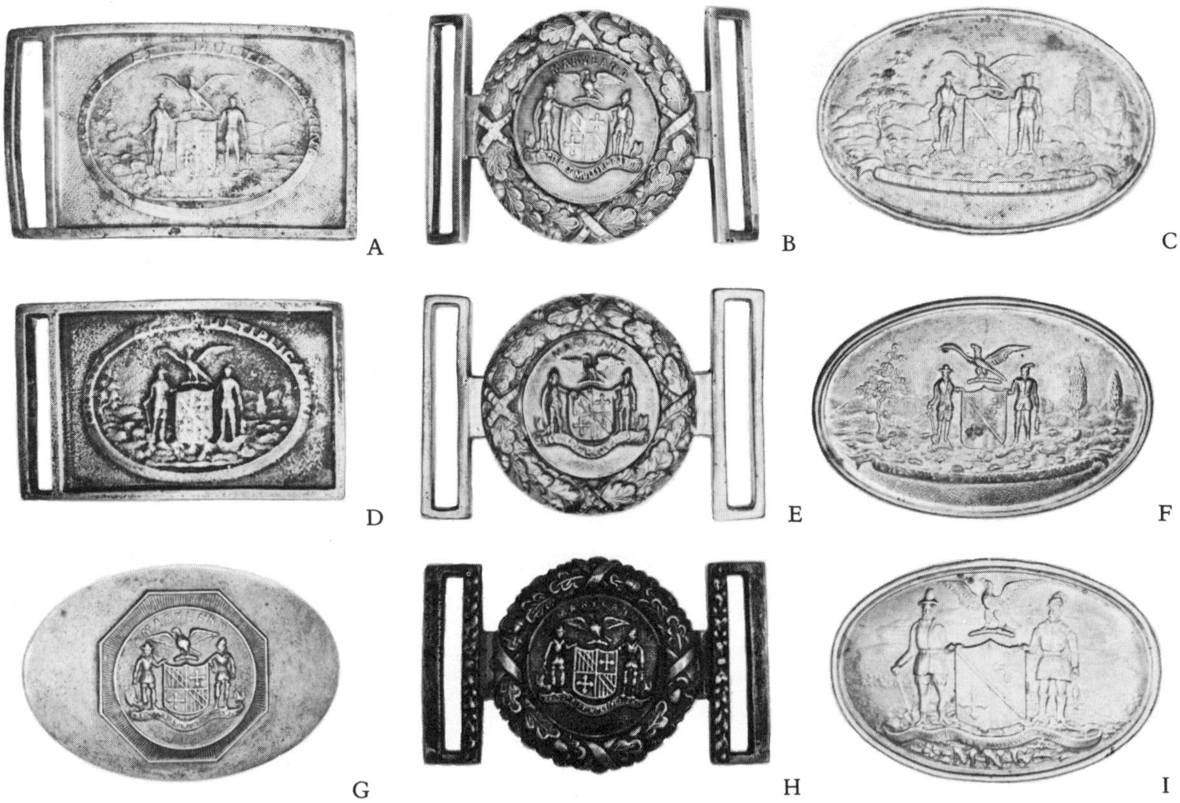

Figure 96. Brass accoutrement plates and belt buckles of the Civil War period, each bearing the Maryland state seal. Specimens A, B, C, F, and I may have been used by pre-war militia companies, but like the state seal buttons, were worn with pride by Maryland Confederate soldiers. Daniel D. Hartzler collection, specimens A, C, D, E, F, G, I; Erick Davis collection, specimen B; John M. Bracken collection, specimen H.

A – A rectangular plate cast in a metal mould, with a belt loop on the right end and a small flat hook on the reverse of the opposite end for attaching a female belt loop. In the center of the plate is the state seal. The lower half of the foreground is turf, with trees in the background, all of which is encircled with a ribbon on which is inscribed "CRESITE ET MULTIPLICAMINI" at the top. None of these plates bears a maker's name, but some are found with numerals on the reverse and have a slightly different hook. This type was worn by Private Henry Holiday, Company A, 2nd Maryland Infantry, and Lieutenant Colonel James Breathed, also a Marylander, who chose to forgo his occupation as a medical doctor for that of a soldier. He fought under the flag of Virginia in Stuart's Horse Artillery. Weight 73.3 grams, dimensions 50 x 90 mm.

B – A high-quality, two-piece buckle, cast in a metal mould. The left side is an extremely well-made hollow-backed wreath. The right side has a sharp die-stamped center with horizontal lines which is soldered to the tongue. Both loops are stamped on the back with Roman numerals and are very concave. This plate was worn by Captain John Eager Howard, Quartermaster, 2nd Maryland Infantry. Weight 82 grams, dimensions 55 x 80 mm. overall, wreath 55 mm. diameter.

C – The Maryland oval also is found as a cartridge-box plate with the name "E. GAYLORD" stamped on the reverse. The front of the plate is identical to specimen F. Probably they were made from the same die. On the reverse, in the leaded back are two small iron-wire hooks for attaching it to the box. All box plates examined are marked with the name "E. GAYLORD". Gaylord made plates and buckles used by both Union and Confederate forces. One plate examined is on an original unmarked box of pre- or Civil War vintage. Weight 128 grams, dimensions 54 x 85 mm.

D – A cast copy of a plate like specimen A. The details on such copies are less distinct. The original, for instance, has clearly-defined horizontal lines in the background behind the seal, whereas the copy does not. An advanced collector also can often use hook details as additional clues. Weight 97 grams, dimensions 48 x 85 mm.

E – An unmarked two-piece, sand-cast plate. The left side is a hollow-backed wreath with leaves, acorns, and crossed bands. The right side has a solid back with the state seal, inscribed ribbon "CRESITE ET MULTIPLICAMIN" below, and the word "MARYLAND" above the eagle. This specimen, in the author's collection, was purchased at a sale in the Shenandoah Valley in the 1930's. This buckle is thought to have been cast in the South during the War. Weight 88.2 grams, dimensions 56 x 87 mm., wreath 54 mm. diameter.

F – A die-stamped, lead-backed oval plate with the state seal in the center. The lower half has turf in the foreground, trees in the background, and at the bottom a ribbon inscribed "CRESITE ET MULTIPLICAMINI." The top portion displays the same horizontal lines in the background as specimen A. This plate has two studs, one hook, and is unmarked. The fasteners are identical to those found on large oval plates of the pre-war period. Captain Warner G. Welsh of Libertytown, commander of Company D, 1st Maryland Cavalry, wore one of these buckles. Weight 125.5 grams, dimensions 55 x 86 mm.

G – A plain oval, with a shako-type state seal in the center, soldered on and held by four brass wires. Above the eagle is the word "MARYLAND" with the farmer, shield, and fisherman, below a ribbon inscribed "CRESITE ET MULTIPLICAMIN". The concave back is not filled, and the three brass hooks are attached with solder. Weight 35 grams, dimensions 55 x 80 mm.

H – An unmarked, two-piece, metal-mould-cast plate. The left side is a flat, solid-backed wreath with differently placed leaves, acorns, and with plainer bands than specimens B and E. The right side has a die-stamped center containing the same state device as specimens B and E. This decorated looped buckle may be post-war. One can be found in the relics of William P. Zollinger, Captain of Company H, 5th Maryland Militia, in 1867, and Colonel in 1877. During the Civil War he was a 2nd Lieutenant in Company A, 2nd Maryland Infantry, C.S.A. Weight 86.4 grams, dimensions 53 x 87 mm., wreath 53 mm. diameter.

I – An unmarked, die-stamped, lead-backed oval plate displaying a large state seal. At the bottom is a ribbon inscribed "CRESITE ET MULTIPLICAMINI," below which are found the letters "N.M.G." surrounded by flowers and corn. To the right of the planter, in the background, is a village. To the left of the fishermen are ships. This plate has two studs and a hook of Civil War period style. The initials "N.M.G." probably refer to Maryland National Guard and this plate may be of the post-war period. Weight 128.2 grams, dimensions 55 x 86 mm.

Figure 97. Secession cockade, referred to as the Maryland blue cockade, made of blue silk, with an Extra Quality state seal cuff button. Daniel D. Hartzler collection.

Figure 98. Union patriotic Maryland cockade, made with red, white, and blue silk and a Waterbury Button Co. state seal cuff button. Daniel D. Hartzler collection.

Figure 99. Silver cross worn by Private Gresham Hough who mustered in Company H, 1st Maryland Infantry, C.S.A., for one year. Upon being discharged he reenlisted in Company A, 1st Maryland Cavalry on 15 June 1862. On 15 June 1864 he became a member of Company D, Mosby's Rangers. Erick Davis collection.

Figure 100. Silver cavalry badge presented to Leonard H. Greenewald of the 1st Maryland Potomac Home Brigade Cavalry, U.S.A. He entered Cole's Cavalry as a private, was promoted to Lieutenant 21 April 1864, and to Captain 13 February 1865. Daniel D. Hartzler collection.

Biographical Data for the Arms Makers of Maryland

The section which follows contains biographical data for 535 men or firms engaged in arms manufacturing in Maryland and Washington, D.C., during the 17th, 18th, and 19th centuries. In pursuing the research, emphasis was placed upon those who made flintlock and percussion longrifles, pistols, and related arms. In the longrifle period there were a few families in which a number of men were occupied at the gunsmithing trade. In such instances, family relationships are particularly important to the student of antique arms, so genealogical diagrams, or "family trees," have been included. This list is as complete as it was possible to make in the time available, but as with any research of this type, small additions will come to light in future years.

The information presented here is not solely the result of my own efforts, for Dr. George Shumway has made an appreceable contribution. My list initially contained about 300 entries, and no attempt had been made to include Washington, D.C., in the research. Dr. Shumway felt it desirable to include Washington, and with help from his small staff, worked through the old directories for the city and some of the old census records. He also did intensive studies of some longrifle gunsmiths and gunsmithing families, notably Peter White, John Armstrong, the Hawken family, and the Rizer family.

Both the author and the publisher have a desire to receive additional information about Maryland's early gunsmiths and their arms, and they would appreciate letters from readers who have supplemental information to contribute.

A

AARON, GEORGE W. — Washington, D. C.

Aaron, George W.

1858	h 733 New Jersey Ave.	gunsmith

[Boyd's Washington and Georgetown Directory]

ALEXANDER, FREDERICK W. — Baltimore

Organizer and captain of the Baltimore Light Artillery U.S.A. during the Civil War. At the time of this cannon patent he was living at 272 Lexington St. with no occupation listed.

Patent 45009 November 15, 1864
Be it known that I Frederick W. Alexander, Captain of the Baltimore Battery of Light Artillery of the City of Baltimore and State of Maryland has invented a new and improved form of bayonet.
[U. S. Patent Office]

Patent 74,478 February 18, 1886
Be it known that I, Frederick W. Alexander of Baltimore city, State of Maryland, have invented a new Apparatus to be Applied to Muzzle-Loading Rifled and Smooth-Bore Guns, which I call a "Calibre Diminisher," for the purpose of diminishing the calibre of guns after they have been loaded, so that any windage of the ball is prevented . . . my invention consists in inserting a hollow cylinder of steel or other metal into the muzzle of the gun.—
[U.S. Patent Office]

Our captain had invented a breech loading cannon, an experimental gun with a one inch bore. This was brought to camp. Our modern breech loading guns have a screw in the breech that swings back on a hinge. The charge is inserted, the screw brought into position and a half a turn locks it. The above mentioned gun had a screw on the top, and when opened there was a hole thru the screw, thru which the charge was inserted and half a turn locked it. Several shots had been fired at a target, so I fortunately moved forward and to the left of the muzzle. With my eyes on the target, I heard the report more violent than the previous ones, and looking back thru the smoke I saw three or four men on the ground and heard the cries and moans of the wounded.
There was one however who was silent. A drummer boy of the 8th Maryland regiment who had his skull crushed by a fragment of the cannon. The most serious of the wounded was one of our boys whose leg was shattered half way above the knee, so that the bone protruded thru his pants; one had a rib broken; the others were not seriously injured.
[MEMOIRS & HISTORY OF CAPT. F. W. ALEXANDER'S BALTIMORE BATTERY, p. 19]

ALFORD, A. G. & CO. — Baltimore

Alford, A. G. & Co.

1883	5 South Howard St.	gunsmith
1884	116 W. Baltimore St.	gunsmith
1885	116 & 129 W. Baltimore St.	gunsmith
1886	129 W. Baltimore St.	gunsmith
1887-89	221 E. Baltimore St.	gunsmith

[Baltimore city directories]

ALFORD, NELSON E. — Baltimore

Alford, Nelson E.	1880	206 W. Pratt St.	gunsmith
	1881-86	59 N. Gay St.	gunsmith
	1887	151 N. Gay St.	gunsmith

[Baltimore city directories]

ALISCS, JOHN J. — Washington, D. C.

October 11, 1864
Be it known that I John J. Aliscs, of Washington, in the District of Columbia, have invented a new & useful Improvement in Ball-Screws for Fire-Arms . . . the interior of a gun-barrel after several discharges, insomuch that in battles guns often become entirely useless until cleaned, which can be very quickly & effectively performed by the use of these gouges.
[U. S. Patent Office, #44,586]

ALLEN, ZACHARIAH

Council of Safety September 13, 1777
That the said Treasure pay to Zachariah Allen five pounds for a Gun.
[AM, p. 375]

ALLSWORTH, SAMUEL — Anne Arundel County

October 13, 1769
Ran away, last Night, from the Subscriber, living on Elk-Ridge, a Convict Servant Man, named Samuel Allsworth, by trade a Gunstocker, about 20 years of age . . .

Samuel Poole

[*Maryland Gazette*]

ARMENDT, GEORGE — Baltimore

Armendt, George	1851	272 Bond St.	gun & locksmith
	1853	274 Bond St.	gun & locksmith
	1856	228 S. Broadway	lock & gunsmith

[Baltimore city directories]

ARMENDT, JOHN E. — Baltimore

1866
The undersigned very respectfully petition your Honorable body to grant him permission to place a sign (a large wooden Gun) in front of his premises No. 262 S. Broadway . . .

John Armendt

[Baltimore City Correspondence Manuscripts, 1866, #783]

Armendt, John E.

1867	261 South Broadway	gunsmith
1868-69	262 South Broadway	gunsmith
1874	261 South Broadway	gunsmith
1875	269 South Broadway	gunsmith
1876	20 South Chester St.	gunsmith
1877-86	80 Thames	gunsmith

[Baltimore city directories]

ARMSTRONG, JAMES Emmitsburg

James was one of the four sons of gunsmith John Armstrong of Emmitsburg, according to Williams' HISTORY OF FREDERICK COUNTY, p. 805; it is mentioned that all four sons were trained in gunsmithing. Nothing is known of James' career or of his work.

ARMSTRONG, JOHN Emmitsburg

John Armstrong stands out as the most famous long-rifle maker of Maryland. His relief-carved and beautifully engraved rifles were made in abundance over a long period of time and many have survived. Research on the details of Armstrong's life is difficult because records of his death are not known and because the potentially valuable property tax assessment records for Emmitsburg were stolen from the Frederick County Court House in the 1960's and have not been recovered. Nevertheless the essential details of Armstrong's life can be determined from the information given below.

John Armstrong was born in the period between 1770 and 1774 according to Federal Census data for 1800, 1810, 1820, 1830, 1840; the census data show age categories, but the categories differ and leave this four four-year interval as the period of birth. The exact date of John Armstrong's birth is provided by an unusual document discovered in June 1975, only a few months before this material was put in final form for printing. Albert M. Sullivan, Sr., a dedicated collector of rifles made by Armstrong with a corresponding interest in the old gunsmith, acquired a document from James Roberson who purchased it at public auction in Gettysburg, Pennsylvania. The document was among a group of old papers that belonged to a mid-nineteenth century justice of the peace in the region. Roberson recognized the value of the paper and passed it along to Sullivan who made it available for use here. This document is a connecting link that ties together two men, from boyhood until old age and death – John Armstrong and Samuel Louden.

Md Frederick County

On the 22d day of August 1828 personally Appeared John Armstrong before me the Subsriber & Justice of the Peace in and for Frederick County aforesaid – and made the following oath and declaration – Deposeth and sayeth that he is personally acquainted with a certain Samuel Louden a citizen of Liberty Township Adams County, Pa. – that he does think to the best of his Opinion that the said Samuel Louden was born in the year 1776 – or before that year – and that he has good reason for forming that opinion respecting the age of the said Samuel Louden – they being boys together – but the said deponent being somewhat the eldest – being born in the year 1772 the 5th day of September – therefore does make oath and declare that the said Samuel Louden is at least fifty two years of age – and further sayeth not.

Sworn before William [unclear]

John Armstrong, the first, the great grandfather of the children of Thomas Hayes, came from England and settled in the Cumberland Valley, Pennsylvania, and had a son John, who was a gun and white smith, a master mechanic. He settled near Emmitsburg, Maryland, the property is now owned by Mrs. Lucy Bean, and married a Miss James, an only child, who lived near Emmitsburg. Her parents were of Welsh descent and her father was a manufacturer of woodenware. John Armstrong the second had seven children, four sons and three daughters. The sons were William, Robert, Samuel and James Armstrong, all gunsmiths. The daughters were Anna, Jane and Elizabeth. William was the only son that married. He was an inspector of guns at the government works at Harper's Ferry for many years. Anna married James Manson; Jane married Samuel Heard; Elizabeth Armstrong married Thomas Hayes. The first gunsmith shop built by John Armstrong is still standing at the present time [1910] and is in a good state of preservation, over one hundred and twenty years. One of John Armstrong's apprentices [Nathaniel Rowe] still lives at the present time, hale and hearty, about ninety years of age and bids fair to live many more.

[WILLIAMS' HISTORY OF FREDERICK COUNTY, p. 805]

Property holder 1808 Emmitsburg John Armstrong, gunsmith. John Armstrong was early in the town, as his name is on the plot of 1808 as owner of no. 1 & 2 lots; his reputation as a gunsmith was good; he made rifles & shot guns; dying, the business was continued by his former apprentice, Nathaniel Row, who retained Armstrong's reputation; his brother Samuel worked with him until he went west . . .

[HISTORY OF EMMITSBURG, MARYLAND]

The Federal Census for 1790 lists one John Armstrong as head of a household in Frederick County, Maryland. The household was made up of three males between 16 and 26 years of age, one male under 16, and two females. This listing is enigmatic, because at this time the young gunsmith would have been 18 years of age, and other public records suggest that there was no other man named John Armstrong living in Frederick County at this time. The one possible explanation is that the young gunsmith at age 18 or 19 was head of his own household. This in turn suggests that Armstrong was not bound into an apprenticeship at this time and instead was working on his own or with others.

There are many indentures recorded in the name of John Armstrong in the Frederick County Court House, for the period between 1797 and 1842, and some in the Hall of Records in Annapolis. Apparently for most of this period there was no other person named John Armstrong in Frederick County other than the gunsmith, for all of these indentures seem to relate directly or at least indirectly to the gunsmith. All of the many real estate transactions concern land in or near the town of Emmitsburg. Only one of the deeds mentions that John Armstrong was a gunsmith, but this transaction ties in with a number of others, proving that they were made by the gunsmith. The deed of 1 April 1822 yields the name of John's wife, which was Agnes. Her name is not mentioned in any other known document or account.

This indenture, a bill of sale dated 20 September 1797, shows that John Armstrong of Frederick County bought from Jacob Harsh of Frederick County for ten pounds, three items in the possession of Conrad Harsh. "I the said Jacob Harsh have put the said John Armstrong in full possession by delivering him one iron hooped barrel in the name of all the said goods and chattels . . ." "Schedule of all and singular the goods, household stuff, and goods and chattels conveyed or intended to be conveyed by the said Jacob Harsh to the said John Armstrong in analy the within Instrument of writing to wit one Blacksmith anvil, one Bellows one iron hooped bucket."
[Frederick CCH, Liber W. R. 15, p. 625]

Armstrong, John 1798 1 front & 2 back lots Emmitsburg ground rent 7 shillings 6 pence, value of land $54, value of improvements 7 pounds 10 shillings
[Property Assessment Book, Frederick County, 1798, Hall of Records #19268]

Tiler Wickham, Son of Margaret Wickham Aged Nineteen the 3rd Day of January next being an Orphan Child is this day with with the Consent of his said Mother bound an Apprentice unto John Armstrong of Frederick County Gunsmith until he arrives at the Age of Twenty One Years the said John Armstrong engaging to find his said Apprentice sufficient Boarding and Cloathes and all other necessaries the said term to cause his said Apprentice to be Taught the said Trade of a Gun Smith & Gun Lock making which he now useth as far as shall be in his power During said term Given under the hands & seals of the Subscribers two of the Justices of the Peace of Frederick County the 11th of March 1799 William Emmitt Henry Williams
[Frederick County Apprentice Indentures, Hall of Records, p. 211]

By this deed, dated 26 August 1799, John Armstrong of Frederick County bought from William Emmit of Frederick County, for 53 pounds, land in the town of Emmitsburg, "viz. three Lotts and part of a Lott, known in a platt made by William Kenworthy . . . by numbers 180, 182, 183, & part of 181, the said part of a Lott lays joining the Stone quarry. Beginning at a post standing at the North East corner of Lott Number 180, and running thence North 16° East three perches and one half perch, then North 61° West eleven perches and four feet, then South 16° West six perches, then with a straight line to the beginning. Containing sixty square perches of land. The Lott Number 182 lays joining a sixty foot street called Church Street, and the Lotts 180 and 183 lays joining a twenty feet alley . . ."
[Frederick CCH, Liber W. R. 19, p. 20]

Armstrong, John 1799 paid 5 shillings for Inquisition held by Wm. Emmitt, J. P.
[Frederick County Levy Book, Hall of Records #19257]

In 1800 the household of John Armstrong in Emmitsburg held one male between 26 and 45 years, one male 16 to 26, three males under 10, one female 26 to 45, and one female under 10.
[1800 Federal Census, Frederick County]

By this deed, dated 26 April 1800, John Armstrong of Frederick County bought from Jacob DeHoff of Frederick County, for 135 pounds, "A Lott of ground with a house thereon in Emmitsburg . . . known by Number two . . . running North 73° West 60 feet, North 17° East 165 feet, South 73° East 60 feet . . . containing 9900 square feet . . ."
[Frederick CCH, Liber W. R. 19, p. 510]

This release from annual payment of ground rent is dated 29 November 1800. ". . . I William Emmit of Frederick County Maryland for and in consideration of the sum of Ten Dollars to me paid by John Armstrong of the said County in consideration of the Annual Ground Rent of one Dollar on his front Lot in Emmitsburg Number Two on which he now lives have remised released & forever quit claim . . . unto the said John Armstrong . . . from any further ground rent for his said front Lot . . ."
[Frederick CCH, Liber W. R. 20, p. 357]

By this deed, dated 14 February 1801, John Armstrong of Frederick County bought from Robert Fleming of Frederick County, for 45 pounds, part of a tract of land called Carrolsburgh in the town of Emmitsburg, "beginning for said part at a stone planted at the Southern Alley of Emmitsburgh and running thence with the back lotts of the said Emmitsburgh . . . containing one Acre of land . . ."
[Frederick CCH, Liber W. R. 21, p. 8]

This release dated 6 April 1801 appears to be a repeat of the release of 29 November. ". . . I William Emmit of Frederick County . . . for . . . the sum of Ten Dollars . . . paid by John Armstrong of the county and state aforesaid . . . I the said William Emmit do hereby acknowledge which sum of Money . . . is to me . . . paid for the express purpose of releasing unto . . . John Armstrong . . . the ground Rents forever of the following lot of ground in Emmitsburgh the said Lot being known in the plat of the said Emmitsburgh by number two . . ."
[Frederick CCH, Liber W. R. 21, p. 361]

This Indenture made the thirty first day of August 1801 Witnesseth that Peter Piper hath this day placed and bound his son George Piper Aged Eighteen Years or thereabouts unto John Armstrong of Frederick County Maryland Gun Smith & Lock Maker to learn the said trade mystery & occupation of a Gun Maker and Gun Lock Maker which the said John Armstrong now useth & will serve him as an Apprentice to Dwell continue

& serve from the day of the date hereof untill the full end and Term of two years & two months from thence next ensuing during all which term of two years & two months the said master well and faithfully shall serve his secrets keep his lawful commands Gladly do & obey, hurt to his s[d] Master he shall not do nor wilfully suffer to be done by others but of the same to the utmost of his power shall forthwith give notice to his said Master the Goods of his said Master he shall not embexxle or waste nor lend them without his consent to any at cards dice or any other unlawful game he shall not play Taverns or ale Houses he shall not frequent fornication he shall not commit matrimony he shall not contract from the service of his Master he shall not at any time depart or absent himself without his said Masters leave but in all things as good & faithful Apprentice Shall & will demean & behave himself toward his said Master and all his during the s[d] Term.

And the said Master doth hereby promise & Engage in the s[d] Trade of a Gun & Gun Lock Maker which he now useth, with all things thereunto belonging shall and will Teach & Instruct after the best way and manner he can and shall and will also find and allow his s[d] Apprentice meat Drink washing and Lodging & all other necessaries fit & convenient and give him one Good Rifle Gun worth Sixteen Dollars and allow him one week in harvest to work for himself or pay him in cash Six days Harvest wages. Witness whereof the said parties hereunto set their hands & seals the day & year first within written.

William Emmit

his
Peter X Piper
mark

John Armstrong

[Frederick County Apprentice Indentures, Hall of Records, p. 312]

Armstrong, John 1802 juror 1 day paid 5 shillings for Inquisition held by Wm. Emmit, J. P.
[Frederick County Levy Book, Hall of Records #19257]

John Blackburn Aged Seventeen years and five months the seventeenth day of March Last by and with the Advise and consent of his father Alexander Blackburn hath placed and bound himself an Apprentice to John Armstrong of Frederick County Gun Smith for the Term of Three years from the sixteenth of this instant during which time the said apprentice his said Master well and faithfully Shall serve and in all respects demean himself as a good and faithful Apprentice ought –

And the said Master shall teach and Instruct or cause to be taught and Instructed to his said Apprentice the art and mystery of the trade of a Gun Smith in the best way he can finding to the said Apprentice meat drink and Lodging and will allow his said Apprentice two weeks in every Harvest during said term for his own use and at the expiration of the Term aforesaid will pay his said Apprentice the sum of Twenty Dollars Current Money. Witness whereof the parties have set their hands hands and seals the Fifth day of April 1804. John Blackburn

John Armstrong
Alexander Blackburn

[Frederick County Apprentice Indentures, Hall of Records, p. 425]

By this deed, dated 6 August 1806, John Armstrong of Frederick County bought from William Porter I son of Nathaniel of Scott County State of Kentucky, for $60, "one lot of Ground in the Town of Emmitsburgh , . . distinguished by lot number one beginning for said lot at a post or corner & running thence North 73° West 60 feet, North 17° East 165 feet, then South 73° East 60 feet . . . containing 9900 square feet . . ."
[Frederick CCH, Liber W. R. 29, p. 291]

By this deed, dated 29 August 1807, "John Armstrong of the Town of Emmitsburgh in the County [of Frederick] . . . Gunsmith" bought from John Troxal Sr. of Frederick County, for $60, a "Lott of ground or parcel of land No. 179 . . . in Frederick County aforesaid being part of a tract of land called Silvers Fancy adjoining a Lott already the property of said John Armstrong No. 180 . . . containing one acre of ground and to square with said John Armstrong other Lott with a stone planted at the South eastern corner of said Lott No. 179 and adjoining to Lott No. 175 the property of said John Troxal the stone marked with the Letters IA on the one side and IT on the the other side . . ."
[Frederick CCH, Liber W. R. 32, p. 244]

By this deed, dated 3 December 1808, John Armstrong of Frederick County bought from William Emmit of Frederick County, for $5, "a part of an alley so far as the said Alley is joined on both sides by the out lots of the said John Armstrong which lots are known and distinguished . . . by the numbers 179 and 180 on the one side and 182, 183, 184 on the other side the said Alley extending from Church Street to a steep Bank of Flat Run and being the most southern alley leading through said Lotts . . ."
[Frederick CCH, Liber W. R. 34, p. 395]

Indenture made March 20, 1809 [release] I, William Emmit of Frederick County, State of Maryland for and in consideration of the sum of $10 to me paid by John Armstrong of the said County in consideration of the annual ground rent of one dollar on his front lot in Emmitsburg #2 on which he now lives have promised release.
[Frederick CCH, Liber W.R. 20, p. 357]

Armstrong, John 1809 juror Inquisition held by Wm. Emmit, J. P.
[Frederick County Levy Book, Hall of Records, #19258]

Indenture made March 19, 1810 [deed] between Robert Floming of Frederick County and State of Maryland of the one part and John Armstrong of the same county and state . . . Part of the tract of land called Carrolsburgh beginning for said part at a stone planted at the southern alley of Emmitsburgh and running thence with the back lotts.
[Frederick CCH, Liber W.R. 21, p. 8]

Both a John Armstrong and a John Armstrong, Jr., maintained households in Emmitsburg in 1810. The household of John, Jr., probably was that of the gunsmith; it held one male between 26 and 45 years, one female between 26 and 45 years, and one male under 10. The household of John, Sr., contained one male over 45, two males between 16 and 26, one male 10 to 16, one female over 45, and two females 26 to 45.
[1810 Federal Census, Frederick County]

The following Bill of Sale was executed 1 October 1810 and recorded in the Court House at Frederick on 6 October 1810.

". . . I Thomas Brady of the County of Frederick . . . in consideration of the sum of $400 . . . to me in hand paid by John Armstrong of the sd County . . . doth . . . sell . . . unto the sd John Armstrong . . . one Horse, one mare, one cow, four feather Beds & four bed steads, one case of drawers, one Cot, one Chest, Three Tables, one Cupboard, five benches, Two doe chests, one Kettle, four Iron pots, eight pewter dishes, twelve plates, twelve tins, one Lanthorn, Two coffee pots, four candlesticks, one Bucket, one tea Kettle, one pair of hand Irons, one pair of tongs, one ten plate stove, one cart and two pair Harness, one cutting box for straw, one womans saddle, eight bags, one scythe, two barrels, Two Tubs, one tub for meat, Two dozen of knives and forks, one ax, one wood saw, two coffee mills, four shovels, one spade, two Ladles, one flesh fork, one spinning wheel, one reel, five chaff beds, and fourteen sheets, five blankets, three coverlets, Two bed quilts, one carpet, Twenty yards of Cotton stripe & twenty yards of Linsey."
[Frederick CCH, Liber W. R. 38, pp. 327, 328]

By this deed, dated 9 February 1811, John Armstrong of Frederick County bought from Charles Robertson of Frederick County, for $70.68, "part of a Tract of Land called Fox Town . . . containing two and one fourth acres and seventeen square perches of Land."
[Frederick CCH, Liber W. R. 39, p. 240]

Armstrong, John 1812 juror paid 66 2/3 cents for helping with Inquisition
[Frederick County Levy Book, Hall of Records, #19258]

By this deed, dated 15 August 1812, John Armstrong of Frederick County bought from John Hughes of Frederick County, for $500, "part of a tract of Land called Douthers Chance situate and lying in Frederick County . . . adjoining the lands of a certain John Troxel, Patrick Reid, Patrick Bradley and a road leading from Emmitsburg to James McCreary's known and distinguished by Lot No. 10 in the plat of the lots lately laid off on the farm of the Revd Patrick Davidson . . . containing ten acres of land . . ."
[Frederick CCH, Liber W. R. 43, p. 546]

By this deed, dated 16 April 1813, John Armstrong of Frederick County bought from Robert L. Annan of Frederick County, for $2700, "part of a tract of land called Alexanders Prospect, part of a tract called the Resurvey on Douthers Chance and a tract of Land lying between the two aforesaid tracts lately taken up as being vacant by a certain Samuel Caldwell . . . containing 27¼ acres . . ."
[Frederick CCH, Liber T. B. 1, p. 348]

By this deed, dated 21 April 1819, John Armstrong of Frederick County mortgages some of his property to Samuel Loudonn of Adams County, Pennsylvania "John Armstrong by his bond or obligation duly executed bearing date the 12th day of April 1819 stands bound unto the said Samuel Loudonn . . . in the sum of $1692.18 . . . now his indenture witnesseth that the said John Armstrong in consideration of the said debt or sum of $846.09 owing to the said Samuel Loudonn . . . doth . . . sell . . . unto the said Samuel Loudonn . . . the two following described lots of land lying near Emmitsburg. . . that is to say lot number nine being among the lots laid off by the reverend Patrick Davidson . . . containing ten acres of land, the other of said lots being also one of the lots laid off by the Revd Patrick Davidson out of a tract of land called "Douthers Chance" beginning for the same at a crooked ash marked with seven notches . . . known in the plat of said lots by number 10 and containing 10 acres . . ."
[Frederick CCH, Liber J. S. 9, p. 486]

In 1820 the household of John Armstrong of Emmitsburg held one male over 45 years, two males between 16 and 26, one male 10 to 16, one male under 10, one female over 45, one female 16 to 26, and one female 10 to 16.
[1820 Federal Census, Frederick County, Fifth Election District]

By this deed, dated 4 May 1821, John Armstrong of Frederick County mortgaged to Samuel Lowden of Adams County, Pennsylvania, his 27-acre tract. ". . . whereas . . . John Armstrong by his Bond of Obligation . . . bearing date this 4th day of May 1821 stands bound unto the said Samuel Lowden . . . in the sum of $1230 . . . with legal interest for the same on or before the 4th day of May 1823 . . . this Indenture witnesseth that the said John Armstrong . . . doth . . . sell . . . unto the said Samuel Lowden all of the tract of land . . . in the County of Frederick . . . and conveyed to him the said John Armstrong by the deed of Robert L. Annan on 20 December 1814 and recorded in Liber T. B. No. 1 folios 348 & 349 & 350 . . . containing 27 acres . . ."
[Frederick CCH, Liber J. S. 13, p. 196]

By this release, dated 4 May 1821, John Armstrong redeems his twenty-acre tract mortgaged to Samuel Lowden in April 1819. "This Indenture made this 4th day of May 1819 Between Samuel Lowden of Adams County . . . and John Armstrong of Frederick County . . . whereas . . . John Armstrong by his Indentures of Mortgage bearing date on or about the twelvth day of April 1819 duly . . . recorded in Liber W. R. No. 39 and 43 folios from 544 to 548 . . . for . . . the sum of $950.44 . . . [did] sell . . . unto . . . Samuel Lowden . . . two . . . portions of Land lying . . . in Frederick County . . . called Douthers Chance known and distinguished by Lots Nine & Ten in a division of said tract as made by the Revd Patrick Davidson . . . containing 20 acres . . . and whereas the said John Armstrong hath fully satisfied . . . Samuel Lowden . . . the said Samuel Lowden doth agree to execute this Instrument of Writing as a full release of the above mentioned two Lots of Land . . .
[Frederick CCH, Liber J. S. 13, pp. 211, 212]

By this deed, dated 1 April 1822, John Armstrong of Frederick County sold a total of five tracts of land, for the sum of $2450, to Samuel McNair of Adams County, Pennsylvania, as follows: "the following described parcel of land situate near Emmitsburg . . . first, part of "Fox Town" containing two acres one quarter and 17 perches conveyed by Charles Robertson to . . . John Armstrong by deed . . . 9 February 1811, secondly, lot Number 9 containing ten acres of land, among the lots laid out by the Revd Patrick Davidson and conveyed by James Hughes to . . . John Armstrong by Deed bearing date 15 August 1812, fourthly, part of a tract of land called "Alexanders Prospect" part of a tract called the "Resurvey on Douthers Chance" and a piece of land lying between the two

aforesaid parcels, containing together 27¼ acres as the same was conveyed by Doct. Robert L. Annan to the aforesaid John Armstrong by deed bearing date 20 December 1814, and fifthly, all that part of "Alexander's Prospect" which was conveyed by Henry Need to the aforesaid John Armstrong . . . viz. beginning at a stone on the North Side of the great road leading from Emmitsburg to Gettysburg and running thence with the said road north . . . making an aggregate of fifty six acres three quarters and 14 perches . . ." "Also at the same time came Agnes, wife of the aforesaid John Armstrong and relinquishes her right of dower in the aforesaid . . ."
[Frederick CCH, Liber J. S. 15, p. 399]

By this Release, dated 30 March 1822, Samuel Lowden of Adams County, Pennsylvania, returns full title to John Armstrong of Frederick County for the 27-acre tract which Armstrong bought of Robert L. Annan on 20 December 1814, Armstrong having repaid in full the debt of $1230 which the property secured.
[Frederick CCH, Liber J. S. 15, p. 402]

By this deed, dated 2 April 1822, John Armstrong of Frederick County bought from John Black of Frederick County, for $1200, ". . . all that lot of ground lying and being in Shield's addition to Emmitsburg known and distinguished on the plot of said addition by number 12 containing one third of an acre more or less . . ."
[Frederick CCH, Liber J. S. 15, p. 701.

By this deed, dated 2 April 1822, John Armstrong mortgaged his lot number 12 in Shield Addition to John Black to secure a debt of $323, probably part of the purchase price.
[Frederick CCH, Liber J. S. 16, pp. 434-437]

Indenture made September 9, 1822 [deed] between John Armstrong of Frederick County . . . in the State of Maryland of the one part and [to] John Black of Frederick County State of Maryland of the other part Property in Emmitsburg.
[Frederick CCH, Liber W.R. 15, p. 625]

On 21 October 1822 John Armstrong of Frederick County sold the tools of his gunsmith shop, for $28, to Jacob Harner by the following Bill of Sale. "Know all men by these presents, that I John Armstrong of Frederick County State of Maryland, for and in consideration fo the sum of twenty eight Dollars Current money to me in hand paid by Jacob Harner of the same County, at or before the sealing and delivery, of these presents, the receipt whereof I the said John Armstrong do hereby acknowledge have granted, bargained, and sold by these presents do grant, bargain and sell unto the said Jacob Harner his executors, administrators and assigns, all the goods and implements herein mentioned viz., forty seven gun stocks, one pair Smiths bellows, one anvil, two vices, one large pair of sheers, three directors for rifling together with all the other tools and implements, necessary for carrying on the gun smith trade – all which said goods & implements are now in my possession, to have and to hold all and singular the said goods and implements mentioned or intended so to be, to the said Jacob Harner, his executors, administrators and assigns, forever and the said John Armstrong for myself, my heirs, executors and Administrators, all and singular the said goods and implements unto the said Jacob Harner his executors, Administrators, and assigns, against me the said John Armstrong my executors and administrators and against all and every other person and persons whatsoever shall and will warrant and forever defend by these presents.
[Frederick CCH, Liber J. S. 16, pp. 748, 749]

It should be noted here in the chronology that John Armstrong seems to disappear from the public records of Frederick County between October 1822 and January 1828. This fact, coupled with the sale of his gunsmithing tools and other equipment in his shop, strongly suggests that Armstrong left town for a period of as much as five years. Where he went is unknown. As a middle-aged man, freed from the responsibilities of supporting his children, he may have gone into a retirement of sorts to pursue other interests. At this time in Pennsylvania and Maryland the gunsmithing trade was in a serious depression. He may have gone west or elsewhere with one of his four gunsmith sons, and in this regard Hellman's (1906) cryptic comment that John Armstrong's "brother Samuel worked with him until he went west," deserves consideration. The use of too many pronouns leaves it unclear whether Samuel or John went west, but in any case, somebody apparently did; also, the Samuel referred to probably was John's son rather than a brother.

In early 1822 Armstrong owned more than a dozen pieces of property. Five of these properties were sold as a group to Samuel McNair on 1 April 1822, but the others, including his residence, apparently remained in his possession during the period of absence. The entry below needs explanation. Possibly Armstrong had tenants on his properties, and their names instead of his are present on the property assessment list. It is also possible that he did, indeed, sell all of his property but there are no other deeds recorded in his name to show this. Deed recording at that time was at the option of the purchaser. There apparently were nine other pieces of property that he owned in 1822, in addition to the five that he sold to McNair. It seems highly improbable that all of these could pass to new owners without any of them being recorded.

In 1825 a property assessment list was prepared for Frederick County. Owners of lands and lots in the town of Emmitsburg are clearly designated. The name of John Armstrong is not found on this list nor is there any other person with the same name of Armstrong listed there.
[Property Assessment Book, Frederick County, 1825, Hall of Records #19269]

Articles of agreement made . . . this 21st day of January 1828, by and between Thomas Hays of Frederick County . . . and John Armstrong of the County and State aforesaid . . . Witnesseth that the said Thomas Hays [son-in-law to the gunsmith]

. . . doth hereby rent unto the said John Armstrong for the term of one year from the date hereof one ten plate stove and pipe; and the said John Armstrong doth . . . promise to pay . . . two dollars and fifty cents.
[Frederick CCH, Liber J. S. 28, p. 654]

In 1830 the household of John Armstrong of Emmitsburg held one male between 50 and 60 years, one female between 50 and 60, and one male between 15 and 20. Presumably this was the gunsmith, his wife Agnes, and a youth who may have been an apprentice, a son, or some other relative.
[1840 Federal Census, Frederick County]

Armstrong, John 1833 paid $8.00 as witness
[Frederick County Levy Book, Hall of Records, #19259]

In 1835 a property assessment list was prepared for Frederick County. Owners of lands and lots in the town of Emmitsburg are clearly designated on one list, and the owners of personal property other than real estate are given on a second list. No mention is made on either list of John Armstrong or of any other Armstrong. Lots.No. 1 &2 in Emmitsburg, containing a "house & shop" are shown to be owned by John Zimmerman.
[Property Assessment Book, Frederick County, 1835, Hall of Records, #19271]

The above entry is of note. A Frederick County deed of 24 March 1865 concerns the sale of Emmitsburg Lots 1 & 2 by Joshua Motter, Executor of the Estate of John Zimmerman, to Jacob Eckenrode. A search through the deed index does not show a deed in which Zimmerman purchased the property from Armstrong. But the above entry shows without doubt that at some time prior to 1835 John Zimmerman came into possession of John Armstrong's residence and shop.

The possibility exists that Armstrong spent some time living in nearby Liberty Township of Adams County, Pennsylvania. The state, county, and township line lies only two or three miles north of Emmitsburg, and it was in Liberty Township that Armstrong's lifelong friend, Samuel Louden, lived. A check of property tax assessment records for Liberty Township for a 25-year period beginning in 1818 showed that William Lowden was there throughout the period, listed as a single man, and his occupation is given as shoemaker in a listing for 1825. The name of John Armstrong does appear on the Liberty Township tax lists, as follows:

1832 Single Men: John Armstrong
1833 Single Men: John Armstrong Occupation Tax $0.70
1837 John Armstrong 1 horse 3 cattle
1838 John Armstrong 2 horses 4 cattle
[Property Tax Assessment Book, Adams County, Pennsylvania, Adams County Historical Society]

These listings must be viewed with caution, however, for they are not at all conclusive. There was another man named John Armstrong in the region at the same time. He was a tanner and farmer, apparently associated with the family of Quintain Armstrong who also was a tanner and farmer. Cumberland Township of Adams County, Pennsylvania, lies adjacent to and northeast of Liberty Township, and includes the town of Gettysburg. The Cumberland Township tax lists for 1825 and 1826 include the name of John Armstrong, Tanner, along with the single men, and the name appears again in 1828, 1829, and 1830, but without an occupation being given.

A percussion rifle in the collection of Albert M. Sullivan, Jr., made by John Armstrong, holds his signature and the date 1836 written in ink on the wood underneath the butt plate (Fig. 32). This piece proves that John Armstrong was building rifles at this late period in his life.

John Armstrong took Nathaniel Rowe as an apprentice to the gunsmithing trade about 1837. Rowe was born in 1821 and would have been 16 years of age at that time. He lived until 1915, and in 1908 gave an account of his memories of early Emmitsburg to the local weekly newspaper, *The Emmitsburg Chronicle.* The newspaper account is quoted at length under Nathaniel Rowe's name in this section, and it suffices to note here that Rowe said, "I was apprenticed to a gunsmith named Armstrong. We used to buy the barrels and make the stocks and other parts . . ."

By this deed, dated 6 March 1838, John Armstrong of Frederick County bought from James Wood and his wife Ellen, for $300, one fourth Lot in Shields Addition to Emmitsburg . . . distinguished on the Plat of said addition to Emmitsburg by lot number seven . . ."
[Frederick CCH, Liber H. S. 7, p. 38]

By this deed, dated 5 October 1838, "John Armstrong of the Town of Emmitsburg" mortgaged to Samuel Louden of Adams County, Pennsylvania, his newly acquired lot number seven. "Whereas the said John Armstrong by his note of hand duly executed bearing date on the 19th of May in the year 1838 stands bound unto the said Samuel Louden . . . in the sum of $300 . . . Now this Indenture Witnesseth that . . . John Armstrong . . . doth . . . sell . . . unto . . . Samuel Louden . . . that tract . . . of . . . land . . . in Shields Addition to Emmitsburg and being known and distinguished on the Plat of said Town as Lot No. 7 in the aforesaid addition to Emmitsburg, and fronting on the main street in said Town . . . and the buildings erected on said lot are a One Story Brick House and a new one story Wooden Shop . . ."
[Frederick CCH, Liber H. S. 8, p. 251]

In 1840 the household of John Armstrong of Emmitsburg held one male between 60 and 70 years, one female between 40 and 50, one female between 10 and 15, one female between 5 and 10, and one female under 5 years.
[1840 Federal Census, Frederick County]

The above information suggests that John's wife, Agnes, was not living by 1840. In 1838 Freedom Township was erected out of the eastern end of Liberty Township in Adams Co., Pa., just north of Emmitsburg. The 1840 Federal Census for Freedom Twp. lists a John Armstrong there as head of a household containing 1 male 30-40 years, 1 female 60-70 years, and 1 female 10-15 years; obviously not the household of the gunsmith.

By this deed, dated 12 January 1841, John Armstrong of Frederick County took an additional mortgage on his lot number seven. "Whereas . . . John Armstrong, by his note of hand . . . bearing date 16 April 1840 stands bound to . . . Samuel Louden . . . in the sum of $100 with the legal interest for the same on or before the 16th day of April 1841 Now this Indenture Witnesseth that the said John Armstrong in consideration of the said debt or sum of $100 owing . . . Samuel Louden as aforesaid, and for better securing the payment thereof . . . doth . . . sell . . . unto the said Samuel Louden . . . a Lot of ground . . . in . . . Emmitsburg called "Shields Addition" and known . . . as lot No. 7 . . . and it is the true . . . meaning of these presents . . . that if the said John Armstrong . . . shall well and truly pay . . . the full sum of $100 on or before the 16th day of April 1841, then and from thenceforth these presents and every matter and thing therein contained, shall cease and be utterly null and void . . ."
[Frederick CCH, Liber H. S. 12, p. 217]

By this deed, dated 16 October 1841, John Armstrong of Frederick County took yet another mortgage on his lot number seven. "Whereas . . . John Armstrong . . . stands bound to . . . Samuel Louden . . . in the sum of $100 . . . on or before the 16th day of October 1842 . . . the said John Armstrong hath sold . . . unto . . . Samuel Louden . . . a lot of ground . . . in that part of Emmitsburg called "Shields addition" and known . . . as lot No. 7 . . . and it is the true intent . . . of these presents . . . that if . . . John Armstrong . . . shall . . . truly pay . . . unto the said Samuel Louden . . . the said full sum of $100 . . . on or before the 16th day of October 1842 . . . these presents . . . shall cease and be utterly null and void . . ."
[Frederick CCH, Liber H. S. 15, p. 175, 176]

By this Bill of Sale, dated 24 December 1841, John Armstrong of Frederick County sold to Samuel Louden, for $60, a number of household possessions. "Know all men by these presents that I John Armstrong of Frederick County . . . for . . . the sum of $60 . . . do . . . sell unto the said Samuel Louden . . . 2 Ten [plate] Stoves, 2 Cupboards, 2 Tables, 2 Feather beds & bedsteads, 1 Clock, 20 yards Carpeting, 1 Bureau, 1 Stand, 1 Trunk, 1 Wheelbarrow, 7 Chairs, 25 Gun Stocks, 1 Box Bees, 1 Chest, 1 Large Meat Barrel & 1 Iron Vise, all Which Property is now in my possession . . ."
[Frederick CCH, Liber H. S. 15, p. 203]

The above Bill of Sale is the last known document for John Armstrong recorded in the Frederick Co. Court House. Presumably Armstrong died within a few years after December 1841, and probably soon after that date. At the end of 1841 he may have owned considerable real estate, namely Emmitsburg Lots 17, 179, 180, 182, 183 and part of 181, an acre of land, Lot 12, and one-quarter of Lot 7. There are no records of these properties being sold in John Armstrong's name prior to 1841. One alternative is that the property passed on to John's three married daughters.

There seems to be no record of John Armstrong's death in the Frederick Co. Court House. There is no record of a will and there is no record of the administration of his estate. And there is no record of an equity case in his name.

Edward R. Flanagan of Thurmont, Md., reports that he has made a thorough search for John Armstrong's grave in Emmitsburg, and that it is not to be found. However, he reports that a senior citizen of Emmitsburg claims that both John and his daughter Elizabeth Hayes are burried in the old First Presbyterian Church cemetary there, John's gravestone being long ago destroyed or removed.

One avenue of research remains which might shed some light on John's late years and perhaps his death. This would involve a title search of the properties owned by John but for which sales were not recorded. It would involve a title search of the properties working back from the present day. An attempt was made made at this for Armstrong's Lots 1 & 2 and it led to a dead end with a John Zimmerman who owned the property in the 1840's.

* * * * * *

A few days before going to press we learned about the important items presented below, which were brought to light by Albert M. Sullivan, Sr. Shortly after 1800 a Catholic religious community was established in Emmitsburg with Mother Elizabeth Ann Seton, America's first and only native-born Saint as its leading light. St. Joseph's Provincial House, a part of this community, used the services of John Armstrong as a locksmith and mechanic from time to time from 1813 until 1841. A total of nine invoices/receipts from Armstrong, written by his hand are preserved in the old records. His signature is present 12 times, and his initials 2 times; they closely resemble the signatures on his rifles and gunlock plates. Sister John Mary Crumlish, D.C., Archivist for St. Joseph's Provincial House kindly provided copies of the original documents, reproduced here on page 278.

Rec^{d} February the 3rd 1813 from Mrs Efeton the Sum of Sixteen Shillings and Nine pence Halfpenny in full of all Demands p^{r} Me — John Armstrong
[Receipts, 1812-1822, St. Joseph's Prov. House, Emmitsburg]

1825 June the 8th the House of St. Joseph to John Armstrong — D
to Repairing two pad Locks & two Cuboard Locks $ = 75 Paid
[Old Bills, 1809-1840, St. Joseph's Prov. House, Emmitsburg]

24th November 1825 St Josephs Sisterhood
to John Armstrong Dr
to a Casteel Eye in a Shuttle $ = 37½
Feby 19th to Mending a key for a Door Lock = 18 3/4
$ = 56¼
Please pay the Bearer My Son Samuel the above Acct and this will be your Rect in full of all Demands to this Date — May 20th 1825 — Paid in full
[Souvenirs, Old Bills, St. Joseph's Prov. House, Emmitsburg]

the House of St Joseph to John Armstrong — Dr
1828 to repairing a Door Lock — $ = 10
please pay Mr. James Stovin the above amount and this shall be your Discharge in full up till this Date June the 19th 1828
pd Me John Armstrong

the House of St Joseph — to Jno Armstrong Dr
for Repairing a Door Lock — $ = 12½
14th October 1835 — then Recd payment in full By the hands of Mr J. Yerk pr m J A
[Expenses of the Infant Community 1809—, St. Joseph's Provincial House, Emmitsburg]

the House of St Joseph — to John Armstrong Dr
for Repairing Screws for a getar, & key for a pieana $ = 62½
October the 1st 1838 Received the above Bill in full of all accounts up till this Date J A
[Souvenirs, Old Bills, St. Joseph's Prov. House, Emmitsburg]

28th of October 1828 the House of St Joseph to John Armstrong Dr to mending Snuffers & Sharpening Scisors & one Clever $ = 31¼ to Dressing a flower Stamp $ = 25 the 6¼ Cents over my agreement is for a screw in one pair of the Scisors — if you do kind as to pay my Son Robert this shall be your Rect in full up till this Date — pr me John Armstrong
[Old Bills, St. Joseph's Provincial House, Emmitsburg]

the House of St Joseph to John Armstrong — Dr
for mending a key — $ = 12½
Received payment in full up til this Date
August the 12th 1839 pr me John Armstrong
[Souvenirs, Old Bills, St. Joseph's Prov. House, Emmitsburg]

ARMSTRONG, ROBERT Emmitsburg

Robert was one of the four sons of gunsmith John Armstrong (see above, also Williams' HISTORY OF FREDERICK CO., p. 805); it is mentioned that all four were trained in gunsmithing, but nothing is known of his career. One rifle has been reported bearing his name.

ARMSTRONG, SAMUEL Emmitsburg

Samuel was one of the four sons of gunsmith John Armstrong (see above, also Williams' HISTORY OF FREDERICK CO., p. 805); it is mentioned that all four were trained in gunsmithing, but nothing is known of his career. One rifle has been reported bearing his name, signed in script on the barrel. It is relief-carved and closely follows the lines of John Armstrong's rifles, but neither the carving nor the engraving is as well done.

So far, no documents have been found listing Samuel Armstrong as a gunsmith, and it is difficult to find him in the public records. During the 1820's there were two men named Samuel Armstrong living in Frederick Co. One of these, who most surely was not John's son, lived in Fredericktown, and signed documents with his "X" mark; this man had a wife named Rebecca who acknowledged a land sale in a deed of 10 Oct. 1829.

The other Samuel Armstrong, who may have been John's son, married Catharine Rife, daughter of Jacob Rife of Emmitsburg. The relationships are given in an equity case. However, Williams (p. 805) stated that Samuel did not marry.

This equity case was filed in 1831 by Samuel Armstrong and his wife, Catharine, of Frederick County, to force the sale of two houses and lots in Emmitsburg, and to have the proceeds distributed. Jacob Rife of Emmitsburg died in 1816, leaving three daughters, Elizabeth, Catharine and Susan. The property of concern consisted of Emmitsburg lot No. 4, with a house on it, close to John Armstrong's residence on Lot. No. 2, and lot No. 17, on the "north side of the main street leading from Nicholson Gap in the South Mountain toward Baltimore, containing 9900 square ft."
[Frederick CCH, Equity Liber J. S. 11, p. 250]

ARMSTRONG, WILLIAM Washington, D.C.

It is probable that William was born about 1797 ± 3 years, the oldest son of John Armstrong. William's career and his abilities as a gunsmith and armorer are all but unknown. Documentation of his work at Harpers Ferry, as reported by Williams remains to be found.

William Armstrong was one of four sons of John Armstrong, gunsmith of Emmitsburg. Of the sons, he was the only one who married. William and his brothers Robert, Samuel, and James all were trained at gunsmithing. William was an inspector of arms at Harpers Ferry for many years.
[WILLIAMS' HISTORY OF FREDERICK COUNTY, p. 805]

Armstrong, Wm. 1822	dwelling w side 8E near the barracks, N. Yd.	master armorer at marine barracks

[Directory, Washington, D.C.]

ARNOLD, W. H. Washington, D. C.

November 15, 1859[
Be it known that I, W. H. Arnold of Washington city, in the county of Washington & District of Columbia have invented a new & useful Improvement in Breech-Loading Fire-Arms . . . fire-arm is designed to be used with my tail-guide projectile.
[U. S. Patent Office, #26,076]

AXER, JOHN Baltimore

Axer, John	1845	130 Potter St.	gunsmith

[Baltimore city directory]

B

BACON, AUGUSTUS M. — Washington, D. C.

July 31, 1866

Be it known that I, Augustus M. Bacon of Washington city, District of Columbia, assignor to himself & George E. H. Day of the same place, have invented new & useful Improvements in Fire-Arms to be applied to guns, rifles, carbines, or pistols . . . for the purpose of cleaning the chamber of the oscillator & forcing out the copper case of the cartridge.
[U. S. Patent Office, #56,846]

BALD, FREDERICK W. — Baltimore

Bald, Frederick W.

1847-51	89 Pennsylvania Avenue	gunsmith
1856-60	91 Pennsylvania Avenue	gunsmith
[no directories for 1861-65]		
1887-88	915 Pennsylvania Avenue	gunsmith

[Baltimore city directories]

Inventory of Goods carried away and destroyed Sunday 21st of April 1861 [from the shop of Frederick Bald]

17 Guns single barrel at an average of $5. each		85.00
3 pairs of Allens Patent Pistols $2 per pair		6.00
2 double barrel Guns	$10. 20. each	30.00
2 Rifles	$6 each	12.00
5 Large shot Pouches 4lb	$1.25 each	6.25
3 fine shot Pouches 3lb	$1.00 each	3.00
3 large fine Powder flasks	$1.25 each	3.75
3 Patent Top Powder flasks	50 cts each	1.50
3 Pistol flask	20 cts each	.60
1 doz Pistol ramrods	12½ cts each	1.50
3 doz Bullet molds	12½ cts each	4.50
2 Boxes of Cooper Cartridges	87½ cts each	1.75
1 lot Percussion Caps Elys Walker Hichs & others		4.00
1 lot Fishing Takle		2.00
9000 Percussion caps 40 cts a 1000		3.60
3 Cannister of Powder	40 cts each	1.20
3 Horse Pistols	2.00 each	6.00
5 or more Bags of Balls	12 cts per lb	15.00
damage of Show Case and door		6.00
1 finished Rifle barrel		2.00
1 finished Rifle barrel		1.50
1 Cane 3 Joint Fishing Rod		1.00
		$200.15

[Baltimore City Correspondence Manuscripts, 1862, No. 3725]

BALL, WILLIAM — Baltimore

A number of beautifully-made silver sword hilts bearing the touch of William Ball are known. These include pieces with short-beaked eagle-head pommels, pillow pommels, and bird-head pommels. But of course, as with most silversmiths of his day, most of his work consisted of making fine items for the table. Among silver collectors his work is highly regarded. The essential details of his life have been known for many years, and it serves well to reproduce here what was published about him in 1930:

William Ball was born in 1763 and settled in Baltimore some time prior to 1790. Family records show that he was the son of William and Rebecca Ball, and he is said by a descendant to have come here from Philadelphia. It is therefore probable that he was the son of William Ball, the Philadelphia goldsmith, who advertised, August 30, 1759, in the *Pennsylvania Gazette* as then located on Front Street; and again, May 28, 1761, "that he intends going to England"; and finally in 1765 that he had returned to Philadelphia. It would therefore appear that our Baltimore silversmith was born in England, if his birth year was 1763.

William Ball was an excellent craftsman, in fact the writer is inclined to rank him as the outstanding Baltimore silversmith. A very large number of pieces made by him during the twenty five years he worked there are known. The first record of him in Baltimore is a notice, dated September 25, 1790, in the *Maryland Journal* announcing the dissolution of the firm of Johnson & Ball, followed by a statement that "William Ball, Gold and Silver Smith, Begs Leave to inform the Public in general, and his Friends in particular, that he has removed his shop to the house formerly occupied by Messrs Abraham Usher and Co., opposite Mr. William Wilson's Boot and Shoe Manufactory, on the North Side of Market Street a few doors below Calvert Street." Of Johnson, the partner of Ball, little has been learned, nor is it known how long the partnership had been in existence when it was dissolved in 1790. He may be identical with Israel H. Johnson, a silversmith working in Easton, Maryland, in 1793, and whose name appears in the census of 1790 as living in Baltimore.

Ball advertises in the *Baltimore Daily Repository* for May 13, 1793, his removal from his shop on Market Street between South and Calvert to the house lately occupied by Mr. George Dowig [the silversmith] "At the Sign of the Golden Urn" in Market Street, seven doors above the Branch Bank and nearly opposite Tripolet Alley. He advertised on November 1, 1794, in the *Baltimore Daily Intelligencer* from the same address, offering silver mounted swords and epaulets for sale. The directories from 1796 to 1815 give his address at 60 (or 62) Market Street in the same shop with Charles Tinges, the watch maker (q.v.), doubtless the shop to which he removed in 1793, and he is listed variously as silversmith, goldsmith, and jeweller. He appears in the land records as having purchased in 1795 a lot on High Street near Low, and in 1802 he and his wife Elizabeth sell land on the Old York Road adjoining Darley Hall to James Rice. In July, 1795, Ball advertises in the *Federal Intelligencer* for apprentices to learn the trade of silversmith. In 1812 he appears in another partnership as Ball & Heald, which lasted only about a year. Of Heald nothing has been learned. For a while Ball also ran a packet line from Baltimore to Alexandria. A year or two before his death he appears to have become involved in financial difficulties, as there is little doubt that he was the William Ball to whom the General Assembly in 1814 extended the benefits of the insolvency laws.

William Ball, who according to a family record was born on January 16, 1763, married by license dated October 21, 1790, Elizabeth Dukehart, a Baltimore Quakeress, the daughter of Valerius Dukehart, and had by her three sons and four daughters born between 1791 and 1811. A brother, Jacob Ball lived in Wilmington. William Ball died on June 2, 1815, and was buried in the Quaker burial ground, Aisquith Street (Cemetery record). The notice of his death in the *Baltimore Patriot* states that he was in his fifty-fourth year, which makes him a year or two older than does the family record. The latter is doubtless correct.

[J. Hall Pleasants and Howard Sill (1930) MARYLAND SILVERSMITHS 1715 - 1830; p. 89, 90]

BAKER, ISAAC Frederick & Washington Co.

The work of Isaac Baker is unknown at present, and as a documented gunsmith he has remained all but unknown up to the present time. In the course of doing research on the arms makers of Virginia, Wallace Gusler came across a deed mentioning him as a gunsmith from Frederick Co., Md. A search of records in Frederick and Washington Counties in Maryland turned up an abundance of deeds bearing his name. They show that he was a good friend of the Maryland gun barrel maker, Thomas Worley, that his wife was named Catherine, that he had a married daughter named Mary Swearington, and that he also lived in Virginia and Tennessee.

Isaac Baker and his brother, Joshua, who was a gunsmith and gun barrel smith, were the sons of Joshua Baker, a gunsmith of Lancaster, Pa., who died in 1754, leaving a will which mentions the two sons. The deeds show that Isaac spent most of his working life in Maryland, but that he lived for a while in the southern end of the Valley of Virginia in the region of Bristol. The deed for 1782 mentions Sullivan Co., North Carolina, which presently is the county in eastern Tennessee just south of Bristol. The deeds also show that Isaac was a man of moderate wealth, for in 1779 he sold a tract of land for 8,000 pounds, a large sum of money for the times.

Isaac Baker probably was born in the late 1720's or early 1730's, for his direct mention in the will of 1754 implies an age of 21 at that time, and by 1758 he was established as a gunsmith in Frederick Co., Md. His last known deed is dated 1792, though possibly there are later deeds in his name in Washington Co., Virginia.

Rifles or other arms by Isaac Baker may have survived, but I am not aware of any. Hopefully,with the information presented here available, some of Baker's work may come to light. It seems probable that he would have worked on muskets for Maryland during the Revolutionary War period, and considering the time and place in which he worked, it seems probable that he made longrifles.

I Joshua Baker of the Borough of Lancaster in the Province of Pennsylvania Gunsmith . . .' do . . . declare this my last will . . . I give to my loving Wife Rebecca Baker the Sum of 50 pounds . . . together with all my household furniture, Plate, Negro Slaves, and what ever other chattles I may die possessed of . . . I give to my Son Isaac the sum of 10 pounds . . . together with my two best wearing coats & a silk jacket . . . I give to my Daughter Mary wife of the Reverend Mr John Elder the sum of 5 shillings having already given the said Mr Elder a Bond for 30 pounds in Lieu of his wifes part of my Estate . . . I give to my son Joshua the two-third Parts of what shall remain of my estate . . . and the other third to be Equally Divided between my two Daughters Ann Daugherty and Eleanor Woods . . . I have hereunto set my hand . . . this [?] Day of May 1753 . . . Will proved 3d July 1754.

[Will Book, Lancaster CCH, Penna.]

The above will was brought to the attention of Wallace Gusler by Samuel E. Dyke, and to George Shumway by Wallace Gusler.

By this deed dated 7 Aug. 1758 Isaac Baker of Frederick Co., Md., Gun Smith, rents from Joshua Baker of the town of Winchester in Frederick Co,, Va., Gun Smith, a tract of land in Frederick Co., Va., on the Potomac River and Warm Spring Run containing 391 acres, for the sum of 5 shillings, and an annual rent of one peppercorn. Joshua Baker had bought the tract on 15 & 16 May 1755.

[Deed Book 5, p. 166-168, Frederick CCH, Winchester, Va.]

By this release dated 7 June 1763 Isaac Baker of the Province of Maryland returns full title to Joshua Baker of Frederick Co., Va., for the tract of 391 acres which he rented from him on 7 August 1758.

[Deed Book 8, p. 485-487, Frederick CCH, Winchester, Va.]

By this deed dated 27 June 1763 Isaac Baker and John Prisen attorneys in trust for Edmond Carlliage of Georgia sell to Joseph Chapline of Frederick Co., Md., a tract of land called Hickory Tavern containing 200 acres.

[Deed Book H, p. 534, 535, Frederick CCH, Maryland]

By this deed dated 30 April 1764 Isaac Baker of Frederick Co., Md., sells to Thomas Bowlus of Frederick Co a tract of land called Blew Haunch.

[Deed Book J, p. 426, 427, Frederick CCH, Maryland]

By this deed dated 17 December 1764 Isaac Baker of Frederick Co., Md., buys from Evan Shelby of Frederick Co., Md., part of a tract of land called Resurvey on the Mountain of Wales.

[Deed Book J, p. 991, 992, Frederick CCH, Maryland]

By this deed dated 21 March 1769 Isaac Baker of Frederick Co., Md., gunsmith, buys from Evan Shelby of Frederick Co. part of a tract of land called Resurvey on the Mountain of Wales.

[Deed Book M, p. 131, 132, Frederick CCH, Maryland]

By this deed dated 20 June 1770 Isaac Baker and Catherine Baker his wife of Frederick Co., Md., sell to John Cary, Merchant, a parcel of land called Pleasant Bottom containing 226 acres. This land was left by will to Isaac Baker and Catherine Baker by John Postlewhaint of Lancaster Co., Pa.

By this deed dated 20 June 1770 John Cary, Merchant of Frederick Town in Frederick Co., sells to Isaac Baker, farmer, of Frederick Co., a tract of land called Pleasant Bottom containing 226 acres.
[Deed Book N, p. 246, 247, Frederick CCH, Maryland]

By this deed dated 1 April 1779 Isaac Baker of Washington Co. sells to Joseph Sprigg of Washington Co., Gentleman, for 8,000 pounds, all that tract of land called Resurvey on Pleasant Bottom and part of the Resurvey on the Mountain of Wales, containing 886½ acres plus 60½ acres. Catherine Baker, wife of Isaac Baker, approved the transaction.
[Deed Book B, p. 7-10, Washington CCH, Maryland]

By this deed dated 1 Sept. 1779 Isaac Baker of Washington Co., Md., Gun Smith, sold to Jane McClalahan and her son James McClalahan a parcel of land, reference being made to a parcel of 109 acres which Isaac Baker sold to Alexander McClalahan in a deed dated 22 April 1772 and recorded in the records of Frederick County, Md.
[Deed Book B, p. 122-125, Washington CCH, Maryland]

By this deed dated 13 September 1779 Isaac Baker of Washington Co., Md., Gunn Smith, sold to Thomas Worley of Washington Co., Md. Blacksmith, for 100 pounds, three tracts of land -- Barren Hill containing 50 acres, a parcel of 18½ acres, and part of the Resurvey on Pleasant Bottom and part of the Mountain of Wales containing 4 acres.
[Deed Book B, p. 163-165, Washington CCH, Md.]

By this deed dated 13 Sept. 1779 Isaac Baker of Washington Co., Md., Gun Smith, sells to Mary Swearinger, Daughter of Isaac Baker, for 200 pounds, a tract of land called Resurvey on the Mountain of Wales containing 222 acres.
[Deed Book B, p. 167-170, Washington CCH, Md.]

By this deed dated 13 Sept. 1779 Isaac Baker of Washington Co., Md., Gun Smith, bought from Nicholas Shaver of Virginia all that tract of land in Washington Co. called Little is Lost containing 23½ acres, for 100 pounds.
[Deed Book B, p. 185-187, Washington CCH, Md.]

By this deed dated 10 Oct. 1779 Isaac Baker of Washington Co., Md., Gun Smith, sells to Jacob Kylor of Washington Co., Tanner, for 425 pounds, parcels of land in Washington Co. contiguous to each other -- Alloway containing 23½ acres, and a parcel of 150 acres. The sale was approved by Catherine Baker, wife of Isaac Baker.
[Deed Book B, p. 188, 189, Washington CCH, Md.]

By this deed dated 13 Oct. 1779 Isaac Baker of Washington Co., Md. sells to Thomas Worley for 3000 pounds a tract of land called Hopewell containing 100 acres, also a part of a tract called Nomore Lost which lies on the west side of the Conococheague, and also that tract called Such containing 12½ acres.
[Deed Book B, p. 197, 198, Washington CCH, Md.]

By this indenture Isaac Baker of Sullivan County, North Carolina, granted Power of Attorney to his trusty friend Thomas Worley of Washington Co., Md., in order to sell to Jacob Baltzhoover of Washington Co. a tract of land called Waggonford situated on the Conococheague Creek containing 25 acres; it is dated 5 Oct. 1782.
[Deed Book C, p. 344, 345, Washington CCH, Md.]

By this deed dated 18 April 1786 Isaac Baker of Washington Co., Virginia, sells to Jacob Kibler of Newcastle Co., Delaware, for 10 shillings, a parcel of land called Alloway containing 150 acres; the transaction was necessary to correct an erroneous description when the land was sold on 1 Oct. 1779.
[Deed Book D, p. 748, 749, Washington CCH, Md.]

By this deed dated 17 Aug. 1792 Isaac Baker of Washington Co., Virginia, grants to Peter Shaver Power of Attorney and sells to George Shavour of Washington Co., Md., for 52 pounds, 10 shillings part of a tract of land in Washington Co., Md., known as Resurvey on the Mountain of Wales, which was conveyed to Baker by Evan Shelby, containing 70 acres. The deed is dated 25 Aug. 1792.
[Deed Book G, p. 845, 846, Washington CCH, Md.]

BAKER, JOSHUA

Joshua Baker was a gunsmith and the brother of the Maryland gunsmith Isaac Baker. During the 1750's and 1760's Joshua worked in Winchester, Virginia, and perhaps further north in what now are Berkeley and Jefferson Counties. He owned a large farm on the banks of the Potomac River directly across from Maryland which he rented to Isaac from 1758 until 1763. There is no evidence that Joshua ever worked in Maryland as a gunsmith, but the possibility that he did should be considered. See the listing for Isaac Baker for details.

BANCROFT, CHARLES F. — Baltimore

Bancroft, Charles F.

1882-85	129 West Baltimore St.	gunsmith
1886	22 East Fayette St.	gunsmith

[Baltimore city directories]

BARNARD & McAINN — Baltimore

An 1860 staff and field officers' sword in the author's collection bears the dealer's name "Barnard & McAinn Baltimore Md" on the throat of the blade.

BARNHIZLE, CHRISTOPHER — Frederick

Contractors with the government under Act of July 5, 1798 for Charlesville pattern muskets at $13.40 per stand

(4) Eli Williams Williamsport, Md. 1000 muskets*
(5) Nicholas White, Thomas Crabb, Jacob Metzger & Christopher Barnhizle
Frederick Town, Md. 1000 muskets

(14) Alexander Clagett Hagerstown, Md. 1000 muskets
(25) Mathias Shroyer Taneytown, Md. 1000 muskets
[U. S. ORDNANCE, Vol. I, p. 20]

*By December 28, 1801, they had delivered 548 stands of arms.

Indenture made June 10, 1800 [bill of sale] I Peter Walters of Frederick County State of Maryland for and in consideration of the sum of 75 pounds current money to me in hand paid by Nicholas White, Christopher Barnhizle and Jacob Metzger of Frederick County and state aforesaid the following property—towit 3 iron anvils, 4 vices and all my smith tools of every description, 1 pair large bellows, 2 grindstones and a large wheel that turns.
[Frederick CCH, Liber W.R. 20, Folio 19]

Barnhassel, Christopher Frederick Town District No. 2
1 male between 26 and 45, one female 26 to 45, 1 male under 10
[1800 Federal Census]

BARRY, STANDISH Baltimore

Barry was an outstanding Baltimore silversmith who produced a number of Baltimore-style silver-hilt swords with eagle-head pommels and thin open knuckle-bows. In addition to being a craftsman, he had a long and distinguished military career. Biographical data about him has been in print for many years:

Colonel Standish Barry, one of the outstanding silversmiths of Baltimore and the issuer of the silver coin or token bearing his name, was born here, November 4, 1763. From family memoranda in the possession of Mr. Francis G. Barry of New York, a great-great-grandson, it is learned that Standish Barry was the son of Lavallin Barry, a young Irishman from Dublin who settled in Baltimore before the Revolution. Standish Barry learned his trade as an apprentice of David Evans, the Baltimore watch and clock maker and silversmith. Barry first advertised, November 26, 1784, in the *Maryland Journal* as a watch and clock maker and engraver at the shop which he had opened on the north side of Market Street three doors below Calvert, but does not refer to himself as a silversmith until a year afterwards. Six months later an advertisement appeared, June 20, 1785, in the *Maryland Gazette* announcing that Joseph Rice and Standish Barry, watch and clock makers, had removed their shop to the northwest corner of Market and Calvert Streets; and that they " return their sincere thanks to those who have (during their short continuance in business) favored them with their custom." They also announce that they engrave copper plates, seals, medals, etc., and on November 8 of the same year advertise that they make and repair gold and silver work. The same newspaper contains a notice, June 21, 1787, of the dissolution of their partnership, and the continuance of Rice in business at the old address. The first Baltimore directory, that for 1796, shows Barry in business at 92 Baltimore [Market] Street, where he was located until 1800, and from 1800 to 1808 his shop was at 20 North Gay Street. He usually appears listed in his earlier years as a watch and clock maker, and later as a silversmith, although there is no question he made and sold silver from about 1784 to 1810. Barry changed his occupation at the latter date, appearing in 1810 as a merchant, and from 1814 to 1823 as a grocer or sugar refiner.

Standish Barry had a long military career, of which a full sketch appeared in the *American* for November 27, 1844, shortly after his death. He served first as a volunteer in the Baltimore militia when an attack on the city was feared during the latter part of the Revolution. In 1794 he saw service under Washington in the suppression of the Whiskey Rebellion. In 1798 when war with France seemed probable, Barry was made lieutenant of the military organization known as the Sans Culottes, and its name changed to the Baltimore Independent Blues. In the War of 1812 as major of the Fifth Regiment Volunteer Infantry he took part in the battles of Bladensburg and North Point, and in these engagements his horse is said to have been shot under him on more than one occasion. He afterwards became colonel and commander of this regiment. Barry is thus described: "a strict tactician--upwards of six feet in height, martial and commanding in appearance, straight as an arrow, eye keen as an eagle's, he was every inch a soldier." Mr. Francis G. Barry of New York, a great-great-grandson, has a portrait of Col. Standish Barry in uniform, attributed to Rembrandt Peale, showing him wearing a sword which Mr. Barry also owns. This portrait confirms the contemporary description of his appearance. On the sword is inscribed, "Presented by the late Members of the Baltimore Independent Blues to their former Captain--Standish Barry, Esq.--now Lt. Col. of the 1st Rifle Regiment, M. M., as an evidence of their respect of his character--Baltimore, Sept. 13, 1824."

Standish Barry probably retired from his business as a sugar refiner when he was elected sheriff of Baltimore County, October 4, 1824, for a three-year term. . . His name last appears in the directory for 1831, and it was probably not long afterwards that he moved to Newport, Herkimer County, New York, where his death at the age of eighty-one occurred on November 6, 1844, at the home of his son Standish Barry, Jr.
[J. H. Pleasants & H. Sill (1930) MARYLAND SILVERSMITHS 1715 - 1830, p. 94, 95]

BARSTOW, FREDERICK S. Baltimore

The business was operated by J. H. Wilson Marriott and Barstow.

Marriott & Barstow

1883	8 South Charles Street	gunsmith
1884-86	7 South Charles Street	gunsmith

[Baltimore city directories]

BARTLETT, A. Baltimore

Bartlett, A. 1817 Ruxton Lane gunsmith
[Baltimore city directory]

BAXTER, ARTHUR T. Baltimore

Several single and double barreled shotguns and a heavy single barreled water fowl shotgun have survived from his shop which reflect his fine craftsmanship. They are finely engraved and checkered with "Arthur Baxter Baltimore" in old English script on the barrels and locks; the barrels are beautifully browned.

Baxter, Arthur T.

1827	72 Centre Market space	gunsmith
1829	76 Centre Market space	gunsmith
1831-40	67 Pratt St.	gunsmith

[Baltimore city directories]

BEALE, R. Washington, D. C.

Along with the invention below, on February 20 of the same year he had a similar patent using a percussion lock to ignite cannon.

January 13, 1835
Be it known that I, R. Beale of Washington city, in the District of Columbia have invented a new & useful percussion lock for cannon.
[U. S. Patent Office]

BEATTY, CHARLES Frederick

First indenture (deed) in Frederick County was recorded in 1762 and his last (deed) in 1805. He was a delegate to the committee of arms in the Province in July of 1775. He was empowered to contract with the local smiths and during that summer brought the two to terms. The smiths differed on two points–single instead of double bridle locks and the date of delivering the 650 muskets. Beatty was the most active commissioner of the Frederick Gun Lock Manufactory.

Delegate from Frederick County at Annapolis Wednesday, 26 of July 1775 and on the committee of Arms in this Province.
August 30, 1775
Resolved that Mr. Charles Beatty of Frederick Town be empowered to contract the making and Delivery of 650 good substantial proved Musquets 3½ Feet in the Barrel, and of three Quarters of an Inch in the bore, with good double Bridle Locks, black walnut or Maple Stocks and plain strong brass mounting Bayonets with Steel Blades 17 Inches long, steel Ramrods, double Screws, priming Wires, and Brushes fitted thereto, with a pair of brass Molds for every Eighty Musquet to cast 12 Bullets on one Side and on the other side, to cast shot of such Size, as the Musquet will chamber three of them, for a sum not exceeding Ten Dollars.
[AM, p. 75]

. . . Resolved, that Thomas Johnson Jun. Esqr. be empowered to contract for the making and Delivery of any number not exceeding 1000 good substantial proved Musquets, of the same Dimensions and with the same Accoutrements with those for which Mr. Beatty is empowered to contract and for the same price; and may contract to advance in the same Manner.
September 1, 1775 Advertisement in *The Maryland Gazette*
[AM, p. 77]

September 20, 1775 Beatty to Council
Agreeable to your request to me I have attempted to agree with our gun smiths for the making of Musketts and they will not come into agreement exactly as your Directions is to me they Differ only in two points which is as followeth single Bridle Locks instead of Double, 2/3 of the Musketts to be delivered by the first of March next the remainder by the first of May following instead of 1/3 by the first of January 1/3 by the first of March and the remainder by the first of May everything else they comply with . . .
[AM, p. 81]

October 18, 1775 Johnson to Council
I inclosed you certificates of the Enrollment of Capt. Chas. Beatty and Capt. James Johnson's Companies they hope that you will be pleased to have the several Commissions made out as far as convenient . . .
[AM, p. 84]

Assembly Proceedings
Ordered That the Treasurer of the Western Shore pay to Mr. Benjamin Ford sixty Pounds Currency to be lodged with Mr. Thomas Johnson at Frederick Town or in his Absence, to be left with Mess Charles Beatty and Barker Johnson to be by them laid out in the purchase of Gunlocks.
[AM, p. 108]

February 1, 1776 Council Proceedings
Ordered, That Messrs. John Hanson Junr., Charles Beatty and James Johnson or any two of them be requested to provide and furnish the said Company with good substantial and serviceable Rifles, necessary Bullet-Moulds and powder Horns and to draw the necessary Order on the Council of Safety from Time to Time to disburse the Expence attending the same.
[AM, pp. 127-128]

July 30, 1776 Council to Beatty
Commissioners of Gun Lock Manufactory Gentn. As a considerable time has elapsed since your erecting your Gun Lock factory, we are in great hopes you have made a number for the use of this Province. The Province is in great distress for arms few of the Troops raised for the flying camp are supply'd with them, & the Militia will not lend theirs. There is an absolute necesity therfore that you immediately send them down what Locks you have. We expect Firelocks from John Yost, they may come from thence if you contrive them there. We desire also to know how many Locks you make per week. We have Barrels enough here and in Kent for the Locks you can make.
[AM, p. 142]

December 10, 1776 Beatty to Council
Gent.
The Barer hereof Mr. Boon (the overseer of the Gun Lock factory) waits on you in order to get some money to enable him to carry on the factory, for it cannot be supported any longer without. Mr. Hanson who keeps the books is not at home so that I cannot get to see them in order to draw for the money, the factory is entitled to, but I am clearly satisfied that it must

be about 400 Dollars by the acct rendered, and the locks now in the factory, which sum please to forward by Mr. Boon. We have now in the factory 156 locks completely finished as will appear by the Inclosed certificate for nearly two months past I was compelled to take money on Interest to support the factory or let the hands go idle. It is not in my power to borrow any more for that purpose here.—pray dont fail sending the money or I shall be compelled to sell locks to pay the money that I have taken up and let the servants sift for themselves, which will be a great loss to the publick . . .
[AM, p. 519]

BEATTY, CORNELIUS E. Baltimore

A single barrel, London marked, percussion, half stock shotgun bears a lock marked "Norris & Beatty Baltimore."

Norris & Beatty

1835	212 Baltimore Street	hardware merchants
1837-42	212 Baltimore Street	importers of hardware

[Baltimore city directories]

BEHR, J. J. Baltimore

In the collection of James A. Knowles of Baltimore is a flintlock musket of Brown Bess pattern with walnut stock. The lockplate is marked "J. J. BEHR." The barrel, of .76 caliber and 42 7/8 inches length, is marked "J. BEHR BALTIMORE TOWN." The numeral "10" is marked on the barrel tang. The butt plate is stamped "R. B. N. 2", and "MARYLAND" is branded into the butt stock. No public records or other documents mentioning Behr as a gunsmith or resident of Baltimore have been found.

BEHRENS, CHRISTIAN F. Baltimore

Behrens, Christian F.

1833-41	Lee Street west of Light	black & gunsmith
1842	Lee Street west of Light	blacksmith

[Baltimore city directories]

BELL, FRANCIS H. Washington, D. C.

Bell, F. H.

1858	60 La av	gunsmith

[Washington, D.C., business directory]

Bell, Francis H.

1858	60 La. av, h 305 Va. av	gun & locksmith

[Boyd's Washington and Georgetown Directory]

Bell, F. H.

1860	60 La av	gunsmith

[Business directory of Washington, D.C., Georgetown & Alexandria]

Bell, Francis H.

1860	60 La. ave, h 305 Va. av	gunsmith & locksmith

[Directory, Washington, D.C.]

April 12, 1859
Be it known that I, Francis H. Bell of the city & county of Washington, in the District of Columbia, have invented a certain new & useful Improvement in Automatic Primers for Firearms.
[U. S. Patent Office, #23,545]

7 July 1860 Washington, D.C.

Bell, Francis H.	age	50	born	N.Y.	gun & locksmith
" Sarah B.		40		Va.	
" Jackson W.		21		Va.	ambro
" Thomas H.		18		Va.	gun & locksmith
" Sephi (m)		16		Va.	ambro
" James E.		15		Va.	
" Charles M.		12		Va.	

[1860 Federal Census]

BELL, LLOYD C. Washington, D. C.

Bell, Lloyd

1858	359 D north, h 673 7th west	gunsmith

[Boyd's Washington and Georgetown Directory]

Bell, L. C.

1860	359 D north	gunsmith

[Business directory of Washington, D.C., Georgetown & Alexandria]

BELL, THOMAS H. Washington, D. C.

In 1860 the household of gunsmith Francis H. Bell held, among others, Thomas H. Bell, gun and locksmith, age 18, born in Virginia, according to the 1860 census.

BELL, WALTER Baltimore

Bell, Walter

1847	Eutaw Street off Montgomery	gunsmith
1858-59	158 North Dallas Street	gunsmith
1860	207 North Bond Street	gunsmith
1864-66	336 East Eager Street	gunsmith
1867	58 West Pratt Street	gunsmith

[Baltimore city directories]

BELL, WILLIAM H. Washington, D. C.

Of his three patents, the last two refer to self-priming firearms. Patent #22,618 dated January 11, 1859, was for primers located in the hammers of the locks of rifles. Patent #27,518 dated March 20, 1860, was for a wafer primer that continuously capped the nipples of revolvers.

May 14, 1836
Be it known that I, William H. Bell, of the city & county of Washington, in the District of Columbia, have invented a new & useful cannon tranverse board.
[U. S. Patent Office]

Bell, William H.

1858	h 251 Va. ave	gunsmith

[Boyd's Washington and Georgetown Directory]

7 July 1860 Washington, D.C.

Bell, William H.	age 26	born Va.	gun & locksmith
" Mary E.	21	Md.	
" Willie	2	Va.	
" Thomas	1	Va.	

[1860 Federal Census]

BELLESON, WILLIAM — Baltimore

The first Belleson family to come to this country was the great-grandfather of George W. Belleson, William Belleson. He was a native of England. On the ship coming to this country, he met and fell in love with a German girl, whom he married. They landed in Baltimore, Md., and settled. He was a gunsmith by trade.
[HISTORY OF FREDERICK COUNTY, Vol. II, p. 1500]

BELSCHNER, JOHN L. — Manchester

Belschner, John L. 1878 Manchester gunsmith
[The Maryland Directory, J. Frank Lewis & Co. Publishers]

BENNETT & PACKSON — Kent Island

Kent Island was inhabited by Matapeake Indians and was occupied by settlers in 1629. Most of the businesses on the Island at this period were owned by William Claiborne. Disborough, Peter and John Bennett resided there in this early period.

I must sincerely let yor Honor know that the Gunsmithery on Kent Island of Bennett and Packson is a 1631 licensed trading house wch has not been suppling the Indians wth gunns and ammunitions I fear theyr unlawfully gotten from the Dutch wch settled on the Delaware.
[Public Record Office, London, Colonial Papers No. 26]

BENNETT, F. W. — Baltimore

The carbines noted below were obtained by the city under ordinance No. 22 which was approved April 20, 1861, to arm the citizens against the Northern invasion. Bennett was located at 28 and 30 South Charles Street and advertised as auctioneer and commission merchants.

April 11, 1862
. . . the precise numbers and quantities of which, as nearly as could be ascertained, are given in the two following schedules:
Arms
1217 Carbines, from F. W. Bennett
[Baltimore Correspondence Manuscripts, No. 695]

BIRD, NIGHTINGALE & NEFF — Baltimore

The firm was composed of Henry Bird, Samuel Nightingale and P. Neff. There is a flintlock fowling piece, maple stocked with a British proof marked barrel signed "Bird Nightingale & Neff." From 1824 to 1829 the firm operated under the name of Neff and Bird at the same location.

Bird, Nightingale & Neff
1819-23 108 Baltimore Street fancy hardware merchants
Bird, Henry, of the firm Neff & Bird
1824-29 108 Baltimore Street fancy hardware merchants
[Baltimore city directories]

BITNER, WILLIAM G. — Washington, D. C.

1850 real estate value $2,000.

Bitner, W. G.	age 50	born	Germany	gunsmith
" Elizabeth	44		Washington, D.C.	
" Anne E.	13		Washington, D.C.	

[1850 Federal Census]

Bitner, William G.
1853 N side H n btw 18 & 19 w gunsmith
[Directory, Washington, D.C.]

Bitner, Wm. G.
1858 158 Pa av gunsmith
[Business Directory, Washington, D.C.]

Bitner, William G.
1858 158 Penn av., h 268 H north gunsmith
[Boyd's Washington and Georgetown Directory]

Bitner, W. G.
1860 158 Pa av gunsmith
[Business Directory, Washington, Georgetown & Alexandria]

Bitner, William G.
1862 h 268 H north gunsmith
[Boyd's Washington and Georgetown Directory]

Bitner, William G.
1862 158 Penn Ave. gunsmith
[Business Directory of Washington, D.C., & Georgetown]

Bitner, W. G.
1864 158 Pa. ave Georgetown gunsmith
[Boyd's Business Directory Washington, D.C., & Georgetown]

Bitner, Wm. G.
1865 158 Penn Ave. gunsmith
[Boyd's Business Directory Washington, D.C., Georgetown & Alexandria]

Bitner, W. G.
1866 158 Penn. Ave. gunsmith
[Boyd's Business Directory Washington, D.C., Georgetown & Alexandria]

Bitner, William T.
1867 158 Penn Ave. gunsmith
[Boyd's Business Directory Washington, D.C., Georgetown & Alexandria]

BLACKBURN, JOHN — Emmitsburg

The only documentation for John Blackburn is the apprentice indenture given below. Whether or not he went on to work as a gunsmith in Maryland or elsewhere after completing his apprenticeship is not known.

John Blackburn Aged Seventeen years and five months the 17th day of March Last by and with the Advice and consent of his Father Alexander Blackburn hath placed and bound himself an Apprentice to John Armstrong of Frederick County Gun Smith for the Term of Three years from the 16th of this instant during which time the said apprentice his said Master well and faithfully Shall serve and in all respects demean himself as a good and faithful Apprentice ought –

And the said Master shall teach and Instruct or cause to be taught and instructed to his said Apprentice the art and mystery of the trade of a Gun Smith in the best way and manner that he can and also the art and mystery of Lock making in the best way he can finding to the said Apprentice meat drink and Lodging and will allow his said Apprentice two weeks in every Harvest during said term for his own use and at the expiration of the Term aforesaid will pay his said Apprentice the sum of Twenty Dollars Current Money. Witness whereof the parties have set their hands and seals the Fifth day of April 1804.

John Blackburn
John Armstrong
Alexander Blackburn

[Frederick County Apprentice Indentures, p. 425, Hall of Records]

BLAKSLEE, B. E. — Cambridge

Blakslee, B. E.	1878	Cambridge	gunsmith
Blakslee, B. E.	1880	Cambridge	gunsmith
Blakslee, B. E.	1882	Cambridge	gunsmith
Blakslee, B. E.	1887	Cambridge	gunsmith

[The Maryland Directory, 1878, 1880, 1882, 1887; J. Frank Lewis & Co., Publishers]

BOERSTLER, DANIEL — Funkstown

His first deed in the Washington County Court House, 1814; last in 1832. All of his property was located in the town of Jerusalem, later renamed Funkstown.

Daniel Boerstler who manufactured rifles, gun and common powder there in 1808 . . . early resident of Funkstown.
[HISTORY OF WESTERN MARYLAND, p. 1280]

BOHRER, JULIUS L. — Baltimore

The below-mentioned Murphy was by trade a patenter, so the original inventor was probably Bohrer. Bohrer was in the United States Navy residing at 14 Broadway. Nothing can be found on his company; therefore, it may have been used only for the patent rights.

June 30, 1868
Be it known that I John D. Murphy, of the City and County of Baltimore, and State of Maryland, assignor to Julius L. Bohrer & Company, have invented certain new and useful Improvements in Cannon . . . invention consists in the provision, in a gun, of an additional chamber . . . which, by being expanded by the explosion of the charge, shall fill the bore, and thus prevent the occurrence of a vacuum, and the consequent recoil.
[U. S. Patent Office, Patent #79,377]

BOND, RICHARD — Cecil County

October 30, 1776 Bond to Jenifer, Caecil County
This may inform you that I am setting up a Gunfactory, which I expect will be ready to go about Crismas, where work will be done in the best manner, where gun-barrels may be had ready for stocking—any quantity. Relying upon the public spirit of the Council I never made application, but hope to meet with encurragement, as it will cost a considerable sum to set it a-going. I may be obliged to aply for money to carry it on, in which case Sir., I beg your influence and also in contracting for work. I beg Sir, that you will signify your pleasure respecting it in a letter, and in so doing you will much oblige.
[AM, p. 412]

November 15, 1776 Council to Bond
We have already more Gun Barrels than Locks, therefore we cannot agree with you for Barrels only. But if you will engage to complete the Guns, either with or without Bayonets, we will agree with you for all you can furnish.
[AM, p. 449]

April 17, 1777 agreement between Council and Bond
Richard Bond agrees to manufacture and make for the use of this State one thousand Gun Barrels, three Quarters of an inch in the Bore, and three and an half Feet in the Barrel, well britched and Looped, and otherwise completely fitted, finished and proved, ready for stocking, for which the Governor and Council agree to allow at the Rate of thirty five shillings p Barrel, computing Bar-Iron at the Rate of fifty five pounds p Ton. and it is agreed that in case Bar-Iron should rise, or fall the Price of Barrels to be rated accordingly. Richard Bond agrees to deliver to the order of the Governor and Council not less than sixty two Barrels p month, compleatly fitted as foresaid, the Governor and Council finding Powder for proving said Barrels. The Gun-Barrels are to be delivered at the Head of Elk, the Britches of the said Guns to be so filled, that in reboring them, the screws shall not be touched.
[AM, p. 219]

February 13, 1778 Council to Bond
If you have any musquet Barrels by you, we shall be glad to receive them by this Opportunity; they are wanted and may come in our Boat without any Expence. If you send any, send an Account of their Number.
[AM, p. 495]

July 30, 1782 Assembly proceeding
That the said Treasure pay to Zephaniah Turner Esqr. for the use of Richard Bond three thousand two hundred pounds of Tobacco due him P Accot. passed.
[AM, p. 225]

October 18, 1782 Assembly Proceeding
Commissions issued to Richard Bond elected Shaeriff of Caecil County . . .
[AM, p. 288]

BOONE, SAMUEL — Frederick

Was overseer of the Fredericktown Gun Lock Manufactory from its inception in December, 1776, to the sale of tools November 10, 1778. The bill of sale shows he purchased many of the working tools for himself.

Indenture made October 24, 1767 [deed] between Samuel Boone of Frederick County and Province of Maryland of the one part and [to] Peter Spycker of Berks County in the Province of Pennsylvania of the other part . . . Lott of ground #15 lying in George Town Frederick County.
[Frederick CCH, Liber L, Folio 78]

Indenture made February 15, 1768 [deed] between John Gordon of Prince Georges County Province of Maryland of the one part and [to] Samuel Boone of Frederick County and the Province of Maryland of the other part . . . Ground situate lying and being in Frederick County being that tract called Knavis Disappointment and joining George Town.
[Frederick CCH, Liber L, Folio 166]

Indenture made April 18, 1770 [deed] between Archibald Orms of Frederick County Province of Maryland of the one part and [to] Samuel Boone of Frederick County and Province of Maryland of the other part . . . Tract or parcil of land lying and being in Frederick County land beginning at the original beginning tree of a tract of land called Montross.
[Frederick CCH, Liber L, Folio 630]

Indenture made April 13, 1772 [deed] between Christopher Lowrides of Prince George County in the Province of Maryland of the one part and [to] Samuel Boone of Frederick County and Province of Maryland of the other part . . . Ground situate lying and being in Frederick County being tract called Maidens Bowers.
[Frederick CCH, Liber N, Folio 683]

Indenture made November 25 [deed] between Samuel Boone of Frederick County and Province of Maryland of the one part and [to] William Boone of Berks County and the Province of Pennsylvania of the other part . . . That part of lott #32 in George Town.
[Frederick CCH, Liber W, Folio 340]

Indenture made January 23, 1776 [deed] between Samuel Boone of Frederick County and Province of Maryland of the one part and [to] Archibald Orms of Frederick County and Province of Maryland of the other part . . . Ground situate lying and being in the same County being that tract of land called Montross and part of a tract of land called Roseway on Rich Meadow.
[Frederick CCH, Liber W, Folio 477]

October 8, 1776 Boone to Council
by orders of they supervisers of they Gun lock facktorey in Fridrich Town I have sent you seventy nine locks and have a quontity more all ready to put together, if you must have all they locks sent to you that is made in they facktorey pleas to let me know by the barer George Bear.
[AM, p. 327]

October 12, 1776 Council to Boone
All the Gunlocks you can make are wanted here. You will therefore be kind enough to send them down to this place as soon as you can.
[AM, p. 338]

December 13, 1776 Council of Safety
That the said Treasure pay to Samuel Boone one hundred and fifty Pounds on Account of the Gunlock Manufactory.
[AM, p. 525]

October 2, 1777 Council of Safety
Ordered That the Commisary of Stores at Annapolis deliver Samuel Boone 6 pieces Kendal Cotton 1 piece Osnabrigs 10 Bushels Salt for the use of the Gun Lock Manufactory in Frederick. Ordered That the western shore Treasurer pay to Samuel Boone four hundred and fifty pounds to be delivered over into the hands of the Supervisors of the Gun Lock Manufactory in Frederick on Acct.
[AM, p. 386]

BORDEN, DANIEL — Taneytown

Except for the information given below nothing is known about this person. The apprentice indenture as copied in the record book clearly shows his name to be "Borden," but it is possible that the name was "Border." The court-house record was transcribed from the original document by a clerk who was not infallible. On one occasion Philip Creamer sent his apprentice Isaac Shunk to the store to buy something for him, and on another occasion he sent "D. Border." If this apprentice was named Daniel Border, he is not to be confused with a gunsmith of the same name who worked in Bedford County, Pennsylvania, in the mid-nineteenth century.

Daniel Borden an Orphan aged Seventeen years the fifteenth day of January next is this day with the approbation of his Mother Rachel Null placed and bound apprentice to Philip Creamer of Taney Town Gun Smith until he arrives at the age of Twenty One years during which Term the said Master shall find and provide for his said apprentice meat drink washing lodging, Apparel and all other necessaries to cause him to be Taught to read and write and cast accounts to cause him to be Taught the said Trade or science of Gunsmith which he the said Master now useth as also to make Gunlocks and when free to give him Forty Dollars in lieu of all other freedom dues given under the hands of the Subscribers two Justices of the Peace of Frederick County the 28th day of October 1799.
[Frederick County Apprentice Indentures, p. 236, Hall of Records]

BORDEN, WILLIAM — Baltimore

Borden, William

1833	Market Space	gunsmith
1835	76 Harrison Street	gunsmith
1845-51	101 Madison Street	gunsmith

[Baltimore city directories]

BOSLEY & HOWE — Baltimore

In 1887 the firm operated under the name of Maryland

Sportsman Supply Company; the proprietor was Arthur Bosley.

Maryland Sportsman Supply Company
1887 110 West Lexington Street gunsmith
Bosley & Howe
1888 110 West Lexington Street gunsmith
[Baltimore city directories]

BOSSELL, V. E. Annapolis

An octagonal barrel target rifle of medium quality is stamped on the barrel "V. E. Bossell Annapolis"; at the breech is stamped "Nelson & Co. steel."

BOTELER, JOHN D. Washington, D. C.

Boteler, John D.
1822 N side, D n btw 6 & 7 w gunsmith
[Directory, Washington, D.C.]

Boteler, John D.
1830 D n btw 6 & 7 w gunsmith
[Directory, Washington, D.C.]

BOYD, JAMES St. Mary's County

August 31, 1776 Council of Safety
Ordered That said Treasure pay to Jas. Boyd eight pounds, two shillings and six Pence for Bayonets.
[AM, p. 250]

September 12, 1776 Council of Safety
James Boyd of St. Mary's County contracts with the Council of Safety for making Muskets, completely fitted in the usual manner at 4 pounds 5 shillings each, the Council engaging to take any quantity from him and find Powder for their proof.
[American Archives, 5th Series, Vol. II, pp. 630, 631]

BRADA, CONRAD Baltimore

Brada, Conrad
1859-64 654 West Baltimore Street gunsmith
[Baltimore city directories]

BRIGGS, E. G. & CO. Baltimore

Briggs, E. G. & Co.
1840 8 Lombard Street gunsmith
[Baltimore city directory]

BRIGHT, JACOB Washington, D. C.

Bright, Jacob
1822 10 e btw Vir av & K s N. Yd. gunsmith
[Directory, Washington, D.C.]

BROTHERHOUSE, WILLIAM Baltimore

Brotherhouse, William
1819 37 George's Fell's Point lock & gunsmith
[Baltimore city directory]

BROWNING, ABEL Garrett County

Abel Browning was born in 1854 and died in the early 1940's. He lived north of Oakland almost to Deep Creek Lake. From Oakland it was north on Rt. 219 past Swallow Falls Rd., right to the next county road to the left. The first farm belonged to John Browning.

Abel Browning had his shop in a very small two-story, two-room building. He had a foot-treadle lathe, a forge, a rifling machine, all home-made, as were his hand tools---hacksaw, bowdrill, spoke shave, and the usual carpenter tools.

In 1939 Mr. Browning was still quite active though 84 years old. We spent the afternoon with him and went over the whole operation of building a rifle from iron, brass and wood. He showed us how the barrel was made by welding two strips of iron together, a 3 or 4 inch bit at a time over a mandrel which was driven out and moved each time to form the bore. The barrel was made of soft, very pure wrought iron which worked easily. After the two strips were welded together they were filed down to octagon, then reamed out by a series of reamers until no imperfections were left. The barrel was then rifled (he showed us the method in detail), a breech plug was threaded in and a barrel to hold a nipple was threaded on. He showed us how he made his moulds from blank castings. A cherry was selected suitable for the size of the bore finished up. As the cherry dulled it was sharpened to a smaller size. The tang of the cherry was square for a carpenter's brace. When a rifle was "freshened", which was frequently, a new mould had to be made, so he had a large stock of cherries.

He had a large selection of "furniture" which he had cast at a foundry by giving the last piece to be used to make the mould. He had sheet brass and German silver for patch boxes, thimbles, etc.

The upstairs had an assortment of wood for gunstocks, mostly maple though he gave me a blank of black walnut he said had been up there for fifty years.

His own personal gun he was most proud of, a double barrel of about .38, not overly long and very nice to hold. It was unique in that only one lock was used to fire either barrel. The hammer had two heads; the upper head hit the top nipple while the lower cleared the cap on the bottom barrel nipple. By a motion of the thumb a lever pushed a block over the bottom nipple which was mounted on a spring. When the hammer fell the bottom head hit the block which struck the bottom nipple. The bottom head was flat and the swinging block was cup-shaped to retain a cap flash.

He showed us a target, shot at the turn of the century where he outshot a sport who thought his breechloader was the best rifle that could be made.

He remembered his grandfather, Meshach Browning (see FORTY-FOUR YEARS IN THE LIFE OF A HUNTER, Winston Printing Co., Winston Salem, N.C.), and recalled to us quite a few stories of his youth. He said that the cannonading at the Battle of Gettysburg was heard as a continuous rumble and people gathered all around wondering what was happening.
[Interview recorded by Maurice H. Miller, 19 July 1939]

BRYAN, FRANCIS — Baltimore

Bryan, Francis

1814	Potter Street, Old Town	gunsmith

[Baltimore city directory]

BUDENBAUGH, MARTIN — Hagerstown

This indenture witnesseth that John Budenbaugh of Washington County and State of Maryland has put his son Martin Budenbaugh aged sixteen years four months, apprenticed to George Kreps Gunsmith of the county and state aforesaid to learn his art trade and mystery . . . to serve the said George Kreps from this 3rd day of January 1797 for four years and eight months. [Washington Co. Apprentice Indentures, p. 48, Hall of Records]

BURKHARDT, PHILIP — Baltimore

Burkhardt, Philip

1871-84	566 West Baltimore Street	gunsmith

Burkart, Philip

1884-86	566 West Baltimore Street	gunsmith
1887	760 West Baltimore Street	gunsmith

[Baltimore city directories]

BURT, ED & COMPANY — Washington, D. C.

1865 Ed. Burt & Co. military outfitters
38 Pennsylvania Ave.

[Washington City directory]

BURNETT, GARLAND — Baltimore

Burnett, Garland

1819	Hull's Lane, Old Town	gunsmith
1831-33	Maiden Lane	gunsmith

[Baltimore city directories]

C

CALDERWOOD, THOMAS A. — Baltimore

Calderwood, T. A.

1853	104 Lombard Street	gunsmith
1874	104 Lombard Street	gunsmith

Gun, Pistol & Rifle guns manufactured & imported, guns repaired, restocked, & altered to percussion in superior manner, moderate terms – gun barrels rebored on improved principles, guaranteed not to recoil when discharged.

[Baltimore city directories]

CALHOUN & COWTON — Baltimore

1862

The joint Standing Committee on claims to whom were referred the following claims against the city—towit 16. The petition of Calhoun & Cowton asking to be indemnified for the loss of pistols and cartridges taken by a mob, on the 19th of April 1861 and claiming $248.22.

[Baltimore City Correspondence Manuscripts, No. 504]

CALVERT & PARKHURST — Washington, D. C.

Calvert & Parkhurst

1867	126 Pa. Ave.	gunsmiths

[Boyd's Business Directory Washington, D.C., Georgetown & Alexandria]

CAMPBELL, DANIEL

September 12, 1776 Council of Safety

Ordered That the said Treasure pay to Daniel Campbell three pounds, ten shillings for a Gun.

[AM, p. 267] [Am. Archives, 5th Series, V. II, p. 630]

CAMPBELL, WILLIAM — Annapolis

Served as a captain in the 3rd Regiment during the Revolution, in Capt. Samuel Griffth's company.

December 11, 1780 Agreement

. . . between the Governor and Council of the one part and William Campbell of Annapolis of the other Witnesseth that the said William Campbell agrees to stock in a good sufficient and workmanlike manner all the Musket Barrels, now belonging to the State in the Care of John Shaw . . .

[AM, p. 241]

December 14, 1780 Council Proceeding

. . . and also to Captain William Campbell one hundred Gun Barrels to be by him stocked agreeable to Contract.

[AM, p. 243]

April 17, 1781 Council Proceeding

That the Collector of the tax for Ann Arundel County pay to Captain William Campbell one thousand three hundred and ninety one Dollars to pay for two hundred and fourteen setts of Mounting to compleat the Muskets agreeable to his Contract.

[AM, p. 401]

April 26, 1781 Council Proceedings

That the said Treasure pay to William Campbell a Lame Soldier of the 3d Regiment Twenty pounds on Account.

[AM, p. 415]

CANFIELD & BROTHER — Baltimore

Presentation swords and buttons bearing this firm's name are in the collection of the author. The firm was composed of Ira C. Canfield, William B. Canfield and Joseph H. Meredith.

Canfield & Brother

1835-40	Baltimore & Charles Sts.	jewellers
1842	Baltimore & Charles Sts.	jewelry & fancy military store
1845	227 West Baltimore Street	jewelry & fancy military store
1847-49	227 West Baltimore Street	import manufacturers & dealer in watches, jewelry & fancy goods
1850-53	227 West Baltimore Street (ad) military goods, a full assortment, military Co. furnished complete	
1855-59	229 West Baltimore Street	jewellers

Canfield, Ira C. (firm Canfield & Brother)

1855	dwelling 28 Pleasant Street	

Canfield, William B. (firm Canfield & Brother)

1855

Canfield & Brother & Co.

1860	229 Baltimore Street (ad) I. C. Canfield, W. B. Canfield, Jos. H. Meredith	fine watches, jewelry
1864-66	229 Baltimore Street (ad) military goods, of every description as swords, belts, sashes, spurs, buttons, lances, embroideries, haversacks, pistols, field & marine glasses, etc. Presentation Swords.	
1867-81	229 Baltimore Street (ad) watches, jewelry, silver & plated ware, rich fancy goods, cutlery on hand and made to order at short notice.	

CARTER, THOMAS — Frederick

Inventory of Work tools Belonging to and now in the Gunlock Factory the 19th August 1778 . . . Thomas Carter – Runaway.

[Hall of Records, Red Book IX, p. 273]

CASHELL, R. — Frederick

INFORMATION is hereby given to the gentlemen of Frederick-Town & its vicenity, that they can have their FIRE ARMS OF EVERY DESCRIPTION cleaned & repaired in the neatest manner also swords cleaned, polished, new mounted, gilt, or plated - blades blued equal to the imported. The Cavalry & Rifle companies can have their carbines, pistols, or rifles eleganitly bronzed, blued, or blacked in such a manner that they cannot collect rust, on very moderate terms by applying to the subscriber opposite Mr Francis Mantz's store in Market street, fourth house below the turn pike, Frederick-Town.
September 7 R. Cashell
[*Frederick-Town Herald,* September 24, 1814]

CATTEY, EDWARD — Baltimore

Cattey, Edward

1847	20 McCeans Alley	gun, pistol & rifle maker

[Baltimore city directory]

CHAMBERS, BENJAMIN — Washington, D. C.

July 31, 1849
Be it known that I, Benjamin Chambers, of this city & county of Washington, in the District of Columbia, have invented a new & useful Improved Cannon . . . My improvements have reference as well to the construction as to the mode of using cannon.
[U. S. Patent Office, #6612]

CHAMBERS, WILLIAM — Washington, D. C.

Chambers, William

1862	bds 89 G south	gunsmith

[Boyd's Washington and Georgetown Directory]

CHARRIER, JACQUES — Baltimore

Charrier, Jacques

1812	50 Market Space	gunsmith
1822-1824	16 Albemarle Old Town	gunsmith

[Baltimore city directories]

CHARRIER, JAMES — Baltimore

Charrier, James

1816-1818	22 Water St.	gunsmith

[Baltimore city directory]

CHARRIER, JOHN — Baltimore

Charrier, John		
1822-1824	16 Albemarle Old Town	gunsmith
1827-1829	53 South Street	gunsmith
Charrier, John & Peter		
1831	53 South Street	gunsmiths
1833-1836	37 South Street	gunsmiths

[Baltimore city directories]

CHARRIER, PETER — Baltimore

Charrier, John & Peter		
1831	53 South Street	gunsmiths
1833-1836	37 South Street	gunsmiths
Charrier, Peter		
1837	53 South Street	gunsmith
1840-1841	Balderston Street	gunsmith
Cleaveland & Charrier		
1845	15 Harrison Street	gunsmiths

[Baltimore city directories]

It was a pet weapon of mine [a rifle] I had bought the barrel in Philada and had it stocked by Pete Chanier, a Frenchman, at that time the best gunsmith in Baltimore.
[A SPORTING FAMILY OF THE OLD SOUTH, p. 138]

CHASE, A. C. — Baltimore

Chase, A. C.

1889	2 Light Street	gunsmith

[Baltimore city directory]

CHISHOLM, ARCHIBALD

Possibly in business with John Shaw through 1777. After this, payments for the stocking of gun barrels were made separately.

February 23, 1776 Council of Safety
Messrs John Shaw and Archibald Chisholm agreed with the Council of Safety to stock the Gun Barrels made by Mr. Isaac Harris for the use of this Province at the price the same are done in Philadelphia.
[AM, p. 180]

April 16, 1776 Council of Safety
Ordered That the said Treasure pay to Messrs John Shaw and Archibald Chisholm twenty two Pounds six Shillings eight Pence for stocking Musquets &c. as Pr. Account.
[AM, p. 333]

February 13, 1777 Council of Safety
That the said Treasure pay to Messrs Shaw and Chisholme Forty Eight Pounds and Two Pence.
[AM, p. 133]

May 28, 1777 Council of Safety
Ordered, That the Western Shore Treasure pay to Archd. Chisolm six Pounds, fifteen shillings for Amt of Acct.
[AM, p. 261]

May 2, 1781 Council Proceedings
Ordered that the western shore Treasure pay to Archibald Chisholm Twenty pounds, twelve shillings and six pence per Account passed by the Aud. G.
[AM, p. 422]

CHRISTOPHER, J. — Washington, D. C.

1850 Georgetown

Christopher, J.	age 33	born Germany		gunsmith
" A.	25	Germany		
" E.	5	Washington, D.C.		

[1850 Federal Census]

CLAGETT, ALEXANDER Hagerstown

First indenture in Washington County was recorded in 1778 and his last in 1817. Elected sheriff of the county in 1780. Apparently he was a wealthy man, because the 1790 census listed 22 slaves in his household.

Contractors with the government under Act of July 5, 1798 for Charlesville pattern muskets at $13.40 per stand

(4) Eli Williams, Williamsport, Maryland	2000 muskets
(5) Nicholas White, Thomas Crabb, Jacob Metzger & Christopher Barnhizle, Frederick Town, Maryland	1000 muskets
(14) Alexander Clagett, Hagerstown, Maryland	1000 muskets
(25) Mathias Shroyer, Taneytown, Maryland	1000 muskets
27 contractors	40,200 stands

[U. S. ORDNANCE, Vol. I, p. 20]

Clagett, Alexander Washington County
4 males of 16 yrs.
3 males under 16 yrs.
5 females
22 slaves

[1790 Federal Census]

Alexander Clagett death June 11, 1821 – age 77 Mr. Clagett was a zealous and earnest patriot during the Revolution, and for three years was sheriff of Washington County. He removed from Hagerstown to Baltimore in 1818.
[HISTORY OF WESTERN MARYLAND, p. 1047]

Episcopal Church St. Johns The Vestry of Washington County
Alexander Clagett
1787
1789
1791
1798
1799
1802

[HISTORY OF WESTERN MARYLAND, p. 1078]

CLAGGETT, WISEMAN

September 5, 1776 Council of Safety
Ordered That the Western Shore Treasure pay to Wiseman Claggett three pounds, ten shillings for a Gun.
[AM, p. 257]

CLARK & SNEIDER

The firm advertised from 1873 to 1884 and was composed of Duncan C. Clark; C. Edward, Charles W. and C. Louis Sneider. They produced many breech loading shotguns, utilizing the many patents of the Sneiders. The various grades of their shotguns show good workmanship and are marked "Clark & Sneider Baltimore". From 1868 to 1872 Duncan was a member of the company of Hough & Clark, shipping merchants.

Price List Sneider Gun
Patent Solid Grip, Top Lever, Self-Closing Gun. Real Twist Barrels, Front Action, Rebounding Locks, no Engraving, but Stock Checked – $65.00
Laminated Steel or Damascus Barrels, Front Action, Rebounding Locks, Engraved – 75.00
Laminated Steel and Damascus Barrels, Front Action, Rebounding Locks, Well Engraved – 100.00
Laminated Steel and Damascus Barrels, Front Action, Rebounding Locks, Finely Engraved – 125.00
Double Grip, Top Lever, Self-Closing Gun.
Damascus and Laminated Steel Barrels, Front Action, Rebounding Locks, Well Engraved – $125.00
Damascus and Laminated Steel Barrels, Front Action, Rebounding Locks, according to quality of Barrels and Engraving – 150 to 250.00
Double Express Rifle, Finely Engraved and Finished – 200 to 250.00 Guns of this class with pistol grip stock if desired.

Sneider Breech-Loading Rifle.
No. 22 Calibre for Target Shooting, 100 to 150 yards, with open and Globe Sight – $35.00. Shoots very accurately and with great force; most elegantly finished rifle in the market.

Muzzle-Loaders Changed to Breech-Loaders.
Solid Top Grip Action – $50.00
Double Grip – 75.00
New Barrels on Breech-Loaders – 50.00 to 100.00
Altering Pin Fire Guns to Central Fire – 25.00
Stocks Bent or Straightened – 5.00
Locks Altered to rebounding – 8.00
New Stocks – 10.00 to 25.00
Guns Rebored to Shoot Close – 6.00 to 12.00

Gun Cases, Implements of all kinds, in fact everything pertaining to a sportsman's outfit, at lowest prices.

Clark & Sneider, 214 West Pratt Street
[Clark & Sneider Catalog, 1870; Eleutherian Mills Historical Library, Delaware]

CLARK, DUNCAN C. Baltimore

Hough & Clark (Duncan C. Clark)
1868-72 shipping mercht
Clark & Sneider (Duncan C. Clark, C. Edward Sneider)
1873-84 214 West Pratt Street gun manufrs
[Baltimore city directories]

CLENDENIN, DANIEL Cecil County

Agreement entered into the 3rd day of April 1826, between Capt. N. Baden in behalf of the U. States, of the one part and Daniel Clendenin of Cecil County State of Maryland of the other part, Witnesseth that the said Daniel Clendenin hath this day agreed to furnish the U. States with ten thousand rough musket stocks to be delivered at the Port of Baltimore in the year 1827.
[National Archives, Contracts for Ordnance & Ordnance Supplies, Vol. I, 1813-1828, p. 245.]

CLEAVELAND, GEORGE — Baltimore

In 1835 a George W. Cleaveland was proprietor of the American Hotel and by the late 1840's was a coachmaker on Scott Street. In 1845 when he advertised as Cleaveland and Charrier, his colleague was probably Peter Charrier.

Cleaveland, George
1840 Aisquith Street gunsmith
Cleaveland & Charrier
1845 15 Harrison Street gunsmiths
[Baltimore city directories]

CLENNEY, JOHN — Baltimore

McComas, Alexander	1850	age 29	gunsmith	birth Md.
Mary	1850	24		Md.
Clenney, John	1850	20	gunsmith	Md.
Johnson, Oscar	1850	14	gunsmith	Md.

[1850 Federal Census]

COMPTON, ALEXANDER — Hagerstown

This indenture made this 24th day of March 1794 between John Gonter of Elizabeth Town and county aforesaid Gunsmith of the one part and John Compton placed his son Alexander Compton as an apprentice to said John Gonter to learn the art and mystery of the Gunsmith business.
[Washington Co. Apprentice Indentures, p. 5, Hall of Records]

CONKLIN, ISAIAH B. — Baltimore

Conklin's invention consisted of various improved and new features on a toothed lever-action rifle. Only one of his rifles is known which incorporates his patent and bears his name.

February 16, 1869
Be it known that I, Isaiah B. Conklin, of the city and county of Baltimore, and state of Maryland, have invented a new and improved Breech Loading Fire-Arm.
[U. S. Patent Office, Patent # 86,971]

COOK, CHARLES — Baltimore

Cook, Charles
1847-54 83 Pearl gunsmith
[Baltimore city directories]

COOKMAN, WILLIAM — Hagerstown

GUNSMITHING The subscribers respectfully inform their friends and the public, that they have commenced business in the old stand, lately occupied by Mr. William Hawken, in Jonathan Street, between Washington and Franklin, – where they intend to manufacture *Guns, Rifles* and *Pistols* of every description.

Guns stocked, altered to percussion, & repaired in a style not to be surpassed.

Keys fitted, Locks repaired, and all kinds of light jobbing, done in the best manner, and at the shortest notice.

The subscribers intend to have constantly on hand a large assortment of *Rifles* and *Shot Guns.*
Hagerstown, April 9 [1841] William Cookman
Joseph W. Keller
[*The Hagerstown Mail*, Friday, 28 May, 1841]

CRABB, THOMAS — Frederick

The records below suggest that Thomas Crabb was more a blacksmith than a gunsmith. At least one rifle has survived bearing the script signature of T. Crabb. The 1800 census suggests that Crabb was born between 1755 and 1774. In 1808 he sold his house in Frederick and his name disappears from the Frederick County records.

Crabb, Thomas Frederick County
1 male more than 16 years
1 male under 16 years
[1790 Federal Census]

Crabb, Thomas 1795 juror 5 days paid 1 pound 5 shillings.
[Frederick County Levy Book, Hall of Records #19257]

Crabb, Thomas 1797 juror paid 5 shillings
[Frederick County Levy Book, Hall of Records #19257]

Indenture made December 21, 1797 [deed] between Mountjoy Bayly and Eleanor his wife of Alleghany County and state of Maryland of the one part and [to] Thomas Crabb of Frederick County and state of Maryland of the other part . . . All that house and lott mentioned in the said deed of Mortgage lying in Fredericktown.
[Frederick CCH, Liber WR 16, Folio 96]

Crabb, Thomas 1798 1/8 lott from Mountjoy Bayley value $100 Frederick Town
[Assessment Book, Frederick County 1798, Hall of Records #19258]

Crabb, Thomas 1798 juror paid 5 shillings
[Frederick County Levy Book, Hall of Records #19257]

Contractors with the government under Act of July 5, 1798 for Charlesville pattern muskets at $13.40 per stand

(4)	Eli Williams, Williamsport, Maryland	2000 muskets
(5)	Nicholas White, Thomas Crabb, Jacob Metzger & Christopher Barnhizle, Frederick Town, Maryland	1000 muskets
(14)	Alexander Clagett, Hagerstown, Maryland	1000 muskets
(25)	Mathias Shroyer, Taneytown, Maryland	1000 muskets
	27 contractors	40,200 stands

[U. S. ORDNANCE, Vol. I, p. 20]

Crabb, Thomas Frederick Town District No. 2
2 males between 26 and 45
2 males under 10 years
1 female between 16 and 26
2 females under 10
[1800 Federal Census]

"We the Subscribers Justices for Frederick County do hereby bind out as an Apprentice Rezin Self and Orphan aged seventeen years on the 27th day of September last unto Thomas Crabb Blacksmith until the said Apprentice shall arrive to the full age of twenty one years . . . said Thomas Crabb to Teach . . . the Black Smith Trade in its different branches to furnish and provide for said apprentice sufficient mead Drink washing lodging cloathing and other necessaries fitting for an apprentice to send him to school to learn to read write and cast up accounts and when free to give him a suit of cloathes worth Twenty Dollars or that sum in money at the option of said apprentice. Jany 31 1803
[Frederick Co. Apprentice Indentures, p. 377, Hall of Records]

Crabb, Thomas 1804 juror 10 days paid 5/12/6.
[Frederick County Levy Book, Hall of Records #19257]

Indenture made November 17, 1806 [deed] between Thomas Crabb of Fredericktown Frederick County and state of Maryland blacksmith of the one part and [to] Francis Mantz of Fredericktown Frederick County and state of Maryland shopkeeper of the other part . . . Stone house and lot there unto belonging whereupon the said Thomas Crabb now lives in Market Street Fredericktown.
[Frederick CCH, Liber WR 30, Folio 29]

Indenture made July 6, 1807 [bill of sale] I, Thomas Crabb of the town and county of Frederick and state of Maryland for and in consideration of money paid to me in hand by John Adlum one black and white cow, horse, three anvils, five vices, one pair bellows, etc.
[Frederick CCH, Liber WR 31, Folio 281]

Indenture made March 9, 1808 [release] between Thomas Crabb of Frederick County and state of Maryland blacksmith of the one part and Francis Mantz of the same county and state of Maryland shopkeeper.
[Frederick CCH, Liber WR 32, Folio 435]

Indenture made March 9, 1808 [deed] between Thomas Crabb of Fredericktown Frederick County and state of Maryland of the one part and [to] George Bentz of the same county and state aforesaid . . . all that stone house and lot in which the said Thomas Crabb now lives situate lying and being in Fredericktown.
[Frederick CCH, Liber WR 32, Folio 436]

John Mason to Dr. John Tyler, Frederick Town, Maryland
Office of Indian Trade, Georgetown, June 9, 1808
Some time last spring a Mr Thomas Crabb of your town, Gun Smith, brought me letters from two or three gentlemen which spoke favorably of him and induced me to give him a small Order to make 20 Rifles for the publick, with intention to give him much larger orders from time to time, since which I have not heard from him. I will be very much obliged to you Sir to make the requisite enquiries and to inform me of what character he is, whether in the first place a good workman, if of such property as to make an order for 100 or 200 arms at a time, safe with him, and if he is expected to be an honest and punctual man in his dealings.

John Mason to Dr. John Tyler, Fredericks Town
Office of Indian Trade, Georgetown, August 8, 1808
I was forwarded in due course with your much esteemed letter of the 13th of June and was much obliged by your enquiries of & information respecting Crabb. he was to have delivered me 20 Rifles at a stipulated price on or before the first of July, I made no other advance to him than by letting him have Twenty Gun Locks for the stores of this Department for which he was to account in the price of the Guns.

As I am now satisfied there is no chance of getting the guns from him I must endeavor to recover the Locks and with this view, I take the liberty to enclose you an order for them. I beg leave Sir to ask your good offices to get back these locks & when received be pleased hold them to my order. they are small german Locks not tightly finished with Rollers at the pan. be pleased excuse the liberty I am taking.

John Mason to Mr. Ths. Crabb, Frederick Town
Office of Indian Trade, Georgetown, August 8, 1808
Your failure to deliver the twenty Rifles you contracted to make for this Office in March last by the time appointed day the first of last month, has determined me after waiting one month to call on you for the return of the 20 Rifle Locks the property of the United States, you received of me. these you will please deliver to Dr. John Tyler of Fredericktown & his receipt shall be good.

John Mason to Dr. John Tyler, Frederick Town
Office of Indian Trade, May, 1809
Be pleased inform me wether you have heard anything of Crabb and the [?] and the gun locks lately or the rifles from him, all hope is lost. I should be glad to get back the gun Locks mentioned in mine to you of August last.

CRANDALL, JOHN

September 4, 1776 Council of Safety
Ordered That the said Treasure pay to John Crandall three pounds, ten shillings for a gun.
[AM, p. 256]

CREAMER, PHILIP — Taneytown

Philip Creamer's name is found with various spellings, namely Creamer, Cramer, and Crawmer. Surviving specimens of his excellent gunsmithing work are signed "P. Creamer," and this can be taken as the preferred spelling. Creamer worked in Taneytown and later apparently moved west. He had the ability to make very fine rifles and pistols. One of Creamer's percussion pistols was owned by Andrew Jackson, and it is preserved today at Jackson's restored home, the Hermitage, in Tennessee; it bears the signature "P. Creamer" on the barrel and holds a silver plate engraved with Jackson's name. Creamer was an important gunsmith and hopefully the future years will bring to light more specimens of his work and more information about his life. Research on the gunsmith Creamer is complicated by the fact that there apparently was a yeoman and distiller named Philip Cramer, with a wife named Christena, living in Emmitsburg during the 1790's. Property transactions for both men, if there were two of them, are given below.

By this deed, dated 30 May 1792, Philip Crawmer of Frederick County bought, for 35 pounds, from Peter Crise a lot in Emmitsburg with a square log house upon it, size 60 feet by 165 feet, it being a "front lot on the south side of the great Road leading to Baltimore."
[Frederick CCH, Liber W.R. 10, p. 700]

By this deed, dated 3 August 1793, Philip Crawmer of Frederick County bought from William Emmit, for 3 pounds, a lot in Emmitsburg, No. 122, size 60 feet by 165 feet.
[Frederick CCH, Liber W.R. 11, p. 670]

By this deed, dated 10 June 1794, Philip Crawmer of Frederick County bought from William Emmit, for 22 pounds, a parcel of land in Carrollsburg [in Emmitsburg] known by the name of Popular Ridge, No. 2, containing 10¾ acres.
[Frederick CCH, Liber W.R. 12, p. 462]

By this deed, dated 13 February 1795, Philip Crawmer of Frederick County sold to Patrick Bradley, for 11 pounds 5 shillings, lot No. 124 in Emmitsburg. Crawmer's wife, Christena, acknowledged the sale.
[Frederick CCH, Liber W.R. 13, p. 78]

By this deed, dated 12 September 1795, Philip Crawmer of Frederick County sold to William Gunter, for 175 pounds, lot No. 35 in Emmitsburg. Crawmer's wife, Christena, acknowledged the sale.
[Frederick CCH, Liber W. R. 13, p. 571]

By this deed, dated 24 May 1796, Philip Crawmer of Frederick County bought from Andrew Harner, for 310 pounds, part of a tract of land called Resurvey on Smith Lott, containing 103¾ acres.
[Frederick CCH, Liber W.R. 14, p. 471]

By this deed, dated 3 April 1797, Philip Crawmer of Frederick County bought from Andrew Harner, for 330 pounds, a parcel of land in Frederick County, part of a tract called Resurvey on Smith Lott, adjacent to land owned by Philip Crawmer, containing 110¼ acres.
[Frederick CCH, Liber W.R. 15, p. 367]

By this deed, dated 2 February 1797, Philip Crawmer, Yeoman, of Frederick County sold to Andrew Harner, Yeoman, for 35 pounds, a parcel of land in Frederick County, a part of Carrolsburg, Lot No. 2, containing 10¾ acres. The sale was acknowledged by Crawmer's whife, Christeana.
[Frederick CCH, Liber W.R. 15, p. 431]

Daniel Borden an Orphan aged Seventeen years the fifteenth day of January next is this day with the approbation of his Mother Rachel Null placed and bound apprentice to Philip Crawmer of Taney Town Gun Smith until he arrives to the age of Twenty One years during which Term the said Master shall find and provide for his said apprentice meat drink washing lodging, Apparel and all other necessaries to cause him to be Taught to read write and cast accounts to cause him to be Taught the said Trade or science of Gunsmith which he the said Master now useth as also to make Gunlocks and when free to give him Forty Dollars in lieu of all other freedom dues given under the hands of the Subscribers two Justices of the Peace of Frederick County the 28th day of October 1799.
[Frederick County Apprentice Indentures, Hall of Records, p. 236]

By this deed, dated 4 November 1799, Philip Crawmer, Distiller, of Frederick County sold to Henry Boyer, Taylor, for 80 pounds, a tract of land, part of the Resurvey on Smiths Lott, containing 20 acres. The sale was acknowledged by Crawmer's wife, Christena.
[Frederick CCH, Liber W.R. 19, p. 170]

Taney Town, Frederick County
In 1800 the household of Philip Creamer in Taney Town contained four males between 16 and 26 years of age, one female between 16 and 26, and one female under 10. This appears to be the household of the gunsmith.
[1800 Federal Census]

By this deed, dated 23 May 1800, Philip Cramer of Frederick County bought from Abraham White of Frederick County, for 335 pounds "... that Lot or portion of ground with the buildings, and improvements thereto belonging lying and being in Taney Town ... and distinguished on the plan of the said town by its number to wit Number seventeen ... " It is to be noted that in this document the buyer's name is spelled, consecutively, as follows: Cramer, Cramer, Cramer, Cramer, Cremer, Cramer, Cremer.
[Frederick CCH, Liber W.R. 21, p. 336]

This Indenture made the 18th day of February 1801 between Peter Shunk and Philip Creamer both of Frederick County and State of Maryland Witness that the said Peter Shunk hath this day bound his son Isaac Shunk unto Philip Cramer Gun Smith for the Term of three Years and nine months from this date the Said Cramer engages to learn the said apprentice the Art and mystery of the Gun smith and Gun Lock making Business, to give said Apprentice three months Schooling find him sufficient meat drink Apparel mending washing and lodging and to give him two weeks in every Harvest and when free to give him as good clothes as he now has and Twenty Dollars. In Witness whereof the parties hath hereunto set their hands and seals the day and year written.
Peter Shunk
Philip Creamer

[Frederick County Apprentice Indentures, Hall of Records, p. 236]

The following entries are from the "Day Book [of] John M. Kaleb & Co. Taney Town May 9th 1801," which is owned by Edward R. and Helen Flanagan, 25 Flanagan Rd., Thurmont, Maryland. The number preceeding the name is the account number assigned by the keeper of the day book; Creamer's name is spelled correctly in the first entry and thereafter it is spelled "Cramer," but the account number indicates that it is the same person. In June 1803 the account number was changed from 16 to 99. The last mention of Creamer in the book is dated 22 May 1805, but the book continues on through 14 April 1809, so it appears that Philip Creamer left Taneytown shortly after 22 May 1805.

L s d

1801 May 16 16 Philip Creamer
pr wife 1 lb sugar & chocolate 3 2
1 oz arafactado 1/3 1 gn whiskey 1/6
1 ditto 1/6
1 ditto 1/6
1 ditto 1/6 7 3

1801 June 15 16 Philap Cramer
4 fiels at 2/2 1 quart whiskey 1/6
for wife 2 oz snuff 10 2

1801 July 6 16 Philip Cramer
by Cash 45 Dolars
Dr to 1 g. whiskey by his Brother Danil 5 6

1801 August 1 16 Philip Cramer
to Cash Lint 16 Dollars 5 long & 1 smal fils 12/6 6 12 6

1801 Septmber 1801 16 Philip Cramer
5 fiels at 2/4 11 0
4 lb soap & 2 fizs blue 4/
1 g. whiskey 1/6
4 fiels at 2/6 10/ 15 6 1 7 2

1801 September 27 16 Philip Cramer
1 skein silk 6d oil 10d 1 4
1 lite glass 2/2
ditto 6/6 2 0 4 0

1801 October 24 16 Philip Cramer
by Cash 2 3 1

1801 November 2 16 Philip Cramer
pr Wife 1/4 1 4

1801 November 14 16 Philip Cramer
1 lb coffe 1 lb Candles 4/22 4 2
3 fiels 1/6 2 lb Candles 3/9 5 3 9 5

1801 December 14 16 Philip Cramer
4 fiels 3/4 pins pr wife 6d 3 fiels 7/ 10 10
1 1 Godfreys Cordial 1/6 2 lb coffe 4/4
2 lbs sugar 2/2 1/4 tea 1/ pins 4d 9 4

1802 February 3 16 Philip Cramer
2 g. rum 4/ 2 lb lose sugar 5/4 9 4
1 guir paper to D. Border 1 10

1802 April 22 16 Philip Cramer
pr I. Shunk 2 lb sugar 1/4 tea 1/1 thread 4d 3 3

1802 April 30 16 Philip Cramer
6 fiels 5/4 1 comb 2/4
2 pr mits 0/1 strip linen for wife 14/9 1 10 5
Dr by silk mits returned 5/7
Dr 8 lb Nails 10/ 1 lb leasins 1/ 2 yd linen
7/6 1 8 6

1802 May 8 16 Philip Cramer
8 yd linen 30/ 6 lb nails 7/6 1 17 6

1802 May 15 16 Philip Cramer
2 Mugs 2/9 Colander 5/ crame jug 1/8 9 5
waterboat 1/10 book 5/ 2 yd Calico 7/6 14 10
silk 6d pantaloons & trimings 23/5 1 3 11 2 - 2

1802 May 21 16 Philip Cramer
pr Wife 1 fan 3/9 linen 3/9 thread 7d 8 1
2 7/8 yd Oil Cloth 10/9 2 lbs sugar 2/
1 lb coffe 2/ 14 9 1 2 10

1802 June 4 16 Philip Cramer
sope & blue 3/10 sugar 2/ tea 2/0 linen 1/10
sundries 2/6 cash 15/ 2 lb sugar 1/1 Rosem 6 1 10 11

1802 June 8 16 Philip Cramer
Cash lent 15 D 3 yds linsey 11/3 6 3 9
2 lbs sugar 2/2 1 coffe 2/2 Rum 1/
Wine 1/3 Whisk 8d 7 3 6 11

1802 August 5 16 Philip Cramer
Godfrey Cordel 1/6 sundries 3d Rum 1/10
B. Gloves 3/9 3 g Rum 5/7 2 g Wine 5/
2 lb candles 3/9 Hooks & Leins 3/4 1 5 0

1802 September 6 16 Philip Cramer
1 pr cards 4/6 linen & tape 4/6 9
coffe bag 1/10 Candles 1/10 Niting pins 4d 4 1

1802 October 30 16 Philip Cramer
files 18/4 cups and sausers 1/10 a crean gudge lid 18 4
2 9

1802 November 12 16 Philip Cramer
pr Wife sope & snuf 2/3 2 3
whiskey 1/6 pots 1/3 ½ B salt 3/9 a g whiskey
1/4 7 10

1802 December 9 16 Philip Cramer
pr Wife balance of flanel 5/9 1 pint sand 1/4 1 13 1

1802 December 25 16 Philip Cramer
2 Handcurchefs 14/ 1½ yd Calico 5/7 1 yd v
Riband 1/6 4½ yds Bumbasat 16/10 2 yd
Calico 7/6 1 Hancurchef 7/ 1 s. silk
62 lb [?] 1/10 a Hankerchef & ribon 8/2 3 12 12

1803 January 1 16 Philip Cramer
2 fiels 1/6 1 Do 10d 2 4

1803 January 9 16 Philip Cramer
pr Wife 4½ yds flanel 15/9 1 yd Leas 2/9 10 11

1803 March 15 16 Philip Cramer
pr wife Ribins 3/ 3

1803 April 7 16 Philip Cramer
pr wife 1 pt wine 1/3 2 qt Whiskey 2/4
2 files 2/6 salt 3/9 10

1803 April 20 16 Philip Cramer
2 files 2/9 2 lb coffee 4/4 7 1
2 lb sugar 2/ receivd in part Cash 6/4 4
2 lb sugar

1803 May 19 16 Philip Cramer
1 lb Candles 1/6 1 6

1803 June 7 16 Philip Cramer
pr Wife 1 teapot 3/9 3 pots 1 salad 3/2 6 11
tureen & spoon 12/6 salad 1/8 14 2

1803 June 22 99 Philip Cramer
1 oz fig blue 1 bottle mustard 1/3 1 8
files 2/3 2 3

1803 August 1 99 Philip Cramer
files 12/4 1 oz pink root 1/ 13 4

1803 August 5 99 Philip Cramer
bridle bits 41/3 wire pr wife 2d 2 1 5

1803 September 1 99 Philip Cramer
2 lb sugar 2/2 2 2

1803 September 20 99 Philip Cramer
fiels 8/8 a g whiskey 5/6 14 2

1803 November 18 99 Philip Cramer
½ bushel salt 5/ 5

1803 November 29 99 Philip Cramer
1 corel 1/1 alspis 4d peper 6d ginger 5
sinamon 1/1 Nutmeg 11d Clove 11d peper 10d
Go Cordal 1/6 7 5

Date	Account	Items	£	s	d
1803 December 21	99 Philip Cramer				
		pr sister in law 2 lb of Meselin 5/10			
		1½ blue sarg 4/6 thred 3d		10	7
1804 January 13	99 Philip Cramer				
		Maselin 5/7 1 lb coffe 2/6 thred 10d salt 8d			
		¼ ginger 9d neadles 4d sugar 2/4 tea 2/4		15	
1804 January 20	99 Philip Cramer				
		2 lb sugar 4/2 ¼ lb tea 2/ salt 8d buttons & thred 4 d		5	4
1804 January 31	99 Philip Cramer				
		2 lb sugar 462 Rum 10d			
1804 February 4	99 Philip Cramer				
		2 lb sugr 2/2 ¼ tea 2/ 3 lb cotton 6/6		11	7
1804 February 13	99 Philip Cramer				
		2 yd moselin 4/8 combs 2/4 thread 3d Coffe 2/8			
		4 lb sugar 4/8 spelling book 1/1 tea 2/		10	8
1804 March 24	99 Philip Cramer				
		pr wife teapot 3/9		3	9
1804 April 13	99 Philip Cramer				
		¼ lb tea 2/ 2 lb sugar 2/4 1 Basket 3/		7	4
1804 May 1	99 Philip Cramer				
		1 qt Mulases 1/10 ½ Whiskey 1/4			
		½ lb Allum 6d 1 B. salt 10/ ½B. Do 5/	5	0	0
1804 June 13	99 Philip Cramer				
		1 qt Mullases 1/10		1	10
1804 July 30	99 Philip Cramer				
		1 Bottle B. oil 1/10/2			
1804 August 28	99 Philip Cramer				
		pr wife stuf 13/6 Calico 4/8 trimmings 2/20	1	1	0
		1 lb tobacco to self 1/6		1	6
1804 September 18	99 Philip Cramer				
		1 qt whiskey 1/4		1	4
	99 Philip Cramer				
		¼ lb Alspice & salt 2/ Buckit 5/6		8	2
1804 October 16	99 Philip Cramer				
		3 oz turkey red 4/6 2½ lb Copress 1/8		6	2
1804 October 16	99 Philip Cramer				
		pr Brother 1 oz Bark 4/			
		1 qt Port wine 5/ cloves & Crem tarter 1/		10	
1804 November 1	99 Philip Cramer				
		files 6/6 ½ lb chocolate 1/2		7	8
1804 December 24	99 Philip Cramer				
		3 lb sope 2/9		2	9
1805 January 1	99 Philip Cramer				
		Blue 5d thread 3d neadles 4d pins 1/6		2	6
1805 February 2	99 Philip Cramer				
		2 lb candles 3/9 Do 1/10		5	7
1805 March 2	99 Philip Cramer				
		for Wife Riband 1/8 1 pt Whiskey 9d thred & tape 1/4		3	10
	99 Philip Cramer				
		tea 2/6 sugar 2/4 candles 1/10½ neadles 5d silk 5d		7	7
1805 March 13 void	Philip Cramer				
		to Cash lent Eight Dollars	3	0	0
1805 May 22 void	Philip Cramer				
		tobac 1/10 8 files 15/			
	99 Philip Cramer				
		coffe & screws 4/6		4	6
1805 May 7	99 Philip Cramer				
		wine 1/3 Do 1/4		2	7

The two deeds which follow appear to be of special importance, for it is probable that they refer to the gunsmith, though specific mention of occupation is not given. Assuming that the Philip Creamer mentioned is the gunsmith, these indentures provide the name of his wife, of his father and mother, and mention that Philip had brothers and/or sisters. The date of these indentures is 15 June 1805, which agrees nicely with the time when his entries in the daybook of John M. Kaleb & Co. terminate. Presumably he left town after he completed the transactions concerning the disposal of his father's property. It is to be noted that the family name is spelled correctly in these two documents. Also, note is to be taken that Philip's mother's name is given both as Margaret Crouse and Sarah Crouse; probably Sarah was the correct name, and Margaret got inserted as a clerical error. The second instrument contains an obvious clerical error where the deceased is referred to as "Casper Crouse, Senr."

This deed was made 15 June 1805 between Philip Creamer Eldest Son of Casper Creamer of Frederick Co. deceased, and Michael Funderlinn of Frederick Co. Casper Creamer died intestate owning three tracts of land in Frederick Co., namely Joseph's Chance containing 40 acres, part of the addition to Brooks Discovery on the Rich Lands containing 20 1/8 acres, and the other part of Brooks Discovery containing 1½ acres, . . . whereupon application was made by petition to Frederick County Court for a commission appointing proper persons as commissioners to divide the Real Estate of the said Casper Creamer deceased if the same would admit of Division without injury and loss to all parties interested among the Children of the said Casper Creamer deceased and to lay off the Dower of the widow of the said Casper Creamer . . . and whereas the said Commissioners did certify . . to the court aforesaid that the said Estate would not admit of Division among the Children . . . without injury and loss to all the parties interested and did make their valuation of the same in current money . . at which valuation the said Philip Creamer did Elect to take the Estate aforesaid and to pay to the rest of the Children of the said Casper Creamer deceased their just proportions of the said valuation, which was accordingly adjudged to him as by the record of the proceedings aforesaid . . . and whereas also the commissioners aforesaid did also lay off and ascertain the Dower of Margaret Crouse who was the widow of said Casper Creamer deceased in the lands aforesaid which said Estate in Dower has since been sold and conveyed by Valentine Crouse and the said Margaret Crouse as by Deed of Bargain and Sale thereof bearing even date herewith will more fully appear. Whereby the whole Legal Estate . . . in the several parcels of Land aforesaid is vested in the said Philip Creamer. Now this Indenture Witnesseth that the said Philip Creamer for . . the sum of 350 pounds . . . hath . . . sold . . . unto . . . Michael Funderlinn . . . all those aforesaid three parcels of land . . . At the same time came Margaret Creamer wife of the said Philip Creamer and . . . did acknowledge the foregoing instrument of writing to be her act and Deed . . . [Deed Book W.R. 27, p. 183-186, Frederick CCH]

By this instrument dated 15 June 1805 Valentine Crouse and Sarah Crouse who was the widow of Casper Creamer Sen[r] deceased both of Frederick Co., for the sum of 70 pounds, release to Philip Creamer all claim of Dower in the Estate of Casper Creamer, deceased. . . To have and to hold all the Estate right interest, title and claim of Dower of us the said Valetine and Sarah Crouse and each of us, in and to any of the lands tenaments and heredements of the said Casper Crouse Sen[r] Deceased [sic] . . .
[Deed Book W.R. 27, p. 187, 188, Frederick CCH]

A gunsmith named Philip Creamer, and doubtless the man from Maryland, was at work in Illinois in 1820. For raw materials he used iron and steel. The cost of goods consumed during the year was $200. He employed two men. Special equipment used consisted of "two pairs of bellows & other tools." His production consisted of "guns & pistols" with a market value from $34 to $120. Sales were reported as "good."
[1820 Federal Census, St. Clair County, Illinois]

On 3 November, 1866 a speech made at a meeting held to form the Missouri Historical Society contained the following quotation: "Being torn down is the old Barber's shop on Olive near Fourth [in St. Louis], long known as the gun shop of old Mr. Creamer, a great favorite of the boys of that date."

This quotation was found by Dewitt Pourie of St. Louis, who searched the St. Louis directories for other references to Creamer or to a gun shop at Olive near Fourth and found no references except for an entry of 1859 listing Philip Creamer in a rooming house.

CRISWELL, JOSEPH — Keedysville

Joseph Criswell was a broom maker and gunsmith, carrying on his shop at Keedysville. He also drove a stage for many years between Boonsboro and Shepherdstown, Va. . . . Mr. Criswell died April 25, 1891; his wife was Elizabeth Ellen Kiger Criswell. They were buried in the Keedysville Cemetery . . . they were married in Boonsboro, April 8, 1834, and had twelve children.
[HISTORY OF WASHINGTON COUNTY, Vol. II, pp. 756-757]

CROMWELL, LEVI, OLIVER, and T. ELVIN — Baltimore

Oliver and T. Elvin were sons of Levi.
During the 1861 disturbances in Baltimore, the city ordered from Levi 119 pounds of musket balls, two muskets and four ship pistols. A large, heavy six gauge, water fowl single-barreled shotgun with a back lock is marked on the lock and barrel "L. Cromwell N. P. Baltimore."

Oliver was in the foundry and coach making business in 1850. He petitioned the city of Baltimore on July 17, 1861 for the seizure and removal of $37.50 worth of pistols from his shop by the military under orders from city authorities on April 21 and 22, 1861.

T. Elvin operated the family business from 1872 to 1880.

Cromwell, Levi

1822-24	37 George Street, Fell's Point	gunsmith
1827-40	35 Thames Street	gunsmith
1845-51	134 Thames Street	gunsmith
1852-57	134 Thames Street	gunsmith
1858	263 Anne Street	gunsmith
1859	118 Thames Street	gunsmith
1860	265 Anne Street	gunsmith

Cromwell, Levi Son & Co.

1865-70	271 South Anne Street	gunsmith
1871	271 South Anne Street & 118 Thames Street	gunsmith

Cromwell, Levi Son & Co. (T. E. Cromwell & John E. Bishop)

1872-80	118 Thames Street	mfg. of guns, deck plugs & wedges, and dealers in hardware

[Baltimore city directories]

Cromwell, Oliver 1850 Foundry & Coach maker
[1850 Federal Census]

(ad, 1856) Gunsmith in General, stocking and repairing done at short notice & on reasonable terms.
[Baltimore city directories]

CRUZEN, JACOB — Washington, D. C.

Cruzen, Jacob
1822 7 e btw L & M s, N. Yd. gunsmith
[Directory, Washington, D.C.]

Cruzen, Jacob
1830 7 e btw L & M s, N. Yd. gunsmith
[Directory, Washington, D.C.]

CURLEY, JAMES M. — Baltimore

The joint Standing Committee on claims to whom were referred the following claims against the city – towit 17. The petition of Jas M Curley asking to be indemnified for the loss of fire arms taken by a mob on the 21st of April 1861 and claiming $160.00.
[Baltimore City Correspondence Manuscripts, 1862, #504]

D

DALHOUSE, RICHARD

September 27, 1776 Council of Safety
Ordered that Western-Shore Treasurer pay to Doctor John Archer for use of Richard Dalhouse 100 pounds on account of his Gun manufactory.
[American Archives, 5th Series, Vol. II, p. 637]

DALLAM, RICHARD Harford County

Was a delegate from St. Mary's County at Annapolis on 26 July 1775; during the following year and thereafter his address was Harford County. James May was associated with his factory. Dallam reached the rank of colonel by the close of the Revolution.

July 4, 1776
Ordered That Western Shore Treasure pay to Richard Dallam one hundred and fifty Pounds for Musquets Pr. Agreement.
[AM, p. 549]

July 16, 1776 Dallam to Jenifer
In answer to yours of the 10th Inst. which I received yesterday I inform you I have twenty two muskets finished compleat and fifteen more ready for Stocking six of which will be finished this week.
Harvest & sickness of two of my best hands and the bursting of twelve or thirteen of my barrels in my absence has disappointed my expectation. Twelve of the Guns finished have been proved with two ounces of powder and one ball, the remainder with one ounce of powder and ball, but if this proof will not satisfy I will apply to the genn you mention to prove them, tho' it will be attended with expence & trouble in taking them out of the stocks & fitting them in again.
What other barrels I make I shall apply to Mr. James Harris to see proved before stock'd as he lives handy to us & take his certificate. I will deliver at my factory of Harford town to your order what arms I have or can have made in time, according to agreement, but will not deliver them at Baltimore town at my own risque & expense. As Colo. Hall is in this neighbourhood, will not a delivery of the arms to him answer. I believe I could get slings for the muskets & scabbards for the bayonets made here, if you'dlet me know the price for them.
P. S. I have not the least doubt but my Arms will please and be found as good as any made in Maryland.
[AM, p. 59] [American Archives, 5th Series, Vol. I, p. 363]

July 22, 1776 Dallam to Jenifer
Since my last I had what guns I had proved by gentlemen appointed by the Committee of Observation for which purpose they have past 23 guns ready fitted. I have five more ready and shall have as many more next week as will pay the sum I received and unless you can raise the price to me I must give over. By the parcel I have now made, I shall be near thirty pounds looser. Our Committee have given 4/10/0 for guns worse than mine. If you want quantity of Bayonets made, I will undertake to make a parcel at 10/ .
[AM, p. 93] [American Archives, 5th Series, Vol. I, p. 491]

July 27, 1776 Council to Dillam Esq.
We received yours of the 22 of July and are glade to hear you have made as many Guns as will reimburse the Money already advanced, they may be delivered to Col. Carvill Hall or his order. The price you ask is high for Guns, but we want them at present, and therefore will take any number you can make in six weeks from this time, and give you four pounds ten shillings currency for them compleatly finished.
[AM, p. 128]

September 6, 1776 Dallam to Jenifer, state of New Jersey, Brunswick
James May who is concerned with me in the Gun Factory will in my absence want a sum of money to go on therewith, and as he is unknown to your Board, I have thought it necessary to advise you thereof, and to request your Board to furnish him with what sum Mr. James Harris may draw on you for, on my account, to be paid in guns, part of which I expect is already finished . . .
[AM, p. 260]

March 13, 1777 Council of Safety
That the said Treasure pay to John May for the use of Richard Dallam forty seven pounds five Shillings – Cy.
[AM, p. 171]

DANN, UNDERHILL

This Agreement, made the 20th day of June 1828 between George Bowford Lt. Col. of Artillery, on Ordnance duty in the service of the U. States of the one part, and Underhill Dann of Port Deposit in the State of Maryland of the other part, Witnesseth . . . deliver at Fort Monroe on or before 31st May 1829 suitable timber for one hundred casemate carriages.
[National Archives, Contracts for Ordnance & Ordnance Supplies, Vol. I. 1813-1828, p. 293]

This Agreement, made the 21st day of May 1829, between Wm. H. Bell, Lt. of Artillery for and in behalf of the U. States on the one part, and Underhill Dann and Savin of Port Deposit, Maryland, of the other part, Witnesseth . . . such kinds, qualities, quantities and description of timber and at such time and prices as are expressed for casemate carriages.
[National Archives, Contracts for Ordnance & Ordnance Supplies, Vol. II, 1829-1842, p. 23.]

DAW, RUBEN Washington, D. C.

Daw, Ruben
1834 Bridge St., South side Gun and Lock Smith
[Directory, Washington, D.C.]

1850 Georgetown

		age	born	
Daw,	R.	40	England	gunsmith
"	E. (f)	40	Ma.	
"	M. (f)	14	Washington, D.C.	
"	S. V. (f)	12	Washington, D.C.	
"	A. (f)	7	Washington, D.C.	
"	C. (m)	3	Washington, D.C.	
"	J. (m)	1	Washington, D.C.	

[1850 Federal Census]

Daw, Reuben

1855	83 Montgomery	gunsmith

[Directory, Georgetown]

15 June 1860 Georgetown

	age	born		real estate
Daw, Ruben	54	England	locksmith	$21,000
" E. J. (f)	33	Maryland		
" Annie	15	Washington, D.C.		
" Charles	13	Washington, D.C.		
" Edward	11	Washington, D.C.		

[1860 Federal Census]

Daw, Ruben

1862	102 Bridge Georgetown	gunsmith

[Business Directory Washington, D.C., & Georgetown]

Daw, Ruben

1863	102 Bridge Georgetown	gunsmith

[Hutchinson's Business Directory, Washington, D.C. & Georgetown]

Daw, Ruben

1865	102 Bridge Georgetown	gunsmith

[Boyd's Business Directory, Washington, D.C., Georgetown & Alexandria]

Daw, Ruben

1866	102 Bridge Georgetown	gunsmith

[Boyd's Business Directory, Washington, D.C., Georgetown & Alexandria]

DEMUTH, JOHN — Frederick County

For most of his life John Demuth lived and worked in Lancaster, Pennsylvania. In his book THE PENNSYLVANIA-KENTUCKY RIFLE, Henry J. Kauffman (1960) reported on Demuth, noting that his family owned and operated a tobacco shop on East King Street in Lancaster from the late 18th century until the present time. He found no documentation that John Demuth worked as a gunsmith, though he did note that Demuth is listed as having contributed 54 rifles to General Hand's contract for "Continental Rifles" in 1794. Kauffman illustrated a flintlock longrifle signed in script "J. Demuth" and bearing an engraved four-piece brass patchbox of Lancaster design. Other flintlock rifles with his script signature are known, some with locks stamped with his full name. These rifles do not have decorative carving, and ornamentation is limited to engraved patchboxes. A swivel-breech over-and-under rifle by him, with well-engraved patchbox is known. The apprentice indenture presented below proves beyond any doubt that John Demuth was a gunsmith and that he worked at the trade in Frederick County at least for a short period.

This Indenture made the 16th day of February 1796 Between John Demuth of Frederick County in the State of Maryland Gunsmith of the one part and Henry Koch Son of George Koch of the County and State aforesaid of the other part Witnesseth that the said George Koch aged 16 years 7 months and 14 days as an apprentice to the said Demuth for the Term of four years four months and sixteen days from the above date to be by him Instructed or cause to be Instructed in the Business Trade and mystery of a Gun Smith and to find said Apprentice during his apprenticeship sufficient meat drink washing lodging and wearing apparel fitting for an apprentice and to pay the said apprentice seven pounds ten shillings at the expiration of the said four years four months and sixteen days and no freedom suit.

And the said John Demuth agrees to give his said apprentice two weeks in the Harvest each year during his apprenticeship to work for himself And the said apprentice is strictly to serve and obey his said master in all his lawful commands as becometh an Apprentice agreeable to the Laws of the State aforesaid.

And for the True performance of all and singular of the aforesaid covenants and agreements either of the said parties bind themselves to the other in the just and full sum of Twenty Five Pounds Current Money to be paid by the party failing to the party complying with the above covenants and agreements. Witnesseth whereof they have hereunto set their hands and seals the day and year above written. John Demuth

George Koch

[Frederick County Apprentice Indentures, p. 66, Hall of Records]

DENMEAD, A. & W. & SON — Baltimore

This firm was in the iron works business at North and Monument Streets.

From the above Summary it will be seen that the expenditures for "Arms and Ammunition" reached the large amount of nearly $25,000. The articles purchased include . . . 3285 Iron Spears from A. & W. Denmead & Son . . .
[Baltimore City Correspondence Manuscripts, 1862, #695]

DENSON & BUCK — Baltimore

Advertised as dealers in metals, furs, paper stock, etc., at 100 Light Street.

From the above Summary it will be seen that the expenditures for "Arms and Ammunition" reached the large amount of nearly $25,000. The articles purchased include . . . 407 Halls Patent Rifles from Denson & Buck . . .
[Baltimore City Correspondence Manuscripts, 1862, #695]

DEWITZLEBEN, ARTHUR — Washington, D. C.

February 7, 1865
Be it known that I, Arthur DeWitzleben of Washington, District of Columbia, have invented a new & useful Improvement in Implement for Removing Balls or Projectiles from Firearms.
[U. S. Patent Office, #16220]

DICKINSON, CHARLES S. — Baltimore

In the winter of 1860 the cannon described below was tested before Army officials. The exhibitions proved it could throw four to five hundred two-ounce balls per minute the distance of one mile with precision and effect.

Work on the steam gun was done at Ross Winan's locomotive manufactory and an engraving of the period shows Winan standing in front of the cannon at his factory.

Several points are misrepresented in the letter. Dickinson, a devout Unionist at the time of the letter, wrote that the traitor Marshal Kane seized the gun; the cannon was actually the property of the city of Baltimore. On May 10, when the piece disguised as farm machinery was captured on the way to the Confederacy, one of the three prisoners was Dickinson himself. The three were taken to Annapolis where Dickinson took the oath of allegiance. He was applying for a government contract to make steam cannon for the federal government in the city of New York.

August 4, 1862 Office of Chief of Ordnance, Dickinson to President Abraham Lincoln

We take the liberty of addressing you for the purpose of presenting a proposition to construct for the United States Government, a Steam Gun, the invention of the undersigned Charles S. Dickinson, and patented August 9th 1859 . . . When the rebel sympathizers at Baltimore assumed a hostile attitude, the traitor Marshal Kane seized the Gun with the intention of using it to prevent the passage of loyal troops through the city, and took the inventor prisoner. There was an attempt subsequently to send it to the rebel army in Virginia, but it was captured on the way by a detachment of the force under General Butler. It was then however, or immediately thereafter, found in such a damaged condition as to be unserviceable, and unfit to be repaired. . . . The one designed for the field can be drawn by horses with the same facility as a piece of ordnance. Its weight, when the boiler and tank are filled with water, is about two tons. It is worked by four men, who, while engaged in their duties, are fully protected by the armor.

[Baltimore city correspondence manuscripts, 1866, #783]

DISTELBART, GEORGE — Baltimore

Distelbart, George

1868-77	159 East Street	lock & gunsmith
1878	159 East Street	blacksmith
1879	159 East Street	locksmith

DODGE, PHILIP TELL — Washington, D. C.

Philip was probably a son of William C.; they shared two patents with each other for the locking of shotgun barrels: #118,350, August 22, 1871; #127,683, June 11, 1872.

August 22, 1871

Be it known that we, William C. Dodge & Philip Tell Dodge, of Washington, in the county of Washington & District of Columbia, have invented certain Improvements in Breech-Loading Guns.

[U. S. Patent Office, #118,350]

DODGE, WILLIAM C. — Washington, D. C.

His numerous patents were on shotgun actions. The three final patented inventions were with Philip Tell Dodge. Patents included #45,983, January 24, 1865; #52,547, February 13, 1866; #58,790, October 16, 1866; #112,763, March 14, 1871; #113,408, April 4, 1871; #114,653, May 9, 1871; #114,794, March 14, 1871; #118,350, August 22, 1871; #127,683, June 11, 1872.

September 20, 1864

Be it known that I, William C. Dodge, of the city of Washington & District of Columbia, have invented a new & useful Improvement in Breech-Loading Fire-Arms . . . consists in the peculiar construction of the lever-guard & retractor, & in so combining & arranging the same with the stock & barrels as that a single movement of the lever-guard shall both elevate the rear end of the barrel & remove the cartridge.

[U. S. Patent Office, #44,290]

DOERMER, J. — Washington, D. C.

Doermer, J.

1846	S side Penn av btw 9 & 10 w	gunsmith

[Directory, Washington, D.C.]

DOLLAR, LAWRENCE — Baltimore

Dollar, Lawrence

1851-54	Peirce next C. Poppleton	gunsmith

[Baltimore city directories]

	age	sex	occ.	birth
Lawrence Dollar	27	M	Gunsmith	Germany
Cath	22	F		Germany
Jos	1	M		Maryland

[1850 Federal Census]

DORRIS, THOMAS — Cumberland

Dorris, Thomas

1887	Cumberland	gun & locksmith

[Maryland Directory & State Gazetteer, 1887]

DORSEY, JOHN E. & BROTHER — Baltimore

On Wednesday past (May 6) Mr. John Dorsey contracted with the government to produce 100 thirty two pound guns of the carronadis style at $139.05 each.

[*Federal Gazette,* 9 May 1813]

Dorsey, John E.

1817	11 Spear's wharf	merchant
1822	N side W. Charles	gent. conewago

Dorsey, Allen

1817-19	104 Smith's wharf	iron store merchant
1822-24	104 Smith's wharf	iron & commission merchant

[Baltimore city directories]

DRAKE, MATTHEW Baltimore

An average quality flintlock fowling piece bears the name "M. Drake" on the barrel and lock.

Drake, Matthew

1817-21	3 Ruxton Lane	blacksmith
1822-24	34 Light Street	locksmith & bellhanger
1836	Lombard & Light Streets	hardware store & manufacturer of Guns, Pistols, Locks

[Baltimore city directories]

DRUCK, CONRAD Washington, D.C.

1850	Druck,	Conrad age	32 b.	Germany	gun & locksmith
		Sophia	28 b.	Germany	
		Carl	4 b.	Washington, D.C.	

[Federal Census, 1850, Washington, D.C.]

E

EALER, LEWIS W. — Baltimore

The 1850 Federal Census indicates that Lewis Ealer was born in Pennsylvania in 1791. He and his wife Susanna had two children born in Pennsylvania: Catharine in 1822 and Sarah in 1824. Apparently he worked in Maryland during this time though his family remained in Pennsylvania.

Ealer, Lewis W.

1822	19 George Street, Fell's Point	lock & gunsmith
1827-31	34 Thames Street	gunsmith
1833	2 Market space	gunsmith

[Baltimore city directories]

Three additional children were born in Maryland: Elizabeth (1826), Susan (1828) and Franklin (1832), according to the 1850 Federal Census, which suggests that he must have situated his family in Maryland sometime between 1824 and 1826. The first evidence that he returned to Philadelphia is an 1837 entry in Desilver directory which lists him as a gunsmith at 354 Vine Street. The McElroy directory lists him as a gunsmith in Philadelphia at various addresses from 1839 through 1856, after which time he is no longer listed.

EASTERLEY, CHRISTOPHER — Baltimore

Easterley, Christopher

1840 Lombard Street, 1 door east of Frederick — gunsmith

[Baltimore city directory]

EASTMAN, JOSEPH

August 10, 1776 Council of Safety
That the said Treasure pay to Joseph Eastman four pounds ten Shillings for a Musquet.
[AM, p. 192]

EBERHARD, ________ — Baltimore

Eberhard, ________

1853	Harford Avenue	gunsmith

[Baltimore city directory]

EBERT, BENJAMIN & SONS — Frederick

Benjamin and John Ebert advertised.

Ebert, Benjamin & Sons

1886	67 West Paterick Street	gunsmiths
1895	67 West Paterick Street	hardware & changes

[MILLER'S GENERAL DIRECTORY OF FREDERICK CITY & FREDERICK COUNTY]

EDON, JOHN — Frederick

Inventory of Work tools Belonging to and now in the Gunlock Factory the 19th August 1778 . . . John Edon three months (Run away).
[Hall of Records, Red Book IX, p. 273]

ELLICOTT, ________

24 June 1781 Smith to Lee
It will be Impossible to Equip Coll. Nicholsons Troop of Horse in less than one Month. I have sent Ellicott to work on the swords &c &c He can make about 10 p week. . .
[AM, p. 311]

ENGLE(S), JOHN & PETER SR. — Allegany & Frederick Counties

John Engles and wife Anna Marie with their son Peter (born c. 1755) left Germania, Prussia, migrating to Frederick County, Maryland. They were members of the Frederick German Reformed Church. Peter was confirmed in the church February 10, 1767. Peter changed spelling of family name by dropping the "s." He and his wife Susanna had 12 children, with four sons continuing in the gunsmith trade. In 1796 they were residing in Allegany County and in 1800 Peter returned from Somerset County, Pennsylvania, to dispose of his remaining real estate. He died in 1833. His deeds were recorded in the Allegany County Court House in 1796, 1797 and 1800. His wife also signed.

Prior to the Engles leaving Frederick County, John and Peter made rifles at the Monocacy, got their iron from the old mine near present Jefferson, Maryland, smelted it on the place and rolled and made the barrels along the Monocacy River near Ceresville, Maryland.
[TWO CENTURIES OF BROTHERS VALLEY]

ENGLE, PETER JR., BARNEY, EZRA, CHRISTIAN

These four sons of Peter Engle Sr. all worked in Greene County, Pennsylvania, as gunsmiths. Ezra was born 27 December 1790, according to the records of the Frederick German Reformed Church. Barney died 25 January 1887.

ERNST, JACOB — Frederick County

Ernst, Jacob Frederick County
1 male of 16 yrs.
1 male under 16 yrs.
2 females
[1790 Federal Census]
Indenture made October 5, 1811, between Christian Creager of Frederick County state of Maryland of the one part and (to) Jacob Ernst of the same county and state aforesaid of the other part . . . ground situate, lying and being in the same

county being part of that tract of land called Peter Promise. [Frederick CCH, Liber W. R. 40, Folio 493]

ESCHERICH, ANTON & SON — Baltimore

Anton was an exceptional craftsman. His exquisite engraving and figure relief metal designs reflect his European training. His target rifles, hunting rifles and pistols are completely and beautifully engraved. They are marked on the barrels and occasionally on the locks "A. Escherich Baltimore." A petition to the city for loss of arms on 21 April 1861 was applied for in Anton's behalf by his son Francis.

Escherich, Anton & Son
 1859-67 477 West Baltimore Street gunsmith
Escherich, Anton
 1866 477 West Baltimore Street barkeeper
[Baltimore city directories]

ESCHERICH, CHARLES — Baltimore

Escherich, C.
 1866 284½ 7th west gunsmith
[Boyd's Business Directory, Washington D.C., Georgetown & Alexandria]

Escherich, Charles
 1869-71 251 Hamburg Street gunsmith
 1872-73 251 Hamburg Street locksmith
[Baltimore city directories]

ESCHERICH, FERDINAND — Baltimore

Ferdinand worked with Anton and later for Francis after 1867 under the name Escherich & Son.

Escherich, Ferdinand
 1864-67 477 West Baltimore Street gunsmith
 1868-70 56 West Baltimore Street gunsmith
(ad 1869) manufacturers and dealers in Guns of all descriptions and Hardware.
[Baltimore city directories]

ESCHERICH, FRANCIS — Baltimore

Undoubtedly a son of Anton, Francis became proprietor of Escherich & Son after working with his father. His rifles are marked on the barrel "Escherich & Son Balt° Md" and are equal to Anton's fine engraving. A handsome rifle made by Francis was won at a shooting match by Albert Loenyi and bears his name on the tang.

Escherich, Francis
 1864-67 477 West Baltimore Street gunsmith
Escherich & Son
 1868-70 56 West Baltimore Street gunsmith
(ad 1869) manufacturers and dealers in Guns of all descriptions and hardware.
[Baltimore city directories]

ESGENBACH, HENRY — Washington, D.C.

Esgenbach, H.
 1858 473 11th av w gunsmith
[Business directory, Washington, D.C.]
Esgenbach, Henry
 1858 473 11th west gun & locksmith
[Directory, Washington, D.C.]

EVATT, EDWARD I, COLUMBUS, EDWARD II, JOHN — Baltimore

It is probable that Edward I taught his sons the gunsmith trade, turning the business over to them in the mid 1820's. They operated as Evatt & Brothers until John left the firm followed by Columbus. The shop or shops of Edward I and his sons were one number apart; it is very possible that there was a numbering change in the streets.

Evatt, Edward [I]
 1810-16 36 Light Street whitesmith
 1817-24 36 Light Street whitesmith & gunsmith
 1831 Ellicott Street gunsmith
Evatt & Brothers
 1827 35 Light Street lock & gunsmith
Evatt, Edward [II] & Columbus
 1831-35 35 Light Street lock & gunsmith
Evatt, John
 1827 3 Water Street, head of Cheapside lock & gunsmith
 1829 35 Light Street & 3 Water Street, Cheapside lock & gunsmith
 1831-35 3 Water Street, head of Cheapside lock & gunsmith
Evatt, Edward [II]
 1837-40 Light Street, S. of Lombard lock & gunsmith
 1845-47 33 Light Street lock & gunsmith & bellhanger
 1851-55 33 Light Street gun & blacksmith
 1856-71 35 Light Street gun & locksmith
Evatt, Columbus
 1837-40 56 Light Street lock & gunsmith
 1845 22 South Howard Street lock & gunsmith
 1851 26 S. Liberty Street lock & gunsmith
 1853 26 S. Liberty Street whitesmith
[Baltimore city directories]

Evatt, Columbus	age 37	Locksmith	birth Maryland
Mary	40		Maryland
Marshall, Edwin	18	apprentice	Maryland

Evatt, Edward	42	Locksmith	Maryland
Agnes	42		Maryland
Young, Frederick	25	Locksmith	Germany

[1850 Federal Census]

EWING, THOMAS — Baltimore

Often a Revolutionary period arms inspector gave the gunsmith a certificate for the muskets which proved to be good. Thomas Ewing, however, used a proof stamp for the muskets he passed with approval. A few Maryland muskets of the period have survived bearing a mark somewhat resembling a fleur-de-lis, believed but not proven to be Ewing's mark.

31 January 1776 Council of Safety
Ordered, That Major Gist and Captain Ewing or either of them be requested to employ a skillful person to prove the said muskets or any others that may be hereafter made by the said Keener, Messersmith and Riddick or either of them under their aforesaid Contracts; and that the said Major Gist and Captain Ewing or either of them be empowered to draw on the Committee of Observation for Baltimore County, for powder sufficient for that purpose.
[AM, p. 127]

12 February 1776 Ewing to Council
Hon. Sirs. I now acquaint you that agreeable to your orders directed to Major Gist and myself, I have prov'd all the guns made by the Gunsmiths in Town, as below, in Company with Mr Vanbibber, who has seen them several times try'd in England. Mr Keener after seeing the rest try'd refus'd to have his proved but upon my threatning him a good deal he comply'd. We charg'd all the guns with one ounce of powder and two Balls. I have a proof stamp with which I stamp'd all that prov'd good, which stamp I will not deliver to any without your orders.

Thos. Ewing

1776				
Feb 7	Peter Lydick	72 Guns	64 good	8 Bad
Feb 7	Sam Missersmith	27 "	23 "	4 "
Feb 7	Sam Keener	32 "	13 "	19 "

[AM, p. 155]

EYSTER, JOHN — Sandy Mount

John Eyster, Gun and Gunlock Smith, Sandy Mount, four miles from Westminster on the Baltimore turnpike. Guns of every description constantly on hand and made to order.
[advertising card]

F

FAVIER, PETER A. — Baltimore

Favier, Peter A.

1837	6 Lancaster Street	gunsmith
1840	Lombard Street E of Bond	gunsmith
1845-53	67 West Pratt Street	gunsmith

(ad 1852) Gun, Rifle, Pistol Maker, having constantly on hand a large & general assortment of double & single barrel bird & ducking guns, rifle & pistols made to order which he will warrent, superior made & patent six barrel pistol of very best & most improved style – every article partaning to sportsmans use.
[Baltimore city directories]

FEINOUR, ALFRED

Feinour, Alfred 1847 11 Baker's Court gunsmith
[Baltimore city directory]

FENNER, ALBERT — Baltimore

Fenner, Albert 1851 181 So. Charles St. gunmaker
[Baltimore city directory]

FERGUSON, ALFRED B. — Washington, D. C.

Ferguson, Alfred B.	age 45	born	N. Y.	gunsmith
"	Catharine	45	Va.	
"	Elizabeth	9	Washington, D.C.	
"	Emily	7	Washington, D.C.	

[1850 Federal Census]

Ferguson, Alfred B.
1853 E side N Jay ave btw M & N s gunsmith
[Directory, Washington, D.C.]

FINDLAY, ROBERT & COMPANY — Baltimore

Findlay, Robert & Company
1887 25 Hopkins Place gunsmith
[Baltimore city directory]

FISHER, A. R. — Cumberland

Fisher, A. R. 1887 Cumberland gun & locksmith
[Maryland Directory & State Gazetteer, 1887]

FISHER, G. FREDERICK — Washington, D. C.

Fisher, Frederick
1864 416 7th west gunsmith
[Hutchinson's Business Directory, Washington, D.C., & Georgetown]

Fisher, Frederick
1865 416 7th Ave. gunsmith
[Boyd's Business Directory, Washington, D.C., Georgetown & Alexandria]

Fischer, G. F.
1866 416 7th west gunsmith
[Boyd's Business Directory, Washington, D.C., Georgetown & Alexandria]

FISHER, JAMES — Baltimore

Fisher, James
1817-18 East Street near Calvert gunsmith
[Baltimore city directory]

FISCHER & SHAEFER — Washington, D. C.

Fischer & Shaefer
1867 416 7th Ave. gunsmiths
[Boyd's Business Directory, Washington, D.C., Georgetown & Alexandria]

FLETCHER, WILLIAM F. — Baltimore

By 1880 Fletcher was listed as a cigarmaker and continued to work at the address below with William G. Fletcher.

Fletcher, William F. & William G.
1876-79 100 William Street gunsmith

FLETCHER, WILLIAM G. — Baltimore

Fletcher, William G.

1847-51	109 Cross Street	gunsmith
1855	150 Cross Street	gunsmith
1858-66	137 Cross Street	gunsmith
1867-69	147 Cross Street	gunsmith
1870-75	100 William Street	gunsmith

Fletcher, William G. & William F.
1876-79 100 William Street gunsmith

Fletcher, William G.

1880-88	100 William Street	gunsmith
1889	906 William Street	gunsmith

[Baltimore city directories]

William Fletcher	age 29	sex M	born Virginia	gunsmith
Hestor	28	F	Maryland	
Catherine	6	F	Maryland	
John L.	4	M	Maryland	
Stephen G.	2	M	Maryland	
H. Clay	4/12	M	Maryland	

[1850 Federal Census]

FLOWER, DANIEL

September 10, 1776 Council of Safety
Ordered That Western Shore Treasure pay to Daniel Flower fifty shillings for a Gun.
[AM, p. 263]

FORNSHELL, JOHN — Baltimore

Fornshell, John

1814-16	39 Union Street	gunsmith
1817-18	corner of North & Union	gunsmith
1822	Bridge Street	locksmith & gunsmith
1824	north side of East Union	locksmith & gunsmith
1831	Ensor Street near Porter	gunsmith
1833-35	77 Ensor Street	gunsmith

1845-53	12 Ensor Street	gunsmith

[Baltimore city directories]

FOY, JAMES Baltimore

In 1850 James was a merchant, dealing in hardware, cutlery, locks and guns, age 37 years, born in Maryland and husband of Harriet. His son James H. was only associated with the business by name for two years according to the advertising.

Foy, James

1847-53	68 South Calvert Street	gunsmith & hdw merchant
1856-69	64 South Calvert Street	gunsmith

Foy, James & Son

1870-75	64 South Calvert Street	gunsmith

Foy, James

1872-75	64 South Calvert Street	gunsmith

[Baltimore city directories]

Foy, James	1850 age 37	Merchant	birth	Maryland
Harriet	1850 36			Maryland

[1850 Federal Census]

FOY, JAMES H. Baltimore

Apparently the son of James Foy. Listed in the 1872 city directory but with no occupation given.

Foy, James & Son

1870-75	64 South Calvert Street	gunsmith

[Baltimore city directories]

FREDERICKTOWN GUN LOCK MANUFACTORY Frederick

The building is standing (1973) and being used as a feed mill. This factory supplied locks to Schley, Winters, Yost and other smiths. The tools were sold 10 November 1778, and on the inventory the overseer Samuel Boone lists four runaway employees.

At the request of Charles Beatty, James Johnson, and John Hanson Supervisers of the Gunlock Manufactory erected at Frederick Town the following deed was recorded on the 13 day of December 1775 to wit. This indenture made the 4th day of December 1775 between Christian Steiner of Frederick Town in Frederick County of the one part and [to] Charles Beatty, James Johnson, and John Hanson of the same County, gentlemen of the other part. Whereas it was resolved by the Convention of Maryland held at the city of Annapolis on Tuesday the 7th day of December 1775 that a Gunlock Manufactory should be established in Frederick Town in Frederick County in the neighbourhood thereof and Charles Beatty, James Johnson, and John Hanson were appointed Commissioners . . . all that piece parcell or portion of ground situate lying and being in Frederick Town in Frederick County known and distinguished by #1 in the original plott of the said town . . .
[Frederick CCH, Liber BD2, p. 471]

May 25, 1776 Council of Safety
Ordered That the Commissioners appointed to supervise the Gun Lock Manufactory at Frederick Town, deliver to Major Price on his order all the Gun Locks already manufactured for public use.
[AM, p. 444]

June 19, 1776 Council of Safety
Ordered That the Supervisors of the Gunlock Manufactory in Frederick Town deliver to Captain Stricker all the Gun Locks which are compleat at the said Manufactory.
[AM, p. 499]

June 27, 1776 Council of Safety
Ordered That the Commissioners of the Gun Lock manufactory at Frederick Town, furnish and make for Jacob Schley ten large Gun Locks agreeable to a patent they will receive for that purpose from said Schley.
[AM, p. 526]

July 2, 1776 Smyth to Council
. . . If Mr. Winters cou'd be furnished with Locks from the Frederick Manufactory he cou'd easly compleat 50 stands of Arms every week.
[AM, p. 544]

July 30, 1776 Council to Beatty & others
As a considerable time has elapsed since your Gun Lock factory, we are in great hopes you have made a number for the use of this Province. The Province is in great distress for arms – few of the Troops raised for the flying Camp are supply'd with them, & the Militia will not lend theirs. There is an absolute necessity therefore that you immediately send them down what Locks you have. We expect Firelocks from John Yost, they may come from thence if you contrive them there. We desire also to know how many Locks you make per week. We have Barrels enough here and in Kent for the Locks you can make.
[AM, p. 142]

September 2, 1776 Council of Safety
Ordered that Supervisors of the Gun-Lock Manufactory at Frederick deliver to Captain Thomas Beal four Rifle Gun - Locks.
[American Archives, 5th Series, Vol. II, p. 626]

September 13, 1776 Yost to Council
. . . I was told by the Manager of the Gun Lock Manufactory at Frederick Town, that they forge Gun Locks much faster than they can finish them off; as that is the Case, I should be glad to furnish myself from thence with 300 ready forged Locks, provided the Terms are admissible.
[AM, p. 271]

October 8, 1776 Boone to Council
By order of they supervisers of they Gun lock facktorey in Fridrich Town I have sent you seventy nine locks and have a quontity more already to put together, if you must have all they locks sent to you that is made in they facktorey pleas to let me know by the barer George Bear.
[AM, p. 327]

December 10, 1776 Beatty to Council
The Barer hereof Mr. Boon [the overseer of the Gun Lock factory] waits on you in order to get some money to enable him to carry on the factory, for it can not be supported any longer without. Mr. Hanson who keeps the books is not at

home so that I cannot get to see them in order to draw for the money, the factory is entitled to, but I am clearly satisfied that it must be about 400 dollars by the acct. rendered, and the locks now in the factory, which sum please to forward by Mr. Boon. We have in the factory 156 locks completely finished as will appear by the Inclosed certificate for nearly two months past I was compelled to take money on Interest to support the factory or let the hands go idle. It is not in my power to borrow any more for that purpose here. Pray dont fail sending the money or I shall be compelded to sell the locks to pay the money that I have taken up and let the servants shift for themselves, which will be a great loss to the publick.
[AM, p. 519]

August 16, 1777 Council of Safety
That the said Treasure Pay to Benjamin Galloway five hundred Pounds to be delivered over to Messrs Beatty, Hanson and Johnson Superintendants of the Gunlock Manufactory on Acct.
[AM, p. 335]

October 2, 1777 Council of Safety
Ordered That the western shore Treasure pay to Samuel Boone four hundred and fifty pounds to be delivered over into the hands of the Supervisors of the Gun Lock Manufactory in Frederick on Acct.
[AM, p. 386]

FRIEDERICK, HENRY

Friederick, Henry
1872 200 4½ sw gunsmith
[Columbia Directory]

FRIEND, GABRIEL Frederick

Gabriel Friend may have been Frederick's first gunsmith, for the will below shows that he was a gunsmith there in 1759 and it implies that he died there in 1760. It is not known whether or not Gabriel's brothers Jacob and Charles put the gunsmithing tools to use. In the late 1780's or 1790's various members of the Friend family moved to the extreme western end of the state of Maryland and the town of Friendsville today bears their name. One member of the family who moved to the western end of the state also was named Gabriel Friend. The 1850 Federal Census for Allegeny County lists Gabriel Friend as a hunter, age 99, and born in Pennsylvania. The Washington County court house has records of a number of indentures in the name of Gabriel Friend in the period between 1790 and 1803.

In the Name of God Amen. I Gabriel Friend of Fred. Coty and province of Maryland Gun Smith being very Sick & weak in body but of perfect mind and memory thanks be therefore unto Almighty God. Knowing that its appointed for all men once to die do make and ordain this my last Will & Testament in the following manner & form First & principally I recommend my Soul to the hands of Almighty God who gave it & my body to be buried at the Discretion of my Execr. hereafter named and touching such wordly Goods as it hath pleased Almighty God to bless me with in this life I give Devise and Dispose of in the following manner & form. Imps. Its my Will and I do order that in the first place all my Just Debts & funeral charges be paid & satisfied.
Item I give & Bequeath to my Loving Brother Jacob Friend all and Singular my Lands buildings and Improvements to me belonging or that hereafter by any Just Claim may Lawfully Deemed to be mine to him his heirs and Assigns forever – together with all my wearing apparell.
Item I give and Bequeath to my loving Brother Charles Friend all my and Singular my moveable Estate Cash Goods Chattels and Household Implements to him his heirs & assigns for Ever except my wearing apparell the use of my Cash and Smith Tools I order to my Brother Jacob untill my Brother Charles be of the age of maturity and then the Cash & Tools to be Returned the cash without Interest the Tools all Replaced and in the order he the said Jacob Friend Received them or the Value agreeable to apprasmt. as the Receipt of them.
I constitute and appoint my trusty and loving friend Col. Thomas Prather and Mr. Van Swearinger Junr. whole and sole Executors of this my last Will & Testament. Revoking Disallowing and Disannuling all former Wills. Testament & Legacies by me made Ratifying & Confirming this and no other to be my last Will & Testament. In Witness whereof I the sd. Gabl. Friend do hereunto set my hand & seal this 22n day Nov. 1759.
Gabriell Friend
Signed Seald published pronounced by the said Gabriel Friend to be his last Will and Testament in the presence of us
Edmn Moran Charles Prather his
Thos. M. Hogg
mark
May 14th 1760 Charles Prather & Thomas Hogg two of the subscribing Witnesses to the within Will Solemly Sworn . . . that they saw . . . Gabriel Friend execute the within Will . . . the Extr refuse to act & renounce the Executorship.

[Will Book A, p. 192, Frederick CCH]

FUGITT, THOMAS M. Washington, D. C.

1850 Washington, D.C., 6th Ward

Fugitt, Thomas M.	age	37	born	Va.	gunsmith
" Mary E.		33		Ma.	
" Charles		13		Washington, D.C.	
" Edward		11		Washington, D.C.	
" William H.		10		Washington, D.C.	
" Josephine		6		Washington, D.C.	
" James P.		5		Washington, D.C.	
" Ugene		3		Washington, D.C.	
" Samuel		1		Washington, D.C.	

[1850 Federal Census]

G

GAFFORD, JOSEPH — Baltimore

Several finely made flintlock double barrel shotguns have been examined that came from Gafford's shop. He was an excellent engraver and his barrels are marked "Gafford Baltimore."

Gafford, Joseph

1816-18	70 Cumberland row	gunsmith
1819-29	70 Market space	gunsmith

[Baltimore city directories]

GARRETT, AMOS — Harford County

March 25, 1776 Garrett to Rumsey
I had a line from you about Bayonets, I forged out by a servent I had (since Ranaway) about 60 for one of the Militia Companys . . .
[AM, p. 287]

August 21, 1776 Garrett to Council
. . . I will continue to make the Bayonets, but the fitting these musqets and repairing some other odd guns from our neighbours has kept my man (who is but slow) employ'd.
[AM, p. 230]

GIBSON, THOMAS — Taneytown

First indenture recorded in Frederick County in 1796 and the last in 1827 but no occupation was ever given.

Gibson, Thomas 1799 juror 1 day paid 11 shillings 3 pence
Gibson, Thomas 1800 Talisman 1 day paid 11 shillings 3 pence
Gibson, Thomas 1801 juror March Court 9 days paid 5/1/3
[Frederick County Levy Book, Hall of Records #19257]

July 23, 1801 John Clarke to Governor
In my route through Maryland, I went to Tanney Town, had an interview with Mr. Thos. Gibson and examined the arms he offerred for sale to our Executive. They are of indifferent quality, considerably inferior to those furnished by Mr. McCormick. He informed me that his lowest price for them was twelve dollars per stand . . .
[Calendar of Virginia State Papers]

Gibson, Thomas 1803 juror 7 days paid 3/18/9
[Frederick County Levy Book, Hall of Records #19257]

GIBSON, WOOLMAND — Talbot County

November 8, 1776 Council of Safety
Ordered That Western Shore Treasure pay to Woolman Gibson thirty shillings for a Gun.
[AM, p. 429]

February 7, 1777 Council of Safety
Ordered that the Western Shore Treasure pay to Woolman Gibson Three pounds Seven Shillings.
[AM, p. 123]

March 10, 1777 Council of Safety
John Gibson and Woolmand Gibson of Talbot County having been brought before the Council of Safety in consequence of Information that they had abused and insulded the Officers of this State, while acting in their public Capacity, and they being heard in their own Defence It was Ordered that they be put in Custody of Captain Fulford and confined in the Guard House till Tomorrow.
[AM, p. 165]

August 8, 1777 Council of Safety
Ordered That the western shore Treasure pay to Woolman Gibson one pound twelve shillings & six pence Amt of Acct Passed.
[AM, p. 327]

GLARPS, GEORGE — Washington, D. C.

1850 Clarps, George age 75 born Germany gunsmith
[1850 Federal Census]

Glarps, together with gunsmiths Edward Gropley and Charles Gweeties, was living in the household of John H. Weaner, gunsmith.

GLATT, NICHOLAS — Salisbury

Glatt, Nicholas 1880 Salisbury gunsmith
[The Maryland Directory, 1880; J. Frank Lewis & Co. Publishers]

GMINDER, JACOB — Baltimore

Gminder, Jacob

1860	10½ South Calvert Street	silver plater & manufacturer of coffin trimmings
1864	10½ South Calvert Street	military ornaments, etc.
1865-67	10 South Calvert Street & 117 West Baltimore Street	

(ad 1865) silver plate & manufacturer & dealer in military equipment of all descriptions, embroidered & metal hat & cap ornaments, large letters, corps badges, shoulder straps etc. fine swords of elegant design for presentation furnished at short notice.

1868-79	10 South Calvert Street	silverplater
1880-84	10 South Calvert Street	silverplater, military goods, fancy hardware
1885-89	14 South Calvert Street	military goods

[Baltimore city directories]

GONTER, JOHN — Hagerstown

The minutes of the Lancaster, Pa., Committee of Safety for 14 August 1776 direct a Mr. Crawford to discharge a few soldiers from his detachment for various reasons. One of them, a lad of about 14 years of age, was a gunsmith's apprentice named "John Gouter." This probably refers to the gunsmith John Gonter, with various transcriptions of the name in longhand yielding a "u" where an "n" should be.
[THOUGHTS ON THE KENTUCKY RIFLE IN ITS GOLDEN AGE, p. 125]

The next documentation for John Gonter places him in Hagerstown, Washington County, Maryland, where he apparently worked as a gunsmith for at least eleven years.

Gonter, John Washington County
2 males over 16 years of age
1 male under 16
3 females
[1790 Federal Census]

Indenture made August 27, 1791 [deed] between Thomas Rinehart of Washington County and state of Maryland of the one part and [to] John Gonter of the county and state aforesaid of the other part . . . Tract or parcel of land lying and being in Rohrens addition to Elizabeth Town.
[Washington CCH, Liber G, Folio 482]

Indenture made November 15, 1791 [deed] between Daniel Stull Esq. Sheriff of Washington County and state of Maryland of the one part and [to] John Gonter of the county and state aforesaid of the other part . . . Tract or parcel of land lying and being that half lott or portion of ground in Elizabeth Town distinguished by number 61.
[Washington CCH, Liber G, Folio 541]

Indenture made March 10, 1792 [deed] between Jonathan Hager of Washington County and state of Maryland of the one part and [to] John Gonter of the county and state aforesaid of the other part . . . Tract or parcel of land lying and being in Elizabeth Town all those two lotts.
[Washington CCH, Liber G, Folio 641]

Indenture made March 26, 1792 [deed] between Jacob Katz of Washington County and state of Maryland of the one part and [to] John Gonter of the county and state aforesaid of the other part . . . Being all that half lott in Elizabeth Town distinguished by number 60.
[Washington CCH, Liber G, Folio 642]

This indenture made this 24th day of March 1794 between John Gonter of Elizabeth Town and county aforesaid Gunsmith of the one part and John Compton placed his son Alexander Compton as an apprentice to said John Gonter to learn the art and mystery of the Gunsmith business . . .
[Washington County Indentures, p. 5, Hall of Records]

Indenture made August 17, 1795 [deed] between Martin Bingham of Washington County and state of Maryland of the one part and [to] John Gonter of the county and state aforesaid of the other part . . . Being all that one half of a lot number 97 in Elizabeth Town.
[Washington CCH, Liber I, Folio 307]

This indenture made the 11 day of January 1796 witnesseth that John Whetstone of Washington County and state of Maryland hath bound, placed and put his son John Whetstone and by these presents doth bind, place and put the said John Whetstone apprentice to John Gonter of the county and state aforesaid Gunsmith . . . John Whetstone shall arrive at or attain to the age of 21 years he now being of the age of 16 years 21 day of August last . . .

his
John X Whetstone
mark

[Washington County Indentures, Hall of Records, p. 29]

September 8, 1799
This Indenture witnesseth that William McIntoch hath put his son William McIntoch apprentice to John Gonter Gun Smith to learn his art, trade and mystery after the manner of an apprentice to serve the said Gonter Gun Smith from the day of this date hereof for and during and to the full end and term of 4 years and 3 months . . .
[Washington County Indentures, Hall of Records, p. 105]

By the Orphans Court of the 17 day of April 1800 Samuel Hammell aged 17 years sometime in this year and as his age cannot be accurately aportained the Court binds the said Samuel Hammell for 3 years and 9 months from this day unto John Gonter for the said term to learn the art and mystery of a Gun Smith . . .
[Washington County Indentures, Hall of Records, p. 110]

Indenture made October 4, 1800 [depos] At the request of John Gonter the following deposition was recorded on Mrs. Mary Hager . . .
[Washington CCH, Liber N, Folio 164]

Indenture made December 19, 1800 [deed] between John Gonter of Washington County and state of Maryland of the one part and [to] William Kreps of the county and state aforesaid of the other part . . . Being that number 97 in Elizabeth Town.
[Washington CCH, Liber N, Folio 268]

Indenture made October 14, 1801 [deed] between John Gonter of Washington County and state of Maryland of the one part and [to] Henry Kealhofer of the county and state aforesaid of the other part . . . Tract or parcel of land lying and being in Elizabeth Town distinguished by numbers 54, 55, and 56.
[Washington CCH, Liber O, Folio 85]

Sometime after 1801 Gonter moved to Columbia in Lancaster County, Pa., for the Lancaster County Historical Society has a business ledger of John Hoff, clockmaker, mentioning a clock made for John Gonter, gunsmith, of Columbia (Kindig, 1976, p. 125). The documentation of latest date for John Gonter indicates that he moved from Lancaster to Reading, Pa. in 1823:

John Gonter, Gunsmith recently from Lancaster, herewith informs the residents of Berks and bordering counties that he has begun the Gunsmith business in his workshop on the west side of Penn Street, the second door, above the sign of the Black Horse in the city of Reading. He makes every kind of new weapon, repairs old ones, and especially rifles. The work will be good and the price cheap.
[*The Reading Adler,* 20 January 1824]

Gonter made a fine longrifle, but only a few of his pieces have survived. One is relief-carved, and signed on the barrel "Jno Gonter." The others are incised-carved, and are stamped on the side plate "J GONTER".

GOPPERT, GEORGE Baltimore

Goppert, George	age 29	Gunsmith	birth Germany
Margaret	37		Germany

[1850 Federal Census]

Goppert, P.
1852 232 South Charles Street gunsmith
[Baltimore city directory]

GORDON, GEORGE

April 6, 1776 Council of Safety
Ordered That the Treasure pay to George Gordon four Pounds on Account of the Iron Work for the Gun Carriages.
[AM, p. 314]

June 12, 1776 Council of Safety
Ordered That the sàid Treasure pay to George Gordon twenty Pounds common money for Gun Carriages.
[AM, p. 480]

July 15, 1776 Council of Safety
That the said Commissary deliver to George Gordon 12 Musquets to be repaired.
[AM, p. 47]

July 19, 1776 Council of Safety
Ordered That the Commissary of Stores deliver to George Gordon 24 Musquets to be repaired.
[Am. Archives, 5th Series, V. I, p. 1339]

July 20, 1776 Council of Safety
Ordered that the Treasurer of the Western Shore pay to George Gordon five pounds six shillings six pence for repairing 24 Muskets.
[Am. Archives, 5th Series, V. I, p. 1340]

August 3, 1776 Council of Safety
Ordered that the Treasurer of the Western Shore pay to George Gordon 40 shillings for repairing Fire Arms.
[Am. Archives, 5th Series, V. I, p. 1347]

September 3, 1776 Council of Safety
Ordered That the Treasure of the Western Shore pay to George Gordon and Oliver Whiddon ten pounds eight shillings and eleven pence for mending Guns.
[AM, p. 255]

September 24, 1776 Council of Safety
They also contracted with George Gordon for the making two thousand Cartough Boxes, Bayonet Bolts and Gunslings at twelve shillings & six pence.
[AM, p. 297]

February 6, 1777 Council of Safety
That the said Treasure pay to George Gordon for a Gun L 4.
[AM, p. 121]

February 11, 1777 Council of Safety
Ordered That the Western Shore Treasure pay to George Gordon Eight Pounds for 2 Guns.
[AM, p. 130]

March 4, 1777 Council of Safety
The Humble petition of George Gordon, sheweth that your petitioner having served your Honours in the course of last summer in cleaning and repairing the arms of this state, and hope to your satifaction since which time your petitioner has lingered under a severer sickness which notwithstanding never prevented your petitioner from stepping forth in defence of his country and as your petitioner understands that there are a number of arms to be repaired for this state Humbly hopes your Honours will please employ him in doing the same as he hopes your Honours from experience will find him as capable as any other in this city at least.
[AM, p. 161]

March 6, 1777 Council of Safety
That the said Treasure pay to George Gordon four pounds for a Gun.
[AM, p. 163]

March 11, 1777 Council of Safety
Ordered That the Western Shore Treasure pay to George Gordon four pounds for a Gun.
[AM, p. 167]

March 17, 1777 Council of Safety
Ordered That the Western Shore Treasure pay to George Gordon four Pounds for a Gun.
[AM, p. 179]

April 18, 1777 Council of Safety
Ordered That the Western Shore Treasure pay to George Gordon four Pounds for a Musquet.
[AM, p. 219]

May 1, 1777 Council of Safety
Ordered, That the Western Shore Treasure pay to George Gordon four Pounds for a Gun.
[AM, p. 235]

May 13, 1777 Council of Safety
That the said Treasure pay to George Gordon four Pounds, for a Gun.
[AM, p. 249]

May 17, 1777 Council of Safety
Ordered, That the Western Shore Treasure pay to George Gordon seven pounds, five shillings for two Guns.
[AM, p. 257]

GRANT, JOHN Baltimore

Grant, John
1810 53 St. Patrick's row gunsmith
[Baltimore city directory]

GREENER, W. W. Baltimore

The predecessor to this English firm was William Linthicum who worked at the same address.

Greener, W. W.
1888 2 West Baltimore Street gunsmith
[Baltimore city directory]

GREINER, WILLIAM Hagerstown

September 3, 1802
This Indenture witnesseth that William Criner with the consent of his mother Deloro Criner hath put himself and by this presents doth put himself apprentice to George Kreps of

Washington County and state of Maryland to learn the art, trade and mystery of a Gun Smith and Rifle Smith.
[Washington County Indentures, Hall of Records, p. 176]

. . . *Maryland Herald* we find the following were residents of Hagerstown: 1805-1815
Greiner, William gunsmith
[HISTORY OF WESTERN MARYLAND, p. 1063]

WILLIAM GREINER *GUNSMITH* Takes this method of informing his friends and the public, that he has commenced the above business, at the Shop lately occupied by Mr. George Kreps, Gunsmith, in Franklin Street, Hagerstown, where he intends carrying it on in its various branches. All those who may favor him with their custom, may depend on his best endeavors to give general satisfaction. January 3, 1812
[Maryland Herald and Hagers- Town Weekly Advertiser, 15, 22, 19 January, 5 February 1812]

GRIFFITH, HENRY

August 7, 1776 Council of Safety
That the said Treasure pay to Henry Griffith fifty nine Pounds ten shillings for fourteen Musquets.
[AM, p. 179]

GRIFFITH, JOHN A. Baltimore

Griffith, John A.

1860	Griffith & Griffith	

Griffith, John A. & Co.

1865-80	Baltimore & Charles Streets (J. A. Griffith, Jo Neil, Jos Maguire)	importers & dealers in merchant tailers & clothiers trimmings
1881-82	247 West Baltimore Street (Edward A. & Henry G. Griffith, John J. Murphy)	tailors trimmings
1883	2 Hanover Street	
1884-88	2 & 4 Hanover Street	
1889	8 Hanover Street (Edward A. Griffith)	tailors trimmings

Griffith, John A.

1853-55	29 Fayette Street	merchant tailer
1856-58	187 West Baltimore Street	

[Baltimore city directories]

GRIFFITH, KINZEY Washington, D. C.

Griffith, Kinzey
1860 lower end King gunsmith
[Business directory, Washington, D.C., Georgetown & Alexandria]

Griffith, K.
1867 24 King Alexandria gunsmith
[Boyd's Business Directory, Washington, D.C., Georgetown & Alexandria]

GROPLEY, EDWARD Washington, D. C.

Gropley, Edward age 21 born Germany gunsmith
[1850 Federal Census]

Gropley, together with gunsmiths Charles Gweeties and George Glarps, was living in the household of John H. Weaner, gunsmith.

GROVE, J.

Two fine-quality relief-carved flintlock longrifles with patchboxes and other features suggesting manufacture in the Hagerstown region are known; each is signed in script with the name "J. Grove."
Documentation that a gunsmith named J. Grove worked in Hagerstown or anywhere else in Maryland remains to be found. The indexes to indentures in the Frederick County court house contain no entries for anyone named J. Grove, but the Washington County court house in Hagerstown indenture indexes hold many. In the 1780's there lived in Hagerstown a clockmaker named John Grove who had a wife Catharine. He bought and sold a number of properties between 1781 and 1792. There also was a Jacob Grove who bought and sold property in Washington County in the 1780's and 1790's. His wife also was named Catharine. The name Grove appears to be derived from the more Germanic name Graff or Groff.

GWEETIES, CHARLES Washington, D. C.

Gweeties, Charles age 22 born Germany gunsmith
[1850 Federal Census]

Gweeties, together with gunsmiths Edward Gropley and George Glarps, was living in the household of John H. Weaner, gunsmith.

H

HAFFNER, JACOB — Frederick

Little is known about Jacob Haffner as a gunsmith. He lived in Frederick during the early decades of the nineteenth century and was active buying and selling real estate. In the author's collection is a longrifle with a daisy-head patchbox, somewhat crudely made, with a barrel marked "J. Haffner." It was purchased from a family descendent who knew Haffner to be a gunsmith. In the Frederick County court house there is a will for one Jacob Heffner dated 1804, a second will for one Jacob F. Heffner dated 1811, and a will for a gunsmith, dated 1827. In the period between 1816 and 1825 there are 11 deeds recorded in the name of Jacob Haffner, or Heffner as it sometimes was spelled.

Jacob Haffner, grandfather of the late Samuel T. Haffner, was a native of Frederick County, Maryland. His ancestors, who were German, were among the early settlers of Frederick County. He was a gunsmith by trade. In politics, he was a member of the Old-Line Whig party; in religion he belonged to the Lutheran Church. He was the father of two sons: William W. and Samuel P.
[HISTORY OF FREDERICK COUNTY, Vol. II, p. 800.]

By this deed, dated 30 August 1815, Michael Haffner of Frederick County conveyed to John Haffner, Jacob Haffner, George Haffner, Frederick Haffner, Joseph Haffner, Benjamin Haffner, Elizabeth Haffner, Appello Haffner, Mary Haffner, Anna Haffner, and Catharine Julian (Heirs of Frederick Haffner) late of Frederick County deceased of Frederick County. Whereas Frederick Haffner Sr. by his last will & Testament did divise unto Frederick Haffner, Jr., and Michael Haffner a tract of mountain land called "Rocky Hill" to be divided between them, and whereas on or about the 6th of January 1797 Michael Haffner did convey his share of "Rocky Hill" . . .
[Frederick CCH, Liber J.S. 1, p. 411]

I Jacob Haffner of Frederick County . . . Item 1 . . . I bequeath to my beloved wife Rebecca all my beds bedding and bedsteads one bureau with its contents, half doz. chairs, copper kettle & choice of a table, one Cow. Item 2 It is also my will she shall have all the remaining part of my real estate (which I shall not hereafter impower my Executor to sell) until my son Samuel arrives of age, and after he being of age, it is my will that my real estate which may be left, shall be valued by two disinterested men, and my son William W. to have the first offer to keep it at this valuation, if he does not keep it, then my son Samuel shall have the second choice to keep it at the valuation, including my wife's thirds in the land from that time. Item 3 It is my will that my Executor hereafter appointed shall sell the remaining part of my personal estate at public sale as soon as convenient after my decease as also so much of my real estate as will be necessary to pay all my just debts and claims, in so doing it is my will that my Executor shall sell first my lot No. 3 and my interest in my brother John D. real property and as much of my other property as will be necessary to pay all my debts and claims & lastly I do hereby constitude my beloved friend Samuel Poole to be my Executor . . . 15 June 1827.
[Frederick CCH, Will Book H. S. 3, p. 561]

Though Jacob Haffner's will states that he was in good health and sound mind when he signed the document, he apparently anticipated death, for the will was probated within about two months, on 20 August 1827.

HAGER, JONATHAN — Hagerstown

Jonathan Hager was born in Germany in 1719 and landed in Philadelphia on the ship *Harle* on 1 September 1736. Hager married Elizabeth Kirselner in 1740, moving to western Maryland where he operated a trading post. He was the founder of Elizabethtown which later was renamed Hagerstown in his honor. He was killed while constructing a church and his will was probated 9 April 1795. According to his descendants, certain correspondence, which has not been seen by the author, indicates that in his youth, Hager followed the trade of a blacksmith and gunsmith.

In the Hager House is a longrifle bearing the name "J. Heger" which he supposedly made. In his early days his last name was spelled as such and only in a few indentures is an occupation listed as a farmer, gentleman and later town proprietary.

Indenture made August 19, 1756 [deed] between William Hollow of Frederick County province of Maryland farmer of the one part and [to] Jonathan Hager of the same County province of Maryland of the other part . . . Tract or parcel of land called the Great Hall lying in said County.
[Frederick CCH, Liber F, p. 39]

Indenture made February 27, 1759 [deed] between Peter Rentch of Frederick County and province of Maryland of the one part and [to] Jonathan Hager of the same County and province of Maryland of the other part . . .
[Frederick CCH, Liber F, p. 647]

Indenture made December 11, 1762 [deed] between Jonathan Hager of Frederick County and province of Maryland farmer of the one part and Samuel Grable of Frederick County and province of Maryland aforesaid farmer . . .
[Frederick CCH, Liber H, p. 249]

Indenture made September 16, 1795 [deed] between M. Bringmas [to] George Vrebs both of Elizabeth Town . . .
Jonathan Hagar Town Proprietary
[Washington CCH, Liber I, Folio 342]

Hager, Jonathan   Washington County
1 male of over 16 years of age
4 females
[1790 Federal Census]

HAHN, GEORGE A. Baltimore

Hahn, George A.

1867-73	124 Mullikin Street	gunsmith
1874-80	202 N. Broadway	gunsmith

[Baltimore city directories]

HAILER, L. Washington, D. C.

July 26, 1870
Be it known that I, L. Hailer of Washington City, in the District of Columbia have invented a new & useful Improvement in gun locks.
[U. S. Patent Office, #105,799]

HAIMES, WILLIAM Harvey Towne

March 19, 1688 Proceeding of the Council held at the city of St. Maries at St. Johns
Ordered that what publick Armes shall be or are Conveyed to Mattapany be put to William Haimes Gun Maker at Harvey Towne to be fixed and made fit for service and he to doe noe other business in the way of his trade till those be done amended and finished.And that afterwards the same be sent to the severall Counties againe according to the direccons of the proclamacon in that case issued And that authority be granted to presse the said Haimes to the service aforesaid.

Ordered that the Armes at Coll Pyes be sent to St. Maries to Coll Digges to be fixed as aforesaid And that the sheriff forewith receive them and send them downe accordingly and that his honr Coll Digges have power to presse a smith for the fixing the said Armes And then to be returned to the severall Counties againe.

The hon Coll Digges Informed this board that the sheriffs of Dorchester County had brought over his Lordsps Armes of the said County kept by Capt Tripp and the same being well fixed and kept the said Coll Darnall ordered the said sheriffs to carry them back to the said County and deliver them to Capt Tripp againe which was well approved of by this board.

The time for the sitting of the Assembly drawing neer, the same was taken into Consideration And Ordered that the Assembly be prorogued to the Last tuesday in october next, And that proclamacons issue accordingly. which follows.
[AM, p. 67]

HALBACH, CHARLES Baltimore

A number of flintlock pistols have survived bearing the name of Halbach, or Halbach & Sons. A crude single-edged knife with blade marked "Halbach Balto." also is known. The pistols all seem to be of European or at least non-American design, and collectors may wonder if they were imported. Little information about the Halbachs can be found. The records below do give some insight into Charles Halbach, showing him as both an instrument maker and hardware merchant.

Halbach, Charles

1835	12 Mercer Street	insterment maker

[Baltimore city directory]

Halbach, Charles 1850 age 40 hardware merchant birth Germany
[1850 Federal Census]

Halbach, Charles & Elenor C. Pirtern both of Baltimore married August 3, 1836, by Scheib.
[Dielman File, Enoch Pratt Free Library, Baltimore]

HAMILTON, JAMES Harford County

Resided at Peach Bottom just south of Fawn Grove, Pennsylvania; the Mason-Dixon Line ran through his property. He paid taxes in Harford County and is listed in several Maryland indentures.

James was the son of Hugh Hamilton . . . He enlisted on June 16, 1780 in Samuel Givens Company in Col. George Alexander Regt. He was taken prisoner at the battle of Camden, S. C., until 2/6/81. After the close of the war he returned to York County, Pa., and there learned the gunsmithing trade. Aug. 23, 1799, he married Martha Wallace . . . and located at Peach Bottom, York County, Pa., and continued to live there until 1814 . . . moved to Cabarris County, N. C. . . . died on Oct. 21, 1842 at the age of 79.
[GENEALOGY OF HUGH HAMILTON]

HAMMELL, SAMUEL Hagerstown

By the Orphans Court on the 17 day of April 1800 Samuel Hammell aged seventeen years sometime in this year and as his age cannot be accurately aportained the Court binds the said Samuel Hammel for three years nine months from this day unto John Gonter for the said term to learn the art and mystery of a Gun Smith . . .
[Washington County Indentures, Hall of Records, p. 110]

HANSON, JOHN JR. Frederick

Born in 1715 and raised on a plantation across the Potomac from Mount Vernon. Moving to Frederick, he was an active patriot from 1774 until his death. He was the first President of the United States of Congress assembled before the country elected George Washington, thus making him truly the first President. He died suddenly 22 November, 1783 while visiting relatives on a plantation at Oxon Hill.

February 1, 1776 Council Proceeds
Ordered, That Messrs. John Hanson Junr., Charles Beatty and James Johnson or any two of them be requested to provide and firnish the said Company with good substantial and serviceable Rifles . . .
[AM, p. 127]

HAPWOOD, DAVID T. Baltimore

Harpwood, David T.

1840	41 South Street	gunsmith

[Baltimore city directory]

HARIG, FREDERICK — Baltimore

Harig, Frederick

1845	Charles & Hill Streets	gunsmith

[Baltimore city directory]

HARRIS, ISAAC

A crude and heavy iron band, walnut stocked, flintlock pistol bears "I Harris" on the round barrel. The pistol appears to be of the Revolutionary period with a flat lock plate and goose neck hammer.

April 11, 1776 Council Proceeding
Ordered That the Commissary of Stores deliver to Mr. Isaac Harris all the Musquets now in the Magazine, which are not fitted with Stell Ramrods in Order that they may be fitted therein.
[AM, p. 326]

June 27, 1776 Council Proceeding
Ordered That said Treasurer pay to Isaac Harris forty six pounds, three shillings, in part of his Acct. as Armourer.
[AM, p. 524]

June 29, 1776 Council Proceeding
Ordered That Isaac Harris furnish Mr Samuel Dorsey with the Dimensions, and sizes of Gun Barrels at the mussles, sufficient for his guidance in manufacturing Bayonets.
[AM, p. 535]

July 6, 1776 Council Proceeding
Ordered That Commissary of stores deliver to Isaac Harriss 2w Gun Power for to prove his rifles and Musquets.
[AM, p. 553]

July 16, 1776 Council Proceeding
Ordered that the Commissary of Stores deliver to Isaac Harris half a Faggot of Steel: Also eighty four Musquets to be repaired.
[AM, p. 54]

July 31, 1777 Council of Safety
That the western shore Treasure Pay to Isaac Harris one hundred & fifty seven pounds two shillings & four pence Amt. Acct. passed.
[AM, p. 323]

February 3, 1778 Council of Safety
Ordered That the Western shore Treasure pay to Isaac Harris one hundred and ninety seven pounds, five Shillings and the furhter Sum of twenty nine pounds fourteen Shillings and two pence due on two Accounts passed by the Board of Accounts.
[AM, p. 482]

December 14, 1780 Council Proceedings
Ordered That the western shore Treasure pay to Isaac Harris Forty three pounds, two shillings and six pence of the new Emission in lieu of one thousand seven hundred & 25 Pounds and also the further sum of four pounds of the same . . .
[AM, p. 243]

August 3, 1781 Council Proceedings
That the said Treasure pay to Isaac Harris seventy five pounds eight shillings specie agreeable to the said Act. p Certs settled by the Aud. Gen.
[AM, p. 537]

HARRIS, WILLIAM — Baltimore

Petitioned the City Committee of Claims for the loss of fire arms worth $266 taken by a mob 19 April 1861. Was the predecessor of Edward Melchoir, who acquired the business in 1877.

Harris, William

1852-57	65 South Street	gunsmith
1858-76	116 West Pratt Street	gunsmith

(ad 1852) Large assortment of bird & ducking guns, double & single barrel, six barreled revolvers, rifles made to order, restocking, all articles necessary to sportsman.
[Baltimore city directories]

Harris, William 1850 age 30 Gunsmith birth Maryland
[1850 Federal Census]

HARTOGENSIS, HENRY S. — Baltimore

Hartogensis, Henry S.

1868-84	10 Harrison Street	gunsmith

[Baltimore city directories]

HASLETT, JAMES — Baltimore

In the flintlock period James Haslett was one of the best gunsmiths of Baltimore, and probably he should be regarded as Baltimore's most outstanding maker of guns. About twenty years ago a rather thorough study of Haslett was completed by Richard H. Randall, Jr., currently Director of the Walters Art Gallery in Baltimore. This was published as a two-part article in a periodical dealing with antique arms, THE AMERICAN ARMS COLLECTOR, for April and July 1957. Randall's account of Haslett is presented below, with some minor editing to eliminate references to illustrations that appeared with the article. Permission for the use of this material here has been granted both by Mr. Randall and by Hugh Benet, Jr., editor of THE AMERICAN ARMS COLLECTOR.

In spite of the vast number of records concerning early American gunsmiths, few have emerged as personalities in the scattered publications. Few enough, indeed have had even the characteristics of their work summarized. James Haslett of Baltimore is one instance where existing documents throw interesting light both on his gun business and on the personality of the man, who came to this country specifically to make arms. He was well educated, though probably at home, well trained in his profession, and led a very successful life. Yet in his later years, Haslett gave up his chosen profession to try his luck in other spheres.

James Haslett was born in 1773 in Ireland, the son of a Presbyterian family. We know nothing of the family except that his brother William S. Haslett was also in this country in 1833.[1] James came to America in about 1798 with his wife, Mary, and settled in Philadelphia. He was then a man of twenty-five and had presumably finished his apprenticeship in gunsmithing in England a few years earlier, because he came to America from Ireland.

Haslett was apprenticed to a gunsmith, whom he termed "one of the first gun makers in Europe," in the art of producing both "military and birding guns." Though it is possible that his apprenticeship was in Ireland, it seems for several reasons more likely that it was in England. The phrase "one of the first gun makers in Europe" could hardly be applied to any Irish gunsmith working in the last years of the eighteenth century, while it most surely could have applied to London makers like John Manton, Durs Egg, or H. W. Mortimer. The striking similarity between Haslett's work and that turned out in London shops allows one to surmise, in the dearth of any documentation, that his training was in that city.

Haslett came to America to become superintendent of the arms factory run by Robert McCormick in Philadelphia. McCormick's first contract for 3,000 muskets was let in 1798,[2] and since a letter of 1801 states that Haslett made the model for the McCormick arms "with his own hands" and superintended the making of all arms delivered by McCormick, it is probable that he arrived in this country in 1798 or earlier. Between May and July, 1801, McCormick had been put in jail for bankruptcy, and his factory and tools sold. The muskets turned out under Haslett's direction at the McCormick works were standard 1798 contract muskets of good quality.

After McCormick failed, Haslett was not idle, as a letter from John Clarke to the Governor of Virginia of July 23, 1801, shows:

"In my route through Maryland, I went to Tanney Town, had an interview with Mr. Thos. Gibson and examined the arms he offered for sale to our Executive. They are of indifferent quality, considerably inferior to those furnished by Mr. McCormick. He informed me that his lowest price for them was twelve dollars per stand. . . . Having travelled in quest of workmen to every place in Maryland and Pennsylvania where I understood arms were manufactured, I had an opportunity of seeing specimens of their workmanship, and find that the arms manufactured in and about this city [Philadelphia] are generally of a quality superior to those I have seen elsewhere, the artists being generally more skillful. A Mr. Miles, of this city, who purchased McCormick's implements, materials, and unfinished work at a very low price, I understand, is now with you. There is a man here by the name of Haslett who was brought by McCormick from Ireland. He has had the chief management of McCormick's manufactory, and wishes to be employed at the Virginia Manufactory as under Master Armourer. He showed me some specimens of his work, with which I was much pleased. General Shee recommended him highly, both as a skillful artist and a good citizen. The workmen who formerly worked at McCormick's Manufactory are much attached to him. Mr. Haslett's terms are $15 per week and his rations found. In the hope of being employed in our manufactory he wishes to make 600 stands of arms for our State, to keep him employed until our works are ready for operation. He has spoken to General Shee and myself on the subject, and General Shee requests me to write to you for information whether, if Mr. Haslett will make 600 stands, which he proposes to do on the same terms upon which you contract for all the rest advertised for, he might not be employed to make them. He promises in the event of being employed to make them, that not a stand shall be inferior to the model for the McCormick arms, which model was made by his own hands. If he gets that employment we may hereafter have it in our power to employ him, if his assistance shall be required. On this subject you will please inform General Shee, as I do not expect to return here in less than three or four weeks.

"Should you think proper to give Mr. Haslett the employment he deserves, he will afford employment to workmen lately discharged from McCormick's works, who for want of employment may soon be widely dispersed and so engaged by the person who undertakes the supply of arms advertised by our State, in which case they perhaps could not leave that employment until the completion of the contract made by them. I have seen those artificers; they are willing to go immediately to Richmond on the terms on which they were employed by McCormick, but having at present no employment, they cannot remain here without it until our works are ready for them. I have made no other promise to them than if I could not get other workmen in the Eastern States equally skillful on lowest terms, I would contract with them on my return to this city."[3]

The immediate results of Clarke's letter are apparent in a second letter of September 25, 1801:

"I found the wages of such men lower in Massachusetts and Rhode Island than in any other of the States. I therefore engaged in those states all the workmen of the desired description I could find, and on my return back again to the works I first visited, the workmen were induced to fall in their prices rather than not be employed when I informed them of the low terms on which I had already engaged similar artists, in the two states above mentioned. Nineteen of the above 68 workmen are now employed by Mr. Haslett at Philadelphia in the manufacture of 600 stand of arms undertaken by him for this state at the works which were formerly Mr. McCormick's. The greater number of these men are natives of Ireland; some of them are Pennsylvanians. I engaged them at the same rates which Mr. Haslett is now giving them which are rather lower than the wages formerly given to them by Mr. McCormick. All the others of the 68 I engaged in the New England States; they are native Americans and their wages are rather lower than those I engaged in and about Philadelphia. About half the number of the New England artificers served apprenticeship in the Springfield manufactory of arms, and about half of who are natives of Connecticut."[4]

The work on the Virginia contract must have occupied Haslett into the following year, and it was probably at this time that he was appointed Inspector of Arms to the Commonwealth of Pennsylvania by Governor McKean.[5] In 1803 we find him listed for the first time in the Philadelphia Directory as "Haslett, James, gunmaker above 511 N. Second." There is no known work of Haslett's marked *Philadelphia*, and we can assume, since he moved to Baltimore in June of 1803, that he did not find private gunsmithing very satisfactory in Philadelphia, probably because of the established competition.

During his five year's residence in Philadelphia, Haslett's family had grown with the addition of his eldest daughter Mary and his son James, Jr. This family of four travelled to Baltimore and after finding a house and shop at 64 North Gay Street, Haslett placed his first advertisement in the *Baltimore Gazette and Daily Advertiser* on June 3, 1803. The ad is typical of the times and reviews the gunsmith's background concisely for his patrons.

GUN MANUFACTORY

– – –

JAMES HASLETT, *GUN MAKER*, RESPECTFULLY informs the public, that he has commenced the Gun Business, at No. 64, north Gay-street, where he intends carrying it on in all its various branches.

Having served a regular time to one of the first gun makers in Europe, and been inspector of arms for the commonwealth of Pennsylvania, (which he resigned in consequence of his removal to this city) he flatters himself to be able to execute his work equal to any imported.

Any state in the union desirous of contracting for muskets, may be supplied with any number (the Charlevill pattern) at 10 dollars 50 cents each, being 50 cents less than they have been manufactured for the United States. June 3.

Haslett's statement that his work was equal to any imported was hardly an idle boast, and his patronage soon included men like Edward Lloyd, the Governor of Maryland, who was used to importing his arms from London directly.[6]

In 1804 (April 20, 23, and 27; May 1 and 3) Haslett placed this second ad in the Baltimore papers:

GUN -- MAKER

– – –

JAMES HASLETT, *GUN--MAKER,* begs leave to inform the public, that he has removed from No. 64, North Gay Street to No. 4, Light Street, nearly opposite Mr. Bryden's coffee house, where he requests continuance of that liberal encouragement he has received from the spirited citizens of Baltimore, since his commencement in this city, and assures them that he will endeavor to merit a continuance of their favors.

Haslett's name appears also for the first time in the *Baltimore Directory* for 1804, where the address is incorrectly given as "Light Street, West side--144."

On October 25, 1805, Haslett purchased his first property in the city, a plot 17 by 95 feet on Water Street, and in November of the following year his third advertisement appeared in the *Federal Gazette and Baltimore Daily Advertiser* advising thè public that he had opened his shop at the "Sign of the Golden Gun."

Sign of the Golden Gun

– – –

JAMES HASLETT

GUN--MAKER

RESPECTFULLY informs the Public, that he has removed from No. 4, Light-street, to No. 28, Water-street, within a few doors of Market-space; where he has, at present, an elegant assortment of double and single Guns, and duelling Pistols, both of his own and London manufacture, which he can afford to sell lower than any of the same finish ever offered for sale in Baltimore.

He begs leave to return his sincere thanks to the Citizens of Baltimore, and the Public generally, for the liberal encouragement he has received since his commencement in Baltimore, and assures them, that he will endeavor to merit a continuance of it. November 12

At the time that Haslett established his shop on Water Street, where he was to dwell and work for the rest of his life, Baltimore was a city of between 29,000 and 33,000 inhabitants. It was rapidly expanding, for the population had doubled between 1790 and 1800 and nearly doubled again by 1810. A glance at the city directories for the early years of the century shows, however, surprisingly few entries for gunsmiths. In 1802 only five firms advertised, and by 1810 only nine other names had been added to the list. Some gunsmiths undoubtedly never placed ads in the newspapers or had themselves listed in the directories. But the majority of good quality gun-makers, like John Stewart and Michael Lenz, who worked in the same block as Haslett on Light Street, could be counted in this small number. The major competition was from hardware importers like John Walraven, of Baltimore Street, who continually advertised "from London and Liverpool--fowling pieces, double and single barreled in cases complete from 50 to 90 dollars. Single ditto common and good from 5 to 55 dollars highly finished."[7]

There is little documentation for the years 1806--1810 to throw light on Haslett's gun business. In that time, however, three more daughters, Sarah, Emily, and Eliza, had been added to the family. In 1810 Haslett was still considering the manufacture of military firearms, though he had long since stopped advertising to make them in Baltimore, as the need in the intervening years was not great. On December 29, 1810, he addressed a letter to the Governor of Virginia with proposals for taking over the Virginia Manufactory. The letter indicates an earlier similar proposal and also reveals Haslett's sound business sense:

"I had the honor of addressing a letter to you on the 11th of January ult., with propositions for the manufacturing of Arms for the Commonwealth of Virginia.

"Understanding that a Law was passed the last session of the Legislature authorizing you to receive proposals for the leasing of the Armory, I have taken the liberty of renewing my propositions.

"I will engage to supply the Commonwealth with as many Arms as they require, provided I may have the use of the buildings and Machinery of the Armory, rent free, which I will engage to keep in good repair during the term of the contract, decay, use and accidents excepted, on the same terms as the United States give to individuals for manufacturing different kinds of Arms. I will receive all the Stocks and unfinished parts of work that are now on hand at their proper value, provided I am permitted to work them in as part of the contract. I agree that the Commonwealth may deduct on delivery of each hundred Muskets, Pistols, or other arms the amount for the materials so purchased that have been wrought in said Guns, or Pistols, &c. The balance I will require immediate payment of, however, if there are an over proportion of any one particular part of work now on hand that I may not be obliged to discharge the person or persons employed in such works. I will require the indulgence of the Commonwealth in not deducting the whole of the amount for such articles at the time of delivery. It will be necessary in the prosecution of so extensive a business that for the better providing materials to advantage that may be required, that the Commonwealth upon the delivery of such goods at the manufactory should guarantee the payment of them in conjunction with the subscriber to such person or persons for whom they were purchased.

"The Foundrey and Boring Mill will receive my particular attention, but not having at present an accurate knowledge of the expence of manufacturing Cannon, cannot state with precision what I can make them for. However, I will engage that part of the work on as moderate terms as they can be manufactured in Richmond.

"By a thorough investigation of the United States Factories in the gun way and those of Richmond, I am confident it will be found that the arms, after calculating waste of materials of different kinds, officers, salaries, men on days' work, and incidental expenses that are impossible to ascertain accurately, cost considerably more than the terms I now offer.

"Having served a regular apprenticeship to the Gun-making business in all its various branches, both Military and Birding guns, and having manufactured Arms for the Commonwealth of Virginia on my own account and superintended the making of the whole of the arms delivered by McCormick, I hope it will not be supposed that I am not competant; However, if it should be doubted, I am personally [known] to Edw'd Lloyd, Esq'r, the present Governor of this State, and Rob't Wright, the late Governor (now member of Congress); by a reference to those Gentlemen, they can testify as to my abilities in the gun business. I can produce satisfactory certificates as to my abilities and character from Philadelphia, where I was appointed inspector of Arms by Governor McKean, and continued as such till I removed into this State, and as to my public and private character in Baltimore an investigation into it will be considered an honor."[8]

His overtures to the Governor were, however, of no avail, and he remained in Baltimore.

From his letters and his ads, one feels that Haslett had a desire for greater activity than his small shop afforded. His early experience in McCormick's works and his apprenticeship in London undoubtedly had shown him that there were larger possibilities in the firearms business than merely making fine arms, and the fact that he sold imported weapons shows that he was desirous of more business than his own forge could command.

Already in 1809 Haslett began investing in property in downtown Baltimore. He purchased a lot on Cove Street in 1809 for $475, which he sold again in 1815 for $750.[9] In August of the same year he purchased a lot on Aliceanna Street for $570 which he disposed of in 1811 for $625.[10] This activity in real estate continued throughout his life, and his wisdom in purchasing is apparent from the continual profit he realized on his sales.

The War of 1812 did not, strangely enough, find Haslett setting up an arms factory or advertising to make arms for the States as he had done earlier. In fact there is no entry under his name in the directory for 1812. Instead we find that by 1814 he had risen to the rank of Brigade Major in the 11th Brigade of the Third Division of the Maryland Militia. A letter preserved in the Baltimore archives is dated October 6 1814, and is addressed to Samuel Hollingsworth Esq.:

"Dear Sir -- Colonel Ragan of Genl. Stansburys Brigade can furnish the Committee of Vigilence & Safety with 600 Men for one day, to labor on the works for the defense of the City of Baltimore. James Haslett, Brigade Major, 11th Brigade Milt."

[Publisher's note: The following letter, turned up by Dan Hartzler in the course of his research, was unknown to Richard Randall at the time he did his writing. It shows that in 1813 James Haslett was involved in the inspection and repair of arms.]

James A. Buchanan, Esqr. Baltimore 15th April 1813

Sir -- I have examined the 49 swords at (Mr Lemons,) and broke in proof 8, leaving 41 servisable, which I have altered according to (Mr. Leonard Taylor's) orders. I have examined at (Mr. Wm Hollins's) 238 muskets of a good quality, 9 of them, the Hammers were soft, 34 the breeching projected beyond the touchole, which defects I have remedy'd, one bayonet broke in proof, not yet replaced.

I examined (Mr. Carthaus's) pistols, there is 12 pair that may do with wiping up.

I have overhauled the whole of (Mr. A. Cloppers) muskets, and find only Fifty, that are any way servisable for militia, they are nearly alike wide in the caliber, they have brass pans. I have them in my possession, they want to be wiped up and some trifling repairs done to some of them.

Likewise the whole of (Captn Stiles) Arms they are of the same Quality as Mr. Clopper's. I have selected 30 from his of the same finish as Mr. Cloppers/ Brass pans/. The bayonets of neither sample are as good as they ought to be.

I have examined Mr. Gitt's 10 muskets they make very good fire, are very good for ship use but not calculated for soldurs they differ in caliber and they have lost their bayonet Fastnings. James Haslett

[Baltimore City Correspondence Manuscripts, 1813, No. 549.]

In 1814 Haslett's listing in the directory at 28 Water Street reappeared and it continued in all subsequent directories up to 1827. In 1814 he was a man of 41 years, apparently much respected in the city, and called "Major Haslett," as later newspaper references show. His gun business must still have been profitable, judging from his various investments in real estate. He purchased one lot in 1815, leased or mortgaged four in 1821, and bought two others in 1831 and 1833. In this period he also sold one lot and rented two others.[11]

In the 1822 directory there appears the following entry: "Haslett, James, gunsmith, 22 Water Street, and grocer, 82 Pratt dw. 22 Water." His apparent dissatisfaction with the renumeration of the gun business, in addition to his advance in years, had perhaps prompted Haslett to turn in a new direction. Possibly he felt that gunsmithing was not the ideal life for a city dweller in America, for his son did not follow in his footsteps in the arms trade. The double entries under Commerce or Grocer and Gunsmith continue in the directories of 1824 and 1827. In the later, an extra entry-- "Haslett, Jas. & Son, grocers 82 E. Pratt" --shows that young James Haslett probably had been running the grocery since its opening in 1822, while his father advised him and continued gunsmithing. In the next directory, 1829, there no longer appears a listing under gunsmith, and the lot on Water Street, formerly the Sign of the Golden Gun, is listed as a dwelling.

In 1825, 1827, and 1828 Haslett and his son signed petitions to the city fathers regarding weights, and in 1829 the firm of James Haslett and Son obtained a retail license.[12] The most interesting document of the period for the characterization of James Haslett is a letter to one of the Mayor's Committee members written in 1830:

"The Honbl. Jacob Small -- Sir:
The bearer Thomas Moore has been in the habit of selling Oranges and Lemons opposite my store in Pratt Street. I believe him to be a quiet industrious Citizen, and if it is not contrary to the City Ordinances I have no objection to his continuing to sell for any time you may think proper. Respectfully your obed't Servt. James Haslett."

A new experiment was undertaken in 1831, and James Haslett opened a clothing store at Pratt and South Streets, while James, Jr., continued in the grocery business. In 1832 "Major" Haslett received the honor of being appointed judge of the Sixth Election District of Baltimore during the debated reelection of Andrew Jackson.[13]

The social position of the Haslett family is in part reflected by the marriages of his daughters and his son. The eldest daughter, Mary was married to merchant Thomas Wilson in 1818. The second daughter, Sarah, married George Gillingham, the clerk of the Baltimore Shot Tower, in 1822, and, so to speak, married in the trade. James Haslett's son, James, Jr., married Margaret J. Turner, about whom nothing is known, in 1823. The third daughter, Emily, married one of Baltimore's more prominent citizens, Dr. Adolphus Dunan, in 1824. The last daughter, Eliza, was married in 1833 to Andrew Meyer, about whom nothing is known except that he was a catholic.

In 1833 James Haslett is listed in the directory as merely dwelling at 28 Water Street. On 15 August of that year he died, at his "Estate at Drum Point, Calvert County, Maryland."

Haslett was sixty when he died at his country estate. He was a man who had been trained by the best of his profession in England, manufactured arms for the State of Virginia, tried his skill in two of the east coast's largest cities, making all types of arms from pocket pistols to duellers and flintlock longrifles, and yet retired from gunsmithing to become a grocer and clothing merchant. Throughout he was an able man and a respected citizen, and apparently he died in affluence, if one were to judge by the properties he owned.

* * * * * *

The quality of James Haslett's firearms must have attracted considerable attention in Baltimore. In 1810, for instance, he was not only able to list the Governor of the state, Edward Lloyd, and a senator, Robert Wright, among his customers, but also to count on their friendship for references to the Governor of Virginia. These gentlemen most likely bought hunting guns from Haslett, such as those he advertised to supply in 1806. Edward Lloyd's taste in fine fowling pieces is attested by his purchases from John Manton in London and from Haslett's contemporary, John Stewart of Baltimore. A fine shotgun by the latter is still at the Lloyd home, *Wye House.* To date, none of Haslett's long fowling pieces have come to light, though a short shotgun either for coach defense or to be carried on horseback in a boot is in the collection of the author.

The shotgun is 32 5/8 inches over-all with a 17 1/8 inch barrel, or roughly the size of the normal coaching blunderbuss. The lock resembles those in use in London about the turn of the century. It has a squared plate tip, a safety catch, a roller on the frizzen-spring, and a gold-lined pan with a separate fence. It bears unusual engraved motifs--palms, a sunburst, a trophy of musical instruments, and a tiny landscape with a flying bird, foliage, and a jump. In the center of the plate is a silver inset with raised inscription HASLETT / BALTIMORE.

The heavily-built barrel is sixteen-sided for half its length and round to the muzzle. It notches into the separate tang with a wide sighting notch, the borders of which are engraved with fruits and leaves. Aside from the thin silver band set in at the breech and and a platinum vent, the decoration consists solely of the amusing front sight. The sight is formed as a silver spider with its eight legs inlaid in the barrel and its body acting as the bead front sight. On top of the barrel is stamped in large letters BALTIMORE, and on the side at the breech is the enigmatic Eagle Head proof mark in an oval. This mark does not appear on any other Haslett arms to my knowledge, and though it could have been a private proof that he used at the time he made the shotgun, it might also bear some connection with his inspectorship of arms in Philadelphia. The mark appears on other American arms, but no satisfactory explanation has yet been put forward for its use. C. Meade Patterson stated in a letter (personal communication) that he did not feel the eagle proof was either a U.S. or a state mark. It was used mainly between 1798 and 1816. On the underside of the barrel is a half rib with a steel ramrod pipe to which is attached a swivelled steel rod.

The stock of the shotgun is dark, figured walnut with a cheekpiece and a checkered grip, nearly obliterated through use. Its minor mounts are undecorated silver, but the butt cap and scroll trigger guard are finely engraved. On the strap of the butt cap is a landscape filled with small plants and a proud stag looking over his shoulder. The scroll-end of the trigger guard is engraved with the same motifs as the steel tang, and the forward finial is a pineapple. The bow of the guard bears a musical trophy composed of horn, drum, and music book below a hunting horn through which protrudes a boar's head. Above the horn are a globe and a burgonet with a plume crest.

The motifs attest Haslett's dependence on his English training, and probably also on an English pattern book. The homogenous quality of the engraving on both steel and silver parts shows that Haslett engraved all the parts himself rather than farming the mounts out to a silversmith.

The second arm that has come to us from Haslett's shop is a very fine little pocket pistol in the Samuel E. Smith collection. It is but 5 9/16 inches long with a barrel of 1 9/16 inches. The barrel is engraved at each end with a leaf pattern and is .36 caliber, smoothbore. It has a side hammer and the trigger guard pushes forward as the safety. There also is a rear sight on a raised rib at the breech, a button front sight, and a gold-lined pan. It is engraved very much like the shotgun, with sunbursts and a spray of flowers on top of the frame, a trophy of arms on the trigger guard, and on the left side a second trophy of banners, drums, and cannon balls surrounding a gold oval with a raised inscription HASLETT / BALTIMORE. The walnut butt is checkered, inlaid with patterns in silver wire and a blank escutcheon, and fitted with an engraved silver cap. The barrel and frame are both numbered 1.

The third and fourth examples of Haslett's work that I am able to describe are a pair of converted duelers, formerly in the Richard D. Steuart collection, and now, I believe, in the William M. Locke collection. Each is 14 7/8 inches over-all with a 9 13/16 inch barrel. The barrel is octagonal, of russeted twist steel, and extremely heavy, each pistol weighing four pounds and a half ounce, with a .62 caliber bore. On the underside is a half rib to carry a brass-tipped oak rod, and the top of the barrel is channelled with a long sighting groove and has a small silver leaf front sight. The barrel is fitted with Nock's Patent breech or a slight variation of it, and on top of the breech are inlaid ribbons of gold with the words HASLETT and BALTIMORE stamped on them respectively. On the under-side, each barrel is again stamped HASLETT / BALTIMORE. The tang is separate, as in the shotgun, and is inlaid in gold with bolts of lightning from three engraved clouds. Each tang is stamped No. 17.

The lockplate of blued steel, is engraved with a sunburst on the toe and in front of the hammer, and with a series of acanthus leaves which appear to date from the conversion. In the center of the plate, an insert of gold bears the stamped name HASLET [sic.] The heavy fences are made as integral parts of the plate and were retained when the conversion was effected by adding a percussion drum. The plates are secured by only one screw, the for-end being fitted with a lug which notches behind a fixed screw in the stock.

Each stock is dark walnut and has a flattened butt insert with an oval steel plate engraved with leaves. The butt is finely checkered, the mitring of high quality. The small mounts are plain silver, and the spur trigger guard is blued steel inlaid in silver and gold with a trophy of flags, spears, drum and shield. There is a set trigger.

This pair of pistols seems typical of Haslett's work. A second pair of flints sold through Clapp and Graham in 1929 have many similar features: the heavy octagonal barrels with a half rib underneath, spur trigger guards and flattened butts with a vase-engraved inset. In the October 1929 issue of *Antiquarian* they were described as having "superb gold inlay and . . . finely chased." A third pair of pistols was sold in Denver in 1941. They were saw-handled flints in somewhat worn condition, and described as being "in the English style," according to a letter by W. Keith Neal.

There is a final piece I can mention only briefly. It is a flintlock longrifle by James Haslett, in a private collection, and I have only a meagre description. The rifle is in mint condition, signed in gold on the barrel HASLETT / BALTIMORE. It is entirely silver mounted, and the mounts are engraved with motifs similiar to the pieces described above.

As far as date order is concerned, I see no way of dating the existing nine examples by Haslett except between 1803 when he arrived in Baltimore and 1827, the last year in which the entry in the Baltimore directory lists him as a gunsmith. The proof mark on the shotgun perhaps indicates a date prior to about 1816. Haslett's production must have been relatively large, if his modest fortune at his death was the result of his smithing, as it seems to have been, judging from the existing records.

REFERENCES & NOTES

1. William S. Haslett was his executor in 1833. He cannot have been identical with the merchant of Philadelphia, William Haslett, who came from Dumnicey, Ireland, and was also a Presbyterian, because he died in 1821. It is possible that they were cousins, however, and that William Haslett of Philadelphia arranged for James Haslett to come to Philadelphia to work with McCormick.
2. Hicks, Major James E. (1946) *NOTES ON U.S. ORDNANCE, V. I. SMALL ARMS 1776-1940,* p. 20. For a further discussion of McCormick see: Knode, Harry C. (1954, July) "McCormick Pistols?" *TEXAS GUN COLLECTOR,* No. 48.
3. Satterlee, L. D. (1935, January) *HOBBIES MAGAZINE* (from *THE PENNA. MAGAZINE OF HISTORY & BIOGRAPHY*, p. 236.).
4. Satterlee, op cit., p. 243.
5. See Haslett's advertisement of 3 June 1803 and letter of 1810.
6. In the Lloyd papers at the Maryland Historical Society are several gunsmith's bills, including one from John Manton.
7. *FEDERAL GAZETTE AND BALTIMORE DAILY ADVERTISER,* Nov. 24, 1803.
8. Flournoy, H. W. (1892) *CALENDAR OF VIRGINIA STATE PAPERS, v.X.*
9. Baltimore Courthouse Archives, W.G. 101,388 and W.G. 131,274.
10. Ibid., W.G. 103,238 and W.G. 114,115
11. Ibid., W.G. 135,466; 160,426; 162,174; 162,176; 162,182; 211,325; T.K. 228,498; W.G. 160,416; i71527; 182,49; 190,49; T.K. 235,366.
12. Ibid., 1825, No. 183; 1827, No. 523; 1828, No. 364; 1829, No. 1603.
13. Ibid., 1832, No. 1253.

HATKILL, J. Washington, D. C.

May 21, 1811
Be it known that we W. Thornton & J. Hatkill of Washington city, in the District of Columbia have invented a new & useful Improvement in Fire-Arms.
[U. S. Patent Office]

HAWKEN FAMILY

Author's note: The information presented below for the Hawken family of gunsmiths was compiled and written by George Shumway, who took considerable interest in publishing here as much data as reasonably could be found about the genealogy of the Hawken family.

Hawken is one of the most famous of all early American gunsmithing names. This is primarily because of the reknown gained by the St. Louis gunsmiths, Jake and Sam Hawken, who made sturdy and functionally beautiful percussion plains rifles, the favorite of plainsmen and mountainmen alike. But Sam and Jake were not the only gunsmiths to bear the name. Research to date has turned up fifteen gunsmiths in the Hawken family, and it is likely that still others will in time be known.

A family with fifteen gunsmiths in it most certainly deserves intensive study. Over recent decades attempts have been made to piece together the genealogical relationships within the Hawken family, both by members of the Hawken family interested in their own genealogy, and by various researchers interested in early American gunsmiths. Each has contributed information, but with one notable exception, nobody has attempted to put the pieces together to show the overall picture.

Attempts at researching the Hawken family go back twenty years or more, and doubtless some of them are by people I am not even aware of. The intentions of the various people involved have not all been the same. Some family members have been primarily interested in their own branches. And some arms enthusiasts have been primarily interested in the activities of Jake and Sam's gunsmith work in St. Louis.

It was fortunate that in the course of trying to put the scattered pieces together I got acquainted (by correspondance) with Glenn R. Scott of Lakewood, Colorado. Glenn's wife, Mary Juanita Weiker Scott, is descended from John Hawken of Hagerstown, a son of Christian, Sr. For a number of years Glenn and Juanita have pursued the Hawken family genealogy in a systematic manner, putting together the pieces from previous researchers, and examining the primary sources themselves. With their generous help, the information presented here has been put together.

Most of the accumulated information about the Hawken family has come from examination of the public records for the communities in which the persons lived. This includes federal census data, and military records, from The National Archives, deeds, wills, estate settlements, property tax records, and other records in various court houses, the Hall of Records in Annapolis, Maryland, and the Library of the Church of Latter Day Saints in Salt Lake City, Utah. Mrs. Frank S. (Rachael S.) Schwartz of Hagerstown provided Hawken family data from the Parish Register of St. John's Lutheran Church there. Lt. Col. Vaughn K. Goodwin of Annapolis did some important research on the Hawkens of Hagerstown which was privately published for members only of the Kentucky Rifle Association; with Goodwin's permission and that of the Kentucky Rifle Association, portions of his research were made use of here.

Information about the Hawken family has accumulated in a file at the Missouri Historical Society in St. Louis and we acknowledge here their contribution. Among the information in the Historical Society's file is genealogical data prepared by Hazel J. MacIvar of Town and Country, Mo. A bundle of Hawken family data from the Society's files was loaned to me by Arthur J. Ressel of St. Louis. Paul E. Hawkins of Kansas City, Mo., kindly supplied information from his branch of the family.

Doubtless there have been others who have contributed to the information put together here. In addition there have been others who have taken particular interest in the Hawken family of gunsmiths, including John Barsotti of the Ohio Historical Society Center, Tom Dawson, and Arthur J. Ressel. One of the principal perpetrators of present-day interest in the Hawken rifle has been John D. Baird of Big Timber, Montana, who has written and talked about them and the times of their use for the past ten years or so. John has written and published two books about the rifles, and presented therein some information about the Hawken gunsmiths of St. Louis: HAWKEN RIFLES, THE MOUNTAIN MAN'S CHOICE (1968), and FIFTEEN YEARS IN THE HAWKEN LODE (1971). Both titles are out-of-print at this writing. In addition, John D. Baird is Editor and Publisher of a monthly magazine titled THE BUCK-

SKIN REPORT which is chiefly concerned with the mountain man of old and his present-day reincarnation. his rifles and accoutrements of old and their present-day counterparts, and his life-style of old as well as its present-day renaissance.

* * * * * * *

The Hawken family appears to have been of Germanic origin, the term being used in the broad sense to include people from Switzerland and other German-speaking countries. The family in Hagerstown were members of St. John's Lutheran Church. Beginning in 1779 the Parish Register for this church lists a number of births, deaths, and marriages for members of the Hawken family. In some of the earlier references the name is spelled "Hachen," which probably was the Germanic spelling. But it also appears as "Hacken," "Hawken," "Hawkens," "Hawkey," and "Hawkes." The first recorded deed for Christian Hawken, Sr. dated 1784, has his name spelled three different ways, "Hockey," "Hawken," and "Hawkin;" the deed is indexed under "Hocky." Allowance must be made for the informality of spelling in this period and for the complexities that resulted in spelling of names when English-speaking and German-speaking cultures got mixed together.

It is hoped that the location of Christian Hawken's birth will be determined from the 1880 Federal Census, for each adult was at that time asked to give the place of birth of his or her parents. Both Samuel and William were living in 1880. William's name was not found in this census in either the lists for Hagerstown or Williamsport. At this writing the 21 reels of microfilm for the St. Louis census have not been searched.

The Hawken family in America, or at least the part we are concerned with here, had its American beginnings in Hagerstown, Maryland. The earliest information about them is the following: "I read an old letter once that one of grandpa's brothers wrote. It said that three brothers came to Hagerstown, Maryland, before the Revolutionary War in 1775. They came from Switzerland. Their names were Hawken. No doubt they served in the Revolutionary War." The "grandpa" mentioned was Thomas Paine Hawkins, and his brother who wrote the letter would have been the gunsmith Henry Clay Hawken of Springfield, Ohio. The writer of the information quoted was Anna Belle Hawkins Diehl.

The information reported above from the letter of Henry Clay Hawken needs careful consideration, for it must be regarded as a secondary source material. However, it is all that we have. The Swiss origin for the family seems like a good possibility. It also seems likely that the name originally was "Hachen," for the Germanic pronounciation of this name would not be much different from the English "Hawken." That there were three brothers involved may have been the case, though we know of only two who possibly or probably were brothers, the gunsmiths Christian Hawken, Sr., (born 1756), and Nicholas Hawken (born before 1755). If there was a third brother, his identity is unknown. It is not proven that Nicholas and Christian, Sr., were brothers, but circumstances suggest that they may have been.

Christian Hawken, Sr.'s mother was living in Hagerstown in 1787. On 27 August 1787 Christian's mother, Catherine Fritz of Washington County, granted him Power of Attorney to act in all legal matters on her behalf. Apparently she had been married twice and widowed twice at the time of this instrument. No further mention of her has been found in the public records.

Whether the Hawkens arrived in Hagerstown in 1775 as suggested in the letter from Henry Clay Hawken, or whether they arrived a few years later is not known. The suggestion that the Hawken brothers served in the Revolutionary War seems to be speculation on the part of Anna Belle Hawkins Diehl, who was working on D.A.R. qualification at the time, rather than information from Henry Clay Hawken. No record of the activities of Nicholas and Christian, Sr., during the war years is known. They may have been present in Hagerstown or Frederick, working at gunsmithing. The first record of the Hawkens in Hagerstown is in the Parish Register of St. John's Lutheran Church which notes that Nicholas and Elizabeth Hachen had a daughter, Juliana, born 28 March 1779. Christian Hawken and his wife Juliana had a son John George, born 19 February 1781. And Nicholas and Elizabeth Hachen had a son, George, born 16 July 1782. Christian and Juliana had a second son, John, born 19 March 1784.

The family of Christian Hawken, Sr., and his wife, Juliann, was large, and the household was stable for the duration of their lives. Christian apparently worked at gunsmithing in Hagerstown until his last days, for his estate inventory lists many tools of the trade. Juliann (as she correctly spelled her name) apparently was well suited to being the mother and grandmother of a large family--a warm and loving person who maintained ties with her partially-scattered family. She must have been well-educated, for she expressed herself well in one letter which has survived and which is in the collection of Hawken family data in the Missouri Historical Society. This was written in 1823, when she was 62 years of age. It is important in a number of ways. It tells something about Juliann's character, including the fact that she could be cheerful and optimistic after having gone through a number of years of sorrow. Her son, Christian, Jr., had died the previous year, her husband, Christian, Sr., had died a year before that, in 1821, her daughter-in-law Rosannah had died the same year, her grandson William L. had died the previous summer, and at some time prior to 1821 she had lost two daughters, Elizabeth and Margaret.

This letter also ties together and helps to confirm relationships within the family. It shows that some of her grandchildren from St. Louis spent time living in Maryland with relatives. And it establishes beyond any doubt that Samuel and Jacob Hawken of St. Louis were her children.

[letter addressed to] Mr. Samuel Hawken
St. Louis, Missouri

Hagerstown
January 26, 1823

To Jacob, Samuel &
Catherine Hawken

My Dear Children ---

I received a letter from Samuel dated December 11, 1822, in which he mentions that he is married. I, with the rest of us wish him much happiness together with his wife -- I hope you will make as good a husband to her as you did to your late amiable wife Rosannah, and I hope your present wife will be a crown to her husband. -- Your dear children are well, and doing well, and grow finely -- they are at present in the country at their Grandmothers. -- I sent word to them that you were married; and do not know what their answer will be. -- As you have mentioned in your letter that you are coming in for them next fall or spring, I think it would be advisable that you should bring your wife with you to assist in taking them out -- We would be glad to see her at all events. You say that she is the daughter of Hamilton Ritchie of Mercersburg, and there stopt -- What Mercersburg is it? Pennsylvania, Missouri, Indiana, or what other state? or dont it belong to any state. I hope in your next you will satisfy us, and I hope your next will be a whole one, and not a half letter like the last. -- As for Mr. Jacob and Catherine, they don't think us worth their notice. I really had great confidence in Catherine? before Jacob came in from St. Louis Catherine would say that he was a very negligent, and should write more frequently, that if she was in his situation she would be more mindful of her friends -- but how is it now. She is there herself, and as careless as Jacob ever was. I wish her to write, and write often, and also Jacob; we like to hear from you often.

I am sorry to inform you that I have entirely lost the sight of my left eye, and I feel the other one getting weaker. -- I suffer very much indeed, the pain is very severe. -- I just received a letter from Harpers Ferry -- our friends there are all well.

William & Leah are well, but no prospects of an increase in number. -- Nancy is going to school and when her quarter is up is going to write herself.

About three weeks ago Mr. Will made sale of all our property except the Harry's Town property. -- That in which I live on I bought myself at $800. That on which Levin C. Willis lives, sold at $1800, which was bought by Samuel Downey and Jacob Swope. The lots were sold at $320. -- Old Stump the Sow gelder bought two and William Hawken bought one. -- William gave $80 for his. There is a purchaser for the Harrys Town property -- the sale for that will take place in a few days. I shall now close this letter by requesting you to answer this letter immediately, and take turn about in writing regularly and often. I want to receive letters from all of you -- you can all write.

You can tell John that his grandfather received a letter from his mother, &c and she and his brother are well.

You will all accept of my kind love, and good wishes for your health and prosperity -- and remember your duty and obligations towards our gracious Lord and your neighbours. --

Do not neglect to serve your God as becometh Christians, and to do unto your neighbour as you would they should do unto you, at least as near as you can.

On account of age and infirmity, in all human probability I shall not be long a dweller upon earth, I will have to travel to that bourne from whence no traveller returns, and therefore as the fountain from whence your existence sprang, I wish you well and hope that you will remember your Creator in the days of your youth, and be kind and loving one to another.

From your affectionate
Mother
Juliann Hawken

P.S. Mrs. Oster and the rest of all their family send their respects to you all. J. H.

* * * *

I must inform Jacob & Catherine that I have another son, it was born the 26th December. I now have three patterns for them. I hope Cate's double chin is fast improving -- let us know in your next -- I wonder if Cate remembers the time she went to Philip Petry's for the pig -- the sleigh didn't upset, did it -- the Indian doctor is well -- you recollect Selser the second --

I was requested to inform you that Sophia Smith is not married; she is now setting here by the stove talking with Nancy, and indeed looks very well -- she has on a plaid frock, which is very flashy, and tipt off other-

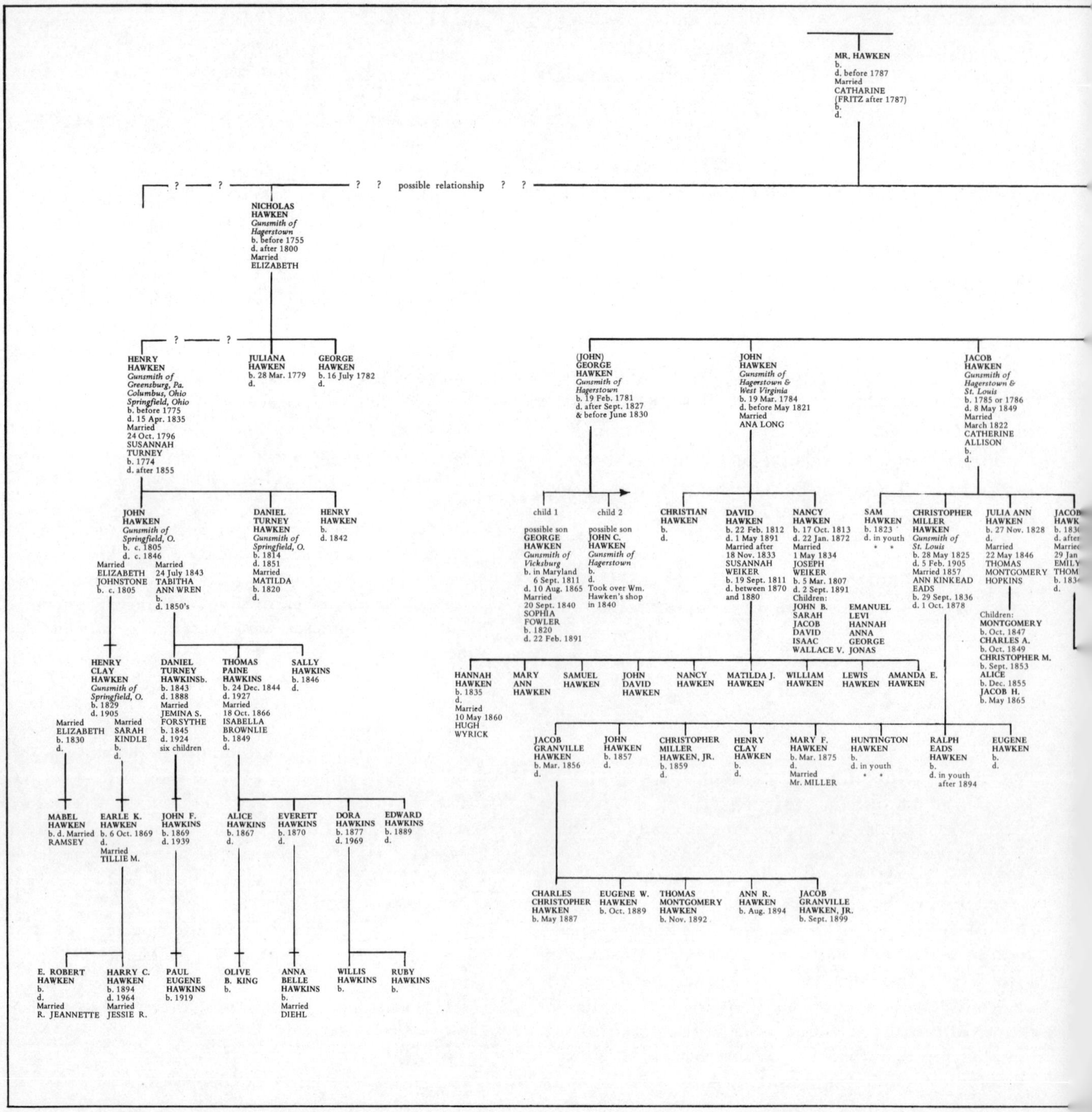

MR. HAWKEN
b.
d. before 1787
Married
CATHARINE
(FRITZ after 1787)
b.
d.
? ? possible relationship ? ?
NICHOLAS HAWKEN
Gunsmith of Hagerstown
b. before 1755
d. after 1800
Married
ELIZABETH
HENRY HAWKEN
Gunsmith of Greensburg, Pa. Columbus, Ohio Springfield, Ohio
b. before 1775
d. 15 Apr. 1835
Married
24 Oct. 1796
SUSANNAH TURNEY
b. 1774
d. after 1855
JULIANA HAWKEN
b. 28 Mar. 1779
d.
GEORGE HAWKEN
b. 16 July 1782
d.
(JOHN) GEORGE HAWKEN
Gunsmith of Hagerstown
b. 19 Feb. 1781
d. after Sept. 1827 & before June 1830
JOHN HAWKEN
Gunsmith of Hagerstown & West Virginia
b. 19 Mar. 1784
d. before May 1821
Married
ANA LONG
JACOB HAWKEN
Gunsmith of Hagerstown & St. Louis
b. 1785 or 1786
d. 8 May 1849
Married
March 1822
CATHERINE ALLISON
b.
d.
JOHN HAWKEN
Gunsmith of Springfield, O.
b. c. 1805
d. c. 1846
Married
ELIZABETH JOHNSTONE
b. c. 1805
Married
24 July 1843
TABITHA ANN WREN
b.
d. 1850's
DANIEL TURNEY HAWKEN
Gunsmith of Springfield, O.
b. 1814
d. 1851
Married
MATILDA
b. 1820
d.
HENRY HAWKEN
b.
d. 1842
child 1
possible son
GEORGE HAWKEN
Gunsmith of Vicksburg
b. in Maryland 6 Sept. 1811
d. 10 Aug. 1865
Married
20 Sept. 1840
SOPHIA FOWLER
b. 1820
d. 22 Feb. 1891
child 2
possible son
JOHN C. HAWKEN
Gunsmith of Hagerstown
b.
d.
Took over Wm. Hawken's shop in 1840
CHRISTIAN HAWKEN
b.
d.
DAVID HAWKEN
b. 22 Feb. 1812
d. 1 May 1891
Married after 18 Nov. 1833
SUSANNAH WEIKER
b. 19 Sept. 1811
d. between 1870 and 1880
NANCY HAWKEN
b. 17 Oct. 1813
d. 22 Jan. 1872
Married
1 May 1834
JOSEPH WEIKER
b. 5 Mar. 1807
d. 2 Sept. 1891
Children:
JOHN B.
SARAH
JACOB
DAVID
ISAAC
WALLACE V.
EMANUEL
LEVI
HANNAH
ANNA
GEORGE
JONAS
SAM HAWKEN
b. 1823
d. in youth
CHRISTOPHER MILLER HAWKEN
Gunsmith of St. Louis
b. 28 May 1825
d. 5 Feb. 1905
Married 1857
ANN KINKEAD EADS
b. 29 Sept. 1836
d. 1 Oct. 1878
JULIA ANN HAWKEN
b. 27 Nov. 1828
d.
Married
22 May 1846
THOMAS MONTGOMERY HOPKINS
Children:
MONTGOMERY
b. Oct. 1847
CHARLES A.
b. Oct. 1849
CHRISTOPHER M.
b. Sept. 1853
ALICE
b. Dec. 1855
JACOB H.
b. May 1865
JACOB HAWK
b. 183
d. after
Marrie
29 Jan
EMILY THOM
b. 183
d.
HENRY CLAY HAWKEN
Gunsmith of Springfield, O.
b. 1829
d. 1905
Married
ELIZABETH
b. 1830
d.
Married
SARAH KINDLE
b.
d.
DANIEL TURNEY HAWKINS
b. 1843
d. 1888
Married
JEMINA S. FORSYTHE
b. 1845
d. 1924
six children
THOMAS PAINE HAWKINS
b. 24 Dec. 1844
d. 1927
Married
18 Oct. 1866
ISABELLA BROWNLIE
b. 1849
d.
SALLY HAWKINS
b. 1846
d.
HANNAH HAWKEN
b. 1835
d.
Married
10 May 1860
HUGH WYRICK
MARY ANN HAWKEN
SAMUEL HAWKEN
JOHN DAVID HAWKEN
NANCY HAWKEN
MATILDA J. HAWKEN
WILLIAM HAWKEN
LEWIS HAWKEN
AMANDA E. HAWKEN
JACOB GRANVILLE HAWKEN
b. Mar. 1856
d.
JOHN HAWKEN
b. 1857
d.
CHRISTOPHER MILLER HAWKEN, JR.
b. 1859
d.
HENRY CLAY HAWKEN
b.
d.
MARY F. HAWKEN
b. Mar. 1875
d.
Married
Mr. MILLER
HUNTINGTON HAWKEN
b.
d. in youth
RALPH EADS HAWKEN
b.
d. in youth after 1894
EUGENE HAWKEN
b.
d.
MABEL HAWKEN
b. d. Married
RAMSEY
EARLE K. HAWKEN
b. 6 Oct. 1869
d.
Married
TILLIE M.
JOHN F. HAWKINS
b. 1869
d. 1939
ALICE HAWKINS
b. 1867
d.
EVERETT HAWKINS
b. 1870
d.
DORA HAWKINS
b. 1877
d. 1969
EDWARD HAWKINS
b. 1889
d.
CHARLES CHRISTOPHER HAWKEN
b. May 1887
EUGENE W. HAWKEN
b. Oct. 1889
THOMAS MONTGOMERY HAWKEN
b. Nov. 1892
ANN R. HAWKEN
b. Aug. 1894
JACOB GRANVILLE HAWKEN, JR.
b. Sept. 1899
E. ROBERT HAWKEN
b.
d.
Married
R. JEANNETTE
HARRY C. HAWKEN
b. 1894
d. 1964
Married
JESSIE R.
PAUL EUGENE HAWKINS
b. 1919
OLIVE B. KING
b.
ANNA BELLE HAWKINS
b.
Married
DIEHL
WILLIS HAWKINS
b.
RUBY HAWKINS
b.

GENEALOGICAL RELATIONSHIPS

The HAWKEN Family of Gunsmiths

Sources: *Public records, Hagerstown, Washington Co., Md.*
Public records, Hall of Records, Annapolis, Md.
Hawken family records, courtesy of Missouri Historical Society
Public records, L.D.S. Library, Salt Lake City, Utah
Family genealogical studies by Mary Juanita Weiker Scott & Glenn R. Scott
Family genealogical studies by Paul E. Hawkins, Kansas City, Mo.
Research by Lt. Col. Vaughn K. Goodwin
Research by George Shumway
Research by Hazel J. MacIvar, Genealogist, Town & Country, Mo., courtesy Missouri Historical Society, St. Louis

Compiled in 1976 by:
*George Shumway, Publisher * R.D. 7 York Penna. 17402*
Glenn R. Scott & Juanita M. Scott, 60 Estes St., Lakewood, Colorado 80226

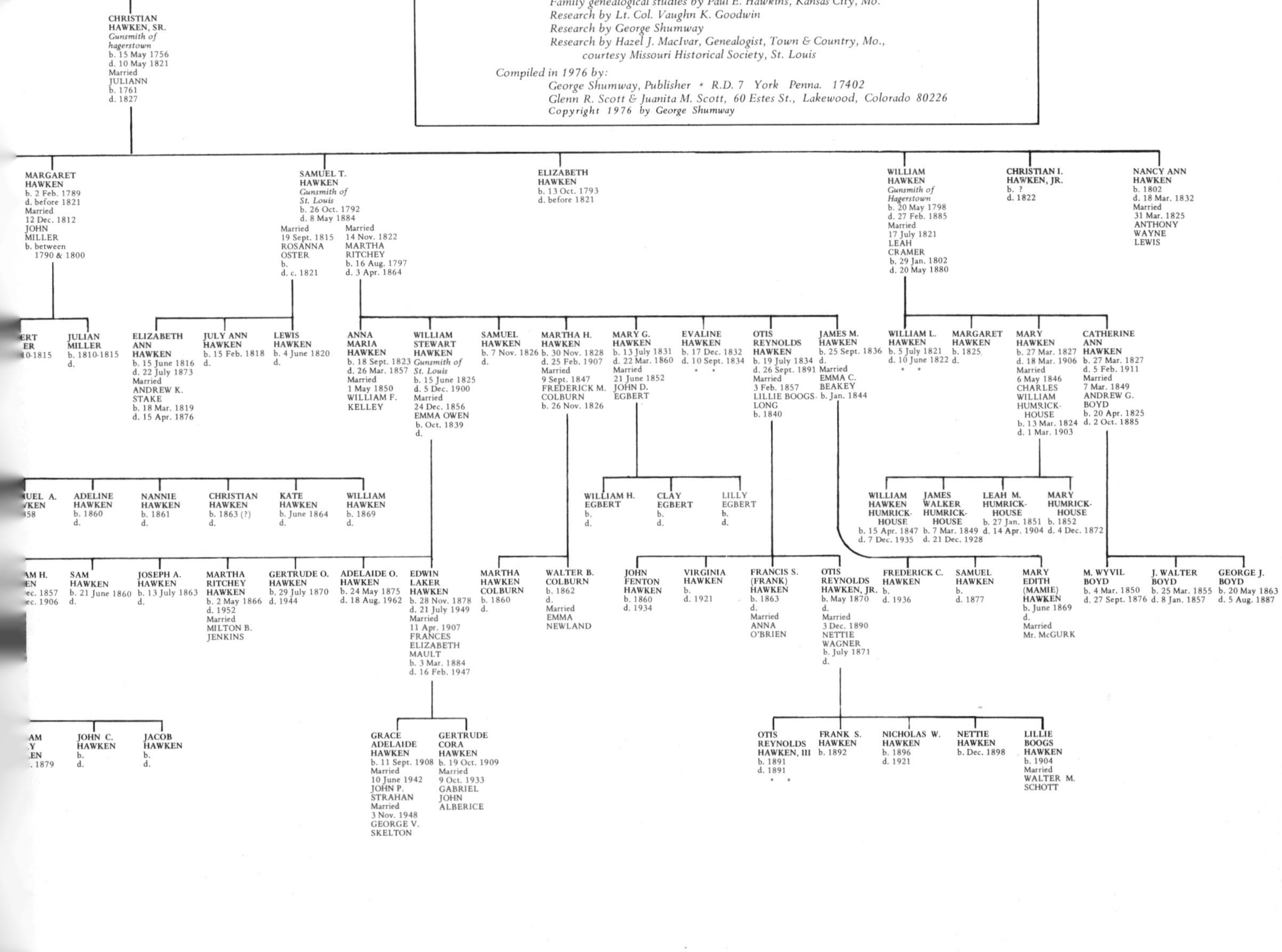

wise right sniptious -- Patience is not here -- Nancy is waiting here with great anxiety for Mr. Griffith from your place -- Patience is gone to her uncle's funeral, Old Nicholas Ridenour is dead -- Nancy & Sophia is here -- they say I must remember their compliments to you which I now do, but if I would say anything about Patience's compliments and a long string of others it would be lies. From your friend

S. Hubert

The letter refers to William and Leah who were Juliann's gunsmith son William of Hagerstown and his wife. It mentions Nancy, most likely a reference to Juliann's youngest daughter who was 20 or 21 years old at the time. And it mentions a letter from "friends" at Harpers Ferry. Earlier in the letter she uses the term "friends" to mean relatives, and it is probable that the reference to friends at Harpers Ferry means relatives there. Harpers Ferry lies about 25 miles south of Hagerstown, and in the 1820's was an active place employing many gunsmiths. It is possible that her son John had been employed there, but at the time of this letter he had been dead for at least a year or two.

Henry Hawken, gunsmith, who possibly was a son of Nicholas, may have had his beginnings in Hagerstown. Henry was married in Greensburg, Pennsylvania in 1796 and lived there until the War of 1812 was underway. After a brief period as a Pensnylvania militiaman during 1812, he and Susannah moved to Ohio where they became the progenitors of the Ohio branch of the family. Their eldest son, John, who was a gunsmith and who resided in Springfield, married twice. He married his second wife, Tabitha Ann Wren in 1843, and within a few years had three children. He died in 1846 and the children were raised in his wife's family; with nobody to defend the name of "Hawken" the children ended up with last names of "Hawkins."

Christian and Juliann Hawken's famous gunsmith sons, Jacob and Samuel T., settled in St. Louis. It is not the purpose of this present account to treat the details of their gunsmithing activities in St. Louis, but it is desirable to show how they tie in with and relate to the other members of the family.

Jacob Hawken was born in Hagerstown in 1786 and probably learned the gunsmith trade from the family without formal apprenticeship. He became 21 years of age in 1807. By 1821 he was established as a gunsmith in St. Louis, but he may have been there for a number of years before that date. It is said that he opened his gunsmith shop there in 1815, or 1819, but definite proof of his location prior to 1818 remains to be found. At some time after the death of his father, Christian, Sr., in May 1821, Jacob returned to Hagerstown, and he was present later in the year at the sale of his father's estate. In May of 1822 he married Catherine Allison there in St. John's Lutheran Church. Juliann's letter shows that he and Catherine had returned to St. Louis by January of 1823.

Samuel T. Hawken was born in 1792 and like his brothers doubtless grew up to the gunsmithing trade. He was drafted into the militia at Hagerstown in July 1814 and mustered out in January 1815. In September 1815 he married Rosanna Oster in Hagerstown. Early the next year they settled in Xenia, Ohio, where Sam did gunsmithing. Sam returned to Hagerstown after his father's death in May 1821 and was present at the estate sale later in the year. By this time Sam and Rosanna had three small children. In the spring of 1822 Sam moved to St. Louis to work with Jacob, his wife having died a few months earlier. Juliann's letter of January 1823 shows that Sam's children were left behind in Hagerstown when he went out to St. Louis. Shortly after Sam's arrival in St. Louis in June 1822 he met Martha Ritchey, and by November they were married. During their long residence in St. Louis they had eight children.

William Hawken, born in 1798, was the fifth son of Christian and Juliann, assuming that Christian, Jr., was born a year or two later. William lived in Hagerstown and worked at gunsmithing throughout most of his long life. He and his wife Leah are mentioned in Juliann's letter of 1823. They were married in July 1821, just two months after Christian's death.

Much less is known about the activities of the other three sons of Christian and Juliann. The eldest son, George, was born in 1781. He worked as a gunsmith in Hagerstown until his death between 1827 and 1830. George was married and had at least two children, though their names are not known. Juliann's will of 1827 provides very specifically that her son George should receive only the interest money from his share of her estate. The implication is that George suffered an incompetancy of some sort.

John Hawken, the second son of Christian and Juliann, was born in 1784. Apparently he learned the gunsmithing trade in the family and moved to Virginia not long after coming of age. His activities in Virginia are unknown, but it is possible that he worked at Harpers Ferry for a number of years. He married and had two sons and a daughter. By 1821 he was dead.

Christian Hawken, Jr., is the Hawken son that we know the least about. He died in 1822, apparently a single man. A record of his birth or birth year is not known. His modest estate was distributed among his mother, his brothers and sisters or their heirs, so he must have been at least 21 years of age at the time. Being named after his father, he might have been the eldest son and therefore born before 1780, but is seems more likely that he was born in the period between William's birth in 1798 and Nancy's birth in 1802.

There were two gunsmiths in the Hawken family whose relationships to the rest of the family are unknown. There is a good possibility that they were the sons of George. One of these men was John C. Hawken, who advertised in a Hagerstown paper in 1840 that he had taken over the gunsmith shop of William Hawken. The other unattached Hawken was George Hawken of Vicksburg, Mississippi, born in Maryland in 1811 and at work as a gunsmith in Vicksburg in the period between 1840 and 1860.

The Ohio branch of the Hawken family stems from Henry Hawken who possibly was a son of Nicholas Hawken of Hagerstown. Henry was born in the period 1770--1775 most likely. By 1796 he was established in Greensburg, Pa., where he married that year. He worked as a gunsmith there until 1812 when he joined the Pennsylvania militia. At some time after his his discharge in 1813 he moved to Columbus, Ohio, and later to Springfield, Ohio, where he died in 1835. Henry had three sons, John, Daniel Turney, and Henry, Jr. John and Daniel T. worked at gunsmithing in Springfield. And John's son, Henry Clay Hawken of Springfield worked as a gunsmith.

The Hawken family genealogy and other data which we present here certainly is not complete or final, but it is the most complete synthesis of data that ever has been put together. It now is possible to see clearly where the missing links and the gaps in the data lie. Some of these gaps will remain always because old records have been destroyed. But some gaps can in time be filled through diligent research efforts.

To date we have found that fifteen members of the Hawken family were gunsmiths. It is likely that still others worked at the trade, at least part-time. And for the sake of completeness, mention must be made here of a particularly skilled gunsmith of the present-day, Tom Dawson, who is a descendent of the the Hawken family.

Hawken, Nicholas of Hagerstown, b. pre-1755
Hawken, Christian, Sr., of Hagerstown, b. 1756
Hawken, Henry of Greensburg, etc. b. pre-1775
Hawken, George of Hagerstown, b. 1781
Hawken, John of Virginia, b. 1784
Hawken, Jacob of St. Louis, b. 1786
Hawken, Samuel T. of St. Louis, b. 1792
Hawken, William of Hagerstown, b. 1798
Hawken, John of Springfield, O., b. c. 1805
Hawken, George of Vicksburg, b. 1811
Hawken, Daniel T. of Springfield, O., b. 1814
Hawken, John C. of Hagerstown, b. pre-1820
Hawken, Christopher M. of St. Louis, b. 1825
Hawken, William Stewart of St. Louis, b. 1825
Hawken, Henry Clay of Springfield, O., b. 1829

The following biographies are those of Hawken family members who were gunsmiths and who relate to Maryland; they are arranged chronologically in order of birth.

HAWKEN, NICHOLAS Hagerstown

Nicholas Hawken lived in Hagerstown for the last two decades of the 18th century, and probably worked as a gunsmith throughout this period, though documentation as a gunsmith is limited to a single deed dated 1798. Surviving examples of his work are unknown. He was the earliest-born of all the Hawken family gunsmiths, being born prior to 1755. It is probable that he was an older brother of Christian Hawken, Sr. The relationship is not proven, but the circumstances of name, occupation, age, town of residence and a family story mentioned above under the Hawken Family heading, argue in favor of it.

The Parish Register for St. John's Lutheran Church in Hagerstown shows that Nicholas Hachen and his wife Elizabeth had children born as follows:

Juliana born 28 March 1779
George born 16 July 1782

Circumstances suggest that the gunsmith Henry Hawken, was a third child of Nicholas and Elizabeth, born prior to 1775. This relationship is not proven.

Only three indentures have been found in the Washington County courthouse recorded in the name of Nicholas Hawken. The first two are curious, for they are dated only three weeks apart in 1792, they both involve household articles, and the second one involves Christian Hawken as well as Nicholas. In searching for clues to relationships, these two bills of sale might be interpreted as documents in which two brothers are involved in the sale and disposition of the household effects of their deceased mother. This is pure hypothesis--but it is known that in 1787 Christian's mother granted him Power of Attorney to handle all of her affairs.

I Nicholas Hawkey of Washington County and State of Maryland for . . . the sum of twenty pounds . . . paid by Wintell Shecter and George Shaver of the County and State aforesaid . . . sell . . . all the household stuffs, Implements and Furniture hereinafter particularly mentioned . . . one Feather Bed and Bed Stead, one ten plate stove, one Cow, one Table, one Chest, one large Iron Kettle, two small Iron Potts, one Copper Tea Kettle, Three Chairs, one pewter dish, two pewter Basons, four pewter Plates, one Watering Pott, one Tin Bucket, one tin Strainer, one frying pan, one small coffee mill all and singular which said articles are now remaining or being in a certain house and lott situate in Elizabeth Town Washington County and now or lately in the occupation of the said Nicholas Hawken . . . In Witness whereof I have hereunto set my hand and seal the 15th day of August 1792.
[Deed Book G, p. 798, 799, Washington CCH]

I Nicholas Hawken of the State of Maryland for and in Consideration of the following Articles, [sic--apparently the recording clerk left out a portion of the document] one Twenty four Our Clock, one Smith Bellace, one New anvill and two Large Vizes and two Hand vizes, one Table, one Cradle, three Bedsteads and two Beads, one Large Iron Pot and two Small Iron Potts, and one frian pan and Skillet, three Chairs and two Chests, and one Stitch Hammer, one Hand Hammer, and three Rivets hammers, and one Shovil and Dung fork, and pair of Hand Irons, a Copper Tea Kettle, and Two pot Rakes, and four pewter plates, and one Large pewter plate, and one Hoge, delivered in the Hands of Christian Hawken and Daniel Wibb of Washington County . . . I have hereto set my Hand and seal this 5 day of September 1792.
[Deed Book G, p. 823, Washington CCH]

By this deed dated 30 June 1798 Nicholas Hawken of Washington Co., Md., Gunsmith, buys from Jonathan Hager of Washington Co. for four pounds Lot No. 312 in Elizabethtown, 82 feet wide by 240 feet long, with annual ground rent of 4 shillings sixpence.
[Deed Book L, p. 321, Washington CCH]

1790 Nicholas Hocky
1 male over 16 years
4 males under 16 years
2 females
[Federal Census 1790, Washington Co.]

1800 Nich[s] Hawken Hagerstown
2 males over 45 years
4 males under 10 years
1 female over 45 years
1 female between 26 & 44 years
1 female under 10 years
[Federal Census 1800, Hagerstown, Washington Co.]

The last documentation for Nicholas Hawken in Hagerstown is the 1800 census. He left no will in Hagerstown and nothing is known about the location or year of his death.

HAWKEN, CHRISTIAN, SR. Hagerstown
Of all the Hawken family gunsmiths, Christian stands out as the patriarch to whom all the others relate in one way or another. Throughout his adult life he lived in Hagerstown and worked as a gunsmith. Documentation for his work as a gunsmith is abundant from 1784 until his death in 1821.

Christian Hawken doubtless produced a large number of rifles as well as some other guns during his long working career. Fortunately, a number of these have survived, ranging from beautiful relief-carved rifles to plainer pieces with no carving. Typically these are signed in script on the barrel with the name "C. Hawken" or "C[n] Hawken." One pistol signed "C. Hawken" is known. At least one rifle has survived bearing the inscription "C. & J. Hawken," which surely must refer to Christian and one of his sons, either John or Jacob.

A tombstone formerly in the graveyard of St. John's Lutheran Evangelical Church in Hagerstown, but now destroyed, contained the information that Christian Hawken was born 15 May 1756 and died 10 May 1821. Family stories mentioned above under the Hawken Family heading suggest that Christian and one or two brothers came to America from a Germanic country, possibly Switzerland, in the 1770's and settled in Hagerstown.

The earliest known record of Christian's presence in Hagerstown is the Parish Register for St. John's Lutheran Church which shows that Christian Hawken and his wife Juliana had children born as follows:

John George	b. 19 Feb. 1781
John	b. 19 Mar. 1784
Margareth	b. 2 Feb. 1789
Elizabeth	b. 13 Oct. 1793

This suggests that Christian and Juliann were in Hagerstown in 1780. He may have been there for a few years prior to that. Where Christian learned the gunsmithing trade is not known--possibly in Europe and possibly in Maryland.

Christian Hawken's mother, Catherine Fritz, lived in Hagerstown, possibly throughout the 1780's. In 1787 she granted Power of Attorney to her son Christian to handle all of her business matters.

Christian Hawken's wife was named Juliann, as is proven by the letter of 1823 written in her own hand. However, in most public records it appears as "Juliana." She was a warm, loving, capable and literate woman, and doubtless a great help to her husband and his gunsmith sons.

Christian and Juliann Hawken had nine children, six sons and three daughters. The famous St. Louis gunsmith, Samuel T. Hawken, was one of the sons. In a newspaper interview given when he was over 90 years of age to the Globe-Democrat of St. Louis, Sam said: "All of my father's boys, except one, were gunmakers. There were six brothers of us..." Christian and Juliann's family consisted of:

(John) George	b. 19 Feb. 1781	Elizabeth	b. 13 Oct. 1793
John	b. 19 Mar. 1784	William	b. 20 May 1798
Jacob	b. 1786	Christian, Jr.	b. ?
Margaret	b. 2 Feb. 1789	Nancy Ann	b. 1802
Samuel T.	b. 26 Oct. 1792		

The first real estate purchased by Christian Hawken that we know about was a town lot in Green Castle, Pa., about 10 miles north of Hagerstown, in February of 1783 (see p. 197).

Christian Hawken and his young wife Juliann may have taken an excursion to North Carolina in 1783. The evidence for this is the 1880 census in which David Hawken, son of John and grandson of Christian, stated that his father, John, was born in North Carolina. The report is curious if not startling, but it is not inconsistent with other known facts. With the signing of the Treaty of Paris in 1783, and in fact for a numbers of years before then, settlers had streamed down the Great Wagon Road through the Valley of Virginia to reach the lands of Kentucky, Tennessee, and North Carolina. Hagerstown lay at the upper end of the Great Wagon Road, and countless wagons, droves of animals, and thousands of people on foot or horse, passed through the town on their way to the promised lands of the West. It is only reasonable to suppose that the young gunsmith, Christian Hawken, and his wife caught a case of wanderlust, like other young people of times past and present, and took off down the Great Road to see what lay at the other end. Juliann got pregnant in the summer of 1783 and on 19 March of 1784 she gave birth to her second son, John. This could have taken place somewhere in North Carolina. Then later in the spring, when infant and mother were ready to travel, they made the return trip to Hagerstown.

The record shows that in June of 1784 Christian Hawken bought his first piece of property in Hagerstown, and happily, the deed mentions that Christian's occupation was that of gunsmith. It also shows the name spelled three different ways, all in one document, and the deed is indexed under yet another spelling of the name, "Hocky."

By this deed, recorded at the request of Christian Hockey, and dated 16 June 1784, between Jonathan Hager of Washington Co., Md., Gentleman of the one part and Christian Hawkin of the County and State aforesaid, Gunsmith, of the other part, Witnesseth that...Jonathan Hager for...20 pounds... doth...sell unto...Christian Hawken...all that Lott...of ground N° 207 lying and being in Elizabeth Town in the County and State aforesaid containing 82 feet breadth and 240 feet in length...to have and to hold unto him the said Christian Hawken...Christian Hawkin...Christian Hawkin...and paying unto ...Jonathan Hager...yearly...the annual...rent of 4 shillings and sixpence...Christian Hawkin...Christian Hawkin...Christian Hawkin...Christian Hawkin...Christian Hawkin.
[Deed Book C, p. 718, 719, Washington CCH]

The next recorded document concerning Christian Hawken is the Power of Attorney for his mother. This contains yet another spelling of the name, "Hochen."

At the request of Christian Hochen the following Power of Attorney was recorded August 27th 1787...Maryland, Washington County -- ...I Catharine Fritz of County and State aforesaid do hereby assign, ordain, appoint...Christian Hochen my son of the County and State aforesaid to be my true and Lawful attorney...this 27th day of August 1787... Came the within named Catharine Fritz...and acknowledged... Christian Hickey to be her Lawful Attorney.
[Deed Book E, p. 496, 497, Washington CCH]

1790 Christian Hocky
2 males over 16 years
4 males under 16 years
2 females
[Federal Census 1790, Washington Co.]

I Nicholas Hawken of the State of Maryland for and in Consideration of the following Articles, [*sic--apparently the recording clerk left out a portion of the document*] one Twenty four Our Clock, one Smith Bellace, one New anvill and two Large Vizes and two Hand vizes, one Table, one Cradle, three Bedsteads and two Beads, one Large Iron Pot and two Small Iron Potts, and one frian pan and Skillet, three Chairs and two Chests, and one Stitch Hammer, one Hand Hammer, and three Rivest hammers, and one Shovil and Dung fork, and pair of Hand Irons, a Copper Tea Kettle, and two pot Rakes, and four pewter plates, and one Large pewter plate, and one Hoge, delivered in the Hands of Christian Hawken and Daniel Wibb of Washington County...I have hereto set my Hand and seal this 5th day of September 1792.
[Deed Book G, p. 823, Washington CCH]

By this deed, dated 23 June 1795, Christian Hawken of Washington Co., Md. buys from George Doyle of Frederick Co., Md., for 200 pounds, Lot No. 28 in Elizabeth Town, 82 feet wide and 240 feet in depth, with an annual ground rent of 7 shillings six pence.
[Deed Book I, p. 264, Washington CCH]

By this deed dated 23 February 1795 Christian Hawken of Elizabeth Town, Gunsmith, buys from Melcher Beltzhoover and wife Elizabeth, for 128 pounds, two tracts of land in Washington Co., the one called Waggon Ford on the west side of the Conogocheige containing 25 acres, the other called Hard Times adjacent to Hopewell and containing 13 ¾ acres.
[Deed Book I, p. 92, 93, Washington CCH]

Christian Hawken, like most well-established gunsmiths of the time, probably had one or more apprentices working with him throughout most of his career. David Miller, who later worked as a gunsmith in Fayette Co., Pennsylvania, was one of these.

This Indenture Witnesseth that David Miller, son of John Miller of Washington County and State of Maryland Deceased, by and with the advise and approbation of John Fisher and Daniel Miller (each of the County and State aforesaid, Guardians of said David Miller) hath put himself, and by these presents doth Voluntarily and of his own free will and accord put himself Apprentice to Christian Hawken of the County and State aforesaid Gunsmith with him after the manner of an Apprentice----to serve from the day of the date hereof till he arrives at or attains the Age of Twenty One Years, he the said Apprentice being fifteen years old the first day of March last, during all which term the said Apprentice his said Master faithfully shall serve, his secrets keep, his Lawful commands everywhere gladly obey, he shall do no damage to his said Master, nor suffer it to be done by others, without letting or giving notice thereof to his said Master, he shall not waste his said Master's Goods, nor lend them unlawfully to any, he shall not commit fornication, or contract Matrimony within said term, at Cards, dice or any other unlawful game he shall not play, whereby his said Master may have damage, with his own goods or the goods of others during said term without license from his said Master, he shall neither buy nor sell, nor haunt Ale houses, taverns nor play houses, but in all things as a faithful Apprentice he shall behave himself toward his said Master, and all his during said term, and the said Master, and all during said term shall by the best manner or methods he can teach, or cause to be taught the said Apprentice the Art and Mystery of a Gunsmith, and also to provide for the said Apprentice sufficient Meat, Drink, Apparel and Lodging and Washing fitting for an Apprentice during said term, and also to send said Apprentice to school until he be able to read the Bible in the German Language, and Write and Cast Accounts, and at the Expiration of said term he the said Master is to give unto him the said Apprentice one full suit of Apparel, worth ten pounds current money, or ten pounds in current money, at the Option of said Apprentice, and for the true performance of all and singular the said Covenants and Agreements either of the said parties bind themselves unto the other by these presents. In Witness whereof they have hereunto Interchangeably set their hands and seals this Fifteenth day of April 1796.

his
David X Miller
mark
Christian Hawken

[Washington Co. Indentures, p. 36, Hall of Records]

Militia units were established in the various counties of Maryland as well as in the other states of the new young nation known as the United States during the 1790's. The 24th Regiment of the Maryland Militia was created in Washington County. A book of Militia Appointments (v. 2, p. 126), preserved in the Hall of Records, shows that on 18 June 1794 Christian Hockey was commissioned an Ensign. He was promoted to Captain on 19 May 1800. He resigned his commission two years later, on 8 July 1802. Reference to this title was made by Scharf in his HISTORY OF WESTERN MARYLAND, v. 2,, p. 1062, "Among the residents of Hagerstown and its immediate vicinity from 1800 to 1805 inclusive were... Hawken, Captain Christian, gunsmith, Hawken, George, gunsmith, Hawken, William, gunsmith." The statement is in error as far as William Hawken is concerned, for he was born in 1798.

1800 Christian Hawken	Hagerstown
1 male 45 or more	1 female 45 or more
1 male 26 - 44	1 female 26 - 44
3 males 10 - 15	1 female 10 - 15
3 males under 10	1 female under 10

[Federal Census 1800, Hagerstown, Washington Co.]

Washington County. I DO hereby certify that CHRISTIAN HAWKEN brought before me, a justice of the peace for said county, as a trespassing stray, a BAY MARE, about three years old, fourteen hands high, with a small star in her forehead, & her hind feet white -- no brand or other marks perceivable. Given under my hand the 8th day of June 1807. ADAM OTT
The owner of the above described stray is desired to come and prove property, pay charges, and take her away. June 10, 1807

CHRISTIAN HAWKEN

[Maryland Herald and Hagers-Town Weekly Advertiser, 12 June, 1807]

List of Letters remaining in the Post Office in Hagerstown . . . Christian Hawken . . . George Kreps . . .
[Maryland Herald and Hagers-Town Weekly Advertiser, 3 July 1807]

Christian Hawken, and George Kreps too, were members of the Hagerstown Fire Company, as the following advertisement shows. This activity carried down from father to son in the case of Samuel T. Hawken of St. Louis, who for many years was an active member of the fire company there.

FIRE COMPANY AGREEABLY to a resolution of the said company, . . . Christian Hawken . . . George Kreps . . . are directed to attend at the Court-house Square on Saturday the 2nd day of July next precisely at 3 o'clock in the afternoon, for the purpose of working the fire engines; those who do not attend or furnish persons in their places, will be fined 25 cents each, agreeably to said resolution. The Engine Directors are requested to attend under penalty of one dollar each.
June 23, 1808 Wm. HEYSER, Director General

[Maryland Herald and Hagers-Town Weekly Advertiser, July 1, 1808]

By late 1804 Christian Hawken had serious intentions of moving to Ohio. He clearly stated this intention in a newspaper advertisement dated in November 1804, in which he attempted to sell all of his property. But the property did not get sold and the large Hawken family did not move to Ohio as far as we know. It is curious that at this stage in Christian's life he would consider a major move.

FOR SALE The subscriber intending to move to the state of Ohio, offers for sale the following property, to wit:

That valuable lot and two story dwelling house where he now lives, opposite Mr. Ragan's Tavern, on the Main Street, in Hagers-town, leading to the Western country. This being a corner lot, on which, besides the before-mentioned two story house, which contains three rooms below and three above, there are also a convenient brick kitchen, a good stable, and a well of excellent water. This lot is eighty two feet in front and two hundred and forty feet in depth, well pailed in. The advantageous situation this lot lies in for any kind of business makes it nearly as valuable as any other in town.

ALSO, one half-lot on the same street, with two small log houses, a convenient stable and a well of excellent water on the same, a few doors above Mr. Ragan's Tavern.

ALSO, a good out-lot, covenient to town sowed down in wheat.

LIKEWISE, Thirty-nine acres of well improved land, six miles from Hagers-town, at the Broad-fordings, on Conococheague Creek, on which there are at least twelve acres of good meadow, a convenient house and stable, a never failing spring, and a most excellent seat for a distillery. For terms apply to

CHRISTIAN HAWKEN
November 12, 1804

[The Maryland Herald and Hagers-Town Weekly Advertiser, 4, 11, 18 Jan. 1805]

By this deed dated 7 April 1811 Christian Hawken of Washington Co., Md. sells to Adam Ridenour of the same place for $800 two tracts of land in Washington Co., the one called the Waggon Ford on the west side of Conococheague Creek and adjacent to Hopewell containing 25 acres, the other called Hard Times, adjacent to Hopewell and containing 13 ¾ acres. The sale was acknowledged by Juliana Hawken, wife of Christian Hawken.
[Deed Book W, p. 515-517, Washington CCH]

1810 Chris. Hawken
1 male 45 or older
3 males between 10 & 15
1 female 45 or older
1 female between 16 & 25
1 female under 10
[Federal Census 1810, Hagerstown, Washington Co.]

By this deed dated 31 March 1813 Christian Hawken of Washington Co., Md., sells to the President, Directors and Company of the Hagers Town Bank, for $5000 Lot No. 99 in Elizabeth Town, being 82 feet by 240 feet in size. The sale was acknowledged by Juliana Hawken, wife of Christian Hawken.
[Deed Book Y, p. 624, 625, Washington CCH]

By this deed dated 22 March 1813 Christian Hawken of Washington Co., Md., buys from Isaac Watson of the same place for $300 Lot No. 69 in Maryannas Town, being 82 feet by 240 feet in size.
[Deed Book Y, p. 627, 628, Washington CCH]

By this deed dated 8 August 1814 Christian Hawken of Washington Co., Md., buys from Samuel Rohrer of the same place for $31 two lots in Rohrer's Addition to Hagers Town (formerly Elizabeth Town), No. 45 & 46, each 82 feet by 240 feet in size.
[Deed Book Z, p. 643, Washington CCH]

By this deed dated 20 October 1814 Christian Hawken of Washington Co., Md., buys from Thomas Martin and Susanna his wife of the same place for $100 the northwesternmost half of Lot No. 81 in Elizabeth Town.
[Deed Book AA, p. 83, 84, Washington CCH]

By this deed dated 1 August 1815 Christian Hawken of Washington Co., Md., buys from John Peter Herr of the same place for $1201 Lot No. 198 in Elizabeth Town, being 82 feet by 240 feet in size.
[Deed Book AA, p. 509-511, Washington CCH]

1815	Christian Hawken	$ 2.20
	To O. H. Williams	7.63
	" Wm. Lewis	.66 2/3
	" Cepi Henry Ridenour	.68 2/3
	" Upton Lawrence	3.33 1/3
	" Cash in part	53.91
	" Cash pd him	305.88 1/2

[Tax Book, Hagerstown, Washington Co., Hall of Records]

1820 Christian Hawken
1 male 45 or older
1 male between 26 & 44
1 male between 16 & 25
2 males between 10 & 15
1 female 45 or older
2 females between 16 & 25
Number of persons engaged in manufacturing - 3
[Federal Census 1820, Hagerstown, Washington Co.]

Christian Hawken, Sr., died on 10 May 1821, five days short of his 65th birthday. His passing may have been sudden and unexpected, for he left no will and apparently made no effort to dispose of his real estate or other property. He probably was active as a gunsmith until his death, for his estate inventory, detailed below, shows many gunsmith tools as well as supplies. Furthermore, the 1820 census shows that he and the two other males in his household were engaged in manufacturing at that time -- though it may be a presumption to consider that he had to be one of the three.

A brief obituary for Christian Hawken was published in the Hagerstown newspaper which confirms the date of his death and the year of his birth as given in the Lutheran Church cemetary records. Also, it provides information about his character.

Departed this life on Thursday last in the 65th year of his age, MR. CHRISTIAN HAWKEN, an old and respected inhabitant of this town. The deceased through his life supported the character of an obliging neighbor and an honest man, and was much respected by a very extensive acquaintance. On Saturday following his remains were interred in the Lutheran burial ground in this town, attended by a large concourse of relatives and friends.
[Maryland Herald and Hagers-Town Weekly Advertiser, Tuesday May 15, 1821]

As a consequence of Christian's death intestate (i.e., without a will), his real estate went to his wife and daughters. Then in order that his sons could receive their share of the estate it was necessary for them to file an equity case in the court. Because of inadequate indexing and storage of old equity cases in the Washington County Court House, the equity case for Christian Hawken is not readily findable. Lt. Col. Vaughn Goodwin made a dogged effort to locate this equity case and eventually he triumphed in his efforts. He reported his findings in an informal publication for the benefit of members of the Kentucky Rifle Association in January 1968, intended for members only. In 1971 the major part of the equity case summary was published in John D. Baird's book FIFTEEN YEARS IN THE HAWKEN LODE (1971, The Buckskin Press), the material having been given to Baird by William Reisner, Jr. In the course of doing research for this book the case also was located by Daniel D. Hartzler. My own efforts to locate it, with the help of two employees from the Clerk's office, were not successful. The following is Goodwin's transcription of Equity Case No. 941.L.373, presented with his permission and that of the Kentucky Rifle Association.

Claiments: Samuel Hawken, Jacob Hawken, George Hawken, and William Hawken
Defendents: Julian Hawken, Nancy Hawken, and other heirs of Christian Hawken, deceased.

Be it remembered that heretofore, to wit, on this 30th day of May in the year of our Lord 1822, Samuel Hawken, Jacob Hawken, George Hawken and William Hawken by David G. Yost their solicitor filed in court here a certain bill which is the manner and form of the tenor following to wit.
To the Honorable Court of Washington County setting as a Court of Equity humbly complaining showeth unto your Honours. Your orator Samuel Hawken, Jacob Hawken, George Hawken that the said Christian Hawken late of Washington County died intestate leaving a widow named Julian Hawken about 60 years of age and a daughter named Nancy a minor and the following grandchildren namely Christian Hawken, David Hawken and Nancy Hawken all minors and children of John Hawken deceased son of said Christian who died before his father Christian, and Robert Miller and Julian Miller minor children of Margaret Hawken deceased daughter of said Christian which said Margaret had intermarried with a certain John Miller and died before the said Christian her father. The said Christian left one other son named Christian Hawken who died since the deceased of the said Christian without issue. All of said defendents reside in Washington County, Maryland, except Christian Hawken son of said John Hawken deceased. The said Christian Hawken minor as aforesaid does not reside in the State of Maryland and your orator further shows that the said Christian Hawken deceased at the time thereof was seized in fee the following lots or portions of ground situated, lying and being in Hagerstown and the addition thereto Washington County, to wit, one lot of ground in Hagerstown distinguished on the plan of said town by Number 198 containing 82 feet in breadth and 245 feet in length -- three lots of ground in Rohrers addition to Hagerstown.

The suit was advertised in the *Maryland Herald* 30 July – 14 Aug. 1821(see Addenda, p. 197). Juliann Hawken responded to the suit and agreed to the claims of the sons on 5 Aug. 1822; the grandchildren, through their guardian agreed to it on 18 Aug. Additional property was involved.

Alexander Neill bonded as Administrator of the estate of Christian Hawken deceased, 22 Oct. 1821.
[Administration Book C, p. 61, Washington Co., Hall of Records]

Alex. Neill, Admin of Christian Hawken D^{d}		
Balance due on passing 1st Acc^{t}		$4,400.83
Distribution as follows, to wit		
The Widow		1,466.94
The Heirs of John Hawken		366.73
The heir of Margaret Miller		366.73
Geo. Hawken		366.73
Jacob Hawken		366.73
Samuel Hawken		366.73
William Hawken		366.73
Christian Hawken		366.73
	$ 2,933.89	
Oct. 3rd 1822	$ 4,400.83	

[Distribution Book, p. 222, Washington CCH]

This document is curious because of the date, which must be correct, and the listing of Christian [Jr.]. After the widow and the heirs of John and Margaret, there are listed the five sons of Christian, Sr. The first four, George, Jacob, Samuel, and William are listed in order of their birth. The fact that Christian is listed last may be an indication that he was born after William in 1798 and before 1801. The youngest daughter, Nancy, born in 1801 or early 1802, was a minor at the time of this distribution, and apparently because of this was not listed as an heir. In the distribution of the estate of Christian, Jr., dated the same day, 3 Oct. 1822, there is a listing of "Geo. Hawken, Jacob Hawken, Samuel Hawken, William Hawken, and Nancy Hawken" in order of decreasing age.

Because Christian Hawken, Sr., died without a will, his personal property was sold at a public sale which took place on 14 November 1821. The sale was advertised in the local newspaper.

Public Sale --- I will sell on WEDNESDAY the 14th day of November next, at the late residence of Christian Hawken, deceased, in Hagers-town, on a credit of six months, All the personal effects of said dec'd. Consisting of a variety of Household and Kitchen Furniture, amongst which are an eight-day Clock elegantly cased, Bureaus, Chests, Tables, Chairs, Beds, Bedsteads, several ten and Six-plate Stoves, with pipes, several Guns, with a general assortment of Gun-smith's Tools, - one valuable Horse, one Cow, one Cart, about one hundred and fifty cart loads of quarried stone, a quantity of Fencing Plank and Locust Posts, and a variety of other articles too numerous to insert. Sale will commence at nine o'clock A.M. when attendance will be given, by

ALEXANDER NEILL Adm'or
Oct. 25

[Maryland Herald and Hagers-Town Weekly Advertiser, 6, 13 Nov. 1821]

A complete account of the sale has been preserved, in Estate Sales Book G, p. 143-147, located in the Hall of Records. The document was recorded 11 December 1821. Numerous gunsmith tools and supplies are on the list, which suggests that Christian, Sr., was actively at work until the time of his death. This is essentially confirmed by the fact that on 29 May 1821, only 19 days after Christian's death, his son William advertised in the newspaper that he "carried on the [gunsmith] business at the old stand (late the residence of Christian Hawken, dec'd) in west Franklin Street." For details of this advertisement see the heading for William Hawken.

At the sale, Jacob, Samuel, William, and Juliann Hawken were among the buyers. William bought most of the gunsmithing tools and supplies.

Acct Sales of Personal Property of Chris. Hawken Dec^{d}

William Hawken	1 Griddle	$ 0.26
Daniel Oster	1 Dutch oven & lid	0.40
Saml Dayly	1 Lanturn	0.20
Jacob Miller	1 Lot Tin ware	0.27
John Schnebly	1 frying pan	0.55
Christian Feckty	1 Basket with lime	0.07
Peter Smith	1 wire riddle	0.25
Jacob Miller	2 arm^{d} chairs	0.19
Samuel Hawken	2 Split bottom chairs	0.15
Henry Dilman	1 chair	0.31
William Lewis	1 Lot Crocks	0.17
Christian Fechty	1 Lot Crocks & 1 jar	0.25
John Schnebly	1 Slaw Cutter	0.50
William Lewis	1 ¾pt Pewter	2.11
William Hawken	1 small table	1.00
Saml Hawken	1 D ditto	0.25
Samuel Hawken	1 D ditto	0.45
John Hartman	1 Kitchen ditto	0.50
William Hawken	1 Washing Tub	0.90
William Hawken	1 Crow bar	1.00
Daniel Long	1 sett boring tools	1.60
Peter Smith	1 Bridle	0.25
Saml Hawken	1 pair saddle Bags	1.40
William Hawken	1 wool whirl	0.50
Jacob Moyer	2 Barrels	0.13
John Hartman	2 Ditto	0.14
Jacob Kirshner	1 Shovel & Hook	0.50
John Harry of Jacob	1 Dung Fork	0.26
John Hartman	1 Shovel	0.15
Peter Smith	1 pick & spade	0.20
Jacob Moyer	5 setts window sash	1.87
William Dittihunt	1 old Chest	0.24
William Hawken	1 Chest old iron	3.75
Peter Figely	1 Six plate stove	6.00
Samuel Hawken	1 old Tub	0.12½
William Dittehunt	1 lot chains	0.25
Saml Hawken	1 Spindle	0.85
Daniel Oster	2 Barrels with flax seed	0.26
Isaac Ridenour	1 Lot lumber	0.50
John Harry	1 sett gears	1.06
William Hawken	1 sett cart Gears	1.10
John Schnebly	1 Grind Stone	0.45
John Hartman	1 Cutting Box	0.15
Saml Hawken	1 Lot Lumber	0.05
Daniel Oster	1 Box old Iron	0.25
William Hawken	1 Wheel	0.05
Saml Hawken	1 Chest with old gun barrels	2.05
John Harry of Jacob	1 old chest & hoops	0.34
William Hawken	1 Lot of Straw	1.25
Jacob Myers	1002 feet of fancy plank	11.22
William Hawken	1 Wheel barrow	2.40
Jacob Myers	2 Boxes	0.14
Francis London	24 Locust Posts	5.04
Peter Smith	1 Lot Plank	0.30
John King	1 Shovel plough	0.14
Saml Hawken	1 pair steel yards	0.31
John Schnebly	the wood work of a cart	8.50
John Grumbaugh	1 Cart	16.12½
Saml Hawken	1 Table	2.50
Saml Hawken	1 sett Castors	0.40
Barney Oldwine	2 Coffee Pots	0.64
Amos Gaylord	1 lot of dishes, etc.	1.25
Charles Humrickhouse	1 double-barreled Gun	8.00
Isaac Ridenour	1 Single Do Do	6.50
Jacob Hawken	1 rifle Gun	16.25
David Lippy	6 Chairs	2.50
Saml Hawken	2 Wheels & 1 reel	2.80
" "	6 Chairs	4.50
" "	1 Mahogany Stand	4.00
" "	1 Walnut Table	3.00
Saml Hawken	1 Bread Basket 12 Waters	0.25
John Harry of Jacob	1 Looking Glass	1.00
Jacob Miller	1 Ten plate stove & pipe	13.12½
Amos Gaylord	1 Desk & Drawers	16.00
John Hartman	1 Lot of Books	0.50
Danl Long	1 Lot of German Do	0.30
Saml Hawken	4 Chairs	1.50

Name	Item	Value
John Kiry	1 Womans saddle	0.12½
Fred^{k} Tickehoof	1 Corn Cupboard	17.00
Saml Hawken	1 powder horn & shot pouch	0.40
David Bower	1 Ax	0.25
Francis London	1 meat Tub	0.50
William Hawken	1 Cross cut saw	0.75
Amos Gaylord	2½ Bushels Potatoes	0.80
Doctor Fred^{k} Doirsy	1 Lot Quarried Stones	20.00
William Bursking	1 Cow	10.25
John Crumbaugh	10 Bushels Rye	4.00
Mrs. Hawken	1 Ten plate stove & pipe	12.00
" "	1 pair shovels & Tongs	1.50
" "	1 Dutch Chest	2.00
" "	1 Breakfast Table	2.00
" "	6 Black chairs	3.75
" "	1 Eight Day Clock	55.00
" "	1 Looking Glass	1.50
" "	6 Pictures	1.50
" "	17½ yds Carpeting	3.50
" "	1 Bedstead Bed & Coverings	20.00
" "	1 German Bible	1.00
" "	1 maple Bureau	9.00
" "	1 Bedstead Bed & Coverings No. 1	12.00
" "	1 Do Do Do No. 3	8.00
" "	1 Do Do Do No. 4	5.00
" "	1 Dough Trough & Scraper	0.50
" "	1 Cotton wheel	1.00
" "	1 Tight Barrel	0.50
" "	1 Meat Tub	0.75
" "	1 Bread holder	0.30
" "	1 Iron Kettle	0.75
" "	1 Iron Pot	0.30
" "	2 Do Do	0.20
" "	1 Skillet	0.12½
" "	1 Copper Tea Kettle	0.20
" "	1 Pot rack & Chain	0.62½
" "	1 Dripping pan	0.50
" "	3 Ladles & 1 flesh fork	0.18
" "	1 Copper Cover	0.06
" "	1 flat Iron	0.20
" "	1 pair Brass candle sticks	0.30
" "	1 Bench	0.06
" "	575 Shingles	2.15
" "	wearing apparel (of the dec^{d})	4.00
" "	13¼ Bushels potatoes	3.31
" "	1 Silver Watch	7.00
" "	1 Spade	0.75
William Hawken	2 Rifle rods	0.50
Do Do	11 rifle Bits	1.50
" "	1 Cast Anvil	0.50
" "	1 Sett Stocking Tools	1.50
" "	1 Do Do Plains	0.62½
" "	1 pair pincers & old Iron	0.50
" "	3 Vices	1.50
" "	1 Iron Stake	1.25
" "	2 Braces	0.50
William Hawken	1 German seyth	1.00
Jacob Hawken	1 Grey Horse	43.00
Mrs. Hawken	68 feet plank	0.14
William Hawken	1 Screw plate & screws	1.50
" "	1 String hoop & muzzle plate	0.62½
" "	1 small Box with old Iron	1.00
" "	2 oil stone	1.00
" "	3 small Boxes with old screws	0.25
" "	1 Drill stock	0.62½
" "	6 files	0.62½
" "	5 Hammers	1.00
" "	4 Saws	2.25
" "	1 vice & Bench No. 1	4.00
" "	1 vice & Bench No. 2	5.00
" "	a Box with old Iron	0.62½
" "	1 Spindle & screws	1.50
" "	2 Bows	0.25
" "	1 Box of old rifling Balls	0.20
" "	1 Bellows	8.00
" "	1 Grind Stone	2.00
" "	1 Lott Hammers	1.00
" "	26 Gun Stocks	3.25
" "	3 Do Do	0.18¾
" "	1 rifle Bench	1.00
" "	a Glue Pot	0.25
" "	7 Bottles & 1 Pitcher	0.50
" "	1 fish Barrel	0.20
		$469.05

Rec^{d} 11th Dec, 1821

HAWKEN, HENRY   Greensburg, Pa., Columbus, O. The Ohio branch of the Hawken family stemmed from Henry Hawken who worked as a gunsmith first in Greensburg, Pa., and then in Columbus and Springfield, Ohio. It seems probable that Henry originated in Hagerstown, but there are no direct relationships that can be proven. The Hagerstown gunsmith, Nicholas Hawken, had a daughter, Juliana, born in 1779, and a son, George, born in 1782. It is possible that Henry was an earlier son of Nicholas, born within a few years before 1775. Family information passed along by descendents of Henry Hawken mention Hagerstown as their place of origin, and this makes the case have some substance.

This family information consists of the following statement: "I read an old letter once that one of grandpa's brothers wrote. It said that three brothers came to Hagerstown, Maryland, before the Revolutionary War in 1775. They came from Switzerland. Their names were Hawken. No doubt they served in the Revolutionary War." The "grandpa" mentioned was Thomas Paine Hawkins, a grandson of Henry, and his brother who wrote the letter was the gunsmith Henry Clay Hawken.

Henry Hawken served in the War of 1812, and from his Bountyland claim it is learned that on 24 Oct. 1796 he married Susannah Turney in Greensburg, Pa. She was a sister of Daniel Turney, a lawyer and politician in Greensburg. Thus Henry's birth year would have been prior to 1775, but probably not many years before.

Property tax returns for Hempfield Township in Westmoreland Co., where Greensburg is located, hold the following listings, with the list for 1802 being the earliest list available for viewing:

1802	Hawken, Henry	gunsmith
1804	Hawken, Henry	gunsmith & innkeeper
1805	Hawkey, Henry	gunsmith & innkeeper
1807	Hawken, Henry	Armorer
1808	Hawkey, Henry	
1809	Hawken, Henry	
1810	Hawken, Henry	gunsmith
1811	Hawken, Henry	gunsmith
1812	Hawkey, Henry	gunsmith
1813	Hawkey, Henry	gunsmith
1814	Hawken, Henry	gunsmith
1815	Hawkey, Henry	gunsmith

The above information was supplied by Robert McAfee after an associate of mine made two unsuccessful attempts to locate the old tax records in the court house in Greensburg.

Henry Hawken's Bountyland Claim states that in August 1812 he volunteered for duty in the Pennsylvania Militia and was in the company commanded by Captain John B. Alexander at first and later by Captain Peter Drum. He held the rank of Sergeant. The last Company Pay Roll for Henry states that he deserted 19 January 1813, but it also shows that he was paid for the period 1 January through 31 May 1813.

After 1815 Henry Hawken's name no longer appears in the tax records of Westmoreland Co., and it is presumed that he moved to Ohio. He located first in Columbus where he worked as a gunsmith, and in his later years was in Springfield. He was located in Columbus as late as 1828, as is shown by the following deed:

By this deed dated 28 May 1828 Henry Hawken and Susannah his wife of Franklin Co., Ohio, sell to John Hawken, son of said Henry and Susannah . . .
[Deed Book 7, p. 323, Franklin CCH, Ohio]

Henry Hawken and Susannah had three children -- John, who was born about 1805, who died in 1846, and who, like his father, was a gunsmith; Daniel Turney, born in 1814 and died in 1851, and also a gunsmith; and Henry, born after 1814, apparently, who died in 1842.

Henry Hawken died in Springfield, Ohio, on 15 April 1835 according to the Bountyland Claim his widow filed twenty years later. Among the administration papers on file relative to his death is a record of the sale of his estate, which was administered by his son John. Two guns and various gunsmithing tools were included in the list of items sold, the tools being bought by his sons John and Daniel T. A partial list of the items sold is as follows:

Item	Purchaser	Amount
1 wheel barrow	James Watson	$ 1.00
1 Barrel	John Lindell	.18¾
1 pair andirons	William Watson	.75
Large German bible	Daniel Kienborts	1.75
Curled maple plank	John Hawken	2.75
1 Gun	John Lindell	17.26
1 Gun	Wm. Tilson	14.00
Screw plate	John Hawken	3.62¾
1 set gunsmith tools	John Hawken	15.00
Desk	John Hawken	.27½
Gun stocks	John Hawken	1.22
1 Set gunsmith tools	John Hawken	10.00
Stove and pipe	John Hawken	6.00
Guide	D. T. Hawken	.50
Press drill & shears	John Hawken	.50
Lot of ramrods	D. T. Hawken	.18¾
New frame	Thomas Arnote	21.00

[Letters of Administration A626, Clark CCH, Ohio]

HAWKEN, GEORGE, OF CHRISTIAN Hagerstown
George Hawken was the oldest of the children of Christian Hawken, Sr., and Juliann. The Parish Register of St. John's Lutheran Church in Hagerstown shows that Christian and Juliana Hawken had a child named John George Hawken born 19 Feb. 1781. Among Germanic people in the 18th century it was common to give a first name of John although it was not used.

Very little is known about George Hawken as a gunsmith or as a person. A few flintlock rifles have survived bearing the signature of "G. Hawken." One piece known is of relatively good quality in workmanship, with incised carving, a brass patchbox with daisy-head finial, and a club-shaped extension to the rear ramrod pipe.

Documentation proving that George Hawken was a gunsmith is not abundant. George's gunsmith brother, Samuel T. Hawken of St. Louis, stated in a newspaper interview late in his life that five of Christian's six sons were gunmakers; of the six, it appears that Christian, Jr. was not a gunsmith. Additional evidence that George was a gunsmith in Hagerstown is found in Scharf's HISTORY OF WESTERN MARYLAND, which lists him together with Christian Hawken and William Hawken as gunsmiths in Hagerstown for the period between 1805 and 1815. William was born in 1798, so he was in error in listing William for that period, though William definitely was a gunsmith there after he came of age.

The best documentation for George's work as a gunsmith is an advertisement he placed in the Hagerstown newspaper from December 1809 through March 1810.

GEORGE HAWKEN *GUNSMITH* Respectfully informs his friends and the public that he has lately commenced business in the Shop next door to Mr. John Kausler, and the third door below the bank, in Washington street, Hagerstown; where he is ready to receive, and will be thankful for any orders in his line-- He will always be careful to execute his work in the neatest and best manner, and has no doubt but he shall be able to give full satisfaction to all who may please to employ him.

December 12, 1809

[Maryland Herald and Hagers-town Weekly Advertiser, 13, 20, 27 Dec. 1809, 3, 10, 17, 24, 31 Jan., 7, 14, 21, 28 Feb., 7, 14 21 March 1810]

The picture of Christian's son George in Hagerstown is confused by a paucity of data and by the fact that the Hagerstown gunsmith, Nicholas Hawken, had a son named George born in 1782. Apparently there were two men of the same name and almost the same age who grew up and worked in Hagerstown. It is possible that George of Nicholas also was a gunsmith.

George of Christian was married and had at least two children, and possibly more. His wife's name is not known. There were two Hawken gunsmiths whose attachment to the Hawken family is unknown, and it is possible that these were sons of George. One of these was a gunsmith named George Hawken who was born in 1811 in Maryland and who worked as a gunsmith in Vicksburg, Mississippi, from 1840 until his death in 1865. The other was a gunsmith named John C. Hawken who advertised in a Hagerstown newspaper that he had taken over the gunsmith shop of William Hawken in 1840.

The two men named George Hawken were listed in the census of 1810 for Hagerstown as follows. There is no person named George Hawken listed in either the census for 1800 or for 1820.

1810 Hawkn, George — Hagerstown
- 1 male between 26 & 44 years
- 1 male between 16 & 25
- 1 female between 16 & 25
- 1 female under 10

1810 Hawken, George — Hagerstown

1 male 45 or older	1 female 16 - 25
1 male 10 - 15	1 female 10 - 15
1 female 26 - 44	2 females under 10

[Federal Census 1810, Hagerstown, Washington Co.]

List of Letters Remaining in the Hagerstown Post Office . . .
George Hawken . . . 1 Jan. 1811.
[Md. Herald & Hagers-Town Weekly Advertiser, 16 Jan. 1811]

George Hawken of Christian apparently did not own property, for there are no deeds recorded in his name. After Christian, Sr., died, George was listed with his brothers on a document dated 3 Oct. 1822 showing how Christian's estate was distributed; his share was $366.73. George again was listed on a document showing how the estate of George's brother, Christian, Jr., was distributed.

George's mother, Juliann Hawken, made her will in March 1827. It states in part:

It is my will and I hereby authorize and empower my Executor to sell my house & Lot in Franklin Street at such time after my decease as his judgement may best direct and on such terms as he may describe, and empower him to execute a Deed or deeds to the purchaser or purchasers, and the proceeds of said sale together with all the residue of my property not already devised, I give and bequeath to my son Geo. Hawken & my Daughter Nancy Lewis share and share alike. It is my will and it must be so understood that the proportion of my estate that is by this will devised to my son George, that my Executor is to invest it in Bank Stock, and the interest only to be paid him during his life -- and at his death to be equally divided amongst his children or the survivors of them."

This implies that George was not capable of handling his own financial matters at this time. Juliann had died by 7 September 1827, but her estate's distribution was not recorded until 8 June 1830. By this time George was dead, for the distribution does not list him but does list "George Hawkins children."

HAWKEN, GEORGE (II) Vicksburg, Miss.

The 1850 and 1860 Federal Census hold the name of a gunsmith, George Hawken, born in Maryland. In view of the known Hawken family genealogy, it is possible, if not probable, that this man, designated here as George (II), came out of the Hawken family in Hagerstown. It is suggested that he may have been a son of the gunsmith George (I) Hawken of Hagerstown; the circumstances of name, occupation, age, and state of birth favor this idea. The census listing was called to our attention by Robert McAfee. Cyril O'Neill of Vicksburg kindly dug into the public records of Vicksburg to supply most of the information given below.

George (II) Hawken died in Vicksburg on 10 August 1865 from enlargement of the heart according to his death record. He then was 53 years, 11 months, and 4 days old, which would place his birth date as 6 September 1811. The death record shows that his funeral expenses of $75 were paid by a Mr. H. Hammett and a Mr. Grammar. The early marriage records of Vicksburg show that on 20 September 1840 he married Sophia Fowler in Vicksburg. At that time she was either 20 years of age (1850 census) or 25 years of age (1860 census).

1850	Hawken, George	age 43	b. Md.	Gunsmith
	Sophia	30	b. Ohio	
	Sharpe, -	60	b. Md.	

[Federal Census, 1850, Vicksburg, Miss.]

1860	Hawkin, George	age 48	b. Md.	Gunsmith
	Sophia	35	b. Ohio	
	Catherine	7	b. Miss.	
	J. C. Throwbridge	26	b. Ohio	Barkeeper
	Joseph Quitzer (?)	37	b. Switzerland	
	Wm. Fagan	42	b. Miss.	Gentleman

[Federal Census 1860, Vicksburg, Miss.]

1860 Hawkins, George residence south side
1st St. between Walnut & Monroe
[Vicksburg City Directory, 1860]

1877 Hawkins, Sophia, widow 37 1st East Street
south side
1886 Hawkins, Sophia, widow 37 1st East Street
south side
[Vicksburg city directories, 1877, 1886]

1851 Hawkins, George 1 free white person
1853 Hawkins, George 1 free white person, 1 clock
[Personal Tax Assessment Records, Vicksburg, Miss.]

Death records for Vicksburg show that Sophia Hawken died 22 February 1891 of dropsy, at age 71. Her funeral expenses of $35 were paid by a Miss Katie Hawkins, presumably a daughter and the same person as the Catherine Hawkin listed in the census for 1860.

HAWKEN, JOHN Hagerstown

John Hawken was the second son of Christian Hawken, Sr., of Hagerstown. Relatively little is known about him, for he apparently moved to Virginia sometime after he came of age in 1805, and his location there has not been firmly established. See p. 198 for some additional information.

The Parish Register for St. John's Lutheran Church in Hagerstown shows that Christian and Juliann Hawken had a son, John, born 19 March 1784. It is curious that John's son, David, reported in the census for 1880 that his father was born in North Carolina. It is not inconsistent with the facts for Christian and Juliann Hawken to have been in North Carolina early in 1784. Like other young people both past and present, Christian and Juliann may have gotten the wanderlust and taken an extended excursion. The Great Wagon Road ran through Hagerstown, headed down the Valley of Virginia to the lands of Kentucky, Tennessee, and North Carolina. In 1783 the Treaty of Paris was signed and settlers streamed down the Great Road. It would have been natural for an adventurous couple to want to get in on some of the action. If David Hawken's report about his father's birth place is correct, we may assume that in the spring of 1784, after infant and mother were ready to travel, that the family returned to Hagerstown, for in June 1784 Christian bought his first piece of property in Hagerstown.

John Hawken married a girl named Ana Long, but the date and place of the marriage is not known. Glenn R. Scott reports that there is one source which mentions that Ana Long Hawken was born at Harpers Ferry, Va. (now W. Va.).

The gunsmithing activities of John Hawken remain unknown, but in time more may be found out about his work, for he has not yet been thoroughly researched in the public records of northern Virginia and West Virginia. It is possible that he may have spent time working at Harpers Ferry.

Wallace Gusler brought to my attention the following mention of one "John Hawkins" working as a gunsmith in Hampshire County, W. Va. in 1815. This is an excerpt from a letter written by the Hampshire County gunsmith, Frederick Sheetz, concerning militia arms.

> After waiting untill noon for his arrival Mas^{t} Sargent and Cap^{n} Kenny (who commands the other rifle company of the reg^{t}) appointed Mr John Hawkins a gunsmith of the neighborhood to act on the part of the Commonwealth . . .
> [Virginia State Library, Executive Papers, Box 223, Folder May 18-31, 1815]

A search for deeds and other related documents in the name of John Hawken (or Hawkins) in the Hampshire County court house in Romney, W. Va., yeilded two deeds in the name of John Hawkins. The first concerns a property of 219 acres and is dated 1796. At that time John Hawken was 12 years of age and living in Hagerstown. The second names "John Hawkins, Senr" and is dated 1810. Thus neither deed is applicable to John Hawken of Hagerstown, and in addition, they cast doubt that the 1815 letter refers to John Hawken from Hagerstown. Census data for 1810 and 1820 likewise rule out John Hawkins, Sr., and Jr., of Hampshire Co.

A date for the death of John Hawken of Hagerstown is not known. His name is not included with those of his brothers and sister in the distribution lists for the estates of Christian, Sr., and Christian, Jr., dated in 1821, and it is mentioned in the equity case for the estate of Christian, Sr., that John had died before his father. So all that we know is that he had died before May 1821.

HAWKEN, JACOB Hagerstown & St. Louis

Jacob Hawken and his brother, Samuel T. Hawken, were the famous gunsmiths of St. Louis, Missouri, who made the percussion rifles so highly regarded by the plainsmen and mountainmen of the West. Jacob was the third son of Christian Hawken, Sr., and Juliann. He was born within the year preceeding 8 May 1786 according to his obituary (see below). His birth was not recorded in the Parish Register of St. John's Lutheran Church in Hagerstown. Most certainly Jacob grew up in Hagerstown and learned gunsmithing with his father and other members of the family. If 1785 is the correct year of his birth, he would have attained the age of 21 in 1806.

Apparently Jacob worked in Hagerstown as a gunsmith for a number of years after he reached 21, probably in his father's shop. He did not immediately get married, and he did not buy property in Hagerstown. In Scharf's HISTORY OF WESTERN MARYLAND v. 2, p. 1062, is the statement, ". . . the following were residents of Hagerstown during the years from 1805 to 1815, inclusive. . . Jacob Hawken." Jacob apparently did not serve in the War of 1812 as a member of any Maryland militia unit, though his younger brother, Samuel T. did serve. The times grew tense during the year 1814 when British troops invaded Maryland and Washington, D.C., and most able-bodied men served for a brief period. Possibly Jacob had moved elsewhere by that time.

Jacob Hawken may have left Hagerstown for the West shortly before 1815. In any case, his location for a number of years cannot be accounted for. Jake's brother, Samuel, and his new wife, Rosanna, left Hagerstown for Xenia, Ohio, early in 1816.

The first definite knowledge we have of Jacob's presence in St. Louis is a deed dated 2 October 1818.

By this deed dated 2 Oct. 1818 Jacob Hawkins buys from David Deshler of St. Louis for $350 half of a certificate issued by Frederick Bates to any quantity of land not exceeding 160 acres in the name of Gabriel Hunot in the town of New Madrid, Certificate No. 427, dated 21 August 1818. [Deed Book H, p. 132, City Hall, St. Louis]

Scharf's HISTORY OF ST. LOUIS numbers and lists a number of early frame and log houses built in St. Louis from 10 March 1804 to May 1821 (p. 154). It describes House No. 152 as a two-story log structure on n. Second above Cherry, north of the town in Connor's blocks, built in 1820 and owned by Lakenan and Hawken (Jacob). In May 1821 it was occupied by William Hughey, laborer. Another house, No. 10, built in 1810 and owned by Antoine Recontre, made of logs and located at 214 N. Main St., St., extension above town in block 56, was occupied by Jacob Hawken in May 1821 (p. 151). And Scharf (p. 360) mentions that in 1821 Jacob Hawkin was a taxpayer in St. Louis, with property valued at $530. The earliest known directory for the young little city of St. Louis is for the year 1821.

1821 Jacob Hawken gunsmith 214 North Main
[Paxton's Directory of St. Louis, 1821]

On 10 May 1821 Jacob's father, Christian Hawken, Sr., died. Jacob returned to Hagerstown during the summer of 1821. Most likely this was to help in settling the estate, to have a reunion with his family, and possibly to see a certain young lady. At the sale of Christian's estate, Jacob bought only two things, "1 rifle Gun $16.25," and "1 Grey horse $43.00."

The Parish Register of St. John's Lutheran Church in Hagerstown shows that Jacob Hawken married Catherine Allison in March 1822. Apparently Catherine was a girl from Hagerstown, whom Jacob had known before he left Hagerstown, for Juliann's letter of 26 January 1823 to Jacob, Sam and Catherine in St. Louis holds the comment, "As for Mr. Jacob and Catherine, they don't think us worth their notice. I really had great confidence in Catherine; before Jacob came in from St. Louis Catherine would say that he was very negligent, and should write more frequently, that if she was in his situation she would be more mindful of her friends -- but how is it now. She is there herself, and as careless as Jacob ever was."

Jacob and Catherine apparently left Hagerstown to return to St. Louis shortly after their marriage. Samuel also left Hagerstown for St. Louis at about the same time, and it would be reasonable to suppose that they made the trip together. Sam went out alone, leaving his three small children behind; his wife Rosanna, probably died the winter before.

Jacob Hawken and his brother, Sam, soon were at work together, building the rifles for which they became famous. Details of their life and work while in St. Louis comprise a separate study and no attempt is made here to cover such ground.

It will suffice to note that Jacob and Catherine had four children: Sam, born in 1823, who died in his youth; Christopher Miller Hawken, born 28 May 1825, who worked as a gunsmith, and who died 5 Feb. 1905; Julia Ann, born 27 Nov. 1828, who married Thomas Montgomery Hopkins; and Jacob Hawken, Jr., born in 1830, who died after 1902.

A cholera epidemic hit St. Louis early in 1849 and it is said that Jacob Hawken died as a result of this disease. His obituary, however, does not state this, though adjacent obituaries on the same page do mention death from cholera. In any case, Jacob was 63 years of age at the time of death at 7:30 o'clock on 8 May.

DIED At Half past 7 o'clock on the 8th instant, JACOB HAWKEN, Sen'r, in the 63rd year of his age.

His friends and acquaintances are requested to attend his funeral this afternoon, at 3 o'clock, from the residence of his brother, Samuel Hawken, No. 156 Sixth street, between Morgan street and Franklin Avenue.
[The St. Louis Republican, 9 May 1849]

HAWKEN, SAMUEL T. Hagerstown, Xenia, Ohio
St. Louis, Mo.

Samuel T. Hawken and his brother Jacob were the famous gunsmiths of St. Louis, Missouri, who made the percussion rifles so highly regarded by the plainsmen and mountainmen alike of the West. Samuel was the fourth son of Christian Hawken, Sr., and Juliann, of Hagerstown. He was born 26 October 1792, a date established by newspaper obituaries after his death in 1884 and by a number of other records.

Sam Hawken doubtless learned gunsmithing under his father, and in association with his brothers George, John, Jacob, and William. As he approached maturity, however, the War of 1812 evolved, and Sam was of suitable age to get involved. When he applied for a pension in 1872 he supplied the following information.

On this third day of June, A.D. 1872, personally appeared before me, Clerk of the County Court of St. Louis County . . . Samuel Hawken, aged 79 years, a resident of St. Louis, County of St. Louis, and State of Missouri, who being duly sworn according to law, declares that he is a widower, that his wife's name was Rosa Aster to whom he was married at Hagerstown, Md., on the 19th day of September 1815; that he served the full period of sixty days in the military service of the United States in the War of 1812; that he is the identical Samuel Hawken who was drafted in Captain Geo. Shryock Company, 1st Maryland Militia Regiment, Comd by Col. Ragan Brigade. . . at Baltimore in the State of Maryland, on the _____ day of May or June 1814, and was honorably discharged at Annapolis Md. on or about 1st day of January, 1815; that after the regiment was organized it was sent to Washington, D.C., and participated in the Battle of Bladensburg, Md., then returned to Baltimore, and supported the battery of Comodore Rodger at the Battle of Baltimore, after this, the Regiment was sent to Annapolis, Md., where it remained until mustered out of service. Affiant has received land warrant for 160 acres, but cannot remember the number, that his discharge was filed with his application for Land Warrant. . .
[Declaration of Surviving Soldier for Pension, Military Service Records, National Archives]

Samuel Hawken and Rosanna Oster were married in Hagerstown on 19 September 1815. Sam was 22 years of age at the time. His wife's first name is variously spelled in the records as "Rosa," "Rosena," and most often Rosanna; her last name appears both as "Aster" and "Oster," but Oster seems to be the correct early spelling.

Early the next year, 1816, Sam and Rosanna moved to Xenia, Ohio, where Sam worked as a gunsmith. When Sam was over 90 years of age he was interviewed by a reporter for the Globe-Democrat of St. Louis who wrote an article about him in which he is quoted as saying ". . . April 1, 1816, I went to Xenia, Ohio, and kept a gun store there, and on June 3, 1822 I arrived in St. Louis."

Sam and Rosanna Hawken purchased a small lot in the town of Xenia in 1818. Sam retained ownership of it for a number of years after he moved to St. Louis, finally selling it in 1839. They also purchased Lot No. 71 in Xenia, but the deed apparently was not recorded.

By this deed dated 12 May 1818 Samuel Hawken of Greene Co., Ohio, buys from Joseph Johnson and Hannah his wife of the same place for $275 a fragment of Lot No. 26 in the town of Xenia, it being 21 feet six inches in front and of full lot length.
[Deed Book 6, p. 331, 332, Greene Co., Ohio]

Sam and Rosanna's first child, Elizabeth Ann, was born 15 June 1816, not long after they arrived in Xenia. A second daughter, July Ann, was born 15 February 1818, and son, Lewis, was born 4 June 1820.

Samuel Hawken is listed on both the federal population census for 1820 and the industrial census for 1820. The population census data suggest that there may have been a third daughter who did not survive long. The industrial census is of some importance in showing the rate of rifle production and the amount of other work done in Sam's shop. If an annual production of 21 rifles was typical, he may have produced over 100 flintlock rifles during his five-year stay there. Not one of these is known to have survived, but a surviving specimen or two may show up sometime.

1820 Samuel Hawken Xenia Greene Co.
1 male under 10 years
1 male between 26 & 44
3 females under 10 years
1 female between 26 & 44
[Federal Census 1820, Greene Co., Ohio]

1820 Samuel Hawken, $200 invested in the business, employing one hand at $20 per month. For one year he produced 21 rifles valued at $420, and made other items and repairs worth $300.
[Census of Manufactures, 1820, Greene Co., Ohio]

Sam's father, Christian, Sr., died 10 May 1821, and shortly after receiving word of this near the end of May he must have packed up and taken his wife and children back to Hagerstown.

Sam was present at the sale of his father's personal property and gunsmith equipment held in the fall of 1821. He bought an assortment of furniture and household items, and "1 chest with old gun barrels" (see entry for Christian Hawken Sr., for details).

The year of 1821 was doubly tragic for Samuel Hawken, for it probably was during this year that his wife, Rosanna, died. She gave birth to her son, Lewis, in Xenia on 4 June 1820, and apparently was dead before Sam left for St. Louis early in the spring of 1822.

We are fortunate to have in Sam's own words a little information about his arrival in St. Louis, as reported by the Globe-Democrat in the interview already mentioned: ". . . on June 3, 1822, I arrived in St. Louis. My first shop was on the levee, near where Olive Street is."

Not long after his arrival in St. Louis, Sam Hawken apparently met Martha Ritchey, daughter of Hamilton Ritchey of Mercersburg, Pa. According to a sheet of Hawken family and Ritchey family genealogical data preserved in the Missouri Historical Society, Hamilton Ritchey moved to Missouri in May 1822 and died in St. Louis in 1833. Sam and Martha were married on 14 November 1822. The first of their eight children, Anna Maria, was born 18 September of the following year.

When Sam left Hagerstown early in 1822 he left his three small children behind, in the care of their grandmothers and perhaps other family members. Sam's mother, Juliann, wrote a letter to him and Jacob on 26 January 1823 (see entry for Hawken Family for full content of letter) and commented, "Your dear children are well, and doing well, and grow finely--they are at present in the country at their grandmothers. -- I sent word to them that you were married; and do not know what their answer will be. -- As you have mentioned in your letter that you are coming in for them next fall or spring, I think it would be advisable that you should bring your wife with you to assist in taking them out. -- We would be glad to see her at all events."

It is not within the scope of this study to cover in detail the life and work of either Samuel or Jacob Hawken after they were established in St. Louis. And it suffices here to note that Samuel T. Hawken lived a long life in St. Louis, dying there on 8 or 9 May, 1884, at the age of 91 years. He had three children by his first wife, Rosanna, and eight children by his second wife, Martha. Genealogical data for the children are included in the accompanying Hawken Family genealogical chart.

Specimens of rifles made by Samuel Hawken are known from his period of work in St. Louis, during the percussion era. Most of these are the classic half-stock percussion rifles marked "J & S HAWKEN ST. LOUIS." Some are marked "S. HAWKEN" having been made in the years after 1849 when Jacob died. But Samuel Hawken also worked during the flintlock era, from about 1813 until the transition period of the 1830's, first in Hagerstown, then in Xenia, Ohio, and after 1822 in St. Louis. It is surprising that there are no known rifles bearing his name that have survived and which without doubt were made in the flintlock period. Probably one or more will show up at some time in the future. Using a conservative number of 20 rifles per year completed, in the flintlock period, it is likely that in the 20-year period Sam Hawken made at least 400 rifles.

HAWKEN, WILLIAM — Hagerstown

William Hawken was the fifth son of Christian Hawken, Sr., of Hagerstown. He was a gunsmith like his father and brothers. But unlike his older brothers, he chose to remain in Hagerstown and nearby Williamsport throughout his life. He made both flintlock and percussion rifles, a few of which have survived. One of these, in the collection of Daniel D. Hartzler, is illustrated in this book. It is a percussion piece with a four-piece engraved brass patchbox, probably made during the 1830's, and exhibiting what could be considered average workmanship for the period.

William Hawken was born 20 May 1798 and died 27 February 1885, according to records obtained from St. John's Lutheran Church graveyard, which later was moved to Rose Hill Cemetary, southeast. It may be presumed that William learned gunsmithing under his father and in association with his brothers. As a lad of 16 years during the summer of 1814, he was too young for militia service during the British invasion of Maryland. Scharf's HISTORY OF WESTERN MARYLAND, v. 2, p. 1062, mentions that "William Hawken, gunsmith" was "among the residents of Hagerstown and its immediate vicinity from 1800 to 1805 inclusive." The dating obviously is in error, for William was but a child at the time mentioned.

William reached the age of 21 in May 1819, but he is not listed as a head-of-household in the 1820 census. He was unmarried at the time and possibly was living at his father's house. On 17 July 1821 William married Leah Cramer of Hagerstown. She was born 29 Jan. 1802, and died 20 May 1880, according to a record from St. John's Lutheran Church graveyard. It should be noted that this marriage took place about two months after the death of William's father. The date of the marriage is given on page 95 of a record of Washington County marriages in the courthouse of Washington County.

Apparently William both lived and worked in his father's house at the time of Christian's death on 10 May 1821. Nineteen days later, on 29 May, William placed an advertisement in the Hagerstown newspaper stating that he was carrying on the gunsmith business at his father's old shop.

William Hawken, GUNSMITH, RESPECTFULLY informs his friends and the public in general, that he carries on the above business at the old stand (late the residence of Christian Hawken, dec'd) in west Franklin Street, Hagers-town; where all orders in his line will be thankfully received and promptly attended to. His charges will be moderate to suit the times. From his knowledge of said business he hopes to give general satisfaction. May 29, 1821
[Maryland Herald and Hagers-Town Weekly Advertiser, 29 May, 5, 12 June 1821]

By the following spring William found it best to move his shop to another location, and accordingly advertised his removal.

REMOVAL -- WILLIAM HAWKEN, *GUNSMITH,* RESPECTFULLY informs his friends and the public generally, that he has removed his shop to Mr. John Smith's house, nearly opposite the dwelling of Mr. John Currey, Cabinet Maker, in Franklin Street, where he is ready to execute all orders in his line with neatness and dispatch. He flatters himself that from his knowledge of the business to give general satisfaction.
Hagerstown April 12
[Maryland Herald and Hagers-Town Weekly Advertiser, 15, 22 April, 7 May 1822]

Early in the fall of 1821 the estate and the gunsmith tools and supplies of Christian, Sr., were sold at public sale. William was present at the sale and bought almost all of the tools and gunsmith supplies. A detailed accounting of the sale has been preserved and is reproduced here under the heading for Christian Hawken, Sr., so it need not be repeated here.

On 7 December 1822 William Hawken mortgaged a number of household items, and a formal bill of sale was recorded for the transaction:

By this mortgage instrument dated 7 Dec. 1822 William Hawken of Washington Co., Md., gives title to Joseph Welty of the same place, for $55.67, for the following items: One bureau, six chairs, one Dining Table, one candle stand, one Chest, one rocking Cradle, one bed stead and furniture, one small Table, one Dutch Oven, one Kettle, two pots and one pan, three tubs, one tight barrel and one bucket. Ownership is to revert to William Hawken provided he repays the sum, with interest, within one year.
[Deed Book GG, p. 20, Washington CCH]

William continued to keep the public informed of his gunsmithing work when he advertised in 1823 for those who had left guns for repair to claim them and pay for them or they would be sold for expenses.

Old Guns, Pistols, &c. ---

To all whom it may concern.
There are at my shop, in Franklin street, Hagers--town, several old Guns, Pistols, and Locks, which were left a long time since (some two years) to be repaired. They have been carefully repaired according to order, since which time the owners have not made their appearance --- I therefore give this public notice, that if they do not want their fire arms, I do want and will have my money for repairing them; and am determined that all old Guns, Pistols, and Locks that are not called for previous to the first Monday in October next, shall be sold to the highest bidder.

The Gunsmith business is carried on as usual.
July 15, 1823 WILLIAM HAWKEN
[Maryland Herald & Hagerstown Weekly Advertiser, 15 July 1823]

After the death of Christian Hawken, Sr., William purchased one of his lots. He sold this in 1825.

By this deed dated 18 Dec. 1823 William Hawken of Washington Co., Md., buys from Alexander Neill, Trustee appointed to dispose of the real estate of Christian Hawken, deceased, for $80, Lot No. 392 in Elizabeth Town, being 82 feet by 240 feet in size. The sale originally was made 8 Dec. 1822 by Neill to William Hawken.
[Deed Book GG, p. 533, 534, Washington CCH]

By this deed dated 30 June 1825 William Hawken of Washington Co., Md., sold to Jacob Crist of the same place for $50 a lot in Elizabeth Town (now called Hagers Town), known by the No. 392, 82 feet by 240 feet in size, part of the real estate of Christian Hawken deceased, which was conveyed to William Hawken by Alexander Neill Trustee for the sale of the real estate of Christian Hawken deceased. The transaction was acknowledged by Leah Hawken, wife of William.
[Deed Book HH, p. 518, 519, Washington CCH]

When William's mother, Juliann Hawken, wrote to her sons Jacob and Sam in St. Louis on 26 Jan. 1823 she passed on whatever news of the family she could write about (see complete letter under Hawken Family heading). William and Leah were mentioned in the letter:

William & Leah are well, but no prospects of an increase in number. . .About three weeks ago Mr. Will made sale of all our property. . . The lots were sold at $320. -- Old Stump the Sow gelder bought two and William Hawken bought one. -- William gave $80 for his.

William and Leah Hawken had a son, William L., who died an infant on 10 June 1822. Apparently Juliann was anxious to see them produce a grandchild for her, which they did in 1825 when a daughter, Margaret, was born. A second daughter, Mary, was born 27 March 1827, and a third daughter, Catherine Ann, was born on 21 Feb. 1830.

1830 Hawken, Wm. Western District of Hagerstown
1 male between 60 & 70 years
1 male between 30 & 40 years
1 male between 15 & 20 years
2 females between 30 & 40 years
2 females under 5 years
[Federal Census 1830, Washington Co.]

1840 Hawken, William Hagerstown District
1 male between 70 & 80 years
4 males between 60 & 70 years
3 males between 50 & 60 years
4 males between 40 & 50 years
6 males between 30 & 40 years
11 males between 20 & 30 years
1 male between 5 & 10 years
1 male under 5 years
4 females between 70 & 80 years
4 females between 60 & 70 years
3 females between 50 & 60 years
4 females between 40 & 50 years
5 females between 30 & 40 years
7 females between 20 & 30 years
1 female between 15 & 20 years
3 females between 10 & 15 years
1 female between 5 & 10 years
1 female under 5 years
[Federal Census 1840, Washington Co.]

The unusually large number of poeple listed under William Hawken's name in the 1840 census suggests that William and Leah were involved with a boarding house or hotel. Early in 1840, or at some time prior to that William apparently went out of the gunsmithing business, and his shop was taken over by one John C. Hawken.

GUNSMITHING John C. Hawken would respectfully inform the citizens of Hagerstown and the surrounding country, that he has taken the Shop of Mr. William Hawken, in Jonathan Street (opposite the 'Mail' Printing Office) where he will continue to carry on the Gunsmith Business in all its branches. No pains will be spared to merit the patronage of the public. Feb. 7
[The Hagerstown Mail, Friday 14 Feb. 1840]

John C. Hawken apparently worked in William's shop for little more than a year, and then it was taken over by two other Hagerstown gunsmiths, as is shown by the following advertisement, the full text of which appears under the names of Cookman and Keller.

GUN--SMITHING The subscribers respectfully inform their friends and the public, that they have commenced business in the old stand, lately occupied by Mr. William Hawken, in Jonathan Street . . . William Cookman
Hagerstown, April 9 Joseph W. Keller
[The Hagerstown Mail, Friday 28 May 1841]

A third newspaper advertisement makes it clear that five years later William Hawken re--opened his old gunsmith shop and went back into the business.

GUN--SMITHING The subscriber would inform the public that he has re--opened his *Gun--Smith Shop,* in Jonathan Street immediately opposite the Methodist Church, where he will be prepared to execute every description of work in his line of business, at the shortest notice and upon reasonable terms. William Hawken
N.B. He also wishes to purchase a lot of two inch curled maple plank. Hagerstown, April 10, 1846
[The Hagerstown Mail, Friday 17 July 1846]

When the 1850 census was taken William gave his occupation as gunsmith, so presumably he still was active at the trade at that time.

1850 Hagerstown District 1

William Hawken	age 51	b. Md.	gunsmith
Leah	48	Md.	
William Davis	11	Md.	
Wm. W. Humrichouse	3	Md.	
Hazel Hawken	25	Virginia	

At sometime between 1850 and 1870 William and Leah moved to Williamsport, on the Potomac River about 6 miles southwest of Hagerstown. Apparently they shared their house with their second daughter, Mary, her husband Charles W. Humrickhouse, and their children.

1870 Williamsport

Hawken, William	age 70	Retired	b. Md.
			real estate $4,000
" Leah	67	Keeping House	Md.
Humrickhouse, Charles W.	45	Merchant "Wholesale"	b. Md.
		real estate $59,000	
		personal property $5,000	
" Mary	43	Keeping House	b. Md.
" Walker	21	Student	b. Md.
" Leah M.	19	At home	b. Md.
" Mary	17	At home	b. Md.

William Hawken's name could not be found in the federal census for 1880 in either Williamsport or Hagerstown. This is unfortunate for it would have given information about the birthplace of his father and mother. William's wife, Leah, died on 20 May 1880, which was about the time that the census was being taken. William died in Williamsport five years later, on 27 February 1885. His will, written many years earlier, in 1852, was brief.

. . . I William Hawken of Hagerstown, Washington County . . .Maryland. . . give and bequeath to my beloved wife Leah all of which I may die seized of, during her natural life. . .

At the death of wife all there shall remain it is my desire shall be divided equally between my beloved daughters Mary who intermarried with Charles W. Humrichouse and Catherine Ann who intermarried with Andrew G. Boyd . . . 24 May 1852.
[Will Book G., p. 356, 357, Washington CCH]

HAWKEN, CHRISTIAN, JR. Hagerstown

Christian Hawken, Jr., the sixth son of Christian, Sr., is the one that we know the least about. It is probable that he did not work as a gunsmith, but this does not preclude the possibility that he too was trained in gunsmithing like his five brothers.

The year of birth for Christian, Jr., is not known, but it is determined that he died in 1822. He was not married, he had no children, and he owned no property at the time of his death other than some money left to him by his father the previous year. It is most likely that he was born in 1800 or 1801 and that he died not long after reaching the age of 21. This could readily explain why he had no wife, no children, no trade, and very little property.

In a newspaper interview given after he was 90 years of age, Samuel T. Hawken of St. Louis, who was a brother of Christian, Jr., stated that five of his father's six sons became gunmakers. Apparently Christian, Jr., was the one who did not. But it is very possible that he grew up to the trade like his brothers.

Christian Hawken, Sr., died 10 May 1821, and during the ensuing 18 months the estate was evaluated and sold. A portion of the estate went to Christian, Jr., who must have been 21 years of age at that time to receive property in an estate distribution. The estate distribution was filed 3 Oct. 1822, but the proceeds of the estate must have been distributed some months earlier, probably early in 1822.

Christian Hawken, Jr., died in the spring or summer of 1822 for causes unknown. Letters of Administration were granted for his estate on 3 Oct. 1822, and an accounting of the distribution of his estate was filed the same date.

In the distribution of the estate of Christian, Sr., it may be noted that after the widow and the heirs of dead children, the five sons are listed, and that the first four are named in order of birth, with Christian, Jr., named last, presumably because he was the last in order of birth. In the distribution for Christian, Jr., again the four remaining sons are listed in order of birth, and then the youngest daughter, Nancy, is listed. Nancy apparently came of age in the period between the deaths of Christian, Sr., and Christian, Jr., These distribution lists are the only documents so far discovered that relate to the order of birth of Christian, Jr., among his brothers and sisters, and they do not prove it.

Christian, Jr., received a distribution of $366.73 from his father's estate, and this apparently was the estate that Christian, Jr., possessed at death a few months later. Christian, Jr., had an estate valued valued at $342.68 after expenses.

Alex. Neill, Admin of Christian Hawken, Dd		
Balance due on passing 1st Acct		4,400.83
Distribution as follows, to wit		
The Widow		1,466.94
The Heirs of John Hawken		366.73
The heir of Margaret Miller		366.73
Geo. Hawken		366.73
Jacob Hawken		366.73
Samuel Hawken		366.73
William Hawken		366.73
Christian Hawken		366.73
	$2,933.89	
Oct. 3rd 1822	$$4,400.83	

[Distribution Book, p. 222, Washington CCH]

Alex. Neill, Adminr of Christian Hawken, Jr.	
Balance due the estate on passing 1st Acct	$342.68
Distribution as follows, to wit	
The Widow	42.83
The Heirs of John Hawken	42.83
The Heirs of Margaret Miller	42.83
Geo. Hawken	42.83
Jacob Hawken	42.83
Samuel Hawken	42.83
William Hawken	42.83
Oct. 3rd 1822	$ 342.68

[Distribution Book, p. 222, Washington CCH]

HAWKEN, JOHN C. Hagerstown

Knowledge about John C. Hawken is limited to the one newspaper advertisement presented here. It shows that in 1840 John C. Hawken took over the gunsmith shop of William Hawken. But this was a short--lived enterprise, for a year later William's old shop was taken over by William Cookman and Joseph W. Keller.

GUNSMITHING John C. Hawken would respectfully inform the citizens of Hagerstown and the surrounding country, that he has taken the Shop of Mr. William Hawken, in Jonathan Street (opposite the 'Mail' Printing Office) where he will continue to carry on the Gunsmith Business in all its branches. No pains will be spared to merit the patronage of the public. Feb. 7
[The Hagerstown Mail, Friday 14 Feb. 1840]

The relationship of John C. Hawken to other members of the Hawken family is unknown. It is known that George Hawken, son of Christian, Sr., had two or more children, but their names are not known. It is possible that John C. was one of the children of George.

HAYWARD, BARTLETT & COMPANY Baltimore

This firm was a wholesale stove manufacturer at 24 Light Street with a warehouse and foundries at Pratt and Scott Streets.

. . . From the above Summary it will be seen that the expenditures for "Arms and Ammunition" reached the large amount of nearly $25,000. The articles purchased include . . . 302 Pikes from Hayward Bartlett & Co. . . .
[Baltimore City Correspondence Manuscripts, 1862, #695]

HEIBERGER, FRANCIS J. & COMPANY Washington, D.C.

1862--65 Heiberger, Francis J. & Company 362 Pennsylvania Ave. military outfitters
[Washington city directories]

HEIDER, HENRY Baltimore

Heider, Henry
1851 Hanover & Henrietta Streets gunsmith
[Baltimore city directory]

HEINZ, RICHARD Baltimore

Jacob Pipino's shop was located at this same address until 1871.

Heinz, Richard
1884-86 20 Ensor Street gunsmith
[Baltimore city directories]

HIBNER, JOHN Baltimore

Hibner, John

1837	Gay & Saratoga Streets	lock & gunsmith
1840	Calvert & Saratoga Streets	lock & gunsmith
1845	25 Harrison Street	lock & gunsmith
1853		locksmith & bellhanger

[Baltimore city directories]

John Hibner age	60	sex M	occ Locksmith	birth Maryland
Barbara	59	F		Maryland
Abraham	22	M	Locksmith	Maryland
William	20	M	Locksmith	Maryland

[1850 Federal Census]

HILL, EDMUND Baltimore

Hill, Edmund
1840 Poca & Ross Streets lock & gunsmith
[Baltimore city directory]

HINELINE, THOMAS Washington, D. C.

Hineline, Thomas
1862 h 533 8th east gunmaker
[Boyd's Washington and Georgetown Directory]

HISS, GEORGE R. A. Baltimore

Listed as a merchant, aged 24, in the 1850 Federal Census.
The King & Hiss partnership was terminated by 1860 when they advertised as King and Hupman; by 1872 the firm was known as Henry S. King and Son.

King & Hiss
1847-55 307 Baltimore Street importers & dealers in hardware, cutlery & guns
[Baltimore city directories]

HITZELBERGER, STEPHEN V. Westminster

Hitzelberger, Stephen V. 1882 Westminster gunsmith
[The Maryland Directory, 1882; J. Frank Lewis & Co. Publishers]

HOBBS, JOSEPH

August 7, 1776 Council of Safety
That the said Treasure pay to Joseph Hobbs junr. three pounds ten shillings for a Musquet.
[AM, p. 179]

HODGSON, JOHN W. Washington, D. C.

Hodgson, John W.
1853 W side 6 w btw 6 & H gun & locksmith
[Directory, Washington, D.C.]

HODGSON & THOMPSON Baltimore

The name "Hodgson & Co" can be found in script on the lock of a maple-stocked flintlock pistol, silver mounted, bearing the touch mark of Charles Freeth of Birmingham, England, on the trigger guard, with a brass barrel marked in script "Baltimore." A pair of flintlock, cherry-stocked pistols with brass barrels marked "England" and the locks marked in script "Hodgson & Thompson" have been examined. The most frequently found pistols bear "Hodgson & Thompson" in two lines of script on the lock, nine inch English proof octagon brass mounted with an eagle and 14 stars engraved on the trigger guard. A long flintlock, maple-stocked fowler is marked on the lock "Hodgson & Co." The same marked lock can be found on a 46 inch stamped English proof round barreled rifle, walnut stock, brass furniture with the same type scalloped engraving as the fusel and with the acorn finial at the head of the trigger guard. A British 1796 period infantry sword and an 1812 saber are also marked "Hodgson & Thompson."

Hodgson & Thompson

1800	72 Baltimore Street	merchants, dry goods
	10 North Gay Street	hardware
1802	10 North Gay Street	merchants
1803	215 Baltimore Street	hardware merchants
1804	215 Cross Forrest Street	merchants

[Baltimore city directories]

HOEN, JOHN H. — Baltimore

Walsh, Hoen & Von Kapff

1887-88	27 South Charles Street	gun importers

[Baltimore city directories]

HOEPNER, BENJAMINE — Baltimore

Hoepner, Benjamine

1856	227 South Charles Street	gunsmith

[Baltimore city directory]

HOEPNER, CASPER B. — Baltimore

Hoepner, Caspar B.

1856-59	237 South Charles Street	gunsmith

[Baltimore city directories]

HOFF, DAVID T. — Emmitsburg

According to his business card he was a repairer of watches, clocks and jewelry, also guns and pistols upstairs in the Olser's building at Mechanicstown. The name of the town was later changed to Thurmont.

Hoff, David

gunsmith	Emmitsburg	1886
gunsmith	Emmitsburg	1895

[MILLER'S GENERAL DIRECTORY OF FREDERICK CITY & FREDERICK COUNTY]

. . . David T. Hoff is the only repairer of guns between Frederick and Gettysburg and Waynesboro and Westminster; he is a dandy as well as a No 1 mechanic; very fond of artistic pictures.
[HISTORY OF EMMITSBURG, MARYLAND]

HOFFENBRIDEL, LEWIS — Baltimore

Hoffenbridel, Lewis

1851	70 Camden Street	gunsmith
1852	459 West Baltimore Street	gunsmith

[Baltimore city directories]

HOGAN, JOHN — Washington, D. C.

13 June 1860 Washington, D.C.

Hogan, John	age 45	born	Ireland	gunsmith	real estate $1500
" Ellen	45		Ireland		
" James	21		N. Y.	rifler	
" Frederick	17		Pa.	fifer	
" Ellen	12		Washington, D.C.		
" John	6		Washington, D.C.		

[1860 Federal Census]

Hogon, John

1860	h493 11th east	gunsmith

[Directory, Washington, D.C.]

HOGLAN, ISAAC — Georgetown

Isaac Hoglan can produce battle axes at 95 cents each and boarding axes at 75 cents each.
[Naval Ordinance, 1816, National Archives]

HOLLIDAY, THOMAS — Washington, D. C.

Holliday, Thomas

1830	dwelling I s btw 7 & 'e & yard	gunsmith at Navy Yd.

[Directory, Washington, D.C.]

HOLLINGSWORTH, HENRY — Head of Elk, Cecil County

He was a wealthy man and employed a number of hands. Produced entire muskets but his largest operation was making barrels and bayonets which he supplied to Winter and Reid.

February 6, 1776 Hollingsworth to Council
The subscriber will engage to have made any quantity from two hundred to two thousand muskets, on the same plan, and at the price, they are made in this or the adjacent provinces and finish one hundred by the tenth day of April next, and one hundred per month after, till the quantity engaged are furnished, providing I am furnish'd locks for said muskets at least twenty days before said day, (for which shall be allow'd fifteen shillings for each lock) out of the price of the musket so finished, and providing I am advanced the sum of three hundred and twenty five pounds immediately (or as soon as good security is offered) and a further sum of one hundred pounds, at the delivery of said hundred muskets, towards defraying the expence of the second hundred, and for each hundred after to have the sum of fifty pounds advanced, till the quantity wanted are compleated as witness my hand . . . P. S. Tis meant that said muskets be compleat, with stock Boinet, Ramer, and brass mounting.
[AM, pp. 138-139]

March 7, 1776 Rumsey to Council
I have received a Lre from Lieut. Col. Henry Hollingworth of the Head of Elk in Cecil County relative to the guns in which he informed me that Mr. Johnson had acquainted him that I had some credentials or orders from the Council of Safety to him relating to the making of Arms and praying that if I had

given the matter out, that the workman were dispersed and must be collected again which he was willing to do out of a hearty wish to forward the service.
[AM, p. 211]

May 31, 1776
Ordered That the Treasurer of the Western Shore pay to Col. Henry Hollingsworth of Caecil County five hundred Pounds common Money on Account of his Contract to manufacture Gun-Barrels.
[AM, pp. 454-455]

July 5, 1776
Ordered That Mr. Henry Hollingsworth be requested to deliver to Elisha Winters, from time to time, such Musquet Barrels, and Bayonets, as he shall have manufactured for the use of this Province, not exceeding six hundred of each sort.
[AM, p. 550]

August 24, 1776 Hollingsworth to Council
Capt. Alexander of the Flying Camp having had his company complete for a considerable time past and hearing nothing from his field officers, hath requested me to inform your Honnours, that he only waits your orders to march. The Guns, Bayonetts, and Cartouch Boxes are ready, as also the Blankets. Cloth might have been had for Tents but as no orders were recd. on that head there is none prepared here. There is about 20 canteens prepaired, and I believe the rest may be had here if thought necessary . . . P. S. My Bayonett forgers have been two of them sick for some time, but yesterday began to work again.
[AM, p. 238]

September 28, 1776 Hollingsworth to Council
. . . The quantity of muskett barrels and Bayonetts, that I have made amounts to about the sum red. on that acct. the accts. of which I shall bring down with me as soon as in my power, together with the accts. for forwarding, and arming the troops, it appears to me, that I have a quantity of barrells and bayonetts now by me, that Mr. E. Winter will not want, as I sent him seventy Bbls and ninety Bayonetts the other day. I shall therefore send them by first opportunity to your Honnours . . .
[AM, p. 309]

October 10, 1776 Council to Hollingsworth
By order of the Honorable Convention we return you eighty four Bayonets and Gun Barrel, which Mr. Winter of Chester Town sent here for our inspection, and we are directed to inform you that unless you comply in a short time with yr. contract with the Board, or give us satisfactory reason for the delay, that your Bond will be put in suit.
[AM, p. 330]

November 3, 1776 Hollingsworth to Council
. . . As to the eighty fore Bayonets mentioned I had no expectation that they would all do, but soposed some of them would, I have not been able yet to finde out but one Smith that can forge a Bayonet properly, but on receiving your letter of the 12th of July in which you request very pressingly that 400 Bayonets be made of different sises for the flying Camp Militia, I sett several workmen to work at forging Bayonetts, who at first made them very deficiant, some of the worst of which I kept hear as not fit for use, at any rate, and sent many on which were of the best with the flying camp militia . . . the remainder such as they are I sent to Mr. Winters desireing him at the same time to send thos that were not fit for use Back, as also all Barells that proved not good or unfit for use, to be sent back with his objection to them in wrighting, by which I hoped in a short time to be compleat in our Buisenis, but to my disappointment have never rec'd one, though frequent opportunityes, had the Large Barells mentioned been sent back they might have been long since made small as they were sent back to Mr. Winters the September . . . but I am of the opinion that Mr. Winters hands did not long stand for want of work, for I finde by my Books that on the 30th of the same month, he had 50 Barrells more, which I hear of no fault to, and had the others been sent back, they could have been made smalle in three days, tis said they Bayonetts are too soft which I doubt not, as they are all steeld with American steel (or at least the Smiths had steel of me for them) some of which I found after to be raw Iron, not at all converted into steel, which at that time I was a Strangure too, as it was sold me for good steel . . .
[AM, p. 419]

January 10, 1777 Hollingsworth to Council Baltimore
I am now hear settling my bill for publick service for the Virginia State with Congress fully intending to have waited on your Honnours in hope of getting credentials to the committee of Safety at Philadelphia or Lanckester, in order to obtain some of the prisoners, to work at my gun factory, if any workmen should be amoung them that would sute for either finishing, or lock making . . .
[AM, p. 38]

April 24, 1777 Council of Safety
Ordered That Henry Hollingsworth deliver to Robert Reid of Chester Town one hundred and fifty Gun Barrels, provided he has a sufficiency, wherewith to supply the other Works.
[AM, p. 226]

May 14, 1777 Council to Hollingsworth
We are very glad to see by yours of the 10th Inst. that your Manufactory of Gun Barrels goes on to your Satisfaction. We are very desirous of having 50 of the large ones, full Inch in the Bore, those sent down were very bare Inch, well finished and sent forward to Gen'l. Washington as soon as possible, he is desirous of having them and we think, if they are well executed, they will be very usefull. The first 20 or 25 may be sent forward, and the rest of the 50, as they are done, recommend them to the Care of your Brother, or some Body at Philadelphia who will push them along. We have not heard, for some Time past, from Mr. Winters; we would, nevertheless, have you drive at the Barrels as we have a good many Locks and expect more soon; if you can get any number of your Barrels stocked and finished well, without interfering with your making the Barrels, we will send you up some Locks, if not, send us down 100, 150 or 200 Barrels, and we believe we can get them stocked here. If you can contract for Stocking and mounting, let us know the Price . . .
[AM, p. 253]

July 31, 1777 Council of Safety
That Colo. Henry Hollingsworth deliver to Mr. Winters fifty Gun Barrels.
[AM, p. 332]

Hollingsworth Henry Esqr. Cecil County
5 males of 16 years
2 males under 16 years
8 females
23 slaves
[1790 Federal Census]

HORST, GEORGE Baltimore

Horst, George

1851	Harford Avenue extension	gunsmith
1853	48 York Avenue	gunsmith

[Baltimore city directories]

	age	sex	occ	birth
George Horst	58	M	Gunsmith	Germany
Mary	47	F		Germany

[1850 Federal Census]

HOUSE, WILLIAM C. Easton

House, William C.

1887	Easton	gunsmith

[Maryland Directory & State Gazetteer, 1887, The Baltimore Publishing Co.]

HOWARD, HENRY Baltimore County

Many cannon from his furnace were supplied to the Council of Safety as well as to the public.

Any person who understands the casting and finishing of cannon by applying to Henry Howard, at Northampton Furnace about ten miles from Baltimoretown, in the Province of Maryland, will meet with great encouragement.
[*Pennsylvania Gazette,* August 14, 1776]

September 12, 1776 Council of Safety
Ordered That the Commissary of Stores at Baltimore Town deliver to Henry Howard one-half Barrel of Powder to try Cannon.
[AM, p. 266]

HOWARD, WILLIAM

July 29, 1776 Council of Safety
That the Treasure pay to William Howard four pounds for a Musquet.
[AM, p. 134]

HUGHES, DANIEL & SAMUEL

Barnabas Hughes emigrated from Ireland in 1750. He married Elizabeth Waters. He engaged in the iron trade and built Mount Etna and other furnaces near the Black Rocks above Beaver Creek. . . . His eldest son Daniel was born in 1730 and Samuel in 1741. Both the brothers became conspicuous in the Revolution for their zeal in behalf of the patriot cause and from the fact that many of the cannon used by the Continental army were cast at their forges . . . The first furnace of the Hughes brothers was built at Black Rocks, near South Mountain in partnership with Mr. Buchanan of Baltimore before the Revolutionary War. The second furnace, now known as Mount Etna was about a mile lower down. After these were built Col. Daniel Hughes formed a partnership with Col. Wm. Fitzhugh and built a forge and nail furnace, now known as the Old Forge between Hagerstown and Leitersburg on the Antietam. The fourth and last furnace was built by Col. Daniel Hughes and his two sons, Samuel and Daniel Hughes both Lawyers of Hagerstown at Mont Alto, Franklin County, Pa. . . . The Mount Pleasant Furnace, near Havre de Grace, in Harford County, Md. was built by Samuel Hughes . . . There are cannon now in existence in the Boston Navy Yard cast at this furnace during the Revolutionary War.
[HISTORY OF WESTERN MARYLAND, pp. 1011-1012]

Owing to the sacarcity of munitions of war it became necessary to provide them, and contracts were made with Mr. Huges, of Frederick County, to supply the Province with cannon, but he was compelled to decline proposals to furnish the Continental army with them. A foundry was established on the Potomac, about Georgetown, where the first cannon were made in this country. Many were also made at the Antietam works by the Hughes, and at Catoctin Furnace by Jones, and Thomas Johnson's powder mills were also erected in Frederick County.
[MILLER'S GENERAL DIRECTORY OF FREDERICK CITY & FREDERICK COUNTY, p. 127]

February 21, 1776 Council to Fulford
From an Opinion of your Knowledge of Cannon we have thought proper to appoint you to prove those to be supplied this Province by Mess. Sam'l. & Dan'l. Hughes and request you will emmediately repair for that Purpose to their works at Antietam . . .
[AM, p. 175]

February 22, 1776 Barrister Carroll to Council
. . . Mr. Hughes by what he has heard supposes there may be 20 Nine Pounders cast by this time, tho he has not since we saw him at Baltimore Town received any letters from his brother. There will be he tells me 1080 wt of Powder wanting to try them al, 180 wt. for the 20 Nine Pounders, and 900 for the 50 Eighteen Pounders at the weight of the ball in Powder for each Gun.
[AM, p. 180]

March 10, 1776 Baltimore Committee to Council
. . . We are sorry to inform you that 4 out of 5 canon at Mr. Hughes burst & killed poor Matthews . . .
[AM, p. 235]

March 26, 1776 Council to Daniel & Samuel Hughes
We received a Letter yesterday from Mr. Lux advising us of your having cast and bored 1 18 Pounder and that you expected to have 4 others finished by this Time. Our Board have appointed Captain Burgess to prove them, and desire you will forward them to Baltimore, as fast as you can get them ready to send off, as they are greatly wanting there.
[AM, p. 288]

July 5, 1776 Council to D. and S. Hughes
The Convention have directed the Council of Safety to fortify Annapolis immediately, this cannot be done without you comply with your contract with the Province or as much thereof as you possibly can – you will therefore be pleased to send as many of the 18 & 9 pounders to Baltimore, as may be in your power.
[AM, p. 551]

August 7, 1776 Council to Messrs. Hughes
. . . We are now much in want of the cannon you contracted to supply us with; we must therefore request you will with all expedition send down all you have ready and that you will not delay, as quickly as you possibly can complying with the whole of your contract. (You may send one half to Baltimore Town and the others to Annapolis, if the waggoners can be procured so to bring them.
[AM, p. 182]

August 13, 1776 Council of Safety
. . . Ordered That Western shore Treasure pay to Dan'l. & Sam'l. Hughes two thousand Pounds on Acct. of his Contract for a Cannon Foundery.
[AM, p. 198]

April 27, 1780 Hollingsworth to Council
. . . I am ordered by the Navy Board to have a quantity (24 pieces) of Cannon haul'd from Mr Hughes furness to Baltimore and to see them forwarded to Philadelphia . . .
[AM, p. 483]

April 29, 1780 Council to Commissions of Frederick & Washington Counties
We request you will furnish Forage for the Subsistence of the teams employed by Col. Hollingsworth in removing Cannon from Hughs's works to Baltimore Town for the use of the the Continent.
[AM, p. 159]

June 4, 1781 Council to Sprigg
We have purchased of Samuel Hughes Esqr. ten pair of four Pound Cannon which we beg you will have proved and transported to Baltimore Town . . .
[AM, p. 461]

June 17, 1781 Sprigg to Council
. . . Agreeable to Your directions I provd. 20 four pound Guns at Mr. Hughes works . . . they stood the first proof so well it inducd. me to give them something more than the usual proof in the second Charge, I believe them to be Extry Guns.
[AM, p. 298]

June 23, 1781 Letters to Council
. . . The Guns sent p Mr Hughes are very handsome but they weigh between 10 and 1100 each . . .
[AM, p. 311]

July 3, 1781 Council to Hughes
The general Assembly not having any particular Notice of the Contract made with you, leaves us at Liberty and we are willing to release you from it on your paying for the Transportation of the Cannon to Baltimore Town and the expence of proving them, unless the Purchaser from you shall have them again proved, then the Expence approving them to be given up.
[AM, p. 494]

July 13, 1781 Letters to Council
. . . 'Mr. Saml. Hughs has Notified to me that he Intends to Take away the 20 Pieces Cannon I received from his Furnace and Requested me to Inform your Excell. Thereof.
[AM, p. 412]

Hughs Daniel Washington County
5 males of 16 years
2 males under 16 years
6 females
6 other free persons
6 slaves
[1790 Federal Census]

Episcopal Church St Johns The Vestry of Washington County
Daniel Hughes
1787
1789
1791
1807
1813
[HISTORY OF WESTERN MARYLAND, p. 1078]

Death notice – December 5, 1818 Col. Daniel Hughes in his 74 year.
[HISTORY OF WESTERN MARYLAND, p. 1050]

HUNT & GOODWIN Washington, D.C.

1862–65 Hunt & Goodwin military outfitters
390 Pennsylvania Ave.
[Washington City directories]

Presentation swords have survived bearing the name of this dealer. One is etched on the obverse side of the blade "Presented by Co. L 1st N.Y. Vet Cav. as a mark of esteem to C. W. Bryant." To other inscriptions were later added side by side, "Captured March 10, 1864, & Presented by Lt. A. E. Richards to Lt. Col. John Singleton Mosby," "Recaptured Sept. 1864 by the 13th N.Y. Cav. Col. H. S. Gansevoort."

HUPMAN, NICHOLAS A. Baltimore

In 1854 he was listed as a merchant and after separating from the below-named firm, he was affiliated in 1872 with N. H. Bell and Company, who were whip manufacturers.

King & Hupman
1860-71 307 West Baltimore Street gunsmiths
[Baltimore city directories]

HUTCHINS (HUTCHINGS), E. & COMPANY Baltimore

Several styles of underhammer pistols manufactured by A. Ruggles (dated 1836) and the Quinnebaug Rifle Company bear the stamp "E. Hutchings & Co Agents Balto." They handled underhammer rifles that were made by Kindell and stamped as the above, also a percussion side by side rifle-shotgun with back action locks made by Baker in Lansingburgh, New York, and stamped likewise. The walnut half stock is fully engraved with German silver hardware and has a lengthy long rifle patchbox with six piercings.

E. Hutchins & Company

1835	110 Baltimore Street	chemists
1836	186 Baltimore Street	hardware merchants, agents for the mfrs. of the long & pocket rifles & dealers in guns, rifles & pistols of every description

[Baltimore city directories]

HUTCHINSON, E. Baltimore

Hutchins, E. & Co.

1835	110 Baltimore Street	chemists

Hutching & Co.

1837	186 Baltimore Street	hardware MTS.

Hutchinson Brothers (H. M., J. L. & Elias S. Hutchinson, dw. 22 South Stricker)

1867-70	14 Light Street	plumbing, ranges, furnaces & stoves

[Baltimore city directories]

Hutchinson, F.	1850	age 30	Peddlar	birth Pennsylvania
S.	1850	28		Pennsylvania

living at the boarding house of William Slicer

[1850 Federal Census]

HUTTEN, PHILIP Baltimore

Hutten, Philip

1872	90 Camden Street	gunsmith

[Baltimore city directory]

* * * * * * *

ADDENDA

HAWKEN, CHRISTIAN, SR.

In Washington County Court sitting as a Court of Equity
July Term 1822

Samuel Hawken, Jacob Hawken, George Hawken and William Hawken, Juliana Hawken, Nancy Hawken, Christian Hawken, David Hawken, Nancy Hawken, Robert Miller, and Juliana Miller, heirs at law of Christian Hawken, late of Washington County, deceased.

The bill in this case states, that the said Christian Hawken, deceased, departed this life intestate, leaving the above named Juliana Hawken, his widow, and the said complainants and defendants his heirs at law. That at the time of his decease he was seized in fee of the following lots or portions of ground situate in Hagerstown, Washington County, Md., to wit, one lot in Hagerstown No. 198, one undivided moiety in the half of lot Numbered 81, in said town, three lots in Rohrer's Addition to Hagerstown, numbered 45, 46, and S92, also two lots of ground situate in Mary Ann's Town, one of the additions of Hagerstown, one numbered 69 and the other being part of a tract of land called the Resurvey on Grogg.

The bill also states that the said Christian Hawken, deceased, left one other son named Christian, who died since the decease of the said Christian the ancestor; and that Christian Hawken, one of the above named defendants who was a minor and grandson of the said Christian, deceased, is not a resident of the State of Maryland.

The bill further states, that a sale of the above described real estate would be beneficial to all the parties interested therein.

The object of the bill is to obtain a decree for the sale of said real estate, and to have a trustee appointed to carry such decree into effect.

Wherefore, it is ordered and adjudged by Washington County Court, as a Court of Equity, this twenty third day of July 1822, that the complainants cause notice of the substance and object of this bill, to be given to the absent defendant, and to that end, that a copy of this order be inserted in some newspaper printed in Hagerstown, once in each of three successive weeks, before the first day of September next, warning them to be and appear in this court, either in person or by solicitor, on the first day of December next, to show cause, if any, he hath, why a decree should not pass as prayed, otherwise this bill will be taken pro con fesso against him.

O. H. WILLIAMS, Clk.

[Maryland Herald & Hagers-town Weekly Advertiser, 30 July, 13 Aug. 1822]

HAWKEN, CHRISTIAN, SR.

The earliest known real estate transaction for Christian Hawken concerns his purchase of a town lot in the town of Green Castle, Pa., which lies about 10 miles north of Hagerstown. The cost was but 3 pounds. William S. Bowers informed me about this deed, and suggested that it might have been taken in payment for a rifle, which at that time cost about 2 or 3 pounds.

This indenture made the 1st day of February . . . 1783 between John Allison of Antrim Township in the County of Cumberland and State of Pennsylvania and Elizabeth his wife, of the one part, and Christian Hawkey of the other part, that Whereas for . . . the sum of 3 Pounds . . . paid by . . . Christian Hawkey to the said John Allison . . . have sold unto the said Christian Hawkey a certain lot . . . in the Town of Green Castle Number 10 in the aforesaid Township County and State containing 60 feet front on the East side of Carlisle Street, and extending backwards 240 feet . . . Now this Indenture Witnesseth that . . . John Allison for . . . 3 Pounds do grant . . . unto . . . Christian Hawkey . . . the above mentioned lot . . . To have and to hold the said lot of land unto the said Christian Hawkey . . . John Allison & Elizabeth his wife . . . do covenant . . . with the aforesaid Christian Hawkey . . . the above mentioned lot of ground . . . unto the said Christian Hawkey . . . Before me . . . came John Allison and Elizabeth his wife and acknowledged the above Deed of Conveyance to Christian Hawkins as their act . . .

[Deed Book, p. 575, Franklin Co., Pa. Court House]

HAWKEN, GEORGE

George Hawken, gunsmith of Hagerstown, probably was living in Hagerstown in 1810; early in the year he advertised in the newspaper there, and he seems to be listed there in the 1810 census. The following listings are worth noting also, however.

1810 Hawken, George
1 male between 26 & 45 years
1 female 16 - 25
1 female under 10
[Federal Census, 1810, Washington Co., Va.]

1810 Hawkin, George
1 male between 26 & 45 years
1 male 16 - 25
2 males under 10
1 female 16 - 25
1 female under 10
[Federal Census, 1810, Washington Co., Va.]

1820 Hawkins, George
1 male between 26 & 45 years
4 males under 10
1 female 26 - 45
2 females under 10
[Federal Census, 1820, Jefferson Co., Va.]

HAWKEN, JOHN

One or more of the following listings may refer to the gunsmith John Hawken of Hagerstown, who moved to Virginia sometime after he came of age in 1805. Harpers Ferry where he may have worked, is located in Jefferson County. The first listing is the most promising.

1810 Hawkin, John C. Harpers Ferry
4 males between 16 & 26 years
1 female 26 - 45
[Federal Census, 1810, Harpers Ferry, Jefferson Co., Va.]

1810 Hawkins, John
1 male between 26 & 45 years
1 male between 10 & 16
1 female over 45
1 female 26 - 45
3 females under 10
[Federal Census, 1810, Jefferson Co., Va.; this listing appears before the Harpers Ferry listings.]

1810 Hawky, Jno.
1 male between 26 & 45 years
2 males under 10
1 female 26 - 45
1 female over 45
2 females 10 - 15
2 females under 10
[Federal Census, 1810, Loudoun Co., Va.]

1820 Hawkins, John Lee Town
1 male between 26 & 45 years
1 female under 10
2 females 10 - 15
1 female 16 - 26
1 female over 45
[Federal Census, 1820, Jefferson Co , Va.]

1820 Hoggins, John
1 male between 26 & 45 years
2 females under 10
1 female 26 - 45
[Federal Census, 1820, Berkeley Co., Va.]

J

JACOB, DAVID C. — Washington, D. C.

Jacob, David
1862 h 581 L north gunsmith
[Boyd's Washington and Georgetown Directory]

Jakob, D. C.
1866 568 L north gunsmith
[Boyd's Business Directory, Washington, D.C., Georgetown & Alexandria]

Jacob, David C.
1867 568 L north gunsmith
[Boyd's Business Directory, Washington, D.C., Georgetown & Alexandria]

JOHNSON IRON FURNACE — Frederick

Johnson Iron Furnace built in 1774 by the four Johnson Brothers – Thomas, Baker, Rodger and James. During the Revolutionary era these furnaces were in action operation and proved of great benefit to the patriot cause; and in furnishing munitions of war for the Continental army . . . located in the Urbana area near the Monocacy.
[HISTORY OF WESTERN MARYLAND, p. 598]

JOHNSON, JAMES — Frederick

August 16, 1777 Council of Safety
That the said Treasure Pay to Benjamin Galloway five hundred Pounds to be delivered over to Messrs Beatty, Hanson and Johnson Superintendants of the Gunlock Manufactory on Acct.
[AM, p. 335]

JOHNSON, OSCAR — Baltimore

			birth
McComas, Alexander	29	gunsmith	Maryland
Mary	24		Maryland
Clenney, John	20	gunsmith	Maryland
Johnson, Oscar	14	gunsmith	Maryland

[1850 Federal Census]

JOHNSON, RICHARD — Maryland Heights

Richard Johnson . . . was employed for more than forty years in the U. S. Armory, (Harpers Ferry) as a gunsmith. He was there at the time of John Brown's famous raid. Mr Johnson was well educated and when the Armory was destroyed, he began teaching in the schools of Washington County. His education was obtained in the common schools and at college. After he gave up teaching, he lived on a small farm which he owned, on Maryland Heights. He served for some time as a commissioner of the county. Mr Johnson was prominent both in political and social affairs, in politics, he was an ardent Democrat. In religion, he was inclined to the Episcopal faith. He died on his farm. Richard Johnson was married to Annie Yourtee and had five children.
[HISTORY OF WASHINGTON COUNTY, Vol. II, p. 1320]

JOHNSON & REAT — Baltimore

An excellent, eagle head pommel sword presented by the state of Virginia to Presly O'Bannon for his deeds during the Tripolitan War bear their mark. The silver hilt has a gold picture in the grip of a man raising a flag with a sword in his other hand. The sword shows superb workmanship and the curved blade is etched with the presentation.

Johnson & Reat Baltimore 1810 silversmith
[OLD SILVER & OLD SHEFFIELD PLATES]

JOSEPH & ANDREW

August 13, 1777 Council of Safety
That the said Treasure pay to Joseph & Andrew two Frenchman, fifteen Dollars for a Musquet, a broken musquet and three Cutlasses.
[AM, p. 332]

JOTUL, CARL — Washington, D.C.

1858 Jotul, Carl lock & gunsmith h 55 E south
[Business Directory, Washington, D.C., 1858]

K

KAIDEL, HENRY — Baltimore

Kaidel was associated with J. D. Moritz

Moritz & Kaidel

1868-79	10 South Charles Street	gunsmith
1880	6 & 8 German Street	gunsmith
1881-84	48 German Street	gunsmith
1885	48 & 50 German Street	gunsmith
1886	48 German Street	gunsmith

[Baltimore city directories]

KAYSER, JOHN C. — Georgetown

Kayson, J. C.
1855 57 Congress gunsmith
[Georgetown Directory]

Kayser, John C.
1863 57 Congress Georgetown gunsmith
[Hutchinsons Business Directory, Washington, D.C., & Georgetown]

Kayser, John C.
1864 57 Congress Georgetown gunsmith
[Boyd's Business Directory, Washington, D.C., & Georgetown]

Kayser, John C.
1865 57 Congress Georgetown gunsmith
[Boyd's Business Directory, Washington, D.C., Georgetown & Alexandria]

Kayser, John C.
1866 57 Congress Georgetown gunsmith
[Boyd's Business Directory, Washington, D.C., Georgetown & Alexandria]

Kayser, John C.
1867 57 Congress Georgetown gunsmith
[Boyd's Business Directory, Washington, D.C., Georgetown & Alexandria]

KEENER, GEORGE P. — Baltimore

Was a third generation gunsmith, born in 1824, son of John (I).

Keener, George P.
1852-60 3 Water Street gunsmith
[Baltimore city directories]

Keener, George 1850 age 26 gunsmith birth Maryland
[1850 Federal Census]

KEENER, JOHN (I) — Baltimore

According to the 1850 census, John (I) was born in 1771. With his father, Peter, he made rifles, muskets, fowling-pieces, and pistols. John (I) had two gunsmith sons, John (II) and George. By 1840 John (I) was located at 8 Lombard St., and from 1845 to 1851 at 3 Water St.

Keener, John & Peter

1796	Oldtown near Green Street	gunsmith
1799	Green & North Streets	gunsmith
1802-06	106 Green Street	gunsmith

Keener, John

1810-24	8 Water Street near Bridge dwelling Green near Bridge	gunsmith

(ad 1831) . . . has for sale a large and extensive assortment of double and single barreled duck guns, also double and single barreled bird guns with flint and percussion locks . . .
[Baltimore city directories]

Keener, John	79	gunsmith	birth	Maryland
Margaret	60			Maryland
John	29	gunsmith		Maryland
George	26	gunsmith		Maryland

[1850 Federal Census]

KEENER, JOHN (II) — Baltimore

Keener, John	79	gunsmith	birth	Maryland
Margaret	60			Maryland
John	29	gunsmith		Maryland
George	26	gunsmith		Maryland

[1850 Federal Census]

KEENER, PETER — Baltimore

As a Revolutionary War gunsmith Peter Keener must have been born prior to 1755, and as the father of the gunsmith John (I) Keener who was born in 1771, Peter must have been born prior to 1750. In 1795 the father and son advertised that they made rifles, muskets, fowling-pieces, and pistols.

A flintlock pistol is known with brass octagon barrel marked "Extre Proof" and a flat lockplate marked "Keener," but which Keener is not known.

September 17, 1777
That the said Treasurer pay to Peter Keener 187 pounds 5 shillings on his Contract for making Arms.
[AM, p. 377]

September 17, 1777 Council of Safety
Peter Keener agrees with the Governor & Council to make and in Workmanlike manner compleat and finish one hundred Musquets fixed with Bayonets, steel rammers, swivels, priming wires and brushes at three pounds fifteen shillings each the Mounting to be of brass or Iron the breech pin to be let into the solid so that on reboring the threads of the screws shall not be injured, that he will enter on the said Work within two weeks from the Date hereof and within six weeks from the Date hereof will deliver fifteen and so from Month to Month thereafter at least fifteen of the said Musquets compleatly finished as aforesaid the said Deliveries to be made at Baltimore Town to Gerrard Hopkins or such other person as shall be appointed by the Governor and Council. The Governor and Council to furnish Locks and such Barrels and Bayonets as were furnished to Nicholas White, and also to advance one hundred and eighty seven pounds ten shillings part of the price on giver Bond with good security to comply with this Contract.
[AM, p. 377]

October 31, 1777 Council of Safety
That the said Treasure pay to Peter Keener sixty pounds twelve shillings and six pence per Acct. passed.
[AM, p. 406]

Peter Keener & Son (John) Gunsmiths Respectfully inform their friends and the publick in general that they carry on the Gunsmith business in all its different branches at their shop in Green Street, Old Town, between Griffith's bridge and the Hay-Scales, where they make and repair Rifles, Muskets, Fowling-Pieces, Pistols and all manner of Gun Work executed in the best and neatest manner. N. B. They have several neat Rifles for sale on reasonable terms.
[*Federal Intelligencer & Baltimore Daily Gazette,* 4 April 1795]

Keener, John & Peter

1796	Oldtown near Green Street	gunsmith
1799	Green and North Streets	gunsmith
1802-06	106 Green Street	gunsmith

[Baltimore city directories]

KEENER, SAMUEL — Baltimore

Samuel may have been a brother of Peter Keener, and apparently was less skilled as a gunsmith.

January 31, 1776 Council of Safety
The Council called on Messrs Keener, Messersmith, and Reddick to know what Quantity of Musquets they had made for the use of this Province in Consequence of their Contracts with Mr. Alexander. Mr. Keener reported that he had made thirty. Mr. Messersmith that he had made Twenty Seven, and Mr Riddick that he had made seventy.
[AM, p. 127]

February 7, 1776 Council of Safety
Samuel Keener 32 guns 13 good 19 bad
[AM, p. 155]

February 12, 1776 Ewing to Council Baltimore
. . . I have prov'd all the guns made by the Gunsmiths in Town . . . Mr. Keener after seeing the rest try'd refus'd to have his proved but upon my threatning him a good deal he comply'd. We charg'd all the guns with one ounce of powder and two balls. I have a proof stamp with which I stamp'd all that prov'd good.
[AM, p. 155]

KELLER, JOSEPH W. — Hagerstown

Fairly plain, sparsely engraved, brass furnished percussion longrifles bear the name "J. W. Keller" in script on the octagon barrel.

GUN-SMITHING – The subscribers respectfully inform their friends and the public, that they have commenced business in the old stand, lately occupied by Mr. William Hawken, In Jonathan Street, between Washington and Franklin, – where they intend to manufacture *Guns, Rifles* and *Pistols* of every description. Guns stocked, altered to percussion, & repaired in a style not to be surpassed.

Keys fitted, Locks repaired, and all kinds of light jobbing, done in the best manner, and at the shortest notice. The subscribers intend to have constantly on hand a large assortment of *Rifles* and *Shot Guns.* William Cookman
Hagerstown, April 9 [1841] Joseph W. Keller
[The Hagerstown Mail, 28 May 1841]

KENNER, ALEXANDER — Hancock

Kenner's first recorded deed in washington Co. was in 1834 and his last in 1893.

Washington County, District No. 2 ½Hancock]

1850	Kenner, Alexander age	50 b.	Md.	gunsmith
	Maria	30	Md.	
	James	4	Md.	
	George	5/12	Md.	
	Robinson, Jane	36	Md.	

[1850 Federal Census, Washington Co., Dist. No. 2]

KENSTER, JOHN — Georgetown

Ran away from the subscriber, last night, two indent'd sevantsman vis . . . John Kenster, born in London about 35 years old . . . He is a gunsmith by trade. John Yost or Thomas Ridgen
[Maryland Journal & Baltimore Daily Advertiser, 15 Apr. 1777]

KETTNER, FRANCIS G. — Baltimore

1840 Kettner, Francis G. gunsmith Fleet St. west of Bond
[Baltimore city directory]

KING & HISS — Baltimore

Henry S. King and George R. A. Hiss formed the first partnership; later it became King & Hupman.

1847-55 King & Hiss 307 Baltimore St. importers & dealers in hardware, cutlery & guns
[Baltimore city directories]

KING & HUPMAN — Baltimore

This firm followed King & Hiss. Henry S. King and Nicholas Hupman apparently concentrated on firearms because they were listed as smiths. Henry S. then worked with his son Henry at the same location.

King & Hupman

1860-71	307 West Baltimore Street	gunsmith

King, Henry S. & Son

1873-80	307 West Baltimore Street	gunsmith
1881-82	53 South Sharp Street	gunsmith
1883-86	53 Hopkins place	gunsmith

[Baltimore city directories]

KING, ANDREW L. — Baltimore

King, Andrew L.

1847	79 North Gay Street	gunsmith

[Baltimore city directory]

KING, JOHN — Talbot County

An English style, flat lock, American walnut stock,

four ramrod piped musket bearing the marks "I King" and stamped "33" has survived.

Committed to the sheriff of Baltimore County on the 15th, day of July, as a Runaway, one John King, who says he is a servant to Joseph Spencer, in Talbot County living within two miles of Mr Mathew Tilgman, he says he is a Gunsmith by trade but lately following Sawing. William Young.
[*Maryland Gazette,* 31 July 1755]

KING, NICHOLAS — Frederick County

This indenture made 16th day of May 1797 witnesseth that Nicholas King an orphan boy aged sixteen years this 5th day of March last hath put himself and by these presents doth voluntarily and his own free will and accord put himself apprentice to Nicholas White gun smith with him after manner of an apprentice to serve untill he shall arrive to the full age of twenty one years . . .
[Frederick County Indentures, Hall of Records, p. 135]

KING, NICHOLAS — Baltimore

It has not been established that this Nicholas King is the same man who served an apprenticeship under Nicholas White in 1797, but it is reasonable to assume that it is.

King, Nicholas
1835 Fleet Street east of Eden gunsmith
[Baltimore city directory]

KIRK, E. CLARENCE — Baltimore

The improvement described below was mainly on the magazine loading tube of the Spencer carbine. Kirk was a gold and silversmith with the firm of Samuel Kirk & Sons at 187 North Calvert Street.

July 9, 1867
Be it known that we, E. Clarence Kirk and E. Sneider of the city of Baltimore, in the county of Baltimore and State of Maryland, have invented certain new and useful Improvements in Magazine Fire-Arms . . .
[U. S. Patent Office, Patent #66,596]

KLEIBACKER, CHRISTIAN B. — Baltimore

He became associated with David B. Trimble after the firm of Poultney & Trimble was dissolved.

Trimble & Kleibacker
1877-80 200 West Baltimore Street gunsmith
[Baltimore city directories]

KLEINHENN, AUGUST — Washington, D. C.

Kleinhenn, August
1867 496 6th west gunsmith
[Boyd's Business Directory, Washington, D.C., Georgetown & Alexandria]

Kleinhenn, August
1868 496 6th west gunsmith
[Boyd's Business Directory, Washington, D.C., Georgetown & Alexandria]

KLEINHENN & DUDLEY — Washington, D. C.

Kleinhenn & Dudley
1866 448 G south gunsmiths
[Boyd's Business Directory, Washington, D.C., Georgetown & Alexandria]

KLOKE, ANTON I — Baltimore

Advertised as a smith in 1840, 1847, 1852 and 1854 at 12 North Green Street; 1855 and 1857 at 10 North Green Street.

Anton Kloke	age 42	sex M	occ. lock & gunsmith	birth Germany
Catharine	42	F		Germany
Eliza	13	F		Maryland
Mary	10	F		Maryland
Anna	8	F		Maryland
John	6	M		Maryland
Anton	3	M		Maryland

[1850 Federal Census]

KNEL, PHILIP — Baltimore

Knel, Philip
1847 51 South Charles Street gun & locksmith
[Baltimore city directory]

KNIGHT, J. W. — Westminster

Knight, J. W.	1878	Westminster	gunsmith
Knight, J. W.	1880	Westminster	gunsmith
Knight, J. W.	1882	Westminster	gunsmith

[The Maryland Directories, 1878, 1880, 1882; J. Frank Lewis & Co. Publishers, Baltimore]

KOCH, GEORGE — Frederick County

Several longrifles signed "H Koch" have survived and Henry Koch, a gunsmith, was working in Tyrone, Cumberland County, Pennsylvania, in 1820.
A deed and bill of sale, by George Koch who was disposing of his holdings, is found in the Frederick County Court House. No occupation is given.

This indenture made the 16th day of February 1796 between John Demuth of Frederick County in the state of Maryland gunsmith of the one part and Henry Koch of the county and state aforesaid of the other part witnofseth that the said George Koch aged sixteen years, seven months and fourteen days as an apprentice to the said Demuth for the term of four years, four months and sixteen days . . .
[Frederick County Indentures, Hall of Records, p. 66]

KOONS, HENRY — Frederick County

There was a gunsmith named Henry Koons who apparently worked in northern Frederick County in the 1810-1830 period, judging from the style of flintlock rifles that have survived bearing his name. Kindig (1960) illustrated one of these and speculated that Henry Koons worked in southern Adams County, Pennsylvania, which is adjacent to northern Frederick County. Research is difficult, for members of the Koons family, because the name was a common one and was spelled a number of different ways. The original Germanic spelling was Kuntz, but in early America the name changed to become Kunz, Koons, Coons, Kuhns, etc. For example, the well-known Philadelphia gunsmith Jacob Kuntz changed his name early in his career to Kunz; his will mentions his grandson, Henry Koons.

There are a number of indentures recorded in Frederick County for various men named Henry Koons during the period in question, but unfortunately none mentions the occupation of gunsmith.

Henry Koons came to the U. S. from the Netherlands in 1730 and after a short stay in New Jersey came to Frederick County (now Carroll County) for a home. He became acquainted with Colonel Bruce, who owned a large tract of land near Bruceville, then a part of Frederick County, and also with Major John Ross Key, who owned a large section of land in the same neighborhood. John Ross Key was the father of Francis Scott Key, and the father-in-law of Roger B. Taney, Chief Justice of the U. S. These two men desiring the services of Koons who was an expert blacksmith, made it easy for him to secure a tract of land in their neighborhood.

Henry Koons had five sons. John during his early manhood settled near Sugar Loaf Mountain, and when he put his name into English used Koontz.

Abraham lived near Keysville.

Henry near Bruceville.

George, a blacksmith, served in that capacity during the Revolutionary War. In 1780 he secured title to the farm near the Double Pipe Creek bridge, one mile from Middleburg. George died in 1817, married Susannah Shroyer (1760-1848) of near Emmitsburg, and there were born to them eleven children.

Jacob (born 1781)
Henry (1783)
Margaret Mrs. Hyder (1786)
Mary Mrs. Ogborn (1789)
Catharine Mrs. Stimmell (1792)
Peter (1795)
George (1797)
Joseph (1800)
Susie Mrs. Dern (1804)

Peter Koons married Mary Ann Ott, of near Millers Bridge between Rocky Ridge and Detour. There were eight in this family, one Henry of Indiana.

[HISTORY OF FREDERICK COUNTY, MARYLAND, Vol. II, pp. 1613-1614]

Indenture made April 2, 1808 [deed] between Conrad Stultz of Frederick County state of Maryland of the one part and [from] Mrs. Henry Koons and Henry Koons Jr. of Frederick County state of Maryland of the other part . . . Tract of land known and called Bedford Conad--ed.
[Frederick CCH, Liber W. R. 33, p. 217]

Indenture made March 2, 1817 [deed] between Capt. John Trucks of Emmitsburg Frederick County state of Maryland of the first part and [from] Henry Koons of Henry of the county and state aforesaid of the other part... Rich lands beginning at the end of the second line of attract of land called Terra Rubra.
[Frederick CCH, Liber J. S. 6, p. 162]

Indenture made March 2, 1819 [deed] between John Shorb of Frederick County in the state of Maryland of the one part and [to] Henry Koons of Frederick County in the state of Maryland aforesaid of the other part . . . Partial of land being part of a tract called Terra Rubra beginning for said part at the end of the first line of Henry Koons part of Terra Rubra.
[Frederick CCH, Liber J. S. 8, p. 215]

Indenture made May 3, 1819 [deed] between Henry Koons of John of Frederick County in the state of Maryland of the one part and [to] Henry Koons of Abraham of Frederick County in the state of Maryland of the other part . . . Tract of land called Part Heads Industry lying in the county aforesaid and beginning at a stone planted on the southeast side of the main road leading from Taney Town to Big Pipe Creek bridge.
[Frederick CCH, Liber J. S. 8, p. 494]

his
Henry X Koons
mark

Indenture made February 19, 1821 [deed] between Abraham Koons in Frederick County in the state of Maryland of the one part and [to] Henry Koons of the county and state aforesaid of the other part . . . Being part of that land of the addition to Heads Industry.
[Frederick CCH, Liber J. S. 12, p. 424]

Indenture made September 24, 1821 [deed] between John Sex and Henry Koons and Elizabeth Koons (wife of said Henry) of Frederick County in the state of Maryland of the one part and [to] John Shorg of Frederick County in the state of Maryland of the other part . . . Tract called the Survey on Benjamin Good Duck Beginning.
[Frederick CCH, Liber J. S. 14, p. 210]

Indenture made March 18, 1822 [deed] between Abraham Koons, John Koons, William Koons, Isaac Koons, Susannah Shank, Catherine Hendricks of Frederick County state of Maryland all of who are heirs of law of William Koons, deceased, of the one part and [to] Henry Koons of Frederick County state of Maryland one of the heirs of law and devisee of John Koons, deceased, of the other part . . . Land contained in folios 386, 387, 389.
[Frederick CCH, Liber J. S. 15, p. 270]

Indenture made March 4, 1824 [bill of sale] I Abraham Koons of Frederick County and state of Maryland [sell] to Henry Koons . . . House hold goods, set of blacksmith tools, ect.
[Frederick CCH, Liber J. S. 19, Folio 386]

Indenture made October 19, 1824 [bill of sale] I Henry Koons of Frederick County and state of Maryland convey to John McKelly a crop of grain.
[Frederick CCH, Liber J. S. 21, Folio 148]

KRAFT, JACOB Hagerstown

Jacob Kraft ties in with the Resor family of gunsmiths, both in Hagerstown and in Lancaster, Pa. Jacob Kraft married Catherine Resor, who was the sister of the gunsmiths Peter and Jacob Resor. Considerable research on the Resor family has been done by Henry J. Kauffman and by Samuel E. Dyke, and they also have reported their findings concerning Jacob Kraft in their books; EARLY AMERICAN GUNSMITHS 1650 -- 1850 (1952) by Kauffman, and THE PENNSYLVANIA--KENTUCKY RIFLE (1960) by Kauffman, and THE PENNSYLVANIA RIFLE(1974) by Dyke.

Jacob Kraft was working as a gunsmith in Lancaster Boro as early as 1770. When Catherine Kraft's father died in 1771, his son Jacob Resor chose not to take the family homestead; Peter bought it but soon sold it to Jacob Kraft and Catherine. Jacob Kraft worked in Lancaster throughout the Revolutionary War and until 1783 at least. Peter Resor moved to Hagerstown after 1786, and probably Jacob Kraft and wife went at about the same time, though the time of arrival in Hagerstown is unknown.

The duration of Jacob Kraft's period in Hagerstown is unknown. His name appears in the federal census for the years 1800 and 1810, so it is likely that he finished out his life there. Kauffman (1960) noted that there was a gunsmith named Jacob Kraft listed as a single Freeman and a gunsmith in the tax lists for Heidelberg Township in York County, Pa. for the years 1807, 1808, and 1809. Possibly this was a son.

1800 Jacob Craft Hagerstown
one male between 26 & 44
one female over 45
one female between 26 & 44
one female under 10
[Federal Census, Washington Co. 1800]

1810 Jacob Kraft Hagerstown
one male 45 or older
one female 45 or older
one female 16 to 24
[Federal Census, Washington Co. 1810]

The census data for 1800 suggest that Jacob Kraft was born between 1755 and 1774, but this is not likely if he was at work as a gunsmith in 1770; probably the age category for the census was marked in error.

Jacob Kraft was a capable gunsmith who made good relief-carved rifles. One of these, which probably was made in Maryland, was illustrated by Kauffman (1960) in THE PENNSYLVANIA–KENTUCKY RIFLE, p. 278.

KRAUSPE, JULIUS Baltimore

Krauspe, J.

1847	15 Hill Street	gunsmith
1858-60	242 West Lee Street	gunsmith

[Baltimore city directories]

KRELL, JOHN Clayton

Krell, Jno	1880	Clayton, Harford Co.	gunsmith
Krell, Jno	1882	Clayton, Harford Co.	gunsmith

[The Maryland Directory, 1880, 1882; J. Frank Lewis & Co. Publishers, Baltimore]

KREPS, GEORGE, SR. Hagerstown

George Kreps of Hagerstown produced particularly beautiful relief-carved flintlock longrifles, and his products stand among the best made in early Maryland. He used relief-carving generously on his rifles, and his patchboxes featured fine engraving and many piercings. Pistols by George Kreps also are known.

A birth-year for George Kreps is unknown, but the census data place it between 1755 and 1765. His middle initial may have been "F." At two periods in his life George Kreps lived and worked in Frederick Co., Virginia, where the principal town was Winchester, but a detailed search for records of him there remains to be done. He is listed in the 1790 census for Hagerstown, but in 1795 "George Kreps late of Frederick County in the commonwealth of Virginia" bought a piece of property in Hagerstown.

About 1810 or 1811 George Kreps changed his trade from gunsmith to cabinet maker. William Greiner, who had apprenticed with Kreps a few years earlier, took over the gunsmith shop. By 1813 Kreps took on two apprentices to learn cabinet making.

George Kreps, Sr., had a son, George Kreps, Jr., who was bound as an apprentice in 1807 to learn the trade of cordwainer (shoemaker), but he got into gunsmithing later, with indifferent results.

1790 Kreps, George Washington Co., Md.
1 male of more than 16 years
1 male under 16 years
2 females
[Federal Census, 1790, Washington Co., Md.]

By this deed dated 16 Sept. 1795, George Kreps late of Frederick County in the commonwealth of Virginia bought from Martin Bringman of Elizabeth Town in Washington Co., Md., lot No. 259 in Elizabeth Town, it being 1½ lots together with the buildings thereon.
[Deed Book EE, p. 604, Washington CCH]

This indenture witnesseth that John Budenbaugh of Washington County and state of Maryland has put his son Martin Budenbaugh aged sixteen years four months, apprenticed to

George Kreps Gunsmith of the county and state aforesaid to learn his art trade and mystery . . . to serve the said George Kreps from this third day of January 1797 for four years and eight months . . .
[Washington County Indentures, Hall of Records, p. 48]

By this deed dated 13 Jan. 1798 George Kreps of Washington Co. bought from Jonathan Hager of Washington Co., Md., a lot in Hager's addition to Elizabeth Town.
[Deed Book K, p. 681, Washington CCH]

By this deed dated 29 July 1799 George Kreps of Elizabeth Town, Washington Co., Md., bought from Timothy Manahan of Elizabeth Town a lot with buildings in Elizabeth Town, distinguished as Lot No. 258.
[Deed Book I, p. 342, Washington CCH]

1800 Kreps, George Hagerstown
1 male 26 - 45 years
1 male 16 - 25
1 male 10 -15
1 female 26 - 45
1 female 10 - 15
3 females under 10
[Federal Census, 1800, Washington Co., Md.]

The following apprentice indenture actually is for William Greiner who learned the trade well enough to take over his master's shop in 1812 when Kreps turned to cabinet making.

September 3, 1802
This indenture witnesseth that William Criner with the consent of his mother Deloro Criner hath put himself and by this presents doth put himself apprentice to George Kreps of Washington County and state of Maryland to learn the art, trade and mystery of a Gun Smith and Rifle Smith . . .
[Washington County Indentures, Hall of Records, p. 176]

his
William X Criner
mark

By this deed dated 25 July 1803 George Kreps of Washington Co., Md., bought from Peter Heffleigh of the same place a half-lot in Elizabeth Town, No. 50.
[Deed Book P, p. 189, Washington CCH]

List of LETTERS Remaining in the Post Office in Hagers-Town . . . George Fischach, George Flegel, care of George Krebs, Gunsmith . . .
[Md. Herald & Hagers-Town Weekly Advertiser, 17 Jan. 1806]

This suggests that Kreps may have had an apprentice or journeyman named George Flegel. This may have been the same George Flegel (Flegle, Flegal) who was an armorer at the U.S. Arsenal in Philadlephia in 1814, was Master Armorer there in 1819, and who was an independent gunsmith in Philadelphia in the 1820's and 1830's.

Martin Kreps, a joiner and cabinet maker, whose death is noted in the newspaper account below, was the father of George Kreps.

This Indenture made the 28th day of August 1807 between George Kreps of Washington County and state of Maryland . . . and George Shall of the County and state aforesaid . . . Witnesseth that the said George Kreps hath . . . put . . . his son George Kreps Junr as an apprentice to the said George Shall for and during the term of three years and six months from the 15th of August (Inst) to learn the art trade and mystery of a cord wainer . . .
[Washington County Apprentice Indentures, p. 92, Hall of Records]

FIRE COMPANY AGREEABLY to a resolution of the said company, . . . Christian Hawken . . . George Kreps . . . are directed to attend at the Court-house Square on Saturday the 2nd day of July next precisely at 3 o'clock in the afternoon, for the purpose of working the fire engines; those who do not attend or furnish persons in their place, will be fined 25 cents each, agreeably to said resolution. The Engine Directors are requested to attend under penalty of one dollar each.
23 June 1808 Wm. HEYSER, Director General
[Md. Herald & Hagerstown Weekly Advertiser, 1 July 1808]

TO BE SOLD BY PUBLIC VENDUE
On Saturday the 2nd day of June next the Dwelling House and half Lot, where the subscriber now lives in Hagerstown . . . also some Gunsmith Tools . . .
May 15, 1810 George Kreps
[Maryland Herald & Hagers-Town Weekly Advertiser, 16 May 1810]

1810 Kreps, George Hagerstown
1 male over 45 years
1 male 10 - 15
2 males under 10
1 female over 45
2 females 16 - 25
1 female 10 - 15
[Federal Census, 1810, Washington Co., Md.]

In 1811 George Kreps made a formal change of his trade from gunsmithing to carbine making. The following advertisement calls attention to the fact that he had commenced business as a cabinet maker by August 1811. Five months later his former apprentice, William Greiner, advertised in the Hagerstown paper that he had taken over the gunsmith shop "lately occupied by Mr. George Kreps, Gunsmith, in Franklin Street . . ." And by 1813 Kreps had taken at least two apprentices to learn cabinet making.

GEORGE KREPS CABINET MAKER Respectfully informs his friends and the public in general, that he has commenced the above business at the shop formerly occupied by his father Martin Kreps, in Franklin Street, Hagers-town; where he intends keeping a general assortment of CABINET WARE on hand, made in the neatest and most fashionable style, which he intends to dispose of at the most reduced prices. Those who may please to favor him with their custom, may depend upon every exertion on his part to render them complete satisfaction.
N.B. A JOURNEYMAN wanted -- Apply as above.
August 6, 1811

[Md. Herald & Hagers-Town Weekly Advertiser, 18, 25 Sept., 2, 9, 26 Oct. 1811]

CABINET MAKING -- The subscriber takes this method of presenting his grateful acknowledgement to his friends, customers, and a generous public, for past favors, and informs them that he still continues to carry on the above business, at his present stand in Franklin Street; where every article in his line is executed at a short notice and on reasonable terms. He flatters himself, that by his thorough knowledge of said business and the expert workmen in his employ, to be enabled to give full satisfaction to those who may favor him with their custom -- and he hopes by his constant endeavors to furnish good and sufficient work, to merit and receive a share of public patronage.
Hagerstown August 3, 1812 GEORGE KREPS
N.B. He has on hand, and intends constantly to keep, an assortment of Cabinet Ware, which he will dispose of at the most reduced prices.
[Md. Herald & Hagers-Town Weekly Advertiser, 5, 12, 19 Aug. 1812]

CABINET MAKING --- GEORGE KREPS Returns sincere thanks to his friends and customers for past favors, and informs them that he continues to carry on the above business at his old stand in Franklin Street, Hagerstown; where every article in his line will be executed at a short notice and on reasonable terms -- He has on hand an extensive assortment of Mahogany, Wild Cherry and Walnut, which will enable him to execute all orders in his line of business with neatness and dispatch. He also intends keeping an assortment of Cabinet Ware ready made which he will dispose of at the Baltimore prices. He flatters himself that from his knowledge of his business, and the expert workmen in his employ, to be enabled to give general satisfaction to those who may favor him with their custom, and by his endeavors to please, he hopes to receive a liberal share of public patronage.
March 31, 1813
N.B. An apprentice will be taken to learn the above business-- A Boy from 15 to 18 years would be preferred.
[Md. Herald & Hagers-Town Weekly Advertiser, 31 March, 7, 14, 21, 28 Apr., 5, 12, 19 May, 2, 9, 23, 30 June 1813]

A similar advertisement appeared in the Hagerstown paper on 10 Aug. 1814.

By Washington County, Orphan's Court Nov. 4th 1813

Daniel Powles aged 17 years the 10th day of January next, is by this court with the consent of his Mother bound unto George Kreps until he shall arrive at lawful age said Master to find and provide for said apprentice good and sufficient meat, drink, washing, lodging and clothing, to send said apprentice to an English day school four months and permit the said apprentice to attend the church for the purpose of learning the catechism and when free to give said apprentice thirty dollars.
[Washington Co. Apprentice Indentures, p. 387, Hall of Records]

John Buchanan aged 10 years the fourth of December next, is by this court with the consent of his Mother bound unto George Kreps until he shall arrive at lawful age said Master to find and provide for said apprentice good and sufficient meat, drink, washing, lodging, and clothing to learn said apprentice to read write and cypher through the rule of three and when free to give said apprentice ten pounds as his freedom dues. [Undated, but position in book indicates October or November 1813]
[Washington Co. Apprentice Indentures, p. 388, Hall of Records]

This Indenture made this seventeenth day of November . . . 1813 Witnesseth that Jacob Poffenberger of Washington County and State of Maryland, hath put and placed, and by these presents doth put and bind his son Andrew Poffenberger (who on the twenty fifth day of July next will be of the age of nineteen years, as an apprentice to George Kreps to learn the art, trade, or mystery of a Cabinet Maker, the said Andrew after the manner of an apprentice, to dwell with and serve the said George Kreps from the day of the date hereof until the twenty fifth day of July 1816, at which time the said apprentice, if he should be living, will be twenty one years of age, during all of which time or term the said apprentice has said Master well and faithfully shall serve, his lawful commands everywhere and at all times readily obey; from his said Master he shall not absent himself . . . And the said George Kreps . . . doth hereby promise, . . . to teach and instruct the said apprentice . . . the art, trade or calling of a Cabinet Maker by the best way or manner he can and to give said apprentice two weeks in every harvest and shall provide for the said apprentice good and sufficient meat, drink, lodging and other necessaries, clothing excepted, fit and convenient for such apprentice .. and at the expiration thereof shall give unto the said apprentice forty five dollars as his freedom dues.
[Washington Co. Apprentice Indentures, p. 390, Hall of Records]

By the Court March 13th 1816

David Steel aged 17 years on the 10th of May next is by this court with the consent of his Mother present bound unto George Kreps until he arrives at the age of twenty one years, to learn the cabinet making business, said Master to find and provide for said apprentice good and sufficient meat, drink, washing, lodging and clothing fit and convenient to send said apprentice three months to school and give him ten pounds current money when free.
[Washington Co. Apprentice Indentures, p. 30, Hall of Records]

As the following indenture release states, George Kreps made plans for leaving Hagerstown in 1816. He moved to Winchester, Virginia, where he advertised as a cabinet maker.

. . . Whereas Andrew Poffenberger was heretofore bound unto George Kreps to learn the cabinet making business and whereas the said George Kreps contemplates leaving the State of Maryland and by reason whereof he is willing to release the said Andrew from his said Indenture, now know ye that the said George Kreps for and in consideration of one dollar current money of the United States do hereby release . . . the said Andrew from said Indenture and every clause and covenant therein contained and that the said Andrew is henceforth and forever freely and clearly discharged therefrom. In Witness whereof the said George Kreps hath hereunto set his hand . . . this 17th day of April 1816.
George Kreps
It is understood between the parties above named that the said Andrew Poffenberger is to make for the said Geo. Kreps two bureaus for which he is to pay to the said Andrew the sum of five dollars.
[Washington Co. Apprentice Indentures, p. 23, Hall of Records]

George Kreps may have remained in Virginia for no more than a year or two, for the one property tax assessment record book available for Hagerstown shows him paying taxes there for the years 1814, 1815, 1817 and 1818. He may have been taxed as an absentee landlord. However, on 19 Feb. 1818 he sold a piece of property in Washington County to Michael Esterday (Deed Book CC, p. 437].

Hagerstown

Kreps, George	1814	to tax office	$5.68½
Kreps, George	1815		6.46
Kreps, George	1818		0.78
		Taxes in 1817	7.67
		Taxes in 1818	0.58

[Tax Assessment Book, Washington Co., 1815 and following years, Hall of Records]

At a public sale held in Washington County 17 Feb. 1824 to dispose of the property of William Kreps, deceased, George F. Kreps bought the following:

3	Chairs	$ 1.00
½	dozen wood screws	1.50
1	rain hogshead	.50
2	maps United States	.50
1	cage & doves	1.50
1	flax wheel	1.25
1	woman's saddle	2.00
1	hayfork	.21
22	lb. beef & pork	13.44
1	wheel barrow	3.85
3	sheets	3.00
1	passage lamp	3.00
	Bed & furniture	9.00
	Bedspread & window curtains	8.00
1	toilet table & glass, carpet	8.00
1	bed & bedding & 10 common chairs	25.00
1	lot of wool & thread	6.80
2	glasses & curtains	18.00
1	table looking glass & stand	2.00
	dining room carpet	2.40
	closet furniture	7.00
2	fire buckets & bag 87½ pantry 5.	5.87½
1	high chair	.46
1	set knives & forks	1.00
1	carriage body	50.00
	contents of cellar	5.00
	wood saw axe & buck	1.50
	set of Drawers & 2 Bags	1.50
2	bags & scales & weight	1.00
	lot of tubs	.75

[Washington Co. Estate Sales Book H, p. 314, Hall of Records]

George F. Kreps died in 1826. John Kreps, Ann Kreps and Thomas Kennedy were bonded in the amount of 2000 pounds to administer the estate on 23 October 1826.

[Washington County Administration Bond Book C., p. 461, Hall of Records]

KREPS, GEORGE, JR. — Hagerstown

George, Jr., apparently did not have the talent of his father, though he did work as a gunsmith. His indenture as an apprentice suggests that he was born in November 1789.

This Indenture made the 28th day of August . . . 1807 between George Kreps of Washington County and state of Maryland . . . and George Shall . . . Witnesseth that the said George Kreps hath and . . . doth . . . bind out his son George Kreps Jr. as an apprentice to the said George Shall for and during the term of 3 years and 6 months . . . to learn the art trade and mystery of a cordwainer . . .
[Washington County Apprentice Indentures, p. 92, Hall of Records]

John Mason to George Kreps
Office of Indian Trade, Georgetown, May 24, 1811
Your two letters of the 22d April and 13th last have been received – the 10 Rifles you sent down came safely to hand and have been inspected – they are received, but am sorry to say are by no means equal to those shewed here as samples, and for which we pay the same price given you. they may shoot as well but are not as well finished. it is not customary at this office to advance more than once on a contract – and beside I have not the security of the gentlemen you offered for more than the first advance – if however you will send me the security of the same Gentlemen or of any good man, I will advance as you have delivered ten of the guns – sixty or seventy five dollars more.

I must request that you finish the Guns to be delivered in a better manner – we want nothing more than neat plain work but it must be well fitted.

John Mason to William Kreps
Office of Indian Trade, Georgetown, November 13, 1811
I have received your letter of the 9th Inst. I was delayed in my reply to that from Mr Geo. Kreps of the 12th Oct by the Gun smiths engaged to refit some of the guns – and whose prices I wanted that I might close the account. I now enclose you a letter to Mr. Geo. Kreps open for your perusal by which you will find that he will receive the balc due him . . . I am sorry that it is impossible for me in justice to the public to approve his work.

John Mason to George Kreps
Office of Indian Trade, Georgetown, November 13, 1811
On the 18th Ulto I received your letter of the 12th and the two boxes containing 20 rifles. I was not in the office when the waggoner delivered them or I should have paid the carriage, and he did not return.

I am sorry to say that I am even less satisfied with these guns than the first parcel you sent down – there is evident want or even coarse finish in all of them – and I must believe you did not inspect them your self or you could not have turned out from your shop such work – the locks are of many sorts, and some of them very indifferent – and generally badly fitted and wood bound. the draw loops are of brass. they should have been hardened iron – and there are only 3 instead of 4 to a gun – the ferrils at the end of the sticks to receive the wipers are rough and flawy – the moulds are generally rough and badly made. these are some of the leading faults. finding them defective generally and some not fit for service, I determined to have them inspected, and called in two gentlemen well acquainted with Rifles to do so – I enclose their report. they both declared that there was not a gun among them equal to the sample which was shown them. they have, however, passed 12 and determined as you will see that eight of them require alteration and repairs before they are fit for service.

I sent for a Gun smith to know what he would refit them for and on a more strict examination it was found that several moulds would fit no one gun. I was particular enough then to have a ball cast for each and found six in that situation of course useless. 1 wiper was wanting and 19 charges. to renew them I find will cost by statement herewith I find $5.16½ to put on the repairs found requisite by the Inspectors $2.43, in all $7.59½ this deducted from the amount of your account as by statement sent leaves $197.90½. If you choose to receive this the Cashier of the Hagerstown Bank to whom I have written will pay it to you. otherwise I should be glad you would refund what is overpaid you and take back the guns

As we differ so much in our opinion of a good rifle, I think it will be best that you and I should not deal for any more.

[Letters sent by Supt. Indian Trade, The National Archives]

SIXTY DOLLARS REWARD – GEORGE KREPS . . . born in Washington County Maryland, aged 30 years, 5 feet 10 inches high of a dark complexion, dark eyes, black hair and by profession a Gun-smith . . . had on the United States uniform.
J. H. Galp 6 Sept. 1819 Capt. 4th U. S. Infantry
[Md. Herald & Hagers-Town Weekly Advertiser, 28 Sept. 1819]

KUEHLING & SON Washington, D. C.

Kuehling & Son
1862 259 Penn Ave. gunsmiths
[Business Directory of Washington, D.C., & Georgetown]

KUEHLING, J. HENRY

Kuehling, J. Henry
1863 259 Penn gunsmith
[Hutchinsons Business Directory, Washington, D.C., & Georgetown]

Kuehling, J. H.
1864 259 Pa. Ave. gunsmith
[Hutchinsons Business Directory, Washington, D.C., & Georgetown]

Kuehling, J. H.
1865 259 Pa. Ave. gunsmith
[Boyd's Business Directory, Washington, D.C., Georgetown & Alexandria]

Kuehling, J. H.
1866 259 Pa. Ave. gunsmith
[Boyd's Business Directory, Washington, D.C., Georgetown & Alexandria]

Kuehling, J. H.
1867 145 King Alexandria gunsmith
[Boyd's Business Directory, Washington, D.C., Georgetown & Alexandria]

Kuehling, John H.
1867 259 Pa. Ave. gunsmith
[Boyd's Business Directory, Washington, D.C., Georgetown & Alexandria]

Kuehling, J. H.
1868 259 Pa. Ave. gunsmith
[Boyd's Business Directory, Washington, D.C., Georgetown & Alexandria]

KUSSMAUL, WILLIAM F. Baltimore

The object of Kussmaul's invention is to provide a means by which the discharge of a gun from any accidental violence to its hammer when not cocked may be prevented.
He advertised in the 1847 city directory as a gunsmith at 56 Chatsworth Street; his last name was spelled Kesmaul. In 1851 his shop was located at 30 New Church Street and his name was spelled Kausmaul.

Kesmaul, William F.
1847 56 Chatsworth St. gunsmith
[Baltimore city directory]

Kausmaul, William F.
1851 30 New Church St. gunsmith
[Baltimore city directory]

October 15, 1867
Be it known that I, William F. Kussmaul, of the city and county of Baltimore, and State of Maryland, have invented a new and improved Lock for Fire-Arms . . .
[U. S. Patent Office, Patent #69,919]

L

LATROBE, FERINOND C. — Baltimore

His colleagues were J. H. Merrill, Lewen W. and Philip E. Thomas. By 1870 he was an attorney as his father was.

Merrill, Latrobe & Thomas

1855-59	6 Hanover Street	manufacturers & importers of guns, gun locks & rifles

[Baltimore city directories]

LATROBE, JOHN H. B. — Baltimore

He was an attorney and the father of Ferinond.

February 26, 1856
Be it known that I John H. B. Latrobe, of Baltimore, in the State of Maryland, have invented a new and useful Improvement in Firearms . . . The object of my invention is to facilitate the application of the Maynard primer . . . I place the primer in a chamber within the hammer, which chamber also contains the means of protruding the primer and of cutting off, at the proper time, the portion to be exploded upon the nipple.
[U. S. Patent Office, Patent #14,319]

LAUDERMAN, WILLIAM H. — Baltimore

Lauderman, William H.

1864-67	136 Thames Street	gunsmith, locksmith, & bellhanger

[Baltimore city directories]

LAURWING, DERMENICO — Washington, D. C.

Laurwing, Dermenico

1867	3d west n Pa. Ave.	gunsmith

[Boyd's Business Directory, Washington, D.C., Georgetown & Alexandria]

LAVIN, M. E. & Co. — Cumberland

Lavin, M. E. & Co.

1887 Wholesale & Retail Dealers in Hardware, Guns, Pistols, Fishing tackle & Sporting Goods 74 & 76 Baltimore St. and 1, 3, 5 Centre St., Cumberland. Agents for Dupont's Powder

[Maryland Directory & State Gazetteer, 1887, The Baltimore Publishing Co.]

LAYPOLE, JOHN H. — Weverton

Laypole made longrifles which are of fair quality, usually plain with little ornamentation and signed on the barrel "J Laypole." There were several John Laypoles who resided in Washington and Frederick Counties. By 1863 John H. was married to a woman named Mary who was illiterate. He signed various indentures, but sometimes only made a mark; his last transaction was in 1892.

Washington County, Petersville District

Laypole, John H.	age 40	b. Md.	gunsmith	property $200
Julia A.	34	Md.		
Theopholes	12	Md.		
Ann C.	9	Md.		
Susan A.	7	Md.		
Eliza B.	5	Md.		
Martha	8/12	Md.		

[1850 Federal Census, Washington Co., Petersville District]

LEE, DAVID — Harford County

David Lee operated a flintlock manufactory during the Revolutionary War in a two-story stone building which still stands as a dwelling a short distance from the mill (Jerusalem Mill, Jerusalem Town, Harford County). Evidently like other Quaker businessmen of his day he did not enter into hostilities against England because of his religious belief about war but established his patriotism by devoting his capital and talents to making arms for the colonial soldiers.
[Baltimore Sun, 8 Sept. 1935]

LEMMER, FREDERICK — Washington, D. C.

Lemmer, Frederick

1867	329 E north	gunsmith

[Boyd's Business Directory, Washington, D.C., Georgetown & Alexandria]

LEPPER, HENRY G. — Baltimore

Lepper, Henry G.

1847-54	221 South Bond Street	gunsmith

[Baltimore city directories]

LEWIS, A. — Baltimore

Lewis, A.

1831	60 South Street	gunsmith

[Baltimore city directory]

LEWIS, WASHINGTON — Washington, D. C.

Lewis, Washington

1822	N side, D n btw 6 & 7 w	gunsmith

[Directory, Washington, D.C.]

LIBIG, WILLIAM — Baltimore

Libig, William

1884-85	536 North Gay Street	gunsmith

[Baltimore city directories]

LIDDLE, ROBERT — Baltimore

Liddle, Robert

1847-51	52 North Eden Street	gunsmith
1855	8 Fish Market	gunsmith

[Baltimore city directories]

LINTHICUM, WILLIAM H. Baltimore

His successor was the English firm of W. W. Greener in 1888.

Linthicum, William H.
1887 2 West Baltimore Street gunsmith
[Baltimore city directory]

LITTLE, JOHN

October 16, 1776 Council of Safety
Ordered That the Western Shore Treasure pay to John Little forty shillings for a Musquet.
[AM, p. 354]

LODON, BENJAMIN Frederick

Inventory of Work tools Belonging to and now in the Gunlock Factory the 19th August 1778 . . . Benjamin Lodon 2 years Run away . . .
[Hall of Records, Red Book IX, p. 273]

LOCKWOOD, C.L. & COMPANY Washington, D.C.

1865 C. L. Lockwood & Co. 324 Penna. Ave.
military outfitters
[Washington city directory]

LOLLIGER, JOHN Washington, D. C.

Lolliger, John
1858 416 7th west, h. do. gunsmith
[Directory, Washington, D.C.]

Lolliger, John
1858 416 7th w gunsmith
[Business directory, Washington, D.C.]

LONEY, F. B. Baltimore

Born in 1823 in Maryland and was married to Ann E. He was a real estate dealer and later a merchant before the firm of Schaeffer & Loney advertised as gunsmiths.

Schaeffer & Loney
1860 1 & 3 Hanover Street gunsmith
Loney, F. B. & Co.
1868-72 240 West Baltimore Street gunsmith
Loney, F. B. & Co.
1868 also hardware dealers
[Baltimore city directories]

Loney, Francis B. 1850 age 27 real estate & merchant
Ann E. 1850 24 birth Maryland
[1850 Federal Census]

To The Honorable Mayor & City Council of Baltimore
Your petitioners would respectfully represent to your Honorable body that on Sunday, April 21st, 1861 between 11 & 12 oclock A.M. our place of business Nos 1 and 3 Hanover Street, was surrounded by a crowd of some 200 or 300 citizens, clamorous in their demand for arms, and threatening that if we did not accede to their wishes that they would break in the doors and take them. – The policeman present said he was unable to restrain them, and in his opinion they would execute the threat, and that in a very few minutes. Other gentlemen present concurred with the policeman, and under these circumstances we were compelled to accept of one of two ways to satisfy them, either to submit to their forcible enterance, which would have entailed a loss of several thousand dollars (as the nature and character of our stock of goods would have made them especially desirable at that time) or to deliver to the policeman to be distributed outside, a portion of our valuable stock of firearms.
We therefore delivered to the policeman 34 Rifles, which he handed to the crowd outside of the Store, through one of the grated windows as per his receipt, (witnessed by 4 gentlemen) now in our possession, a copy of which accompanies this. We would remark that our entire Stock of arms amounted to 285 pieces, the balance (after this delivery) 251 pieces were on the succeeding day (Monday) packed in 9 cases, and delivered to the Mainhall of Police for safekeeping . . . May 22, 1861
F. B. Loney & Co.
[Baltimore Correspondance Manuscripts, 1861, p. 428]

LUDEKE, FRANCIS JOSEPH Washington, D. C.

19 June 1860 Georgetown real estate $600
Ludeke, Joseph age 30 born Germany locksmith
" Elizabeth 26 Germany
" Mary 22 Washington, D.C.
" Annie 10 Washington, D.C.
[1860 Federal Census]

Ludeke, Joseph
1860 168 Bridge gunsmith
[Business Dir., Washington, D.C., Georgetown & Alexandria]

Ludeke, Joseph
1862 168 Bridge gunsmith
[Business Directory, Washington, D.C., & Georgetown]

Ludeke, F. J.
1864 168 Bridge gunsmith
[Boyd's Business Directory, Washington, D.C., & Georgetown]

Ludeke, F. J.
1866 168 Bridge Georgetown gunsmith
[Boyd's Business Directory, Washington, D.C., Georgetown & Alexandria]

Ludeke, Francis J.
1867 168 Bridge Georgetown gunsmith
[Boyd's Business Directory, Washington, D.C., Georgetown & Alexandria]

Ludeke, F. Jos.
1875 176 Bridge Georgetown gunsmith
[Georgetown Directory]

LYDICK, PETER Baltimore

A slender longrifle has survived bearing "P Lydick" in script on the barrel, brass mounted with a daisy patch box. The rifle is of a much later period than the documentation shown below, signifying he must have remained in the business.

February 7, 1776
Peter Lydick 72 guns 64 good 8 bad
[AM, p. 155]

July 8, 1777 Council of Safety
That the said Treasure pay to Peter Littig seventeen pounds sixteen shillings and six pence due on Accts. passed by the board of Accts.
[AM, p. 312]

September 17, 1777 Council of Safety
That the said Treasure pay to Peter Littig two hundred and eighty one pounds five shillings on Contract for making Arms.
[AM, p. 377]

September 17, 1777 Council of Safety
Peter Littig agrees with the Governor and Council to make and in a Workman like manner compleat and finish one hundred and fifty Musquets fixed with Bayonets, Steel Rammers, Swivels priming wires and brushes at three pounds fifteen shillings each the mounting to be of brass or Iron the breech pin to be let into the solid so that in reboring the threads of the Screws shall not be injured that he will enter on the said Work within two Months from the Date hereof will deliver Twenty and so from Month to Month thereafter at least twenty of the said Musquets Completely finished as aforesaid, the said deliveries to be at Baltimore Town to Gerrard Hopkins or such other person as shall be appointed by the Governor and Council. The Governor and Council to furnish Locks and such barrels and Bayonets as were furnished to Nicholas White and also to the said Peter Littig two hundred and eighty one pounds five shillings part of the price on giving Bond with good security to comply with this Contract.
[AM, p. 377]

September 19, 1777 Council of Safety
That the said Treasure pay to Philip Lydick six pounds Ten Shillings for a Gun purchased of him.
[AM, p. 379]

October 23, 1777 Council of Safety
That the said Treasure pay to Peter Littig one hundred and sixty pounds fifteen shillings per Account passed.
[AM, p. 403]

LYNCH, JOHN — Baltimore

On 8 February 1786 he purchased a property and his occupation was listed as a silversmith. He married Naomi Willey in 1804. Aleysius Owings apprenticed to him in 1800 for five years, six months. Lynch died 15 January 1848 after being in business for sixty some years.

He produced much silver including several fine eaglehead pommel swords in both the Baltimore and Philadelphia styles. Lynch used the pierced knucklebow which was popular with the Baltimore makers.

Lynch, John

1802-04	Franklin Street	silversmith
1807-19	Strawberry Lane & Franklin Street	clock maker & silversmith
1822-42	7 Franklin Street	jeweller
1848	104 Franklin Street	silversmith

[Baltimore city directories]

M

MAEDEL, ADOLF — Washington, D. C.

1860 Washington, D.C.

Matel, Atolf	age	53	born	Germany	engraver
Matel, Chas.		25		Germany	gunsmith
" Caroline		50			
" Adelfenia		30			
" Rosater		22			
" Adeline		18			

[1860 Federal Census]

Note: Directories for 1862 and 1863 list Charles B. Maedel, which is probably the same person listed here, the census-taker having taken liberties with the spelling.

MAEDEL, CHARLES B. — Washington, D. C.

1860 Washington, D.C.

Matel, Atolf		53		Germany	engraver
Matel, Charles	age	25	born	Germany	gunsmith
" Carolina		50			
" Adelfenia		30			
" Rosater		22			
" Adeline		18			

[1860 Federal Census]

Maedel, Charles B.
1862 153 Penn. Ave. gunsmith
[Business Directory of Washington, D.C., & Georgetown]

Maedel, Charles B.
1863 153 Penn gunsmith
[Hutchinsons Business Directory, Washington, D.C., & Georgetown]

MANLEY, DAVID — Baltimore

Manley, David
1822 Potter north of McElderry, Old Town gunsmith
[Baltimore city directory]

MARCH, MORRIS DEC. — Baltimore

March, Morris Dec.

1840	Gay Street 3 doors south of Aisquith	gunmaker
1845	313 North Gay Street	gunsmith

[Baltimore city directories]

MARKER, DANIEL SR. — Middletown Valley

His fine longrifles show excellent quality. The carving is high relief and the profuse use of silver illustrates that he was a very capable craftsman. Daniel's superior engraving, as the rest of his workmanship, closely follows that of the Emmitsburg school. Rifles of this quality are usually signed "Daniel Marker" in script on a brass plate in the barrel or stamped on the barrel with a symbol placed between the two names. He produced many rifles for men of all income brackets. Some guns have an oval patch box, no carving, few inlays and are usually signed on the barrel "D Marker" in script.

Was born in 1774, oldest son of George Marker 1753-1827 who came from Germany about the time of the Revolutionary War. Deed is recorded in Frederick County Court House in 1795 for a 296 acre farm where he farmed and ran a gristmill and/or a power mill near Myersville. George married Mary Mohler about 1773. Mary 1746-1846, when about 14 years old, came to the U. S. as a redemptive slave. There were 8 children born from this marriage, including . . . (5) George and . . . (3) Paul. Daniel married Christina Beckenbaugh of Sharpsburg in May of 1797 . . . made muzzle loading rifles around Sharpsburg. He went to Ohio seeking work leaving his family behind, at South Dixie Highway and the Springboro Pike, southwest of Dayton. Christina died 4 October 1835 and is buried in the Sharpsburg district. From this marriage there were born two sons who never left Maryland.
Daniel Jr. married Catherine Sutten in 1840 and lived near Sharpsburg. Will recorded 19 October 1891.
James married Amelia Naeff in 1832 and had eight children, with a farm near Sharpsburg and made guns.
Daniel remarried Nancy Ann Dowler in August of 1835 and migrated to Preble County [Ohio]. He is buried in Christian Cemetary, in Darke County, along the Stillwater River, southeast of Webster, Ohio. His tombstone is a Government stone without a death date.
[Mrs. Nellie Marker Grillot, unpublished genealogical data]

Daniel was a Captain in the 28th Reg. during the War of 1812, composed of men from the Middletown Dist. named the "Mountain Rangers." They were largely represented in camp and field during the war. He was among those who raised companies and participated in the Battle of Bladensburg. On September 16, 1813, Capt. Marker's fine Company which had been summoned to Annapolis, passed through Frederick enroute home. He was commissioned First Lieutenant September 21, 1807 and promoted to Captain April 28, 1808, assuming command of the Mountain Rangers on June 26, 1812.
[Annapolis Record, 950-M2, p. 130]

Indenture made August 21, 1802 [bill of sale] I James Stokes of Frederick County and state of Maryland [sold] one cow black and white, two sows, and barrow, six pigs, two beds, and all my ketchen furniture to have and to hold the above goods and chattels into Daniel Marker of Frederick County and state of Maryland.
[Frederick CCH, Liber W. R. 23, p. 320]

Indenture made May 24, 1803 [deed] between John Custard of Frederick County in the state of Maryland of the one part and [to] Daniel Marker of Frederick County in the state of Maryland aforesaid of the other part . . . Being in the same County the tract of land called Hogs Neck.
[Frederick CCH, Liber W. R. 24, p. 432]

THE MARKER FAMILY OF GUNSMITHS

Source: *Genealogy of the Marker Family* by Mrs. Nellie Marker Grillot

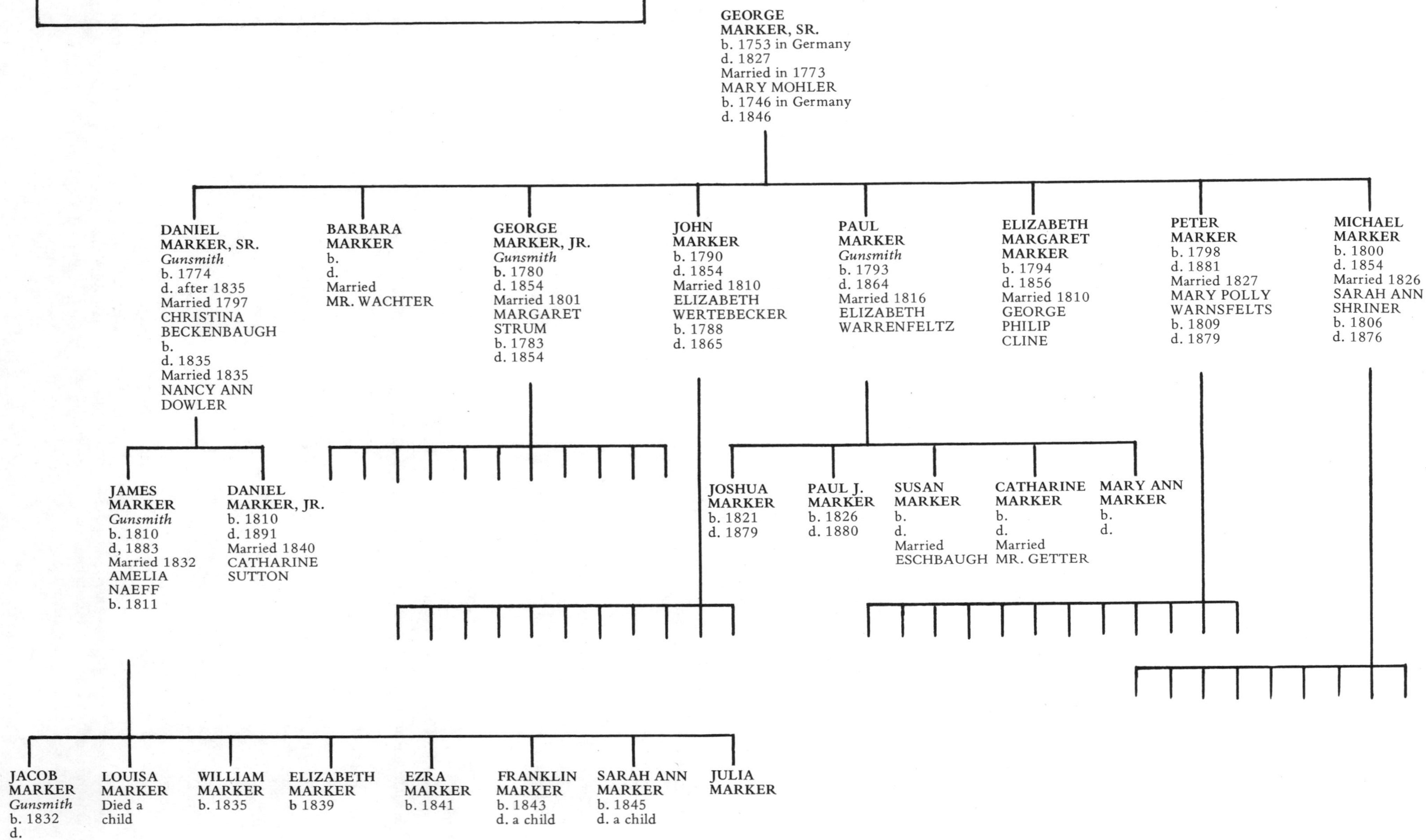

Indenture made March 4, 1809 [bill of sale] I John H. Myers of Frederick County state of Maryland [sell] unto the said Daniel Marker of Frederick County state of Maryland his excutors adminstrators and of size one sorrel stallion four years old last spring, one red and white cow about four years old, two beds and clothing there to as they now stand one chest, and one walnut table all now in my possession.
[Frederick CCH, Liber W. R. 34, p. 415]

I George Gaver of Frederick County . . . have released and forever quit claimed unto Daniel Marker of Frederick County of and from all manner of Goods and chattels mentioned in a Bill of Sale given . . . by the said Daniel Marker to me . . . bearing date the 21st day of February 1814 and whereas . . . Daniel Marker hath fully paid to . . . George Gaver the sum of money mentioned in the aforesaid Bill of Sale for which satisfaction the said George Gaver doth execute this instrument of writing as a full release of the goods & chattels specified in the Bill of Sale . . . Witness whereof I . . . George Gaver hath hereunto subscribed my name . . . this 29th day of February 1816.
[Frederick CCH, Land Record Book J. S. 2, p. 732]

This Indenture made the 29th day of February 1816 Between George Gaver of Frederick County . . . and Daniel Marker . . . Witnesseth that the said George Gaver for . . . $200 . . . paid by . . . Daniel Marker . . . doth . . . sell . . . unto . . . Daniel Marker . . . the following tracts of land lying . . . in the County . . . aforesaid to wit "The Mistake" containg 5 acres of land, part of "Hogs Nest" containing 10 acres, also part of a Tract called "More Stones" and containing 16 acres the above said three tracts . . . of land was conveyed to . . . George Gaver by . . . Daniel Marker . . . bearing date . . . 21st day of February 1814 . . .
[Frederick CCH, Land Record Book J. S. 2, p. 733]

Indenture made March 19, 1817 [bill of sale] I Daniel Marker of Frederick County and the state of Maryland for and in consideration of the sum of $100.00 paid to me in hand by Paul Marker, tinplate stove and pipe, and chest of drawers, one kitchen cupboard, one corner cupboard, one chest, two bedsteads, eight common chairs, two rifles, one red heifer, one black heifer, one black mare and saddle, one set of blacksmith and gunsmith tools.
[Frederick CCH, Liber JS 4, Folio 197]

Indenture made April 26, 1824 [bill of sale] Know all men by these presents that I Daniel Marker of Washington County state of Maryland for and in consideration of the sum of $250 current money to me in hand by George Marker Jr. one leader for cutting rifle barrels, two rifle guns, six rifle gun barrels, fifty gun stocks.
[Washington CCH, Liber GG, Folio 839]

Indenture made March 15, 1825 [deed] between Daniel Marker of Washington County state of Maryland of the one part and [to] David Bowlus and John Herring of Frederick County state of Maryland of the other part . . . Tract or parcel of land lying and being in Frederick County being all that tract called The Mistake containing 5 acres of land, part of a tract of land called Hog Neck containing 10 acres of land, part of that tract of land called More Stones Added containing 16 acres.
[Frederick CCH, Liber J. S. 22, Folio 60]

MARKER, GEORGE JR. Myersville

Did not have the great ability that his brothers Daniel and Paul illustrate. None of his carved longrifles have survived and his engraving is a crude imitation. The top facet of the barrels are signed "G Marker."

George II was born 19 January 1780 and died 29 November 1854. Was the third child of George I . . .
George II married Margaret Strum in May of 1801. Margaret was born 13 March 1783 and died 29 November 1854. They were married in Maryland and were slave owners but lost all the endorsing papers in 1823 when moving to Montgomery County, Ohio, on the Germantown Pike, near Ellerton. Two years later moved to the Michael farm, south of Liberty, Ohio. There were 11 children, all born in Maryland except the youngest. George II is buried in the Church cemetery at Ellerton, southwest of Dayton. Margaret is buried at Sugar Hill cemetery, West Alexandria, Ohio.
[Mrs. Nellie Marker Grillot, unpublished genealogical data]

Indenture made March 28, 1816 [deed] between Leonard Flohr of Frederick County in the state of Maryland of the one part and [to] George Marker Jr. of Frederick County in the state of Maryland of the other part . . . Tract or parcel of land lying and being in Frederick County containing 177 acre and ¾ acre aforesaid tract called The Face Put and Patricks Day in the Morning.
[Frederick CCH, Liber J. S. 2, Folio 536]

Indenture made April 19, 1816 [deed] between George Marker Jr. of Frederick County state of Maryland of the one part and [to] George Peter Brown of the aforesaid county and state of the other part . . . Tract or parcel of land lying and being in Frederick County aforesaid being that tract called La[?] are Bottle.
[Frederick CCH, Liber J. S. 2, Folio 259]

Indenture Made November 23, 1816 [mortgage] between George Marker, Jr., of Frederick County state of Maryland of the one part with George Marker Sr. and Michael Wachter of the aforesaid county and state of the other part . . .
[Frederick CCH, Liber J. S. 3, Folio 653]

Indenture made May 22, 1817 [bill of sale] I Barnet Shields of Frederick County and the state of Maryland [sell] unto the said George Marker Jr. of Frederick County and the state of Maryland . . .
[Frederick CCH, Liber J. S. 4, Folio 477]

Indenture made April 15, 1818 [deed] between George Marker Jr. of Frederick County state of Maryland of the one part and [to] George Delawter of the aforesaid county and state of the other part . . . Tract or parcel of land lying and being in Frederick County aforesaid being part of that tract called Markers Delight.
[Frederick CCH, Liber J. S. 6, Folio 331]

Indenture made March 2, 1819 [deed] between George Marker Jr. of Frederick County state of Maryland of the one part and [to] Christian Geesey of the aforesaid county and state of the other part . . . Tract or parcel of land lying and being in Frederick County being part of that tract called Markers Delight.
[Frederick CCH, Liber J. S. 8, Folio 223]

Indenture made August 23, 1819 [deed] between John Gaver of Frederick County state of Maryland of the one part and [to] George Marker Jr. of the aforesaid county and state of the other part. Tract or parcel of land lying and being in Frederick County being part of that tract called New Germany with all buildings.
[Washington CCH, Liber 9, Folio 373]

Indenture made June 6, 1821 [mortgage] between George Marker Jr. of Frederick County state of Maryland of the one part and Michael Wachler and Michael Heffner of the aforesaid county and state of the other part . . . Become securities or endorsers for the said George Marker Jr. on his note payable in the bank of Westminster for the sum of $3000. current money and on his note payable in the farmers bank of Maryland.
[Frederick CCH, Liber J. S. 11, Folio 176; 15 May 1821, Frederick CCH, Liber J. S. 13, Folio 273]

Indenture made August 21, 1821 [deed] between George Marker Jr. of Frederick County state of Maryland of the one part and [to] Henry Leatherman of the aforesaid county and state of the other part . . .
[Frederick CCH, Liber J. S. 14, Folio 85]

Indenture made October 20, 1821 [release] between Michael Wachler and Michael Heffner of Frederick County state of Maryland of the one part and George Marker Jr. of the aforesaid county and state of the other part . . .
[Frederick CCH, Liber J. S. 14, Folio 328]

Indenture made November 19, 1821 [deed] between George Marker Jr. of Frederick County state of Maryland of the one part and [to] Henry Crenise of Adams County state of Pennsylvania of the other part . . . Tract or parcel of land lying and being in Frederick County containing 177 acre and ¾ acre aforesaid tract called The Face Put[?] and Patricks Day in the Morning.
[Frederick CCH, Liber J. S. 14, Folio 485]

Indenture made April 26, 1824 [bill of sale] Know all men by these presents that I Daniel Marker of Washington County state of Maryland for and in consideration of the sum of 250 current money to me in hand by George Marker Jr. one leader for cutting rifle barrels, two rifle guns, six rifle gun barrel, fifty gun stocks.
[Washington CCH, Liber GG, Folio 839]

Indenture made February 18, 1829 [bill of sale] Know all men by these presents that I George Marker of Frederick County and state of Maryland do owe and stand justly indebted to Jacob Stottlemire aforesaid two notes of hand for $160.
[Frederick CCH, Liber JS 31, Folio 119]

Indenture made October 25, 1838 [deed] between George Wachter of Frederick County state of Maryland of the one part and [to] Peter Marker and George Marker of the same place of the other part . . . Tract or parcel of land lying and being in Frederick County aforesaid being part of land called Millers Delight, part of land called Catch It If You Can, part of land called Fullers Delight, part of land called Neighbors Accommodation, part of land called Hognest, part of land called We Have Got The Bird In The Cage.
[Frederick CCH, Liber HS 16, Folio 66]

MARKER, JACOB — Sharpsburg

After working under his father, Jacob operated the gun business on his own in 1855. Jacob and James were collectors of state taxes in the Sharpsburg district in the early 1870's. By 1875 Jacob was elected Sheriff. His last indenture recorded in Washington County was in 1890. James signed his full name on rifle barrels; the barrels stamped "J. Marker" are of better quality and are thought to have been made by Jacob.

Washington County, Sulberow #2

	age	sex	occ.	
James Marker	40	M	Gunsmith	prop. $4500. birth Maryland
Amelia	39	F		Maryland
Jacob	18	M	Gunsmith	Maryland
William	15	M		Maryland
Elizabeth	11	F		Maryland
Ezra	9	M		Maryland
Franklin	7	M		Maryland
Sarah	5	F		Maryland

[1850 Federal Census]

MARKER, JAMES — Sharpsburg

Many of his rifles have very large and heavy octagonal barrels making the lines of the rifles not too graceful. He did not possess the great ability of his father and is not known to have done any carving; apparently accuracy was his main objective. In the author's collection is a long flintlock longrifle with many simple inlays and little engraving, but it is a fine shooter. His rifles are signed "James Marker" in script on the barrels. Many of his father's rifles were probably reworked and restocked by James, thus accounting for the quality of some rifles that bear Daniel's name.

James was the son of Daniel and Christina Beckenbaugh Marker, was born and spent his entire life around Sharpsburg. He married Amelia Naeff in 1832 and they had eight children: Jacob, Louisa, Sarah Ann, William, Ezra, Elizabeth, Franklin, and Julia. (Louisa, Sarah Ann, and Franklin all died as small children.) James owned a large farm which borders the town of Sharpsburg where he made muzzle loading rifles as his father did. He died at the age of 75 years and is buried in the Sharpsburg Methodist cemetery. His will was probated May 15, 1883, Washington County Court House, Liber G, Folio 228.
[Mrs. Nellie Marker Grillot, unpublished genealogical data]

Washington County, Sulberow #2

James Marker age 40	sex M	occ. Gunsmith	prop. $4500.	birth Maryland
Amelia	39	F		Maryland
Jacob	18	M	Gunsmith	Maryland
William	15	M		Maryland
Elizabeth	11	F		Maryland
Ezra	9	M		Maryland
Franklin	7	M		Maryland
Sarah	5	F		Maryland

[1850 Federal Census]

Indenture made February 28, 1833 [deed] between Paul Marker of Washington County in the state of Maryland of the one part and [to] James Marker of Washington County in the state of Maryland of the other part . . . Tract or parcel of land lying or being that portion of ground in Sharpsburg Washington County aforesaid it being distinguished by lot number 160.
[Washington CCH, Liber NN, Folio 824]

Indenture made September 22, 1841 [deed] between Ervin Spong of Washington County in the state of Maryland of the one part and [to] James Marker of Washington County in the state of Maryland of the other part . . . Tract or parcel of land lying or being in Washington County situated upon the Chesapeake and Ohio Canal known as a tract of land owned by the heirs of Matthias Spong dec. containing 116 acres.
[Washington CCH, Liber YY, Folio 598]

Indenture made July 9, 1844 [deed] between Margaret Hedrick of Washington County in the state of Maryland of the one part and [to] James Marker of Washington County in the state of Maryland of the other part . . . Tract or parcel of land lying or being in the town of Sharpsburg known on the platt of said town as lot number 88.
[Washington CCH, Liber OHW2, Folio 322]

Indenture made January 21, 1845 [deed] between David Bruner of Washington County in the state of Maryland of the one part and [to] James Marker of Washington County in the state of Maryland of the other part . . . Tract or parcel of land being in the aforesaid county being called Hickory Tavern or otherwise Mount Pleasant.
[Washington CCH, Liber OHW2, Folio 664]

Indenture made June 29, 1846 [deed] between Harrison Lowry of Washington County in the state of Maryland of the one part and [to] James Marker of Washington County in the state of Maryland of the other part . . . Tract or parcel lying or being on the South side of the front street in the town of Sharpsburg being distinguished by half of lot number 128.
[Washington CCH, Liber IN1, Folio 896]

Indenture made 1864 [bill of sale] John Grice to James Marker of house hold goods.
[Washington CCH, Liber IN2, Folio 148]

Death of James Marker – one of the prominent men of our county . . . James Marker was born about the year 1808, near Myersville, Frederick County, Md., and came to this county with his father who was a gunsmith when he was a boy. This trade he learned and at the age of manhood took charge of the business in Sharpsburg and continued in it a good portion of his life. Not only did the rifles and firearms manufactured by him obtain a wide reputation but he himself was noted not only in the county but through the surrounding country as a famous marksman excelling all his competitors in that practice. At his business, he acquired in connection with other operations a handsome competency and in 1855 he turned over the gun-making to his son Jacob, he having become the owner of real estate and to farming of this and buying and selling of property.
[The Hagerstown Mail, 18 May 1883]

MARKER, PAUL — Sharpsburg

He was an able carver as his rifles show both fine raised and incised carving. He enjoyed using silver and brass at any place that enhanced its beauty. The glorious engraving alone makes his longrifles masterpieces. The head of the patch boxes are often a beautiful bird. The rifles are signed "P. Marker" in script on the barrel and sometimes the name is in block letters in a brass plate.

Paul was born in 1793, the fifth child of George 1753-1827 who came from Germany about the time of the Revolutionary War. . . .
Paul married Elizabeth Warrenfeltz in December of 1816 in Frederick County where he made muzzle loading rifles. There were five children born to this marriage and the entire family moved to Dayton, Ohio, where he farmed and later had a gunshop, with his brother Daniel at south Dixie Highway and the Springboro Pike, southwest of Dayton. Paul died in November of 1864 and his wife in 1862. Both are buried in the Church cemetery at Ellerton, Ohio.
[Mrs. Nellie Marker Grillot, Genealogy]

Indenture made March 19, 1817 [bill of sale] I Daniel Marker of Frederick County and state of Maryland for and in consideration of the sum of $100.00 paid to me in hand by Paul Marker, tinplate stove and pipe, and chest of drawers, one kitchen cupboard, one corner cupboard, one chest, two bedsteads, eight common chairs, two rifles, one red heifer, one black heifer, one black mare and saddle, one set of blacksmith and gunsmith tools.
[Frederick CCH, Liber JS 4, Folio 197]

Indenture made May 16, 1832 [deed] between John Hedrick of Washington County in the state of Maryland of the one part and [to] Paul Marker of Washington County in the state of Maryland of the other part . . . Tract or parcel of land lying or being all that portion of ground in Sharpsburg, Washington County and state of Maryland aforesaid it being distinguished by lot number 159.
[Washington CCH, Liber NN, Folio 291]

Indenture made May 16, 1832 [deed] between John Hedrick of Washington County in the state of Maryland of the one part and [to] Paul Marker of Washington County in the state of Maryland of the other part . : . Tract or parcel of land lying or being all that portion of ground in Sharpsburg Washington County and state of Maryland aforesaid it being distinguished by lot number 160.
[Washington CCH, Liber NN, Folio 292]

Indenture made February 28, 1833 [deed] between Paul Marker of Washington County in the state of Maryland of the one part and [to] James Marker of Washington County in the state of Maryland of the other part . . . Tract or parcel of land lying or being that portion of ground in Sharpsburg Washington County aforesaid it being distinguished by lot number 160. [Washington CCH, Liber NN, Folio 824]

Indenture made July 17, 1833 [deed] between Paul Marker of Washington County in the state of Maryland of the one part and [to] George Edwards of Washington County in the state of Maryland of the other part . . . Tract or parcel of land lying or being that portion of ground in Sharpsburg, Washington County aforesaid it being distinguished by lot number 167. [Washington CCH, Liber OO, Folio 359]

Indenture made January 11, 1834 [deed] between Paul Marker of Washington County in the state of Maryland of the one part and [to] Elizabeth Wade of Washington County in the state of Maryland of the other part . . . Tract or parcel of land lying or being that portion of ground in Sharpsburg, Washington County aforesaid it being distinguished by lot number 166.
[Washington CCH, Liber OO, Folio 729]

MARKS, ANDREW — Washington, D. C.

Marks, Andrew
1822 armorer at Marine barracks, N. Yrd.
[Directory, Washington, D.C.]

Marks, Andrew
1843 armorer – W side, 8 e btw 6 and I s nr G
[Directory, Washington, D.C.]

Marks, A.
1855 497 E s gunsmith
[Directory, Washington, D.C., & Georgetown]

MARKS, JACOB — Washington, D. C.

Marks, Jacob
1860 h 12th east & cor E south gunsmith
[Directory, Washington, D.C.]

MARRIOTT & BARSTOW — Baltimore

The proprietors of the firm were J. H. Wilson Marriott and Frederick S. Barstow.

Marriott & Barstow

1883	8 South Charles Street	gunsmith
1884-86	7 South Charles Street	gunsmith

[Baltimore city directories]

MARSH, SAMUEL W. — Washington, D. C.

His second invention of 5 November 1861, #33,655, was a revision of the first patent on lever action sporting rifles.

December 6, 1859
Be it known that I, Samuel W. Marsh, of Washington, in the District of Columbia, have invented a new & useful Improvement in Fire-Arms . . . in falling breech rifles.
[U. S. Patent Office, #26,362]

MARSHALL, EDWIN — Baltimore

Evatt, Columbus	1850	age 37	locksmith	birth Maryland
Mary	1850	40		Maryland
Marshall, Edwin	1850	18	apprentice	Maryland

[1850 Federal Census]

MARTIN, JOHN — Charles County

October 27, 1682 Assembly Proceedings
An account exhibited by John Martin of Charles County Smith for scowing cleaning and fixing Arms amounting unto four thousand one hundred and eighty pounds of tobacco resd and thus andorsed Vist.
[AM, p. 336]

MARYLAND ARMS PURCHASES

Henry Brookes Plte Nichas Cawseene deft
According to the direccon of an order of the 25th of June last made in this Cause the deft this day brought the Gunne in question into Court to bee viewed. And Richard Cole being sworne and examined in open Court upon his oath saith. That the Gunne nowe pduced in Court was about four or five yeares since (as hee remembers) taken by Governor Calvert deceased from the plte upon pmise to redeliver the same to the plte at his returne from Kent. And lt Willm Lewis in open Court averred upon oath, that Governor Calvert acknowledged hee had a Gunne of the pltes and promised to deliver the same againe to the plte at his returne from Kent being about four or five yeares since (as hee remembers) . . .
[AM, p. 40-41

October-November, 1678 Assembly Proceedings
Armes & Ammunition at Matapenny bought by the Rt. honorable The Lor– defence of the Province. Muskets 315; Carbines 101. Blunderbus 1. Armes abroad for the country Service in Charles County Muskets 50. Carbines 40. Jn Somerset County Muskets 50. Jn St Maryes Count muskets 58. Carbines 30. Calvert County Carbines 6. muskets 40. Carbines in officers hand 6. Gunns 797. Powder at Matapenny 1750 in Calvert County 100 in Somerset County 200. Powder in all 2050. shot at Matapenny 6400 at the Gournors 100 in Somerset County 600 in Calvert County 300 in all 7400. Spent since the Lord Proprietaryes going for England about 2000 ls of Powder & 6000 ls of shot & of his Lopps Gunns lost & Broken . . .
[AM, p. 30]

October 12, 1681 from the Council to Lord Baltimore
Wee cannot but take notice thereof unto your Lordship praying and requiring you to cause the same if true to be speedily redressed, and that in the distribution of the Armes and Ammunition (which at the request of your Agent Nicholas Lowe Merchant, Wee have permitted to bee transported for the Defence of that Place) your Lordship do express your trust and confidence in His Majestie's Protestant Subjects by putting the said Armies into their hands.
[AM, p. 300-301]

January 19, 1688 Proclamation by Deputy Governors
. . . And wee the said Deputy Governors doe hereby also further publish declare will require and strictly command all and every the said officers that they use all possible dilligence

Industry and Endeavor for the procuring and getting into each of their respective custodies The publick Armes aforesaid and delivering the same together with such Armes as already they have in their hands to the sheriffs as aforesaid and that imediately after the receipt of the same by the sheriffs They the said sheriffs giveing the severall Officers receipts for what they receive shall imediately presse boats and hands to convey the same to Mattapany to the hono Coll Henry Darnall or to St. Maries to the hon Coll William Digges . . .
[AM, p. 56-57]

October 15, 1691 Order for 100 Barrells of Powder and 200 muskets for Maryland
Upon the Report of the Right Hon the Lords of the Committee for Trade & Plantations. It was this day ordered in Council that Sir Henry Goodrick Knt. and Barrt. Lieut. Generall and the principall Officers of their Majesties Ordnance do cause to be provided and sent to Maryland One hundred Barrells of Powder as of their Majesty's Bounty to that Colony and also two hundred Muskets for their Majestie's service there the value of which muskets is to be reinbursed to the Office of the Ordnance out of the duty of one shilling per hoggshead arising the last year in Maryland . . .
[AM, p. 287]

March 9, 1692 Councill Meeting at the City of St Marys, Maryland
Ordered that Coll Nicholas Greenberry of Ann Arundell County and the rest of the Chief Military Officers of other the respective Countys of this Province be advised & required by Letter from this Board to Press and employ all the Smiths within their respective Precints and Commands for the immediate Cleansing and fixing the publick Arms to them respectively belon which was accordingly done . . .
[AM, p. 470]

December 21, 1692 Governor and Councill of Maryland
Ordered that Capt Edw Burges of Ann Arundell County do forthwith deliver into the hands and possession of the hon Col Nicholas Greenberry or his order upon demand the publick arms & ammunition in his Custody according to the tenor of his receipt passed for the same . . .
[AM, p. 444]

April 1, 1693 Coll Lionel Copley Captain Gen & Gov of Maryland
. . . Upon motion made p Coll David Browne setting forth that severall of the Countrey Arms sent into Somerset County are wanting & Concealed Ordered that praecept issue from this Board cammanding and requiring allmanner of persons having had the custody or keeping the said arms or any of them forthwith to deliver to produce the same & deliver them to the Chief Milatary Officer to be him Distributed & Disposed of as he shall think fit for his Majestys & the Countreys Service, & in Case any Person shall be delinquent or Fail in his Duty in this matter to be bound over to the next Provincial Court to answer . . .
[AM, p. 519]

July 1, 1746 Council Proceedings
. . . Advice of his Lordships Council shall and he is hereby empowered and required to send immediately to the said William Hunt Merchant in London to send into this Province for the Use of the Public the Value of Two Hundred Pounds in such Arms as he the said Governor or Commander in Chief with the Advice aforesaid . . .
[AM, p. 342]

June 4, 1748 Assembly Proceedings
. . . An Act for raising three pence Sterling p Hogshead &c to Mr Adams, Merchant in London, directly; . . . Resolved, That the money be payed by the Naval Officer to the Treasure of this Province, to be by them transmitted directly to mr Joseph Adams Merchant in London.
[AM, p. 103]

December 26, 1754 Sharpe to Dinwiddie
What made me apprehend that the Guns sent to Belhaven were only four pounders was Mr. Lewis's misrepresenting to me their Dimensions and the Diameter of their Chase which he told me he had measured but I have received a Letter from him since advising me that on his reviewing and reexamining them at his Return in Compliance with my Directions he found he had not been sufficiently accurate in his first Measurement and that he had upon discovering his Error writ to you to acquaint you that he found the Diameters of the Bore of those pieces to be 3 Inches & 7/10 & to desire you to order Ball to be made of the same Diameter, in that however he has again erred for the Diameter of the Ball ought to be always 2 Lines at least less than that of the Guns Chase so that instead of having the Ball made 3.7 Inches you will please to order them to be only 3.5 Inches. Be pleased to present my Compliments to Capt. Arbuthnot & acquaint him with the Cause of my Suspicion & that if you approve thereof I will be obliged to him for his Men at that Rate. I should be glad you would get some Shot made for that small Brass-Gun at York & send them up together with the Gun & those Wheels from Williamsburg by the first Opportunity. It was reported with us that 2000 Men were expected over immediately from Europe which I should have been glad to have seen confirmed.
[AM, p. 152]

December 27, 1754 Sharpe to DeLancey
. . . if you have received any Intelligence from Oswego or elsewhere of the March of such numbers you will be kind enough to acquaint me & as we shall in case we proceed to act offensively next Spring stand in great need of a Mortar or two & some Bomb Shells neither of which can be procured in either of these Colonies I hope you will not be averse to sparing us a small Supply of both from your Magazines & any small Cannon that you cannot have immediate use for, till we can get sufficient Number of each sent us from England . . .
[AM, p. 154]

January 2, 1755 Dinwiddie to Sharpe
. . . I have been endeavouring to procure Carpenters & Coopers, but as yet have had no Succes. I sent to Fredericksburg two Smiths with Bellows, Anvil & Tools, with orders to proceed directly for Will's Creek . . .
[AM, p. 155]

February 8, 1756 Sharpe to Lords of Trace
. . . I am to acquaint your Ldps that there are in & belonging to this Province 19 Carriage Guns all Iron but very good 4 of them are 6 pounders and 15 four pounders the last not mounted. In the Magazine is 16 lb weight of 6 lb. shot & 24

half Barrels of Powder; of musquet Ball & Bar Lead we have a considerable quantity & there were last Spring about 500 stand of small Arms belonging to the Publick most of which are in the hands of the men that have been raised for the protection of the Frontiers & some were lost at the Action of the Monongahela. The Militia of this Colony are near 165,000, One third of whom at least are entirely destitute of Arms & many of the Guns that are the property of the Rest are very bad & scarcely fit for use. For want of a proper Militia Law (Which the assembly has been frequently in vain sollicited to make) the people are undiciplined as well as badly armed & cannot be compelled to serve in Defence of the Country. The white Inhabitants in Maryland are 107,963, the Black 46225 . . .
[AM, p. 352-353]

April 24, 1756 Shirley to Sharpe
In your last, you mention'd that you was in want of small Arms. I am now upon my journey to New York, and have only time to inform your Honour that the 10,000 stands of Arms, & proportion of Ammunition sent by his Majesty to Boston for the use of the Colonies, according to the Descretion of the Commander in Chief of His Forces in North America, are now arriv'd, & are under the Care of the comptroller of Ordnance settled by the Board at that place. Of these Arms I have order'd the Comptroller to deliver 300 Stands with a proper proportion of Ammunition to any person who shall be authoriz'd by your Honour's Government to receive and indent for the same on the behalf of the Province; vizt to return the Arms at the End of the present Service, for which they are lent to the Province, into his Majesty's Stores at Boston, pursuant to the Direction of the Board of Ordnance there to be kept by their Officers as a Magazine for the use of the Colonies, which shall want them. The Rule, I have observ'd in the present Distribution of these Arms, with respect to my own & all the other Governments, except Virginia & Connecticutt, is to deliver to each of them Arms for four sevenths of the Men, they now raise: I am oblig'd to furnish his Majesty's four Regiments of Regulars immediately with 1200 Stands; and the 10,000 stands will hold out for no larger a proportion, leaving only 800 stands in the Stores for Extrordinary Contingencies; you will be pleas'd Sir, in case your Government shall raise 500 Men for the Expedition to send for the Arms, when you think proper, to Mr. Turner Comptroler of his Majesty's Ordnance at Boston . . .
[AM, p. 392]

June 9, 1758 Sharpe to Forbes
. . . I dispatched an Express to Annapolis & ordered the Person who had the Care of our Arms to pack them all up that they might be ready to be sent on the first notice to any Place that you should think proper to order them, a few Days afterwards most of them were brought to Frederick Town agreeable to Sir John St Clairs Desire & 500 new Muskets, upwards of 50 old but the best Carbines that I could collect in the Province & 33 Pair of Pistols have been since sent from that place to Winchester to arm the Virginia Forces. I have likewise ordered 50 broad swords Scabbards & Belts 40 Pair of Holsters & as many Blankets & Shoulder Belts with Springs as can be got in Annapolis to be sent thence to Winchester ... .
[AM, p. 197]

December 9, 1758 Pitt to Sharpe
. . . It is His Majesty's Pleasure, that you do, with particular Diligence, immediately collect and put into the best condition, all the Arms, issued last Campaign, which can be, any ways, rendered serviceable, or that can be found within your Government, in order that the same may be employed, as far as They will go, in this Exigency, I am at the same time, to acquaint you, that a reasonable Supply of Arms will be sent from England, to replace such, as may have been lost, or have become unfit for future Service . . .
[AM, p. 309]

August 2, 1776 Council of Safety
There are three Gun Smith in Baltimore Town, one in George Town, four in Fredericktown, one near Fredericktown, two in Hagers Town, one in Jerusalem Town, in the whole amounting to twelve. We are informed that each of these shops can in one month, complete twenty substantial musquets (the dimensions of which may be 42 inches in length, ¾ of an inch clear in the bore 1½ inch diameter at the breech and 7/8 of an inch at the Muzzel) with Steel rammers and Bayonets 20 Inches in length including the Stock. The price of the Musquet with its Bayonet may be about four pounds, the Cartridge Box of 23 charges & sling, the Sling for the Musquets, the Scabbard and belt for the Bayonet will altogether cost about 20/.
[AM, p. 65]

January 12, 1781 S. West to Council.
I think there are State Arms in the hands of Contractors.
Peter Keener of Baltimore
Peter Lettigg I believe of Baltimore
James Mey for 50 delivered 15 qr. of Hartford
Jacob Reasor of Frederick County
Jacob Schley for 10 Rifles pr. Contract Frederick
Nicholas White of Frederick Town
John Yost of Frederick or Montgomery County
[AM, p. 14]

July 31, 1781 Adam Hunter to Smallwood
In my Brothers Absence who has resided at his Plantation in Culpepper ever since our late Alarm of the Enemy coming this way, I have your favour of 29th curr't. by Express, to which am sorry I cannot return a satisfactory answer. We have long discontinued the manufactory of small Arms of which none remain on hand, & the Workmen employed in that Branch, are so dispersed that we believe it would be difficult matter to engage them again. I shall tomorrow transmit your letter to my Brother . . .
[AM, p. 376]

MARYLAND SPORTSMAN SUPPLY COMPANY
Baltimore

In 1888 the name of the firm was changed to Bosley & Howe, who were no doubt the earlier proprietors.

Maryland Sportsman Supply Company
1887 110 West Lexington Street gunsmiths
[Baltimore city directory]

MASLIN, MICHAEL M. — Baltimore

Many American flintlock longrifles are known that hold English-style locks bearing the name "M M MASLIN." These locks appear to be typical Birmingham-made locks of the kind imported in quantity by a large number of American hardware merchants during the early decades of the nineteenth century. The earliest directory listing for M. M. Maslin is 1822. Between 1824 and 1833 he worked in partnership with William McKim as McKim & Maslin. It would be desirable to know the earliest date at which locks marked "M M MASLIN" were imported, but such information is not yet known.

Maslin, Michael M. house of Tiernan & Maslin

1822	186 Baltimore Street	hardware merchants
1824	223 Baltimore Street	

Maslin, M. M. firm of McKim & Maslin

1829-33	248 Baltimore Street	wholesale hardware

[Baltimore city directories]

MASON, JOHN — Georgetown

This Agreement, made this 28th day of July 1827 between George Bowford Lt. Col. of Artillery on Ordnance duty in the service of the United States, acting with the consent and under the direction of the Hon. James Barbour, Secretary of War, on the one part, & John Mason of Georgetown on the other part . . . That the said John Mason shall manufacture and deliver for the Military Services of the United States eighty three 24-pds iron cannon, to be delivered at the Columbia Foundry in Georgetown on or before the 1st day of July 1828.
[National Archives, Contracts for Ordnance & Ordnance Supplies, Vol. I, 1813-1828, p. 268.]

It is hereby agreed the 23rd day of May 1828 that the number of 24-pds iron cannon, mentioned in the contract dated January 29, 1828, viz forty shall be extended, to sixty.
[National Archives, Contracts for Ordnance & Ordnance Supplies, Vol. I, 1813-1828, p. 290.]

This agreement made the 13th day of December 1828 between George Bowford Lt. Col. of Artillery on Ordnance duty in the service of the United States, acting with the consent and under the direction of the Hon. Peter B. Porter, Secretary of War, on the one part, & John Mason of Georgetown on the other part . . . That the said John Mason shall manufacture and deliver for the Military Services of the United States twenty-five 24-pds iron cannon to be delivered at the Columbia Foundry in Georgetown on or before the 1st day of May 1829.
[National Archives, Contracts for Ordnance & Ordnance Supplies, Vol. I, 1813-1828, p. 344.]

This agreement made the 20th day of March 1829 between George Bowford Lt. Col. of Artillery on Ordnance duty in the service of the United States, acting with the consent and under the direction of the Hon. Peter B. Porter, Secretary of War, on the one part, & John Mason of Georgetown on the other part . . . That the said John Mason shall manufacture and deliver for the Military Services of the United States forty 32-pds iron cannon and eight hundred 32-pds cannon balls to be delivered at the Columbia Foundry in Georgetown on or before the 1st day of July 1830.
[National Archives, Contracts for Ordnance & Ordnance Supplies, Vol. II, 1829-1842, p. 14.]

MATTHEWS, GEORGE — Baltimore County

October 14, 1775 Matthews to Purviance, Kingsbury
. . . I could not say anything with any degree of certainty till now as I have fully considered the madder I find that it will require about 200 worth of Apparatus to go into the business of making guns Wch. will be of little or no value for any other purpose, the expence of Marking Boring drilling & finishing compleat will be about 40 Currency p ton & any size or bore from 9 Pounds downwards maybe cast at this Country Furnaces . . .
[AM, p. 82]

January 30, 1776 Council of Safety
Mr. George Matthews, Cannon Founder in the Neighborhood of Baltimore Town, being requested to attend the Council of Safety, and give what Information he could with Regard to casting of Cannon, and the Cannon now in that place, informed that there were two 12 pounders, six 9 pounders, six 6 pounders, eight 3 pounders and two Field pieces, 3 pounders, which in his Judgment and Opinion might be of some service, but not to be.depended on till they were bored again. That they had been a long Time unused and were scaly within side of the Guns. That there 7 or 8 of the Annapolis Guns left by Capt. Stone, 4 pounders, all good except one which had lost a Trunnion, might be made serviceable by having Trunnions fixed on the Outside with an Iron Hoop.
[AM, p. 122]

MAY, JAMES — Harford County

In a letter on 6 September 1776 Richard Dallam stated that May was associated with him in the gun factory and in his absence was in charge.

March 13, 1777 Council of Safety
That the said Treasure pay to James May for the use of Richard Dallam forty seven pounds five shillings – Cy.
[AM, p. 171]

March 14, 1777 Council of Safety
Ordered That the Western Shore Treasure pay to James May one hundred pounds on Acct. of Gun Contract.
[AM, p. 173]

November 13, 1777 Council of Safety
That the said Treasure pay to James May Ninty four pounds twelve shillings on Account passed.
[AM, p. 408]

MAYFIELD, JOHN — Baltimore

Mayfield, John

1824	Poco Street north of Saratoga	locksmith
1835-37	Baltimore Street east of Eden	gunlocksmith

[Baltimore city directories]

1850 Georgetown

Mayfield, J.	age	45	born	England	locksmith
" M. (f)		40		Va.	
" S. (f)		25		Md.	
" Thos		11		Washington, D.C.	

[1850 Federal Census]

Mayfield, —

1853	N side Bridge btw Congress & Wash., G town	gunsmith

[Directory, Washington, D.C.]

Mayfield, John

1855	78 Bridge	gunsmith

[Directory, Georgetown]

MAYNARD, EDWARD — Washington, D. C.

He was a dental surgeon who pursued the tape primer system which he continued to refine. His patents include #8,126, 27 May 1851; #26,364, 6 December 1859; #30,537, 30 October 1860; #48,423, 27 June 1865; #49,130, 1 August 1865; #86,566, 2 February 1869.

The primer which bore his name was submitted to the Board of Ordnance, where after gaining its approval, the government purchased the primer rights.

September 22, 1845
Be it known that I, Edward Maynard, of the city of Washington & District of Columbia have invented a new & useful Improvement of a tape primer for Fire-Arms.
[U. S. Patent Office]

MEIGS, JOSIAH V. — Washington, D. C.

After the war, on 22 May 1866, he also received patent #54,934, an alteration of his first.

October 21, 1862
Be it known that I Josiah V. Meigs of Washington city, in the District of Columbia have invented certain new & useful Improvement in Breech-Loading Fire-Arms . . . to provide a simple & efficacious means for traversing the breech pin or plug back & forth in its socket.
[U. S. Patent Office, #36,721]

MELCHER, EDWARD O. — Baltimore

Melcher, Edward O.

1868-70	3 Constitution & 200 Madison	gunsmith
1871-74	221 North Central Avenue	gunsmith

[Baltimore city directories]

MELCHER, HARMAN — Baltimore

Melcher, Harman

1865	149 North Central Avenue & Frederick Road west of toll gate	gunsmith

[Baltimore city directory]

MELCHOIR, EDWARD — Baltimore

William Harris was his predecessor at the same address.

Melchoir, Edward

1876-83	116 West Pratt Street	gunsmith
1884	114 West Pratt Street	gunsmith

[Baltimore city directories]

MELCHOIR, NATHANIEL — Baltimore

Produced target rifles and longrifles of below average quality. These rifles had patch boxes and crude inlays. He usually used a punch on each letter on the top facet of the barrel which read "N. Melchoir Baltimore." He occasionally used a dye punch on the barrel and lock on which the above was found.

He was called a manufacturer when he died 15 September 1879.

Melchoir, Nathaniel

1840	Mercer & Grant Streets	lock & gunsmith

[Baltimore city directory]

Melchoir, Nathaniel 1850 age 42 cutler birth Germany
Henriette
[1850 Federal Census]

MEREDITH, BENJAMIN — Baltimore

Meredith, Benjamin

1810	6 Water Street near Light	whitesmith & cutter
1812	33 South Liberty Street dwelling Water Street	
1814-15	33 South Liberty Street	whitesmith
1816	corner of Light & Bank Streets	whitesmith & bellhanger
1817-18	Baltimore Street extended near Paca	gunsmith

[Baltimore city directories]

MERRILL, JAMES H. — Baltimore

Born in Washington, D. C., in 1816 and had experience in Europe and America; from his many improvements and patents was one of the most prominent arms inventors of the time. Beginning in 1861 all of his patents were assigned to Merrill Patent Fire Arm Manufacturing Company. His three colleagues were Ferinond C. Latrobe, Lewen W. and Philip E. Thomas. The company received from the federal government $374,804.63 for 14,495 Merrill carbines during the Civil War. They also supplied the Navy Department with 770 Merrill rifles and many arms were purchased by individuals as state companies. Along with the obsolete Jenks carbines, they altered everything from 1851 Springfield musketoons to 1812 flintlock muskets to the Merrill conversion They also manufactured double set trigger, half stock rifles with his actions. Percussion pistols and shotguns can be found with back lock actions marked "J. H. Merrill Balto." The Merrill action guns are marked the same except for various patent dates with the actual manufacturing done in Baltimore.

Merrill, James H.

1840	South Street 1 door from Pratt	gunsmith & hardware store
1845	65 South Street	gun, rifle, pistol manufacturer

Merrill Patent Fire Arm Manufacturing Company

1861-67	239 West Baltimore Street	gunsmiths

[Baltimore city directories]

MERRILL, JEREMIAH L. Baltimore

Was the operator of a livery stable in 1845. Was possibly the father of James H. and William Y. Merrill.

Jeremiah L. Merrill age 70 sex M occ gunsmith birth New Jersey
Phebe
[1850 Federal Census]

Merrill, Jeremiah L.

1851	13 Columbia Street	gunsmith
1864	275 West Lombard Street	gunsmith
1867	145 North Charles Street	gunsmith

[Baltimore city directories]

MERRILL, LATROBE & THOMAS Baltimore

Merrill, Latrobe & Thomas

1856-59	6 Hanover Street	gunsmiths

Merrill, Thomas & Co.

1860-67	239 West Baltimore Street	gunsmiths

(ad 1865) Merrill carbines, sporting rifles – Merrill action – with globe sight & Starr's repeating four shooter.
[Baltimore city directories]

14077 January 8, 1856
The Merrill Patent Fire Arm Manufacturing Company, in offering their Arms to the Public, take this means of informing those not acquainted with the Arms of their Manufacture, that the Merrill Rifle is the result of a great deal of experiment by the Inventor Mr James H Merrill, of Baltimore whose vast experience of Arms, both in America and Europe, has enabled him to give the world this great improvement in small arms which classes him among the most prominent of modern inventors. The simplicity and strength of construction arrived at in the Merrill rifle is only equalled by the extraordinary range and accuracy of fire and the effectiveness of the arm not being depended upon any metallic or Indian rubber cartridge case but using the simple paper cartridge or loose powder and ball and the ordinary percussion cap to be had at any country store or trading station gives them the great advantage over arms depended upon fixe and regular prepared ammunition and makes them justly preferred not only on the frontier but wherever a rifle is required. Having been approved by the U. S. government after being subject to the most severe tests by both army and navy departments and having been ordered then for service, it is entirely confident that they are offered to the public. Another advantage of the Merrill-planned process is that it can be easily applied to the muzzle-loading arms at a small cost without restocking, altering the lock or changing the general appearance or dismantling the strength of the arm. The alteration of the Merrill loader to the Merrill plan of the breech loading has been practicably tested by the U. S. government with the most entire success and has placed in the hands of the troops an arm that cannot be surpassed and also at a low cost. A great advantage is claimed for these arms in being solid in their construction like the regular muzzle-loading arm or saber of ordinary bayonets can be used on them, whereas the breech-loading arm in which the barrel moves cannot stand the shock or charge and are awkward to use when their length is increased by the bayonet. The addition in length of the Merrill rifle makes no difference in handling it with as much ease as without the bayonet. For sporting purposes or target shooting it cannot be equalled as it has attained the highest perfection in every respect. The barrels are of the best material and the chambers are bored out to the exact size of the conical ball which they use so that the plunger which puts the cartridge in its place forces the point of the ball firmly into the grooves or rifling of the barrel, thereby ensuring it going straight to its mark and not turning as conical balls frequently do when fired from other rifles. The sporting rifles are fitted with globe sights the most improved patent which ensures fine shooting. The rifles have been fired 500 times without cleaning and any number of shots may be discharged without any charge taking place in the working of the machinery. This is owed to there being no escape of gas which when escaping causes them to be trimmed or clogged in other breechloaders, but which can never occur in the Merrill rifle as the plunger or breech pin is reamed so as to make the expansion spring and the cavity be filled up with the brass cap upon which the force and heat of the exploding action at the moment of discharge causes the plunger to expand as much as the barrel will allow it or in other words to keep up with the expansion of the barrel or surrounding surfaces and thereby preventing the escape of gas which not only clogs up the working parts of the gun when escaping but causes great loss in power. Thus no gas escapes from the Merrill rifle accounting for its having more penetration than other breech-loaders. Attention is respectfully called to the recommendation on the following page and any further information may be given by applying by mail or in person to the Merrill Patent Fire Arms Manufacturing Company, Baltimore, Maryland.

Patent No.	20954	Dated Jan. 20, 1858	Reissued Mar. 26, 1861
	32032	Apr. 9, 1861	
	32033	Apr. 9, 1861	
	32450	May 28, 1861	
	32451	May 28, 1861	
	33536	Oct. 22, 1861	
	40884	Dec. 8, 1863	

The ordinance office report indicates that they have been purchased from Jan. 1, 1861 to June 30, 1866 14,495 Merrill carbines costing $374,804.63 583 Merrill rifles costing $23,880.50. Many fine patents could be illustrated.
[Advertising pamphlet]

MERRILL, WILLIAM Y. Baltimore

In the 1850 Federal Census James H. was married to Martha and had seven children, one by the name of William who was nine. The relationship of the gunsmith William Y. to James H. was possibly brother. Jeremiah and William Y. were both listed at the same address in 1864. By the late 1840's all three Merrills were residing at 145 Washington Place.

Merrill, William Y.
1864 275 West Lombard Street gunsmith
[Baltimore city directory]

MERWIN & COMPANY Washington, D.C.

1862 Merwin & Co. military outfitters 354 Penna. Ave.
[Washington city directory]

MESSERSMITH, SAMUEL Baltimore

January 31, 1776 Council of Safety
The Council called on Messers Keener, Messersmith and Reddict to know what Quantity of Musquets thay had made for the use of this Province in Consequence of their Contracts with Mr. Alexander. Mr. Keener reported that he had made thirty. Mr Messersmith that he had made Twenty Seven, and Mr Riddick that he had made seventy.
[AM, p. 127]

February 7, 1776
Sam Missersmith 27 guns 23 good 4 bad
[AM, p. 155]

May 31, 1776 Ware to Council Baltimore
The bearer will produce to the board three Gun Locks sent by Mr Messersmith for your inspection, Who tells me he has now twenty five more finished off, and if the public will Encourage him properly he can finish ten or twelve Pr. week I beleave the Encouragement he expects is three dollars per Lock, if you approve of the work and price, please let me know by a line, and I will order him to employ all his hands in that besiness.
[AM, p. 456]

July 17, 1776 Council of Safety
Ordered that the said Treasurer pay to Samuel Messersmith 7 pounds 1 shilling 9 pence for mending Muskets.
[American Archives, 5th Series, Vol. I, p. 1338]

September 12, 1776 Council of Safety
Ordered that Samuel Messersmith be requested to deliver to Captain Forrest all the Guns which were left in his custody, to be repaired.
[American Archives, 5th Series, Vol. II, p. 630]

June 20, 1777 Council to Messersmith
We have a good many Musqet Barrels and unfinished Bayonets by us, belonging to the Public and want them stocked, fitted & finished we would be glad to know whether you would undertake a number of them and at what Price; if you are inclined to do so, it might save some Time in finishing the Agreement by coming down . . .
[AM, p. 296]

October 23, 1777 Council of Safety
That the said Treasure pay to Samuel Messersmith One hundred and fifty nine pounds six shillings and ten pence per Acct. passed.
[AM, p. 403]

METZGER, JACOB Frederick

Jacob Metzger apparently learned gunsmithing in Lancaster during the Revolutionary War years and then settled in Frederick where he spent the remainder of his life. Kindig (1960) noted that an 'Assessment of Non-associators" of the Borough of Lancaster dated 23 August 1777 includes the name of Jacob Metzger as a gunsmith. If this is the same person who worked in Frederick in later years, he was at the time a lad of about 12 years. The pressing need for gunsmiths in 1777 may have provided openings for apprentices younger than usual, and the informalities of the time may have allowed an apprentice to be listed as a gunsmith.

Metzger was born between 1765 and 1767. The 1810 Federal Census shows him to be not more than 45 years of age; the 1830 census shows him to be between 60 and 70 years; and he must have been 21 years old in 1788 for he was issued a deed for land in that year.

Jacob Metzger was an accomplished carver and engraver, and some fine relief-carved flintlock rifles have survived bearing his name. His earlier rifles are signed in script "J. Metzger." He also used a die stamp having the same letters, "J. Metzger," and other stamps, "I. M." and "J. M." with a symbol used repeatedly between the two initials. Kindig (1960) noted that details in his engraved designs resemble some used by Jacob Dickert of Lancaster, and suggested that he may have apprenticed under Dickert.

Indenture made January 15, 1788 [deed] between Jacob Groff of Fredericktown and Frederick County in the state of Maryland of the one part and [to] Jacob Metzger of Fredericktown and Frederick County in the state of Maryland aforesaid gunsmith of the other part . . . Ground situate lying and being in addion to Fredericktown, laid out by a certain Casper Myers, deceased, and on the north side of the street leading to Conicochingue and part of the lott known and distinguished in said platt by number 5.
[Frederick CCH, Liber W. R. 7, p. 663]

Metzker, Jacob Frederick County
2 males over 16 years
2 males under 16 years
1 female
[1790 Federal Census]

Indenture made May 25, 1793 [deed] between Jacob Metzger of Fredericktown, and County in the state of Maryland of the

one part and [from] Jacob Shellman and Henry Line of Fredericktown, and County in the state of Maryland aforesaid of the other part . . . Ground situate lying and being in the additian to Fredericktown commonly wives called Bentztown being part of a tract of land called Long Acre.
[Frederick CCH, Liber W. R. 11, p. 579]

Indenture made June 3, 1793 [deed] between Jacob Metzger of Fredericktown, Frederick County and state of Maryland gunsmith of the one part and [to] George Bowman of Frederick County and the state of Maryland of the other part . . . Land on the north side of the tract leading to Conicocheague . . . at the same time came Christona, wife of Jacob Metzger.
[Frederick CCH, Liber W. R. 11, p. 601]

Contractors with the government under Act of July 5, 1798 for Charlesville pattern muskets at $13.40 per stand

(4) Eli Williams, Williamsport, Md. — 2000 muskets
(5) Nicholas White, Thomas Crabb, Jacob Metzger & Christopher Barnhizle, Frederick Town, Md. — 1000 muskets

[U. S. ORDNANCE, Vol. I, p. 20]

Indenture made November 23, 1803 [deed] between Frederick Reel of Fredericktown, Frederick County state of Maryland of the one part and [to] Jacob Metzger of the same place of the other part . . . Ground situate to Fredericktown aforesaid it being part of that half of a lott of ground distinguished by the half of lott number 1.
[Frederick CCH, Liber W. R. 25; p. 163]

Indenture made May 25, 1816 [deed] between Jacob Metzger of Fredericktown, Frederick County state of Maryland of the one part and [to] Jacob Metzger Jr. of Fredericktown, Frederick County state of Maryland of the other part . . . Two part of lotts of [?] situate in Bentz addition to Fredericktown.
[Frederick CCH, Liber J. S. 3, p. 56]

Indenture made June 10, 1800 [bill of sale] I Peter Walters of Frederick County in the state of Maryland for and in consideration of the sume of 75 pounds current money to me in hand paid by Nicholas White, Christopher Barnhizle, and Jacob Metzger of Frederick County and state aforesaid the following property - towit 3 iron anviles, 4 vices and all my smith tools of every description, 1 pair large bellows, 2 grindstones, and a large wheel that turns.
[Frederick CCH, Liber W. R. 20, p. 19]

Indenture made October 9, 1821 [deed] between Jacob Metzger Jr. of Fredericktown, Frederick County state of Maryland of the one part and [to] Jacob Metzger of the same place of the other part . . . Two parts of lotts of [?] situate in Bentz addition to Fredericktown.
[Frederick CCH, Liber J. S. 12, Folio 419]

Indenture made September 17, 1823 [deed] between Frederick Stover of Fredericktown, Frederick County state of Maryland of the one part and [to] Jacob Metzger of the same place of the other part . . . Ground situate to Fredericktown aforesaid it being part of lott called number two.
[Frederick CCH, Liber J. S. 18, Folio 465]

Indenture made June 3, 1824 [deed] between Jacob Metzger of Fredericktown Frederick County state of Maryland of the one part and [to] Philpipina Metz of the same place of the other part . . .
[Frederick CCH, Liber J. S. 20, Folio 158]

Indenture made May 20, 1825 [bill of sale] I Jacob Hill of Frederick County in the state of Maryland convey to Andrew Low and Jacob Metzger . . .
[Frederick CCH, Liber J. S. 22, Folio 414]

I Jacob Metzger Senr in the State of Maryland . . . I devise and bequeath as follows all my estate real, personal and mixed, to my wife Christina during her natural lifetime for her own use and benefit and after her death and funeral charges are paid the said property to be sold to the highest bidder for the benefit of my children share and share alike but should either of them die before my wife and leave any heirs they shall receive that part of the estate And as Mrs William Bentz, Richard Noxes widow and F. C. Nusz Wife have each of them received Sixty dollars which I desire shall be deducted out of their respective shares . . . I have hereunto set my hand . . . this 19th day of June 1826. Jacob Metzger
On the 10th day of July 1837 came William Shellman, Jacob Riehl, John Elbert the subscribing witnesses to the foregoing last Will & Testament of Jacob Metzger dec'd . . . Probated 9 October 1837.
[Frederick CCH, Will Book G.M.E.,2, p. 272]

MEYER, JOHN C. J. — Baltimore

Shotguns and heavy target rifles bearing "John C. J. Meyer Balto Md" in script on the barrels and sometimes the locks came from his shop. Meyer used silver mountings over brass and his exclusive grade rifles are superiorly engraved, reflecting the Escherich style and engraving.

Meyer, John C. J.

1856-60	14 West Pratt Street	importer & manufacturer of guns, rifles & pistols

[Baltimore city directories]

The joint Standing Committee on claims to whom were referred the following claims against the city – towit . . . 2. The petition of John C. J. Meyer asking to be indemnified for the loss of guns and other property taken from his store by a mob on the 19th of April 1861 – and claiming $1,500.00.
[Baltimore City Correspondence Manuscripts, 1862, #504]

MICHAEL, HENRY — Baltimore

Michael, Henry

1847-51	143 East Fayette Street	gunsmith
1855	155 Fayette Street	gunsmith
1858	Central Avenue, corner of Fayette	locksmith
1860	Central Avenue	hardware

[Baltimore city directories]

	age	sex	occ	birth
Henry Michael	30	M	Gunsmith	Germany
Bertha	25	F		Germany
Josephine	1	F		Maryland
Virginia	1/12	F		Maryland

[1850 Federal Census]

MIDDLEKAUFF, DANIEL Washington County

Indenture made August 24, 1805 [deed] between Daniel Rench of Washington County in the state of Maryland of the one part and [to] Daniel Middlekauff of Washington County in the state of Maryland of the other part . . . Tract or parcel of land lying and being in Washington County aforesaid being that tract of land called Settled in Times.
[Washington CCH, Liber R, Folio 356]

Indenture made August 24, 1805 [deed] between Daniel Middlekauff of Washington County in the state of Maryland of the one part and [to] Daniel Rench of Washington County in the state of Maryland of the other part . . . Tract or parcel of land lying and being in Washington County aforesaid being that tract of land called Locust Bottom.
[Washington CCH, Liber R, Folio 357]

Indenture made October 23, 1805 [bill of sale] Know all men by these presents that I John Wetstone of Washington County in the state of Maryland convey to Daniel Middlekauff of the aforesaid county and state in consideration of the sum of thirty pounds current money of the state aforesaid goods and chattles following one bellows, one boring wheel, two large vises, one letter, three rifle barrels, twenty rifle stocks, one sign, one adze, one saw, one large screw plate, six small screw plates, one drawing knife, two beds, one stove, five planes, one small anvil, five hammers, one brace, one [?], three chairs, one table, twenty eight bundles lath.
[Washington CCH, Liber R, Folio 430]

Indenture made August 19, 1806 [deed] between John Middlekauff of Washington County in the state of Maryland of the one part and [to] Daniel Middlekauff of Washington County in the state of Maryland of the other part . . . Tract or parcel of land lying and being in the aforesaid county being part of that tract of land called Locust Bottom.
[Washington CCH, Liber S, Folio 357]

Indenture made October 22, 1812 [deed] between Walter Boyd, Martin Bachlet, and Daniel Middlekauff all of Washington County in the state of Maryland of the one part and [to] John Rohrer of Washington County in the state of Maryland of the other part . . . Tract or parcel of land lying and being in Washington County 200 acres being part of that land called Locust Bottom.
[Washington CCH, Liber Y, Folio 377]

Indenture made October 22, 1812 [deed] between Walter Boyd, Martin Bachlet, and Daniel Middlekauff all of Washington County in the state of Maryland of the one part and [to] Peter Mong of Washington County in the state of Maryland of the other part . . . Tract or parcel of land lying and being in Washington County being part of that land called Locust Bottom.
[Washington CCH, Liber Y, Folio 380]

April 24, 1824
David Bline apprentice to Daniel Middlekauff to learn the mystery of Hatter.
[Washington County Indentures, Hall of Records]

Death – In January 1852 Daniel Middlekauff, a leading farmer of Western Maryland.
[HISTORY OF WESTERN MARYLAND, p. 1052]

MIER, JACOB Frederick County

Jacob Mier was settled in the Brothers Valley by 1762 and he worked there as a carpenter and gunsmith. There are deeds dated 1762, 1768 and 1773 in his name, with spelling variations of "Myerer" and "Myer." His son, Samuel Mier, learned the trade with his father and worked in Somerset County, Pennsylvania.

MIER, SAMUEL Frederick County

Samuel Mier learned gunsmithing with his father in Brothers Valley, then settled at the southern end of Somerset County, Pa., at Salisbury, where he worked as a gunsmith until the 1870's.

Samuel Mier settled at Salisbury, Pa. He was a son of Jacob Mier and learned the [gunsmithing] trade at his father's workbench. Father and son traded or exchanged parts and craftsmanship and produced guns . . . Samuel Mier made only the stocks and bought the parts, mostly from his father and other smiths of the area. . .
[Cooper, Auston TWO CENTURIES OF BROTHERS VALLEY, p. 49]

MILLER, DAVID Hagerstown

Apparently David Miller moved a little bit west, to Fayette County, Pennsylvania, immediately after finishing his apprenticeship. The name of David Miller with occupation of gunsmith is found in tax assessment records there, in George Township, for the years 1802, 1803, 1804, 1805, 1806, 1810, according to Kauffman (1960) THE PENNSYLVANIA-KENTUCKY RIFLE.

This Indenture witnesseth that David Miller son John Miller of Washington County and state of Maryland deceased, by and with the advice and approbation of John Fisher and Daniel Miller (each of the county and state aforesaid guardian of said David Miller) hath put himself and by these presents doth voluntarily and of his own free will and accord put himself apprentice to Christian Hawken of the county and state aforesaid Gunsmith . . . till he arrives at or attains the age of twenty one years he the said apprentice being fifteen years old the 1 day of March last . . . David X Miller his mark
[Washington County Indentures, Hall of Records, p. 36]

MILLER, ISAAC L. Washington, D.C.

1850	Miller,	Isaac L.	age 59	b. N.J.	gun carriage maker
		Jane S.	47	Wash. D.C.	
		Margaret	17	Wash. D.C.	
		Edward D.	16	Wash. D.C.	
		Sarah H.	13	Wash. D.C.	
		Virginia	11	Wash. D.C.	

[1850 Federal Census Washington, D.C., 7th Ward]

MONK, HENRY Washington, D.C.

1855 Monk, Henry gunsmith 361 Pa av
[Directory, Washington D.C., & Georgetown]

MOORE, EDWARD F. — Baltimore

Edward F. Moore	age 36	sex M occ Gunsmith	birth Caracas S.A.
Rachael	21		Maryland
Rachael	4/12		Maryland

[1850 Federal Census]

Moore, Edward F.

1851	8 Maiden Lane	gunsmith
1855	46 French Street	gunsmith

[Baltimore city directories]

MORITZ & KAIDEL — Baltimore

The proprietors of this firm were J. D. Moritz and Henry Kaidel.

Moritz & Kaidel

1868-79	10 South Charles Street	gunsmith, also hardware dealers
1880	6 & 8 German Street	gunsmith
1881-84	48 German Street	gunsmith
1885	48 & 50 German Street	gunsmith
1886	48 German Street	gunsmith

[Baltimore city directories]

MOSS, JOSIAH — Cumberland

This Indenture made this 18th day of September 1813 Witnesseth that Experience Moss of Bedford County Pennsylvania hath put and placed and by these presents doth put and bind her son Josiah Moss (who on the 31 day of August next will be 16 years of age) an apprentice to George Rizer to learn the trade or mystery of a Gun Smith . . . and the said George Rizer promise to give the said apprentice sixteen dollars as a freedom [dues].

[Washington County Apprentice Indentures, p. 384, Hall of Records]

MOUNT ETNA FURNACE

June 25, 1781 Deeport to Lee

May It Please Your Execellency I Received in the Course of Last Week Eighteen Peices of Cannon [from Mount Etna Furnace] Said to be Four Pounders but are not all of a Bore some of them will Carry a Six Pound Shot, The Guns are Very Fine To Judge from Appearance I think them Equal to any Ever Cast on the Continent But are By no means Proportioned or Proper for Field Peices as they are much too short & Heavy. They are only of the Common Length for Ship Guns, as I was not Informed wether Those are the Guns Intended for the Travelling Carriages I was Ordered to have made. Your Excell'y will Please to Inform me wether Those are the Guns Intended for Field Peices and wether the Whole or What Part is to be mounted on Proper Travelling Carriages. There is 4 Three pounders (which would Answer for Field Peices much Better) Offered in Exchange for 4 of Those Guns.

[AM, p. 315]

MUNCK, C. H. — Washington, D. C.

Munck, C. H.

1855	No. 361 Pennsylvania Avenue, opposite National Hotel, Washington, D.C.	manufacturer & dealer in GUNS Repairing neatly executed

[Directory, Washington, D.C., & Georgetown]

Munck, C. H.

1855	361 Pa av	gunsmith

[Directory, Washington, D.C., & Georgetown]

Munck, C. H.

1858	496 6th w	gunsmith

[Directory, Washington, D.C.]

Munck, C. H.

1858	N. 496 6th St near Penn Ave., Wash. D.C.	manufacturer & dealer in GUNS Repairing neatly executed

[Directory, Washington, D.C.]

Munck, C. H.

1860	496 6th w	gunsmith

[Business Directory of Washington, D.C., Georgetown & Alexandria]

Munck, C. H.

1862	496 6th west, h 476 9th west	gunmaker

[Boyd's Washington and Georgetown Directory]

Munck, C. H.

1862	6th west	gunsmith

[Business Directory of Washington, D.C., and Georgetown]

Munck, C. H.

1863	596 6th west	gunsmith

[Hutchinsons Business Directory, Washington, D.C., & Georgetown]

Munck, C. H.

1864	496 6th west	gunsmith

[Hutchinsons Business Directory, Washington, D.C., & Georgetown]

Munck, C. H.

1865	496 6th Ave.	gunsmith

[Boyd's Business Directory, Washington, D.C., Georgetown & Alexandria]

Munck, C. H.

1866	496 6th west	gunsmith

[Boyd's Business Directory, Washington, D.C., Georgetown & Alexandria]

MYERS, A. — Boonsboro

November 9, 1832

Be it known that A. Myers of Boonsborough, Washington County, and State of Maryland, have invented certain new and useful improvements in gunstock sawing machine.

[U. S. Patent Office]

McAINN (BARNARD & McAINN) Baltimore

These dealers' names along with "Baltimore Md" can be found on an 1860 style staff and field officers' sword in the author's collection.

McCOMAS, ALEXANDER Baltimore

Established himself as a smith in 1843 and became Maryland's largest producer. He produced everything from percussion shotguns, pistols, full and half stock longrifles to hammer and hammerless breech loading rifles and shotguns. Some of his rifles are elaborately decorated with silver inlays and some of his shotguns are high grade with fine engraving. His arms are of various qualities and are stamped "A McComas Baltimore." Alexander produced many guns and was known to have had at least two apprentices, John Clenney and Oscar Johnson.

1850	McComas, Alexander age	29 b.	Md.	gunsmith
	Mary	24	Md.	
	Clenney, John	20	Md.	gunsmith
	Johnson, Oscar	14	Md.	gunsmith

[Federal Census 1850, Baltimore]

McComas, Alexander

1845-47	51 S. Calvert St.	gun, rifle & pistolsmith
1850	51 S. Calvert St. & 53 Cheapside	gunsmith
1852	51 S. Calvert St.	gunsmith
1856-58	47 S. Calvert St.	importer & manufacturer of guns
1859-66	51 S. Calvert St. & 53 Cheapside	gunsmith
1870-84	51 S. Calvert St.	gunsmith
1885-1889	22 S. Calvert St.	gunsmith

[Baltimore city directories]

McCOMAS, Alexander, was born near Bel Air, Harford Co., Md. 27 Feb. 1821. His father, Preston McComas, and grandfather, Alexander McComas, were natives of the same county. His family were among the first settlers of the county. His father was elected High Sheriff of Harford County in 1833, and served for three years, to the entire satisfaction of the citizens. He married Hannah E. Gough, eldest daughter of Harry Gough, a native of England, who came to this country and settled at South Hampton, a well-known mansion near Bel Air. Her brother Harry D. Gough, represented Harford County in the Maryland Legislature for a number of years. He was also Clerk of the county about seven years. Harry Gough McComas, who was killed in the battle of North Point, and with Daniel Wells had the credit of killing General Ross, the British commander, was first cousin to Hanna E. Gough. The parents of Alexander were married October 10, 1809. They had eight children, of whom the subject of this sketch was the fifth, he being the second son. His educational advantages, like those of most of the country boys of that day, were limited to the elementary English branches; and as was the custom then, only availed of in the winter––the remainder of the year being given to work on the farm, or whatever else was needed. At thirteen years of age his father required him to make a selection of his future busi-nesss. He promptly decided to be a gunmaker. His father brought him to Baltimore and placed him with C. C. C. O'Brien, a leading gunmaker, with whom he remained about six years, until Mr. O'Brien's death. He then worked about three years in the same business with another gentleman. At the end of that time he found himself financially no better off than at the beginning; he therefore concluded to begin business on his own account, and rented a small portion of the warehouse, No. 51 South Calvert Street, July 1843, and opened a gunsmithing establishment on a very limited scale. His business career has been one of great prosperity, and is the result of the honorable business policy through which he has maintained and steadily strengthened his reputation. He commenced business at his present stand with a capital of twenty-one dollars. His house is now, and has long been the most notable in that line of business in Baltimore. Both as an importer and manufacturer he has maintained the highest standard in all lines of goods, and has made his house famous as a depot for the best guns and sportsmen's goods known in the world. The rifles and shot-guns manufactured by Mr. McComas have been in use for thirty-five years, and are widely and favorably known all over this country and Europe, and the McComas heavy guns made expressly for buffalo, bear, deer, geese, and duck, are not excelled anywhere in the world. At the Maryland Institute, Mr. McComas took the first premium and the first diploma for guns, a silver medal for gun-locks, a silver medal for rifles and pistols, and a gold medal for double-barrelled guns. His goods were also awarded a silver medal by the Metropolitan Mechanics' Institute. Mr. McComas has travelled extensively in the United States and Canada. Polit-ically, he has been a Whig; during the civil war a decided Union man. His mother died in 1831, when he was but ten years old, and, although nearly fifty years have passed since that event, her dying admonitions stand out before him as if written in letters of living light. He believes if mothers would bestow more attention on their sons in training them in the right way, to be just and honorable, there would be better and truer men in the land. He was married by the Rev. Henry Slicer, of the Methodist Episcopal Church, November 24, 1844, to Miss Mary A. Hahn, second daughter of the late William Hahn. By this union he has three children living, one son and two daughters. His son Harry H. is associated with him in his business.

[BIOGRAPHICAL CYCLOPEDIA, p. 252, 253]

McCOMAS, NICHOLAS A. Baltimore

In the Baltimore directories Nicholas A. McComas advertised that he was a "manufacturer in guns, rifles, pistols of every description."

1847-60 McComas, Nicholas A. gunsmith 44 W. Pratt St.
[Baltimore city directories]

McINTOCH, WILLIAM JR. Hagerstown

William Sr. was a property holder but no occupation is given nor can any trace of his son be found.

Sept. 8, 1799 This Indenture witnesseth that William McIntoch hath put his son William McIntoch apprentice to John Gonter Gun Smith to learn his art, trade and mystery after the manner of an apprentice to serve the said Gonter Smith from the day of this date hereof for and during and to the full end and term of four years and three months . . .
[Washington County Indentures, Hall of Records, p. 105]

McKIM, WILLIAM D. Baltimore

As a wholesale hardware merchant in the late flintlock and early percussion periods, McKim sold many locks marked with his firm's name to gunsmiths on the trade routes from Baltimore. Small flintlock plates often are marked "McK Bros Balto" and other plates large and small are marked "McK & Brother Baltimore" and "McKIM & BROTHER BALTIMORE." Between 1824 and 1833 he apparently was in partnership with M. M. Maslin who imported and sold locks marked with his own name. Percussion lock plates are marked "McKIM & Co."

McKim, William D.

1819	Holliday Street	

McKim & Maslin

1824-27	223 Baltimore Street	
1829-33	248 Baltimore Street	wholesale hardware merchants

McKim & Brother

1835	248 Baltimore Street	hardware merchants
1837-41	248 Baltimore Street	wholesale hardware merchants

[Baltimore city directories]

McLAUGHLIN, MATHEW Baltimore

Article of Agreement made this 31st day of January 1821, between Lieut. Nelemical Bader of the one part on behalf of the United States and Mathew McLaughlin of the City of Baltimore of the other part, Witnesseth that the said Mathew McLaughlin on this part binds himself to Build and complete in a Workmanlike manner for the said United States Eleven 6 pds Gun Carriages after a model pattent furnished by the authorities of the State of Maryland, the said Carriages being intended for the State of Maryland furnished by the United States as a part of their quota of arms.
[National Archives, Contracts for Ordnance & Ordnance Supplies, Vol. I, 1813-1828, p. 144.]

N

NEFF, P. Baltimore

The barrel of a British marked, flintlock fowling piece bears the signature of "Bird Nightingale & Neff." The firm continued to operate and advertised at the same location until 1830 under the name of Neff & Bird.

Bird, Nightingale & Neff

1819-23	108 Baltimore Street	fancy hardware merchants

[Baltimore city directories]

NESTLE, FREDERICK Baltimore

The early evidence of his work is found in heavy German type *Scheutzen* rifles. All these rifles are trimmed in silver and are marked "F. NESTLE BALTIMORE MD" or "NESTLE-Balto" on the large octagonal barrels and usually the lock.

Nestle, Frederick

1867	306 South Bond Street	gunsmith
1868-69	43 Lancaster Street	gunsmith
1870	56 Lancaster	gunsmith
1871-73	332 North Gay Street	gunsmith
1874-75	91 Low Street	gunsmith
1876	180 North Dallas Street	gunsmith
1877-79	125 Low Street	gunsmith
1880-84	132 Low Street	gunsmith

Nestle, Josephine, widow of Frederick

1885-87	132 Low Street	

[Baltimore city directories]

NETZ, JOHN Allegeny County

Netz, John age 35 born Germany gunsmith Allegeny County
[The spelling of the last name is not clear.]
[1850 Federal Census]

NEW YORK CLOTHING HOUSE Baltimore

New York Clothing House

1878-84	184 West Baltimore Street	I. Becker manager
1885-89	102 East Baltimore Street	

[Baltimore city directories]

NICHOL & GEORGE Hagerstown

Early residents of Hagerstown and its immediate vicinity from 1800 to 1805.

Nichol & George – gunsmith

[HISTORY OF WESTERN MARYLAND, p. 1062]

NICKERSON, CHARLES V. Baltimore

January 27, 1852
Be it known that I, Charles V. Nickerson of the city and county of Baltimore, and state of Maryland, have invented a

new and good improvement in Fire-Arms for Loading of the Breech . . . Dividing the stock at the junction of the barrel and breech, and mounting the barrel and that portion of the stock to which it is attached with a sheath or case upon a longitudinal bar or tongue projecting from the butt of the stock, as represented in the drawings, whereby the stock and barrel are allowed to have a movement from the breech for inserting the cartridge into the chamber thereof, and returned and locked by catch to confine them together . . .
[U. S. Patent Office, Patent #8,690]

NIGHTINGALE, SAMUEL — Baltimore

The barrel of a British proof, flintlock fowling piece is signed in script "Bird Nightingale & Neff."

Bird, Nithgingale & Neff

1819-23	108 Baltimore Street	fancy hardware merchants

[Baltimore city directories]

NOCK, JOSEPH

Nock, Joseph

1866	329 E. North	gunsmith

[Boyd's Business Directory, Washington, D.C., Georgetown & Alexandria]

NORRIS & BEATTY — Baltimore

A London marked, single barrel, percussion, half stock shotgun has survived with a lock marked "Norris & Beatty Baltimore."

Norris & Beatty

1835	212 Baltimore Street	hardware merchants
1837-42	212 Baltimore Street	importers of hardware

[Baltimore city directories]

NORRIS, RICHARD, SR. — Baltimore

Norris, Richard, Sr.

1810-16	197 Baltimore St.	hardware
1818-29	272 W. Baltimore St.	hardware

[Baltimore city directories]

NORRIS, RICHARD & SONS — Baltimore

Norris, Richard & Sons

1831-39	216 W. Baltimore St.	hardware

[Baltimore city directories]

NORRIS & BROTHER — Baltimore

Norris & Brother

1840-44	216 W. Baltimore St.	hardware
1845-48	272 W. Baltimore St.	hardware
1849-60	272 W. Baltimore St.	hardware & foreign commission merchants
1864	272 W. Baltimore St.	iron importers

[Baltimore city directories]

Large percussion pistols of nice quality with saw handle grips and octagonal barrels are known with engraved back-action locks marked "Norris & Bro. Maker Baltimore."

NORTHAMPTON FURNACE — Baltimore County

To be sold at Northampton Furnace, in Baltimore County, about 10 miles from Baltimore Town, 12 miles from the navigation of Gunpower River, 20 miles from Otterpoint on Bush River, and 30 miles from the Susquehanna lower ferry, Cannon of any size from two to twelve ponders, (eighteen may be had if applied for sometime before they are wanted). Also swivels, shot, salt pans, and plates, pots, kettles, dutch ovens, tea kettles, waggon-boxes, flat irons, etc. Guns made at this Furnace, have been allowed by the best judges, to be equal in quality to any yet made on the continent.
[Maryland Journal & Baltimore Advertiser, 31 July 1781]

NOTTINGHAM COMPANY FURNACE

July 8, 1776 Council to Charles Ridgely & John Westen
We have immediate occasion, on account of the Province, for some swivels and small cannon, from four-pounders downward and shall (as we are informed your furnace is in blast) be very glad to know whether you can cast them.
[American Archives, 5th Series, Vol. I, p. 114]

July 16, 1776 John Skinner to Council
I am favored with yours of the 8th instant concerning the casting some small cannon for the use of the Province, in answer to which I have to acquaint you that the Nottingham Company's furnace under my care blowed out about five weeks ago. The remainder of this year will be employed in providing stock for another blast, so that we shall not be able to get to work again sooner than sometime next January.
[American Archives, 5th Series, Vol. I, p. 364]

NUNEMAKER, GEORGE — Emmitsburg

A very beautiful relief-carved and inlaid longrifle signed on the barrel "George Nunnamacher 1797" was illustrated and described by Kindig (1960, p. 337-339). This rifle holds design features and details which suggest that it may have been made in northern Maryland, and perhaps in association with the Emmitsburg School.

Nunemaker, George 1798 1 front lot value 80 pounds, Emmitsburg.
[Assessment Book, Frederick County, 1798, Hall of Records, #19268]

O

OBERTEUFFER, CHARLES A. Baltimore

The 1850 Federal Census indicates that Charles Oberteuffer was born in Switzerland in 1811. He evidently moved to Maryland where he worked as a gunsmith and married Mary (born 1816 in Maryland). They had three sons born in Maryland: Charles (1836), Ulrich (1839), and Adila (1846).

In 1850 Charles Oberteuffer was listed as a gunsmith in Kensington. From 1842 through 1845 he is listed in the McElroy directory as a gunsmith in Philadelphia.

Oberteuffer, Charles A.

1837	Monument Street west of Eden	gunsmith

[Baltimore city directory]

O'BRIEN, CHARLES C. C. Baltimore

Alexander McComas was apprenticed to O'Brien at 13 years of age for six years until O'Brien's death.

O'Brien, Charles C. C.

1833	64 South Frederick Street	lock & gunsmith
1835-41	55 South Frederick Street	gun & rifle maker

[Baltimore city directories]

OEHM'S ACME HALL Baltimore

OEHM, CHARLES N. Baltimore

"Oehm & Co. Baltimore Md" is marked on several 1860 style staff and field officers' swords in the author's collection.

Oehm, Charles N.		
1860	150 Light Street	merchant tailor
Oehm, John L.		
1860	338 West Pratt Street	merchant tailor
Oehm, Charles N.		
1864	150 Light Street	merchant tailor
Oehm, Charles N. & Brother (Charles N. & Franz F.)		
1867-74	150 Light Street	merchant tailor & clothier
Oehm, Charles N. & Brother		
1875-77	150 Light Street & 230 West Pratt Street	
Oehm, Charles N. & Brother		
1878-84	230 West Pratt Street	
Oehm & Co. – Acme Hall Clothing House		
1885-86	209 West Baltimore Street	Alfred Haupt, manager
Oehm's Acme Hall Clothing House		
1889	5 & 7 West Baltimore Street	Charles H. & Charles W., proprietors

[Baltimore city directories]

OLIVER, JOHN Baltimore

Oliver, John

1810	37 George Street, Fells Point	gunsmith

[Baltimore city directory]

P

PAGENHARDT, C. A. Westernport

Pagenhardt, C. A. 1878 Westernport guns & pistols
[The Maryland Directory, 1878; J. Frank Lewis & Co. Publ.]

PAGE, GEORGE, & COMPANY Baltimore

In 1860 this firm did manufacturing and machine work, and probably made the pikes mentioned. By 1864 they also advertised naval stores at 111 Smiths Wharf.

. . . From the above Summary it will be seen that the expenditures for "Arms and Ammunition" reached the large amount of nearly $25,000. The articles purchased include . . . 414 Pikes from George Page & Co. . . .
[Baltimore City Correspondence Manuscripts, 1862, #695]

PARK, JOHN

August 27, 1776 Council of Safety
Ordered That said Treasure pay to John Park three pounds, ten shillings for a Gun.
[AM, p. 242]

PARKER, JOHN

August 12, 1776 Council of Safety
Ordered that the Treasurer of the Western Shore pay to John Parker 3 pounds 5 shillings for a Musket.
[American Archives, 5th Series, Vol. I, p. 1352]

September 2, 1776 Council of Safety
Ordered That the Treasure pay to John Parker three pounds, five shillings for a Gun.
[AM, p. 252]

September 23, 1776 Council of Safety
Ordered That W. Shore Treasure pay to John Parker forty five shillings for a Musquet.
[AM, p. 293]

Probably John Park and John Parker are the same man; however, for the sake of historical accuracy they are listed as found in the Maryland Archives.

PARKHURST (see CALVERT & PARKHURST)

PARSONS, HIRAM Baltimore

Parsons was employed by Samuel Colt and was instrumental in perfecting and producing several of the first Colt models in his shop.

Parsons, Hiram

1819	corner of North Eutaw Street near Dutch Alley	gunsmith & locksmith
1822-23	West side North of Franklin	gunsmith & locksmith
1824	corner of Wilk & Granby Streets	whitesmith & gunsmith
1831	Wagon Alley West of Park	brass founder

[Baltimore city directories]

PATTERSON, WILLIAM P. Baltimore

Patterson, William P.

1876-80	178 West Pratt Street	gunsmith
1881-85	168 West Pratt Street	gunsmith
1886-89	38 East Pratt Street	gunsmith

[Baltimore city directories]

PAYNE, E. SCOTT Baltimore

A medium grade, double barrel, hammer, breech loading shotgun has survived bearing the name "E Scott Payne & Bro Baltimore Md" on the rib and "Richard Moore" on the lock plates.

Payne, E. S. & R. K.

1876-88	174 North Gay Street	hardware
1889	366 North Gay Street	hardware

[Baltimore city directories]

PEABODY, JOHN J. Washington, D.C.

Peabody, John J.
1862 60 La. av gunsmith
[Business Directory, Washington, D.C., & Georgetown]

Peabody, John J.
1863 60 La. av gunsmith
[Hutchinsons Business Directory, Washington, D.C., & Georgetown]

Peabody, J. J.
1864 51 La. av gunsmith
[Hutchinsons Business Directory, Washington, D.C., & Georgetown]

Peabody, J. J.
1865 49 La. Ave. gunsmith
1866 414 D north gunsmith
[Boyd's Business Directory, Washington, D.C., Georgetown & Alexandria]

Peabody, John J.
1867 414 D north gunsmith
[Boyd's Business Directory, Washington, D.C., Georgetown & Alexandria]

Peabody, John J.
1868 414 D n gunsmith
John J. Peabody Gun & Locksmith Dealer in Powder, Shot, Balls, Percussion Caps and Cartridges of all kinds. Special attention paid to the repairing of Guns, Locks, etc. Also Bell-hanging promptly attended to at short notice.
[Boyd's Business Directory, Washington, D.C., Georgetown & Alexandria]

Peabody, John J.
1872 gunsmith
1872 683 D nw Guns and Sporting Tackle
1878 Peabody, John J. 619 Louisiana Ave. Washington, D.C. Gun & Locksmith & dealer in Fishing Tackle, Sporting Goods, &c.
[The Maryland Directory, 1878; J. Frank Lewis & Co., Publ.]

PHIPPS, JOHN — Washington, D. C.

1853 Phipps, John gunsmith E sdie 4[w btw M & N s
1855 Phipps, John gunsmith 396 4½
1862 Phipps, John gunsmith h 386 4½
[Washington, D.C. directories]

PIERSON,______ — Baltimore

1835 Pierson,______ gunsmith 66 Centre Market Place
[Baltimore city directory]

PIKESVILLE ARMORY — Pikesville

September 3, 1813 received from J. J. Henry Phila
297 gun screws
September 3, 1813 received from Mr. Derringer
125 tin chargers
150 Prickers & Brushes
7 screw wipers
September 10, 1813 received of Jonathan Waters
40 pair of Holsters
September 10, 1813 received of James Williamson
30 pair of Holsters
September 29, 1813 received of Jonathan Waters
100 Rifle Bayonet Scabbard & Belts
November 2, 1813 received from William Roft
13 Muskets 1 screw lock
November 12, 1813 received of J. J. Henry
722 Muskets at $12.75
110 Swords at $9.50
14 pair of Pistols at $13.50
690 Wipers
110 Screw drivers
100 bullet drawrs
November 15
280 Muskets
108 Swords
November 25
436 Muskets at $13.75
205 Swords at $9.50
93 Pistols at $6.75
25 Bayonets at 0.89
1 Bayonet reamer
4 Stamps letter M
November 29, 1813 received from Mr. Derringer Phila
108 Rifles with long bayonet
60 Pairs of pistols
December 11, 1813 received of Jonathan Waters
500 Cartridge boxes
500 Cartridge box belts
December 20, 1813
5 boxes of military stores
sword 4 of them from Bouchet Philadelphia
December 26, 1813 received from William Allen of N. Y.
2 boxes of Muskets containing 25 each
December 31, 1813 received of Irwin & Saith N. Y.
2 boxes containing 99 pair of pistols
December 31, 1813 received from William Allen N. Y.
260 Artillery swords
86 Horseman do
March 30, 1814 received of Jonathan Hutton
191 rifle pouches
March 31, 1814 received of James Williamson
1000 cartridge boxes with belts
24 bayonet belts
19 scabbard for rifles
April 4, 1814 received from Joseph Henry Philadelphia
29 boxes with 404 muskets
7 boxes with 375 cavalry swords
sold Mr. Henry 300 old gunlocks belonging to the State at $1 each
April 30, 1814 received of Jonathan Waters
100 Shoulder sword belts
May 9, 1814 [received from] Jonathan Hutton
30 Cavalry & Artillery Shoulder belts
[MARYLAND ARMORY BOOK, Vol. 165, 1812-1820, p. 156.]

PIPER, GEORGE — Emmitsburg

George Piper, who was born about 1783, apprenticed with John Armstrong (see p. 119, 120 for full text of indenture) beginning in 1801. After completing his indenture in October 1803 he may have worked in the southwestern part of nearby York County, Pa. At least two rifles bearing his name have survived. One of these is a fine-quality flintlock rifle with incised carving, a checkered wrist, and a nicely engraved patchbox, and the name "G Piper" in script on the barrel. A second rifle is known with "G Piper" engraved in script on the lock plate and "Jacob Sell" engraved on the barrel. This fine flintlock holds fine engarving and relief carving, and does not appear to be the work of Jacob Sell the younger whose name is on the barrel. The first piece mentioned has York County attributes, but the second one relates to the Emmitsburg school.

This Indenture made the 31st day of Aug. 1801 Witnesseth that Peter Piper hath this day placed and bound his son George Piper Aged 18 years . . . unto John Armstrong of Frederick County Maryland Gun Smith & Lock Maker to learn the said trade . . . of a Gun Maker and Gun Lock Maker which the said John Armstrong now useth . . . & serve from the day of the date hereof until the full end and Term of two years & two months from thence . . .And the said Master doth hereby promise . . . and give him one Good Rifle Gun worth Sixteen Dollars . . .
[Frederick Co. Apprentice Indentures, Hall of Records, p. 312]

PIPINO, JACOB — Baltimore

Only two arms have survived bearing his name. A target rifle and a longrifle, both of superb quality showing excellent engraving, are stamped on the barrel and lock "Jacob Pipino."

Pipino, Jacob

1847-54	18 Ensor Street	gunsmith
1856-71	20 Ensor Street	gunsmith

(ad, 1853) Gun, Rifles, Pistols manufacturer, constantly on hand & made to order every variety of above.
[Baltimore city directories]

PIPINO, JOHN L. — Baltimore

Pipino, John L.

1847	16 Mercer Street	gunsmith
1851	46½ Harrison Street	gunsmith
1867	Cross Street near Shop	gunsmith
1868	Peach Alley near Cross	gunsmith

[Baltimore city directories]

PIPINO, LEWIS — Baltimore

He is listed at the address of the shop of Jacob C. Pipino, so there is no doubt of some relationship between the two.

Pipino, Lewis

1860	20 Ensor Street	gunsmith

[Baltimore city directory]

POPPLEN, G*N JR. — Baltimore

In 1833 Nicholas was an importer of drugs and chemicals, later advertising as a druggist. By 1853 he had a factory that manufactured chemicals.
A good percussion, double barrel shotgun has been examined that bears the name "G*N Popplen Jr."

POULTNEY & TRIMBLE — Baltimore

Thomas Poultney and David B. Trimble advertised as partners from 1860 to 1876 at 200 West Baltimore Street. They employed Charles E. Sneider who developed the Smith carbine and exclusive rights were obtained by Poultney. During the War between the States the federal government purchased 30,362 Smith carbines from the firm. The actual manufacturing was subcontracted to three Massachusetts companies: American Machine Works, American Arms Company and the Massachusetts Arms Company. The carbines are marked "address Poultney & Trimble Baltimore U. S. A." A well engraved double barrel shotgun exists bearing the name "Poultney & Trimble."

Poultney & Trimble

1860-76	200 West Baltimore Street	gunsmith

Trimble & Kleibacker

1877-80	200 West Baltimore Street	gunsmith

Trimble

1881	200 West Baltimore Street	gunsmith

Poultney & Trimble

1868	also fishing tackle & Cutlery dealers	

[Baltimore city directories]

February 28, 1865
A small rod passed through the breech of each chamber parallel to and central with the bore, and has a longitudenal movement. It has a knob each side of the breech of the chamber; its forward part passed through the cartridge and explodes the priming situated at the base of the bullet; its forward knob, striking upon the case of the cartridge shell, expels it, its near knob limits the entrance of the pin.
[U. S. Patent Office, Patent #46,612]

May 14, 1867
Be it known that we, Thomas Poultney, of the city of Baltimore, in the State of Maryland, and Silas Crispin, of the city and State of New York, assignors to Thomas Poultney, have invented a certain improvement in the construction of Breech-loading Fire-Arms . . . consists in the peculiar manner of forming the breech-piece receptacle (in arms having a breech-piece hinged to swing upward and forward, and lying in a recess formed in the barrel,) so as to facilitate the insertion of the primed metallic cartridge, and afterward the expulsion of the shell . . .
[U. S. Patent Office, Patent #64,701]

PRATT, WILLIAM — Baltimore

Occupation was listed as a dental surgeon from 1845 to the late 1860's. By 1870 he was an engineer. He died in 1876. He used an 1815 Wickham musket in inventing a breech loading system. The action is marked "W Pratt Balto 1853 patent model." This invention may have been very similar to another patent or perhaps thought not to be successful because it was never patented.

PURUCKER, JOHN W. — Baltimore

Purucker, John W.

1885-86	74 Camden Street	gunsmith
1887	139 West Camden Street	gunsmith

[Baltimore city directories]

R

RABRECK, CHRISTOPHER Baltimore

July 17, 1776 Council of Safety
Christopher Rabreck of Baltimore-Town agrees with the Council to supply the Province with one thousand Priming and Brushes at 7 shillings 6 pence per dozen, to be delivered in two weeks from this day.
[American Archives, 5th Series, Vol. I, p. 1338]

READ, (REED, REID), ROBERT Chestertown

May 23, 1776 Smyth to Council Chester Town
Since my return home I have agreed with Robert Reed of this place blacksmith for 10 musketts for acct. of the Province at 85/ each they are all finished and only wait your Determination to confirm my Contract before they are delivered.
[AM, p. 440]

September 5, 1776 Council to Fiddeman
As your Lieutenant Mr Downes informs the Council that the Arms of your Company want a good many Repairs, we would have you halt at Chester Town until you can get them done. Mr Winter and Mr Robert Read are employed to repair Guns for the Province.
[AM, p. 255]

December 1, 1776 Smyth to Council Chester Town
This will be delivered you by Robert Read a Black Smith of this Town who waits on the Council to sell them 17 muskets which he has made and carries with him, they seem to be strong and substantial, he proposes contracting with your Board for a good many, he bears the character of an honest man, has some real property in this Town and I believe would be pretty punctual.
[AM, p. 498]

December 10, 1776 Council to Reid
Mr Winter has been with us, and is desirous of having 25 barrels that were delivered to you, it would oblige us if you would let him have them.
[AM, p. 517]

April 15, 1777 Council of Safety
Ordered That the Western Shore Treasure pay to Robert Reid eighty four pounds for Amt. of his Acct. on Contract for Gunmaking.
[AM, p. 212]

April 19, 1777 Council to Reid
The Council and I are very unwilling that you should decline the Business of making Arms, we would have you proceed and will if Things continue in their present Circumstances, give you the five Pounds ten shillings under a Confidence of your Diligence, and that you will do the Work faithfully though we look on it a very high Price. Recd. 30 Musquets.
[AM, p. 221]

April 27, 1777 Council of Safety
Ordered That Henry Hollingsworth deliver to Robert Reid of Chester Town one hundred and fifty Gun Barrels, provided he has a sufficiency, wherein to supply the other Works.
[AM, p. 226]

July 29, 1777 Council of Safety
Received of Mr Robert Reid thirty Musquets in Consequence of a former Agreement.
[AM, p. 320]

July 29, 1777 Council of Safety
Ordered That the Armourer Deliver to Robert Reid fifty Gun Bdles to Stock taking his Rect. for the same.
[AM, p. 320]

December 15, 1777 Council of Safety
That the said Treasure pay to Robert Reed one hundred and forty five pounds fifteen Shillings and three pence due to him per Acct. passd. and also the further Sum of fifty nine Pounds, twelve Shillings and six pence to be delivered over to Daniel Fisher due to him per Acct. passed.
[AM, p. 438]

March 5, 1778 Council of Safety
It appearing from Accounts Lodged by Robert Reid that the Public is considerable indebted to him which Accounts cannot now be settled It is ordered that the western shore Treasure pay to the said Robert Reid three hundred Pounds on Account.
[AM, p. 526]

READ, WILLIAM Baltimore

Read, William
1799-1801 between 11 & 13 Water Street gunsmith
[Baltimore city directories]

REASOR [RESOR], JACOB Frederick

September 15, 1777 Council of Safety
That the said Treasure pay to Jacob Reasor one hundred and eighty seven pounds ten shillings on Account of this contract for making Arms.
[AM, p. 376]

September 15, 1777 Council of Safety
Jacob Razor agrees with the Governor and Council to make and in a Workman like manner compleat and finish one hundred Musquets fixed with Bayonets, Steel Rammers Swivels priming wires and Brushes at three pounds fifteen shillings each the Mountings to be of Brass or Iron the Breech pin to be let into the solid so that on reboring the threads of the screw shall not be injured. that he will enter on the said Work within one Month from this Day and within two Months from this Day will deliver at least twelve and so from Month to Month thereafter at least twelve of the said Musquets the Deliveries to be at Frederick Town to Charles Beatty or such other person as shall be appointed by the Governor and Council. The Governor and Council to furnish Locks and such Barrells and Bayonets as were furnished to Nicholas White And also to advance one hundred and eighty seven pounds ten shillings part of the price on giving Bond with good Security to comply with this Contract.
[AM, p. 376]

August 2, 1781 Council to P. Thomas
. . . We are desirous of giving Reasor all reasonable Encourage-

ment and shall be glade to have an Account of what he has done, and a sample of the Arm he professes to make, the Number and the Terms . . .
[AM, p. 535]

August 19, 1782 Assembly Proceedings
Ordered that the western shore Treasure pay to Jacob Reasor a distressed Creditor one hundred & fourteen Pounds. twelve shillings and eleven pence specie in part of his Claim against the State on Account.
[AM, p. 238]

January 2, 1783 Assembly Proceedings
. . . there will be no Necessity for an Armourer; we think however, it would be right to employ Mr Razer to repair the Arms that are now out of order . . .
[AM, p. 335]

This Indenture made this 15th day of August 1785 Between Jacob Razor of Frederick Town and Frederick County . . . and George Jacob Schley of Frederick Town . . . Witnesseth that . . . Jacob Razor for . . . 372 pounds 7 shillings and 1 penny . . . doth . . . sell . . . unto . . . George Jacob Schley . . . a certain part of a Lott of Ground . . . in Frederick Town adjoining Ground belonging to Stephen Brenner . . . 60 féet [by] . . . 45 feet . . . Jacob Reaser
. . . At the same time came Margaret wife to said Jacob Razor . . .
[Frederick CCH, Land Record Book W. R. 7, p. 53]

This Indenture made the 15th day of August 1785 Between Jacob Reasor of Frederick Town Frederick County Gun Smith . . . and Christopher Edelen of Frederick Town . . . whereas the said Jacob Reasor is now justly and truly indebted unto . . . Christopher Edelen in the sum of 260 pounds . . . Now this Indenture Witnesseth that the said Jacob Reasor for the payment of the said sum of money . . . doth . . . sell . . . unto . . . Christen Edelen all that part of a Lott of Ground conveyed by John Jacob Schley to Jacob Reasor lying . . . in Frederick Town being part of Lott N° [blank] adjoining Ground belonging to Stephen Brenner . . . 60 feet . . . [by] 45 feet . . . Provided . . . that if the said Jacob Reasor . . . shall well and truly pay . . . to the aforesaid Christopher Edelen . . . the aforesaid sum of 260 pounds . . . with the legal interest thereon due on the 15th day of August 1787, then these presents and every Claim thereof to be void.
[Frederick CCH, Land Record Book W. R. 6, p. 51,52, 53]

Indenture made August 16, 1785 [bill of sale] between Jacob Reasor of Fredericktown, Frederk County in the providence of Maryland to Dr. Adam Fischer of the same county and providence . . . Sundry small smith tools, one reel and brake troft, one large German Bible, one Testament German, one vice, three bench hammers, one hand vice, one Leder, two rifling rods, two boring bitts for stocking, one iron brazo, five ruff barrels, 12 screw plates large and small, 2 drawing knives, 1 large chest, 1 frying pan, 1 upright trill, 1 ruff bubbard, 3 gun locks, 2 grundstones, 1 small turkey stone, 1 hand saw, 1 pennall saw, 4 plains, 1 bridgo-wredgo, 1 tin half gallon, 2 tin quarts, 1 proof for casting, 1 spade and howe, 1 little garden howe, 10 ruff gun stocks, 1 gluepot, 1 pair jaw clamps, sundry small bench tools, 1 large iron melting pot of nail iron, 1 rough iron melting pot of nail iron, 1 rough forged gun barrel, 1 turnip cutter, 1 cabbage d.d.
[Frederick CCH, Liber WR6, p. 60]

Wanted Immediately, One or two Journeymen Gunsmiths GOOD workmen may have constant employment, and generous wages, from Jacob Resor. Mercersburg, (Penna.)
May 14, 1808
[Maryland Herald and Hagers-Town Weekly Advertiser, 20 May, 17, 24 June 1808]

RIDDICK, _________

A gunsmith named David Riddick worked in Annville, Lebanon Co., Pa., about 1800. The tax list there for 1800 lists him as "David Reduk" according to Kauffman (1960, p. 324). The name also may be found as "Reddick." A beautiful relief-carved flintlock longrifle is known with architecture and decoration closely similar to the work of Peter Berry of Annville, and bearing the signature "D. Riddick." The gunsmith mentioned below may be the same person.

January 31, 1776 Council of Safety
The Council called on Messers Keener, Messersmith and Reddick to know what Quality of Musquets they had made for the use of this Province in Consequence of their Contracts with Mr. Alexander. Mr. Keener Reported that he had made thirty Mr Messersmith that he had made Twenty Seven, and Mr Riddick that he had made seventy.
[AM, p. 127]

REINHARDT, CHARLES C. Baltimore

He made side knives which were very popular during and after the Mexican War. They are marked "Reinhardt Balto."

Reinhardt, Charles C.

1837-40	24 Lombard Street	instrument maker
1842-50	9 Light Street	

Reinhardt, Charles C., & Co.

1851-55	9 Light Street	cutlers, surgical, dental instrument manufacturers
1856-64	7 Gay Street	

Reinhardt & Brothers (Wm. H. Reinhardt, J. Reinhardt)

1865-72	2 Anthony Street	cutlers & instrument manufacturers

[Baltimore city directories]

REMINGTON, E. & SONS Baltimore

Remington, E. & Sons

1876-77	47 North Charles Street	gunsmith

1878-79	47 German	gunsmith
1880	27 North Charles & 23 South Calvert Streets	gunsmith
1881-82	21 South Howard & 97 Lexington Streets	gunsmith
1885-86	7 South Charles Street	gunsmith

[Baltimore city directories]

RESOR, PETER Hagerstown

Peter Resor was one member of a gunsmithing family whose name is variously spelled in the old records: Roesser, Roeser, Roesor, Reeser, Razor, Reasor, Rosser, and Resor. Peter was one of at least three children of Mathias Roesser, a gunsmith of Lancaster, Pennsylvania. Mathias was born in 1708 and apparently learned gunsmithing in Europe. He arrived on a ship from Rotterdam on 1 September 1736, and one of his shipmates was Jonathan Hagar, gunsmith, who founded Hagerstown. Mathias Roesser settled in Lancaster and bought property there in 1740. He had a number of children, the oldest of whom was Jacob. His daughter Catharine married Jacob Kraft, a gunsmith who worked at York, Pennsylvania. Peter Resor was another son of Mathias. Mathias died in 1771 and Peter acquired his father's residence, but he sold it later to Jacob Kraft. Peter worked as a gunsmith in Lancaster during the Revolutionary War period, at least between 1776 and 1784, and his son Jacob Resor, who became a gunsmith, was born there in 1784.

About 1785 Peter Resor and his family moved to Hagerstown, where Peter continued in the gunsmithing trade. His children Johan and Elizabetha were baptized in the Zion Reformed Church there on 22 February 1789 and 17 October 1793 respectively. By 1810 Peter and his family had moved up the valley about 15 miles to Mercersburg in Franklin County, Pennsylvania. His last years were spent in Lancaster where he died in 1823.

The above information about the Resor-Roesser family is abstracted from a longer account presented by Henry J. Kauffman (1960) in his book THE PENNSYLVANIA–KENTUCKY RIFLE.

A few flintlock longrifles signed "P. Resor" or "Peter Resor" are known. Some hold relief carving in the Lancaster style, and most have four-piece brass patchboxes with daisy-head finials and no piercings.

Reser, Peter Washington County
2 males of more than 16 years
4 males under 16 years
3 females
[1790 Federal Census]

Indenture made October 30, 1790 [bill of sale] Know all men by these presents that I Peter Riechter of Elizabeth Town in Washington County in the state of Maryland convey to Peter Resor of the aforesaid in consideration of the sum of 50 pounds, 16 shillings, 9 pence, goods and chattles following one complete set of smith tools, furniture, one cow, 2/3 of wheat crop, etc.
[Washington CCH, Liber G, Folio 271]

Indenture made November 2, 1791 [mortgage] between Jacob Rohrer of Elizabeth Town in Washington County in the state of Maryland of the one part and [to] Peter Rezor of Elizabeth Town in Washington County in the state of Maryland of the other part paid in hand of current money 10 pounds . . . Tract or parcel of land lying and being in Elizabeth Town aforesaid it being two lots distinguished by lot number 45 and 46.
[Washington CCH, Liber G, Folio 536]

Indenture made April 30, 1794 [deed] between Jacob Rohrer of Elizabeth Town in Washington County in the state of Maryland of the one part and [to] Peter Rezor of Elizabeth Town in Washington County in the state of Maryland of the other part . . . Tract or parcel of land lying and being in Elizabeth Town aforesaid it being two lots distinguished by lot number 42 and 43.
[Washington CCH, Liber H, Folio 676]

RETNER & THOMAS Baltimore

A walnut longrifle in the author's collection has a percussion lock which is marked "Retner & Thomas Baltimore."

RETTBERG, JACOB Hagerstown

Rettberg, Jacob	1878	Hagerstown	gun & locksmith
Rettberg, Jacob	1880	Hagerstown	gun & locksmith

[The Md. Directory, 1878, 1880; J. F. Lewis & Co. Publ.]

				born	father b.	mother b.
Rettburg, Jacob	age 64	lock & gunsmith		Hessed	Hessed	Hessed
Louisa	56	wife		"	"	"
Rosa	35	daughter		Md.	"	"

[Federal Census 1880, Hagerstown, Washington Co.]

RETTBERG. LEWIS Hagerstown

Rettberg, Lewis 1882 Hagerstown gun & locksmith
[The Maryland Directory, 1882; J. Frank Lewis & Co. Publishers]

REUTER, A. Baltimore

Reuter, A.
1845 42 Harrison Street lock & gunsmith
[Baltimore city directory]

Andrew Reuter age 41 sex M occ hardware merchant birth Germany
[1850 Federal Census]

REYNALL, RICHARD Baltimore

He apparently came from Philadelphia, where he advertised as a gun maker in 1797. His last name was also spelled Reynell, Reynolds, and Reynells.

THE ROESSER—RESOR FAMILY OF GUNSMITHS

Sources: *Public records of Washington Co., Md.*
Kauffman, Henry J. (1952) EARLY AMERICAN GUNSMITHS 1650–1850
Kauffman, Henry J. (1960) THE PENNSYLVANIA KENTUCKY RIFLE

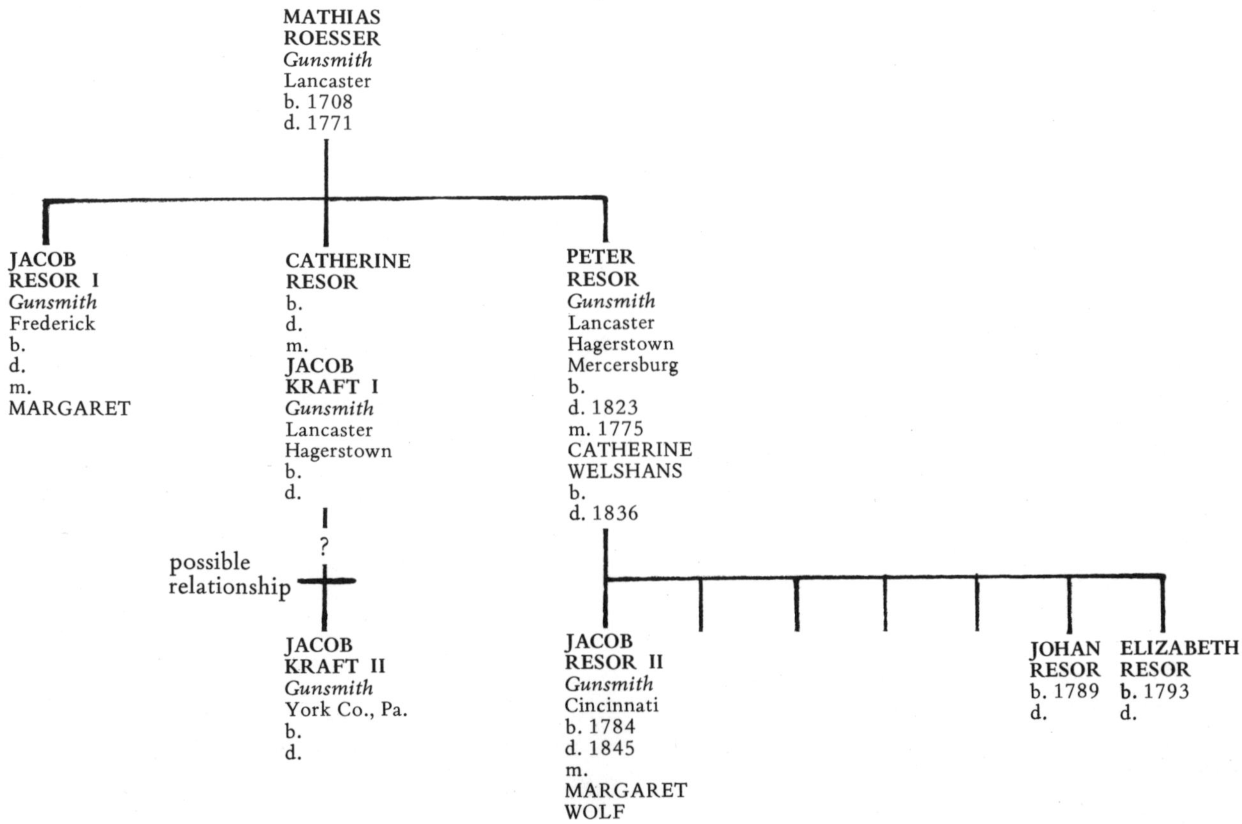

Reynall, Richard

1801	56 Water Street	gunsmith
1816-29	35 Water Street	gunsmith

[Baltimore city directories]

RICHARDS, JACOB — Cecil County

Know all men by these Presents, that it is hereby mutually agreed, by and between the United States by the agency of George Bromford, Lieut. Col. of Ordnance, deputed therefore by the Dept. of War, and Jacob Richards, of Cecil County, Maryland, that the said Jacob Richards shall and will manufacture, and deliver within two years from the date hereof Thirty Five Thousand Musket Stocks for the use of the Armories of the U. States. July 1818.
[National Archives, Contracts for Ordnance & Ordnance Supplies, 1813-1828, Vol. I, p. 75]

RIDGELY, HENRY

August 13, 1776 Council of Safety
Ordered That the Western Shore Treasure pay to Henry Ridgely jurn. four Pounds, ten Shillings, for a Musquet & Bayonet.
[AM, p. 198]

April 3, 1777 Council of Safety
Ordered That the Western shore Treasure pay to Henry Ridgely one hundred and thirty two pounds, ten shillings for Amt. of his Acct.
[AM, p. 198]

RIECHTER, PETER — Hagerstown

Reichter sold "smith tools" to Peter Resor and possibly was a gunsmith.

Indenture made October 30, 1790 [bill of sale] Know all men by these presents that I Peter Riechter of Elizabeth Town in Washington County in the state of Maryland convey to Peter Resor of the aforesaid in consideration of the sum of 50 pounds, 16 shillings, 9 pence, goods and chattles following one complete set of smith tools, furniture, one cow, 2/3 of wheat crop, etc.
[Washington CCH, Liber G, Folio 271]

Indenture made November 11, 1790 [deed] between Peter Riechter of Elizabeth Town in Washington County in the state of Maryland of the one part and [to] Peter Resor of Elizabeth Town in Washington County in the state of Maryland of the other part . . . Tract or parcel of land lying or being in Elizabeth Town one brick house with house hold furnishing and ½ lot adjoining house for 180 pounds paid to me in hand current money.
[Washington CCH, Liber G, Folio 274]

RIGGS, GEORGE W. — Baltimore

Born in Montgomery County on 8 August 1777, the son of Samuel and Amelia Dorsey Riggs. George started his work in Georgetown in 1805 before coming to Baltimore. He retired to Woodville where he died on 26 August 1864. He was a talented craftsman. His silver hilted swords are rare and none has been found under the name of Riggs & Griffith.

Riggs, George W.

1812-14	74 Baltimore Street	jewellery & fancy store
1819-40	260 Baltimore Street	silversmith

Riggs & Griffith

1816-18	74 Baltimore Street	silversmith

[Baltimore city directories]

RILEY, JOHN W. — Baltimore

Riley, John W.

1873	30 South Eutaw Street	gunsmith
1874-75	102 South Eutaw Street	gunsmith
1878	224 Hollins Street	gunsmith

[Baltimore city directories]

RIZER FAMILY

Four members of the Rizer family worked as gunsmiths in Maryland, and it is likely that there were others. The family name generally is spelled "Rizer" in the public records, but sometimes it appears as "Riser," and less frequently "Risor" or "Reiser." There does not seem to be any connection with the gunsmithing family of Resor or Reasor, at least as far as mis-spellings in the public records are concerned.

Our knowledge about the Rizer family begins with Martin (I) Rizer who was living in Martinsburg, Virginia (now W. Virginia) in 1782 when the first surviving tax assessment records were taken. Martin (I) apparently was the father of the Cumberland gunsmiths Martin (II) Rizer and George (I) Rizer, Sr., and he may have been the father of Matthias (I) Rizer and Anthony Rizer. Martin (I) moved to Cumberland in 1793 and died there in 1796. During the 19th century the Rizer family in Martinsburg diminished, but in Cumberland it flourished.

Today there are a few members of the Rizer family still living in Cumberland. In the course of doing this research, contacts were made with some of these family members. With the exception of George Huth Rizer, they had no knowledge of the early history of their family. George Huth Rizer was interviewed in the fall of 1975. He was born in Piedmont on 13 February 1895, the son of Oliver Mortimer Rizer born 16 Nov. 1859, and the grandson of George Washington Rizer born 22 Feb. 1833. But it is not known how this branch of the family ties in with the rest.

In the records presented here, the designation (I) is used with the earliest Martin Rizer. The two gunsmiths of Cumberland are designated as Martin (II) Rizer, Sr., and George (I) Rizer, Sr. Tax records of Martinsburg before 1793 designate Martin, Sr., and Martin,

THE RIZER FAMILY OF GUNSMITHS

Sources: Public records of Berkeley Co., W. Va., Allegany Co., Md., Nelson Co., Ky.
Compiled by George Shumway, Daniel D. Hartzler, Thomas A. Strohfeldt

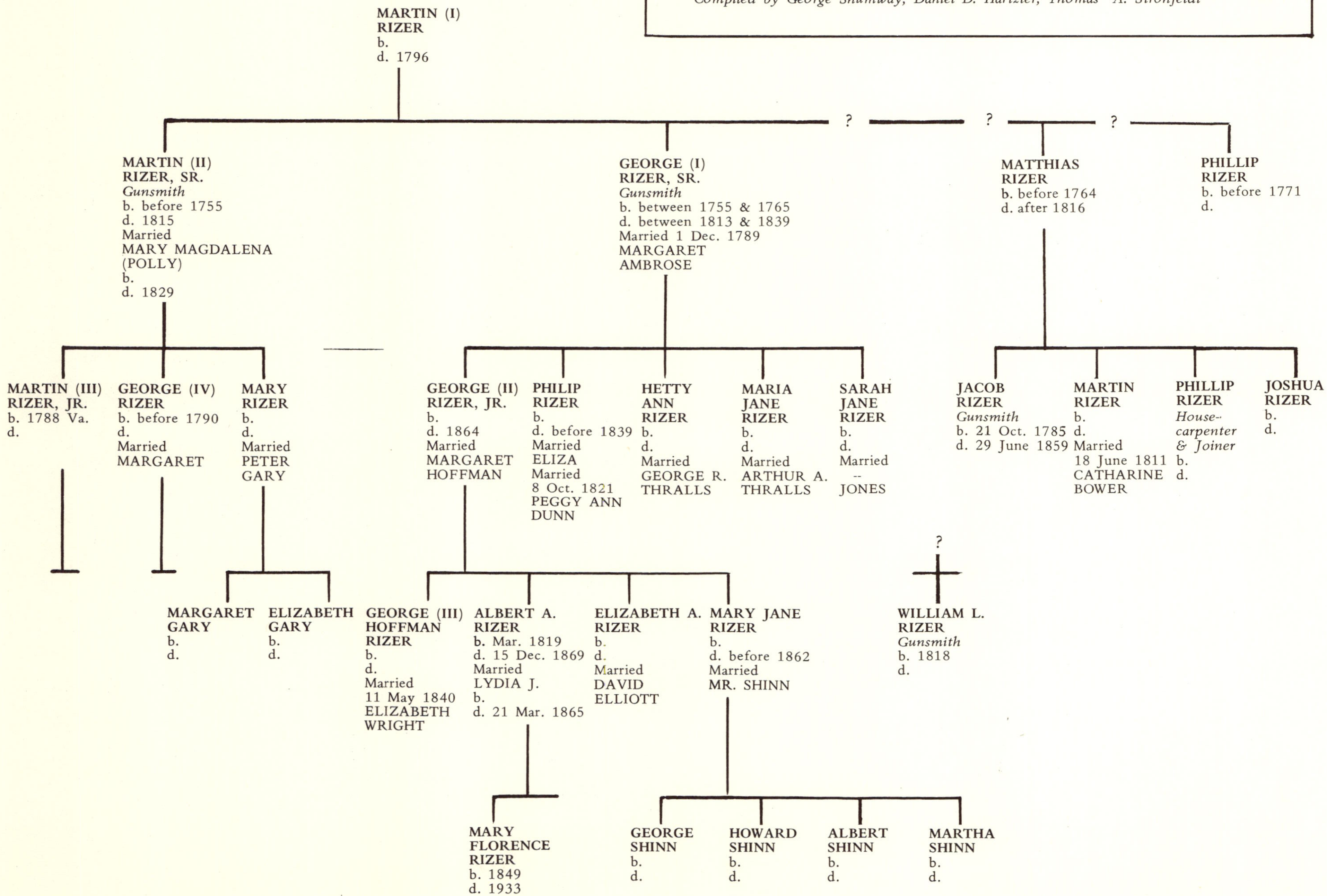

Jr., doubtless referring to Martin (I) and Martin (II) respectively.

RIZER, GEORGE (I), SR. Cumberland

George (I) Rizer, Sr., and Martin (II) Rizer, Sr., were the principal gunsmiths of Cumberland in the flintlock period, but relatively little is known about their gunsmithing activity. A number of fine-quality longrifles by Martin (II) Rizer are known, but signed specimens by by George (I) Rizer are unknown to the author. In May 1800 Jacob Rizer began an apprenticeship under George Rizer which apparently was completed in 1805. It is reported that Ben Troutman began his apprenticeship with George Rizer in 1805 but confirmation remains to be found. And in 1813 Josiah Moss began his apprenticeship with George Rizer.

In trying to learn something about the life and life-spans of the gunsmiths it was necessary to get involved with the inter-relationships of the entire Rizer family, because in the period between 1790 and 1850 there were four men named George Rizer and four men named Martin Rizer to contend with, and the public records often do not distinguish clearly between them. The accompanying chart shows the basic family relationships.

The two gunsmiths, George (I) Rizer, Sr., and Martin (II) Rizer, Sr., apparently were brothers who came to Cumberland from nearby Berkeley Co., Va., in 1793. A contemporary of theirs in Martinsburg and Cumberland, Matthias Rizer, may have been a third brother.

A man who presumably was the father of George (I) and Martin (II) died in Cumberland in 1796. His name was Martin Rizer, designated here as Martin (I), and his estate was administered by George (I) and Martin (II), the first account being filed 12 Oct. 1796. His estate inventory was small and typical for the time, and contained nothing to suggest that he too was a gunsmith. Research in the tax lists of Berkeley Co., W.Va., shows that Martin (I) Rizer paid taxes there from 1783 until 1793, so 1793 probably is the year in which Martin (I) and his sons George (I) and Martin (II) arrived in Cumberland.

The Rizers were active and respected members of the bustling community of Cumberland. George (I) served as a Justice of the Peace, and invested in real estate. The year of birth for George (I) is unknown, and at best it can be surmized that it was in the period between 1755 and 1770. George (I) had five children who survived him: George Jr. (II), who died in 1864; Philip, Hetty Ann who married George R. Thralls, Maria Jane who married Arthur A. Thralls, and Sarah Jane who married a man named Jones. From the existing records it is difficult to determine the year or even approximate year of the death of George (I). He left no will and there are no administration records in his name either in Allegany County or in Washington County. At best it can be stated that he died between 1813 and 1839. The first indenture recorded in Allegany County in the name of George Rizer is dated 1793. Later in his life he apparently owned land in Washington County. George Rizer and others in his family were slave holders.

By this deed dated 13 April 1793 George Rizer bought from Thomas Beall of Samuel, of Allegany County, for 10 pounds, lots No. 213 & 231 in Cumberland, each 60 feet in front.
[Land Record Book A, p. 286, Allegany CCH]

By this deed dated 9 Feb. 1795 George Rizer of Allegany County bought from Thomas Beall of Samuel, of Allegany County, for 30 pounds, part of a tract of land called "Walnut Bottom" containing 3 acres.
[Land Record Book A, p. 522, Allegany CCH]

The First Account of Martin & George Rizer Administrators of Martin Rizer . . . of Allegany County deceased . . . 12 October 1796.
[Allegany CCH, Administration Account Book A, p. 14]

Cumberland Town Hundred:			
1798	George Rizer		117/10/0
1800	George Rizer		-
1801	George Rizer		117/10/0
1802	George Rizer		95/-/-
1803	George Rizer		95/-/-
Fourth District, Cumberland Hundred:			
1802	George Rizer	Long Meadow 68 acres	11/6/8
1803	George Rizer	Long Meadow 68 acres	11/6/8
Third District, Cumberland Town Hundred:			
1798	George Rizer	2 horses	12/10/-
		3 black cattle	11/5/-
		other property	12/10/-

[Allegany CCH, Tax Assessment Book 1798]

By this Indenture dated 13 November 1801 George Rizer of Allegany County sold to Thomas Blaire of Allegany County for 150 pounds a portion of a tract of land called Walnut Bottom containing 3 acres. The deed was acknowledged by George's wife Margaret.
[Allegany CCH, Land Record Book C, p. 411]

I, George Rizer, one of the Justices of the Peace for Allegany County and State of Maryland do hereby certify that on the 8th day of February 1805, George Bruce, Esqr, duly commissioned Register of Wills for Allegany County, took before me the several oaths required by law, and signed a declaration of his belief in the Christian Religion — In Witness whereof I

have hereunto set my hand and seal the day and year above written. George Rizer
[Allegany CCH, Will Book A, p. 151]

Messrs George & Martin Rizer
Cumberland, Maryland
Sir – A letter has been received from the War Department from which it appears that you have communicated to R. Nelson, Esquire your wish to manufacture rifles for the U. States. Our practice is to obtain some persons willing to furnish supplies, their written proposals. If you will send me to Philadelphia for 50 or 100 stands of good plain Rifles fit for the Indian Service, to pass a strict inspection & the time & place you will engage to deliver any parcel or number with certainty, I will file it and apply to you on any further occasion that may offer. My last contract was for $10 Dollars for the rifle complete, to be equal in quality, comparing a Rifle to a musket, with the Charlesville musket and used by our Army and made in Harpers Ferry.

I am, Sir Tench Coxe
196 Spruce St.
Phila. 19 April 1806

[National Archives]

The Second Account of Martin & George Rizer Administrators of the Estate of Martin Rizer Deceased . . . 8 February 1807 . . .

Current money paid to

Catharine Rizer	pounds	4	8	0
Catharine Rizer		11	0	6
Philip Rizer		1	12	6
Christena Rizer		18	6	6
Philip Rizer		8	15	9
Mathias Rizer		6	12	9
Mathias Rizer		15	11	8
Philip Rizer		7	10	0
Mathias Rizer		15	0	0
Philip Rizer		9	7	6
Philip Rizer		7	10	0

[Allegany CCH, Administration Account Book A, p. 98]

In 1810 the household of George Rizer of Cumberland contained 1 male over 45 years, 1 male between 26 & 45, 3 males between 16 & 26, 1 male between 10 & 16, 3 males under 10; 1 female between 26 & 45, 1 female between 16 & 26, 2 females between 10 & 16.
[Federal Census, 1810, Allegheny Co., Md.]

The Fourth Account of Martin & George Rizer Admr of Martin Rizer Senr . . . 13 October 1813.
[Allegany CCH, Administration Account Book A, p. 167]

By this Indenture dated 3 October 1839 (quit claim) "Eliza Rizer widow of Philip who was a son of George, George R. Thralls and Hetty Ann his wife, Arthur A. Thralls and Maria Jane his wife, Sarah Jane Jones of the one part . . . sell . . . to George Rizer of the other part all children of George Rizer deceased . . . tract or parcel of land lying . . . in Washington County . . . containing 294 acres . . .
[Washington CCH, Land Record Book UU, p. 573]

RIZER, MARTIN SR. (II) — Cumberland

Several fine-quality flintlock longrifles have survived which bear the script signature of M. Rizer. The flowing lines and good composition that he used for his relief carving and for his engraved figures show that he was a master craftsman. Apparently he learned the trade in Berkeley County, Virginia, before moving to Cumberland in the early 1790's. For comments about the Rizer family in Cumberland, see the previous listing for George Rizer Sr. and the chart of family relationships.

Martin (II) Rizer, Sr., is no easier to follow through the public records than is George (I) Rizer, Sr., who presumably was his brother, for in the period between 1795 and 1850 there were at least 4 and probably 5 men in Allegany County named Martin Rizer. One of the few clues to identification of the various Martins is the name of each man's wife, which usually was indicated when real estate was sold. The desigantions of Senior and Junior associated with certain Martin Rizers in the public records do not necessarily clarify matters, for it appears that the designations of Junior was applied to two or three of them.

In 1796 George (I) and Martin (II) Rizer jointly administered the estate of Martin (I) Rizer, who presumably was their father. Like other members of the Rizer family, Martin (II) Rizer, Sr., was an active man in the community as a craftsman, as a real estate investor, and probably in community affairs.

Martin (II) Rizer, Sr., the gunsmith, may have been born before 1755, for the 1800 census shows his household to contain one male over 45 years, and none between 26 and 45 years. His son, Martin (III) Rizer, Jr., was born in Virginia in 1788 according to the Federal Census for Cumberland for 1850, and he was therein listed as a cabinet maker. A property tax assessment book for Allegany Co. for the years 1798–1803 lists taxable personal property for Martin Rizer in the Third District, Wills Town Hundred, and for George Rizer in the Fourth District, Cumberland Town Hundred, but in listing real estate, George Rizer and Martin Rizer, Jr., are listed under the Fourth District, Cumberland Hundred. The only way to square these two Martins with other data is to assume that the references are to the same person, it being not many years since the death of the elder Martin (I) Rizer.

In trying to distinguish between the various Rizers named Martin on the basis of the names of their wives, further complications arise. When a man sold property it was customary for his wife to acknowledge the sale, and her name and acknowledgement was recorded as part of the deed. It appears that the wife of Martin (II) Rizer, Sr., was named Mary Magdalena and that she sometimes was called Polly. In the 19th century it was not uncommon for a woman named Mary to use the nickname of Polly. Mary Magdalena Rizer outlived her husband and died in 1829.

Martin (II) Rizer, Sr., and his wife Mary Magdalena had three children, it can be deduced from various deeds plus Mary's will. Martin (III) Rizer, Jr., was born in Virginia in 1788, married Rosanna or Rosina Gephart, and worked as a wagon maker, plough maker, and cabinet maker. It is likely that he made some guns also. A second son was George (IV) Rizer and he married a woman named Margaret. The third child was a daughter, Mary, who married Peter Gary.

Martin(II) Rizer, Sr., died in 1815, without a will. Administration accounts for his estate are available, which include an inventory appraised 4 Jan. 1816, but other than a set of blacksmithing tools there is nothing in it to suggest gunsmithing activity.

The following property tax assessment records for Berkeley County, Va., involve both Martin Rizer, Sr. [Martin (I)], and Martin Rizer, Jr. [Martin (II)]. The father, Martin (I), owned the 100-acre tract, while Martin (II) had the single lot.

1782	Riser, Martin	1 lott	tax 0/2/0
1783	Riser, Martin	1 lott	tax 0/2/0
	Riser, Martin	100 acres	tax 0/2/8
1784	Riser, Martin	1 lott	tax 0/2/0
	Riser, Martin	100 acres	tax 0/2/8
1785	Riser, Martin	1 lot	tax 0/2/0
	Riser, Martin	100 acres	tax 0/2/8
1786	Riser, Martin	1 lot	tax 0/1/6
	Riser, Martin	100 acres	tax 0/2/0
1787	Riser, Martin	1 lot	tax 0/14/0
	Riser, Martin	100 acres	tax 0/3/11
1788	Riser, Martin, Junr	1 lot	tax 0/8/0
	Riser, Martin, Senr	100 acres	tax 0/4/0
1789	Rizer, Martin	1 lot	tax 0/3/0
	Rizer, Martin, Senr	100 acres	tax 0/2/0
1790	Rizer, Martin, Junr	400 acres	tax 0/4/0
1791	Rizer, Martin, Senr	100 acres	tax 0/1/0
	Rizer, Martin, Junr	400 acres	tax 0/4/0
1792	Rizer, Martin, Senr	100 acres	tax 0/1/0
	Rizer, Martin, Junr	400 acres	tax 0/4/0
1793	Rizer, Martin, Senr	100 acres	tax 0/0/8
	Rizer, Martin, Junr	400 acres	tax 0/2/8

[Property Tax Assessment Records, Berkeley Co., W. Va.]

By this deed dated 3 August 1784 Martin Rizer of Berkeley Co., state of Virginia, bought from John Miller a parcel of land in Berkley Co. on a branch of Sleepy Creek, part of a larger tract of 400 acres which was granted to William Barley.
[Berkeley CCH, W. Va., Deed Book 6, p. 278]

By this deed dated 17 March 1789 Martin Rizer of Berkeley Co., Va., bought from Adam Stephen, Esq. of Berkeley Co. . . . land . . . in the town of Martinsburg marked on the general plan of the said Town Number 59 . . . standing in the corner between Burke Street and Spring Alley . . .
[Berkeley CCH, W. Va., Deed Book 6, p. 278]

By this deed dated 14 April 1793 Martin Rizer of Berkeley Co., Va., bought from George Dick of Berkeley Co. a tract of land on the middle fork of Sleepy Creek in Berkeley Co.
[Berkeley CCH, W. Va., Deed Book 11, p. 295]

By this deed dated 16 Sept. 1793 Martin Rizer and Mary his wife sold to George Dicked of Berkeley Co. . . . lot Number 59 . . . in the Town of Martinsburg. Mary signed with her mark.
[Berkeley CCH, W. Va., Deed Book 11, p. 292

By this deed dated 17 June 1794 Martin Rizer and Mary his wife of Berkeley Co. sell to Peter Harbourt of Lowden Co. in Va. the tract of land on the middle south fork of Sleepy Creek which Rizer purchased 14 April 1793. Mary signed with her mark.
[Berkeley CCH, W. Va., Deed Book 11, p. 568]

By this Indenture dated 22 June 1795 Martin Rizer of Allegany County bought from Joshua Scritchfield of Bedford County, Pennsylvania, for 350 pounds part of two four-acre tracts of land, containing 7 acres.
[Allegany CCH, Land Record Book B, p. 48]

The First Account of Martin & George Rizer Administrators of Martin Rizer . . . of Allegany County deceased . . . 12 October 1796.
[Adm. Acct. Book A, p. 14, Allegany Co. CH]

By this Indenture dated 27 December 1797 Martin Rizer Jr. bought from Jacob Corns of Allegany County, for 150 pounds, two pieces of land, Lot No. 3352 of 50 acres and "Corns Addition" of 69 acres.
[Allegany CCH, Land Record Book B, p. 537]

By this Indenture dated 27 December 1797 Martin Rizer sold to Christian Albright, for $750, Lot No. 3352 containing 50 acres. Magdalena Rizer, wife of Martin Rizer, acknowledged the sale.
[Allegany CCH, Land Record Book G, p. 89]

Third District, Wills Town Hundred:

1798	Martin Rizer	24/4/0
1799	Martin Rizer	24/4/0
1800	Martin Rizer	24/4/0
1801	Martin Rizer	24/4/0
1802	Martin Rizer	26/4/7½

Third District, Wills Town Hundred:

1798	Martin Rizer	1 horse	1	15/-/-
		4 black cattle		8/-/-
		other property		12/10/-

Fourth District, Cumberland Heights:

1802	Martin Rizer, Junr	
	part of Prospect, 17 acres	3/8/-
1803	Martin Rizer Junr	
	part of Prospect, 17 acres	3/8/-

Cumberland Town Hundred:

1800	Martin Rizer Junr Lot.No. 5	10/-/-
1801	Martin Rizer Junr Lot No. 5	10/-/-
1802	Martin Rizer Junr Lot No. 5	10/-/-
1803	Martin Rizer Junr Lot No. 5	10/-/-

Allegany CCH, Tax Assessment Book 1798]

In 1800 the household of Martin Rizer [Sr.] of Wills Town [near Cumberland] contained 1 male over 45 years, 1 male between 16 & 26, 2 males between 10 & 16, 2 males under 10; 1 female between 16 & 26, and 2 females between 10 & 16.
[Federal Census, 1800, Allegheny Co., Md.]

By this Indenture dated 12 October 1803 Martin Rizer Junr bought from Peter Gephart, for 5 shillings, a small piece of land containing 5720 square feet, at the corner of Bedford and Liberty Streets in Cumberland Town.
[Allegany CCH, Land Record Book D, p. 89]

By this Indenture dated 12 October 1803 Martin Rizer Junr sold to George Blocker of Allegany County, for $35, parts of tracts of land called Prospect and Ridge.
[Allegany CCH, Land Record Book D, p. 114]

By this Indenture dated 10 November 1804 Martin Rizer sold to Jonathan Cox, for 425 pounds, part of two four-acre tracts called Walnut Bottom, containing 7 acres, the land having been bought from Thomas Beall of Samuel on 11 February 1786.
[Allegany CCH, Land Record Book D, p. 252]

Messrs George & Martin Rizer
Cumberland, Maryland
Sir – A letter has been received from the War Department from which it appears that you have communicated thro R. Nelson, Esquire, your wish to manufacture rifles for the U. States. Our practice is to obtain some persons willing to furnish supplies, their written proposals. If you will send me to Philadelphia for 50 or 100 stands of good plain Rifles fit for the Indian Service, to pass a strict inspection & the time & place you will engage to deliver any parcel or number with certainty, I will file it and apply to you on any further occasion that may offer. My last contract was for $10 Dollars for the rifle complete, to be equal in quality, comparing a Rifle to a musket, with the Charlesville musket and used by our Army and made in Harpers Ferry.

I am, Sir Tench Coxe
Purveyor's Office
196 Spruce St.
[National Archives] Phila. 19 April 1806

Roger Perry, Esqr
Cumberland, Maryland
Sir – I am much obliged by your attention in speaking to Mr. Riser about the Rifles. I now send you by Mr. Isaac Beals one of our Indian Rifles you will please to remark to him that this Gun is not intended as a complete sample but only to show the general manner in which we wish them furnished. The faults in the Gun I send are the Rifles are too shallow, the lock is too large and stiff and the mountains are not well secured as they should be. with a proper amendment of these faults & to carry from 40 to 60 Balls to the pound. I should give him an order for an hundred Guns at $11 each. you will be pleased have this gun returned by some of the first Boats after Mr. Riser has examined it. Very Respectfully I am S'r Your Mo Obt.

John Mason
16 April 1808 Office of Indian Trade
[National Archives] Georgetown, Wash.

Roger Perry, Esq.
Cumberland, Maryland
Sir – I received in due course the letter you did me the favor to write me on the 30th April, at that time, as I had on hand Rifles enough to answer the purpose of this office for the summer, and as I thought Mr. Risers price too high, I declined (bothering) you further on the subject.
As I shall want a parcel of Rifles to send off early in the spring, I will be much obliged to you Sir to see and talk with Mr. Riser again if he will agree to make 100 Rifles, delivered by the middle of March, equal to the sample sent except that they be rifled in a more perfect manner & have tighter locks. I will give him $12 cash – the bores I wish from 40 to 60 Balls in the pound. should there be any difficulty about providing Locks in time, I will furnish Locks which I have on hand at what they cost – more than $12 per gun complete I will not pay. If Mr. Riser should accept this offer, I should be glad he would make a sample gun & send it to me, or if he can come down himself it would be better when the contract could be made . . .

John Mason
20 November 1808 Office of Indian Trade
[National Archives] Georgetown

The Second Account of Martin & George Rizer Administrators of the Estate of Martin Rizer Deceased . . . 8 February 1807 . . .
[Allegany CCH, Administration Account Book A, p. 98]

The Third Account of Martin & George Rizer Administrators of the Estate of Martin Rizer late of Allegany County deceased . . . 10 May 1807.
[Allegany CCH, Administration Account Book A, p. 104]

Houses in Cumberland, 1813:
West side of Mechanic Street: – First Martin Rizer's house; . . .
[HISTORY OF CUMBERLAND, p. 291]

Inventory of Goods Chattels & personal Estate of Martin Rizer Senior late of Allegany County deceased:
[many ordinary items]
1 set Blacksmith Tools $90.00
Appraised 4 January 1816
Martin Rizer Jr. Administrator
[Allegany CCH, Administration Account Book A, p. 345]

By her will, dated 21 March 1829, Mary Magdalena Rizer left to her grand daughter Margaret Gary her bed stead & bedding, and to her grand daughter Elizabeth Gary her coffee mill, tea kettle & skillet. Peter Gary was the Executor, and the will was probated 8 September 1829.
[Allegany CCH, Will Book A, p. 293]

RIZER, MARTIN JR. (III) Cumberland

It is probable that Martin Rizer Jr. (III) made rifles at some time in his life, for he is known to have worked at the trades of cabinet maker, wagon maker and plough maker in Cumberland, and his father, Martin (II), was a gunsmith. Martin Jr. (III) was born in 1788 in Virginia. There is confusion concerning the name of the woman who married Martin Jr. (III). The original Allegany County marriage license records for the early nineteenth century were destroyed by fire, but they had been copied or extracted by Will H. Lowdermilk for his book HISTORY OF CUMBERLAND published in 1878. According to Lowdermilk the following marriage licenses were issued:
Rizer, Martin & Rosina Gephart 4 June 1799
Rizer, Martin of Martin & Catherine Bower 18 June 1811
Rizer, Martin, Jr. & Mary Saylor 12 June 1823

Rizer, George of Martin & Margaret Hoffman 18 June 1811

Something appears to be amiss in the above. A marriage date of 1811 would be most appropriate for a man born in 1788, hence the entry mentioning Catherine Bower seems correct. However, there was a Martin Rizer of Mathias in the county at the same time, a relative of the other Rizers by an unknown relationship; he was born before 1790 and had a wife named Catharine, so it is likely that the above entry should read "Rizer, Martin of Mathias and Catherine Bower." A deed dated January 1810 involves the sale of land by Martin Rizer Jr., acknowledged by his wife Rosanna, and there is a second deed, dated April 1814 likewise involving the sale of land by Martin Rizer Jr. and acknowledged by his wife Rosanna. Perhaps Rosanna died and Martin Jr. (III) married Mary Saylor in 1823. But the marriage application date of 1799 could not be correct if Martin's age, as given in the 1850 census, was 62 years. The estate of one Martin L. Rizer was administered by J. Rizer and M. L. Rizer, and this person probably was Martin (III).

By this Indenture, dated 2 January 1810 Martin Rizer Jr. sold to Jacob Hoblitzell, for $200, lots No. 247 and 248 on Bedford Street. The sale was acknowledged by Martin's wife, Rosanna.
[Allegany CCH, Land Record Book E, p. 526]

By this Indenture, dated 18 April 1814, Martin Rizer Jr. sold to John Seawright, for $50, lot No. 61 in Cumberland. The sale was acknowledged by Rosanna Rizer, wife of Martin Rizer Jr.
[Allegany CCH, Land Record Book G, p. 414]

The First Account of Martin Rizer Junr Administrator of Martin Rizer Senr late of Allegany County deceased . . . 12 August 1818.
[Allegany CCH, Administration Account Book A, p. 14]

The Third Account of Martin Rizer Junr Administrator of Martin Rizer Senr late of Allegany County, Decd . . . 14 August 1822.
[Allegany CCH, Administration Account Book A, p. 282]

The Fifth and Final Account of Martin Rizer Junr Administrator of Martin Rizer Senr . . . 9 March 1825.
[Allegany CCH, Administration Account Book A, p. 307]

Martin Rizer, Jr., plough & wagon maker 1832.
[newspaper in Cumberland, 1832]

1850 Rizer, Martin age 62 cabinet maker born Va.
[1850 Federal Census, Allegany County]

RIZER, M. L.

A flintlock longrifle signed "M. L. Rizer" is illustrated in the book KENTUCKY RIFLES & PISTOLS 1750–1850 by J. R. Johnston and George H. Carroll. This piece holds an engraved brass patchbox, some elaborate inlays, and a abundance of silver or brass wire inlay. Presumably this rifle was made in Cumberland, and presumably it was made by one of the Rizers named Martin. Martin (II) Rizer Sr., died in 1815 and probably did not make this rifle, which appears to be of post-1820 manufacture. Perhaps Martin (III) was the maker.

RIZER, GEORGE JR. (II) — Cumberland

Whether or not George Rizer Jr. (II) worked at gunsmithing is not known, but there is a good possibility that he did, at least on a part-time basis during his early years. He married Margaret Hoffman, daughter of George Hoffman, one of the more prominent citizens of Cumberland. The year of his birth is not known, but probably it was about 1790. George and Margaret Rizer had four children, George Hoffman Rizer who married Elizabeth Wright in 1840, Albert A. Rizer, Elizabeth Rizer who married David Elliot, and Mary Jane Rizer who married a man named Shinn. George (II) died in 1864.

I George Rizer of Allegheny County in the State of Maryland . . . do make . . . my last Will and Testament . . . 1) To my son George Hoffman Rizer of Frederick County I give . . . the house and lot . . . on the east side of Mechanic Street in Cumberland situated opposite to the old banking house and devised to me and my deceased wife by the last Will of her late father George Hoffman deceased . . . 2) To my son Albert A. Rizer of Allegheny County I give . . . the farm on which he resides . . . 3) To my grand daughter Florence Rizer, daughter of my son Albert A. Rizer I give . . . the Bed, Bedding and all the furniture in my room in her father's house. 4) To my daughter Elizabeth A. Elliot,wife of David Elliot, I give . . . the Bed, Bedding and furniture belonging to me in her husband's house. 5) All my real estate not hereinbefore disposed of I hereby devise to my son Albert A. Rizer . . . in trust for the use . . . of my daughter Elizabeth A. Elliot and the four children of my deceased daughter Mary Jane Shinn in equal moiety . . . the one half for the use of . . . Elizabeth A. Elliot and the other half . . . to be equally appurtained among the said four children to wit George Shinn, Howard Shinn, Albert Shinn and Martha Shinn . . . 26th day of October 1862.

George Rizer

On the 9th day of July 1864 came . . . witnesses to the . . . last will . . . of George Rizer late of Allegheny County deceased . . .
[Allegany CCH, Will Book G, p. 97]

RIZER, JACOB — Cumberland

Jacob Rizer learned the gunsmithing trade as an apprentice to George Rizer in Cumberland, and then moved on to Kentucky where he worked at it for many years. The information given below is the result of the thorough research efforts of Thomas A. Strohfeldt, whose article detailing the life and work of Jacob Rizer was published in 1974: *Jacob Rizer: A Bardstown Riflemaker* (1974) THE FILSON CLUB HISTORY QUARTERLY, v. 48, No. 1, pp. 5-15, Louisville, Ky.

Jacob Rizer was born in Berkeley Co., Virginia, on 21 Oct. 1785 according to the Industrial Census for 1850, Kentucky Division. He was the son of Matthias Rizer who owned property there from 1784 until 1812 according to the property tax records. When he was 14 years of age Jacob became an apprentice in gunsmithing to George Rizer of Cumberland who doubtless was a relative and possibly an uncle. It was fortunate that Strohfeldt copied the indenture when he did, for an attempt made in the fall of 1975 to see the old indenture book in the basement of the Allegany County Court House met with no success; the book could not be located.

This Indenture made the 23rd day of May in the year of our Lord 1800 between Matthias Rizer of Berkeley County and Commonwealth of Virginia of the one part, and George Rizer of the town of Cumberland, Allegheny County, State of Maryland, of the other part, Witnesseth that the said Matthias Rizer hath this day put and placed his son Jacob Rizer as an apprentice to the said George Rizer to learn the art trade and mystery of a gunsmith and to serve him, the said George Rizer for and during the full due and term of five years from the date hereof, and the said George Rizer doth on his part promise and obligate himself, his heirs, executors, and administrators, by all necessary and lawful endeavors to learn the said apprentice the trade aforesaid, and to find the said apprentice aforesaid, sufficient meat, drink, washing, lodging and clothes suitable to an apprentice bound to such trades to give him six months schooling, two months of which he is to have in the last year, and when free, which will be when he is of the age of twenty years to give him nine pounds lawful money, or the value thereof in clothes, at the option of said apprentice. In witness whereof the parties have hereunto set their hands and seal the day and year above written.

[signed] Matthias Rizer
[signed] George Rizer

[Apprentice Indenture Book, Allegheny CCH, p. 56]

After completing his apprenticeship in October 1805, Jacob apparently went fairly soon to Bardstown, Kentucky, about 40 miles south of Louisville. Jacob's brother, Phillip Rizer, purchased 104 acres of land in Bardstown in 1806. Jacob married Polly Linebaugh on 27 January 1808. In 1815 he bought Lot No. 44 in Bardstown. He took a second wife, Elizabeth McAkron, on 4 Oct. 1815. Matthias Rizer, who presumably was Jacob's father, bought half of his town lot in 1816. Matthias and Phillip Rizer are listed together in the tax records of Berkeley Co., Va., until 1812, and Phillip is listed there in 1792 and 1793.

Jacob Rizer prospered in Kentucky, apparently working there as a gunsmith until he died on 29 June 1859. His inventory mentions gunsmith shops and gunsmith tools. His estate was valued at $7,631.89½, which was by no means a small sum for the time and place, and it included considerable furniture and household goods and 6 negro servants.

RIZER, WILLIAM L. — Allegany County

What relation William L. Rizer had to George Rizer (I) and Martin Rizer (II) is unknown. It seems unlikely that he was a descendent of George (I). He may have been a grandson of Martin (II), that is, a son of either Martin (III) or George (IV). He also could have been a son of Martin Rizer of Mathias.

A percussion longrifle signed on the barrel in script "W. L. Rizer" has been examined by the author. The brass patchbox has engraving, but the stock is without carving.

1850 Allegany County
Rizer, William L. age 32 gunsmith born Maryland real estate, $1000.
[1850 Federal Census]

ROSE, ADAM L. — Washington, D.C.

1850 Washington, D.C., 6th Ward

Name	Age	Born	Occupation
Rose, Adam L.	35	Washington, D.C.	gun carriage maker
" Mary	33	Va.	
" Ann E.	9	Va.	
" Marah C.	6	Washington, D.C.	
" Mark	3	Washington, D.C.	

[1850 Federal Census]

Rose, Adam L.
1858 h 622 7th east gun carriagemaker
[Business directory, Washington, D.C.]

ROSENFELD BROTHERS — Baltimore

Rosenfeld & Co. (M. Rosenfeld, G. Rosenfeld)
1867 23 & 25 West Baltimore Street clothiers
Rosenfeld Brothers (M. & G. R. Rosenfeld)
1868-72 23 & 25 West Baltimore Street clothiers
Rosenfeld Brothers (Michael & Goody Rosenfeld)
1873-84 Baltimore & Centre Market clothiers
Rosenfeld Brothers (J. Goody & I. Rosenfeld)
1885-89 617 East Baltimore Street clothiers
[Baltimore city directories]

ROTH, HENRY SR. — Frederick County

First deed in Frederick County was in 1749 when his occupation was listed as a farmer. After this no occupation is given, and his last indenture was in 1767. After moving back to Pennsylvania he was the first minister of the old Stony Creek Church.
He supposedly made a plain longrifle that now belongs to a descendant.

Henry Roth, original name changed to: Roads, Rodes, Rhodes,

Rhoads, Rode . . . [Heinrich Roth Sr.] was born in Manheim, Germany, in 1686, and died in Ursina, Pennsylvania in 1774 and is buried in the Rhoads Cemetery, on his tract of 1771, near Brooks Tunnel on the B & O Railroad, near Ursina, in Lower Turkeyfoot Township, Somerset County, Pennsylvania. The father of Henry (Heinrich) was Christian Roth. Heinrich Roth, Senior, was married twice. His first wife was Elizabeth Ulrich, sister to Catherine (Caterina). To this union were born four sons: Peter, who was a minister and on coming to America remained in Lancaster County. The other three sons, in order of their birth were: Frederick, Ulrich and John. To date no date of their births have been discovered by this author. His first wife Elizabeth, and his second wife Catherine . . .
[TWO CENTURIES OF BROTHERSVALLEY, p. 162]

The sixth and last gunsmith to point to in this short treatise was Henry Roth, Sr. He was the first minister of the old Stony Creek Church. He was a blacksmith by trade. He also made rifles, though this was in opposition to the Dunker practices. But Henry Roth had grown up largely in the practices of the Seventh Day Baptists of Ephrata. Nevertheless, Roth moved to Ursina, Pennsylvania, in 1771 and continued to make rifles. He died in 1774 and is buried on this tract there. It is said that several of these rifles were taken to Kentucky in 1785, when young Captain Henry Roth, Jr., led the party of about one hundred souls into the wilderness of the "dark and bloody ground." Not much more than this brief mention of the rifles of Henry Roth is known by this historian.
[TWO CENTURIES OF BROTHERSVALLEY, p. 53]

ROTHANHEAFER, LAWRENCE J. Harmony

L. Rothanheafer gunsmith in Harmony
[HISTORY OF WESTERN MARYLAND, p. 635]

Rothanheofer, L.	1878	Harmony	gunsmith
Rothanheofer, L.	1880	Harmony	gunsmith

[The Maryland Directory, 1878, 1880; J. Frank Lewis & Co. Publishers]

ROWE, NATHANIEL Emmitsburg

The following information is a composite of the research efforts of Edward R. Flanagan, Wm. S. Bowers, George Shumway, and the author. In particular, however, it was the work of Edward R. Flanagan which made this extensive account possible, for it was he who discovered the important newspaper interviews with Nathaniel Rowe and who put together the basic story of his life from these accounts and other data. Flanagan prepared an article on Rowe which was published privately in January 1976 by The Kentucky Rifle Association. With the kind permission of Flanagan and the K.R.A., material from this article has been utilized here.

Nathaniel Rowe was born 8 August 1821 on his father's farm about three miles south of Emmitsburg. His parents were Daniel and Susan Rowe. When he was about 16 years of age he moved to Emmitsburg and became the last apprentice of the aging gunsmith John Armstrong. Upon finishing the apprenticeship at age 21 in 1842, he took over the gunsmithing business of Armstrong who apparently died then or shortly afterward. Rowe listed his occupation as gunsmith in the federal census returns for 1850, 1860, and 1870, but in 1880 he gave his occupation as machinist. As the Industrial Revolution advanced and the demand for muzzle-loading arms declined, Rowe turned to foundry and machine shop work. Being in good health, it is probable that Rowe did some limited shop work while in his eighties. If this included a bit of gun repair work, which is likely, it would be possible to write that Rowe and his master worked in three centuries.

In 1844 Nathaniel Rowe married his second cousin, Elizabeth Rowe, who was born 21 Dec. 1817 and died 24 Nov. 1883. They raised a family of five children, Quincy Edgar, Charles C., E. Kate, Helen J., and Albert S. In 1847 Nathaniel Rowe bought a house and lot on West Main Street two blocks from the square, where he lived for over 70 years. He encountered business difficulties in the late 1870's, and in 1879 his business property was sold at public auction.

Nathaniel Rowe apparently enjoyed good health throughout his long life. When he was interviewed by newspaper reporters in 1908 his health was fine and he mentioned that with the help of his glasses he could read all day. And in 1910 T. J. C. Williams noted:

One of John Armstrong's apprentices [Nathaniel Rowe] still lives at the present time, hale and hearty, about ninety years of age and bids fair to live many more.
[Williams, T.J.C. (1910) HISTORY OF FREDERICK CO., MD., p. 805]

But time caught up with Nathaniel Rowe not many years later, and he died in 1915 on 10 April, at the age of 94 years.

In 1908 Emmitsburg's local weekly newspaper interviewed old residents and published their remembrances of the early days in Emmitsburg. Rowe was one of their favorite subjects, being clear-headed and talkative. The interviews, which are quoted below, are particularly valuable in giving a first-hand account of a gunsmith's life in early Maryland.

A number of rifles made by Nathaniel Rowe have survived to the present day. It is unlikely that Rowe signed any rifles until he completed his apprenticeship and reached age 21 in 1842. This of course was well into the percussion period. During the 1840's he may have made a few rifles with flintlock ignition, but nearly all his production utilized percussion locks. Probably there exist today a number of his rifles made in the early part of his career which never were flintlock until altered in the mid-twentieth century by eager collectors, restorers, and dealers. Rowe's earlier rifles follow the carefully-designed and beautiful profile for rifles worked out and used by John Armstrong, with the flaring butt suitably high to hold large relief-carved designs and generously-sized brass patchboxes of the Armstrong style. But Rowe's percussion arms had to keep in style with the times. The relief-carving gave way to incised carving and then the incised carving was eliminated. A smaller butt with less flare was made. The inlays changed in style, and the patchboxes became more simple in design. It is curious that Nathaniel Rowe signed his name as "Row" on his rifles. The name is pronounced to rhyme with "cow."

Nathaniel Rowe's father, Daniel, was dead by the time Nathaniel finished his apprenticeship in 1842. George Rowe, the gunsmith's grandfather who had been a Revolutionary War soldier, was living at that time however; his will, dated 9 Mar. 1842, mentions Daniel as being deceased and it bequeaths to his grandson Nathaniel the sum of $30 plus additional money unspecified. The will, recorded in Frederick County Court House, mentions that George Rowe died in 1845.

By this deed of 29 March 1847 Nathaniel Rowe bought from Philip Nunemaker and Jane Nunemaker his wife for $420 a house and lot in Shields Addition to Emmitsburg . . . the eastern half of Lot. No. 18.
[Deed Book W.B.T. 11, p. 684, Frederick CCH]

Rowe,	Nathaniel	age 31	sex m	Gun Smith	birth Frederick Co.
	Elizabeth	25[?]	f		Frederick Co.
	Helen J.	5	f		Frederick Co.
	[unclear]	2	m		Frederick Co.
	Daniel	17	m	Apprentice	

[1850 Federal Census, Emmitsburg, Frederick Co.]

By this indenture dated 12 October 1850 Nathaniel Rowe and his wife Elizabeth execute their mortgage deed to Joshua Motter and George W. Rowe, trading under the name of Motter & Rowe . . . Whereas the said Nathaniel Rowe . . . stands bound unto the said Motter & Rowe in the sum of $365.55 . . .
[at margin of page] I hereby release the within mortgage. Witness my hand and seal this 22d day of February 1877 . . . J. Taylor, Adm. of Joshua Motter.
[Deed Book W.B.T. 11, p. 685, Frederick CCH]

Rowe,	Nathaniel	age 39	sex m	Gunsmith	birth Maryland
	Elizabeth	38	f		Maryland
	Helen J.	13	f		Maryland
	Jemey E.	12	m		Maryland
	Albert S.	9	m		Maryland
	Charles C.	6	m		Maryland
	Emma R.	1	f		Maryland

[1860 Federal Census, Emmitsburg, Frederick Co.]

Whereas Joseph Rowe late of Frederick County deceased did by his last will . . . appoint Joshua Rowe his sole executor . . . and whereas the said Executor did . . . about the first day of April 1862 sell unto Elias S. Rowe a house & lot . . . situate on the south side of Main Street . . . in Emmitsburg . . . and whereas the said Elias S. Rowe died before the title was executed for said house & lot & whereas the said Joshua Rowe, James A. Rowe & Sarah his wife, Charles F. Rowe & Ann his wife, Nathaniel Rowe & Elizabeth his wife, Dietrick Zeek & Mary his wife, Peter Hoke & Francis A. his wife, and Sarah J. Troxel are legal heirs . . . of the . . . deceased, Witness therefore that in consideration of . . . $1,000, we . . . do grant unto Catherine A. Rowe all that lot & parcel of ground with the buildings thereon . . . in Emmitsburg . . . designated by number 23 on the plot of said Addition. Dated 10 June 1865.
[Deed Book J.W.L.C. 2, p. 703, Frederick CCH]

By this deed of 10 June 1865 Charles F. Rowe & Ann Eliza his wife, Catherine A. Rowe, Sarah J. Troxell, and Peter Hoke & Frances A. his wife sell for $316.67 unto Joshua Rowe & James A. Rowe the following tracts of land in Frederick County: part of "Carrolsburg" adjoining Shields Addition to Emmitsburg containing 5 acres, also five lots in Shields Addition being back lots and designated by numbers 58, 59, 60, 61, 62.
[Deed Book J.W.L.C. 2, p. 704, Frederick CCH]

Rowe,	Nathaniel	age 49	sex m	Gunsmith	birth Maryland
	Elizabeth	30[?]	f	Keeping house	Maryland
	Helen	24	f		Maryland
	Jainey	21	f		Maryland
	Albert	16	m		Maryland
	Catherine	10	f		Maryland

Real Estate of N. Rowe value $1,500, personal estate $300
[1870 Federal Census, Emmitsburg, Frederick Co.]

By this deed of 11 October 1875 Nathaniel Rowe buys from Daniel G. Adelsberger of Emmitsburg for $4,000 Lots No. 68 & 69 in Emmitsburg with all the buildings & improvements thereon.
[Deed Book T.G. 4, p. 92, Frederick CCH]

By this deed of 22 March 1876 Nathaniel Rowe & Elizabeth Rowe his wife sell for $1,650 to Helen J. Rowe a lot on the South side of Main Street in Shields Addition to Emmitsburg . . . the east half of Lot . . . Number 18.
[Deed Book T.G. 4, p. 653, Frederick CCH]

By this deed made 24 June 1879 by Charles W. Ross & Charles J. D. Levy . . . by a decree of the Circuit Court . . . as a court of Equity passed on this 14th day of January 1878 in the case of the Frederick Town Savings Institution vs Nathaniel Rowe, Elizabeth Rowe his wife and others, being No. 4304 Equity in said Court . . . We the said Charles W. Ross & Charles V. S. Levy were appointed trustees to sell the Land decreed to be sold and have sold the same to J. Taylor Motter, Frances A. Maxwell, & Ezra R. Zimmerman partners trading as Motter Maxwell & Co. for the sum of $1,510 . . . all those lots . . . known and designated as Lots 68, 69 . . . which the said Nathaniel Rowe obtained from Daniel G. Adelsberger on 11 October 1875 . . .
[Deed Book T.G. 12, p. 161, Frederick CCH]

The Fredericktown Savings Institution vs Nathaniel Rowe, Elizabeth Rowe his wife, Helen J. Rowe and Daniel G. Adelsberger . . . Recorded 22 July 1879 . . . Nathaniel Rowe being indebted to Fredericktown Savings Institution for $1,872, executed a mortgage deed for Lots 68 & 69 but neither money nor interest ever was paid . . . Further at the December 1876 Term of Circuit Court J. Taylor Motter administrator of Joshua Motter deceased,

obtained a Judgement in the sum of $367.64 with interest against Nathaniel Rowe. And on4 October 1877 a Judgement was entered to the use of Helen J. Rowe to the **extent** of $213.00. Also at the same court J. S. Auman got a Judgement against Nathaniel Rowe for $136.96. And on 20 September 1877 a Judgement in favor of Daniel G. Adelsberger was entered.
[Equity Record Book T.G. 7, p. 232]

Rowe,	Nathaniel	age 59	sex m	Machinist	birth Maryland
	Elizabeth	62	f		Maryland
	Helen J.	35	f	Teacher	Maryland
	Quincy E.	32	m	Conductor	Maryland
	Albert S.	29	m	Teacher	Maryland
	Charles C.	25	m	Machinist	Maryland
	Emma K.	20	f		Maryland

[1880 Federal Census, Emmitsburg, Frederick Co.]

By this deed of 21.March 1892 Nathaniel Rowe, acting as Executor of the last will of Catherine Rowe, sells the 27-acre tract of land on the road leading from Emmitsburg to Harbaugh's Valley.
[Deed Book J.L.J. 1, p. 281]

The *Chronicle* reporter found Mr. Nathaniel Rowe in his cozy home on Main Street where he has lived continuously for nearly 65 years. The house is one of the oldest in Emmitsburg, having been first constructed of logs like so many of the first buildings in town. It was rebuilt many years ago and is now as comfortable as a modern house could be, while retaining something of the atmosphere of the long ago. In the dining room is a clock Mr. Rowe's grandfather had built one hundred and forty years ago, which is still ticking out,

Forever - Never
Never - Forever

as faithfully as when its time was young.

"They don't make them like that nowadays," said Mr. Rowe. "Things were made solid and good in the old days. They were made for service, not for show —— they had no factories and no trusts then." "Now, Mr. Rowe," said the reporter, "Won't you tell *The Chronicle* about the girls in Emmitsburg, maybe something of the way their grandparents used to live? What was Emmitsburg like when you were a boy? But first tell about the early history of your family in these parts and anything you can remember about them so that *The Chronicle* may make of this interview a part of the permanent history of Emmitsburg.

"I will begin with my great-grandfather who was one of the earliest settlers in this county. He was a German emigrant and, with his wife, came to Pennsylvania in colonial days and lived first near Lancaster, Pa. At the time of the Scotch-Irish movement from Lancaster to Frederick county he came along and took up a large plantation about three milles below Emmitsburg. He had three sons: Michael, Arthur, and George, who was my grandfather. He divided his farms amongst his three sons, and my father, Daniel, was born in the same house I was born in. My grandfather, George Rowe, fought in the War of the Revolution and he often used to tell me of the hardships they endured but I can't remember much about it. I recollect he said they were camped through an entire winter in a valley; the snow was deep and the soldiers had to carry wood for their fires from the Ridge a mile away. When his regiment was in camp near Lancaster, Pa., he often saw the great Pulaski, the Polish patriot who fought in the Revolutionary army and was killed at the siege of Savannah in 1779. My great-uncle, Michael Rowe, was a lieutenant in a Maryland regiment during the War of the Revolution and he also used to tell me stories of his adventures which I wish I could remember for your readers. A good many of the Hessian soldiers, who were hired by England to help her fight the American patriots, settled in Frederick county after the war was over.

"I lived the life of the average country boy in those days. We suffered some hardships but not more than many a boy does now who lives on a farm. It wasn't very comfortable, for example, to get up on a winter morning and step with your bare feet into a pile of snow that had sifted in through the cracks in the walls and roof. I helped about the place, took care of the stock and went to school – a little."

"What did they teach you at school?" asked the reporter.

"Oh, not much of anything. A little grammar, writing, reading and cyphering – that's about all we got."

"Did the master ever flog you?" inquired the reporter.

"Yes, like the Devil," responded Mr. Rowe with fervor, as though he had a lively recollection of the fact. "I remember one teacher named Haas I used to be afraid of. He had a leather strap cut into four or five thongs with a tight knot on the end of each thong. It was called a taws and let me tell you it could bite when it was laid on right. When a boy had been caught at any mischief in school the teacher would throw the taws to the boy and he had to carry it to the teacher who then applied it where it would do the most good. Once he threw the taws to me when I hadn't been doing anything bad. I picked it up, ran out the door with it and went home and never went back to school while that teacher was there. We had another master, William Stone, who turned the tables on us once when we were trying to play a joke on him. It was the custom on the last day of school before the Christmas holidays, for the teacher to give the boys a treat – an apple and a piece of gingerbread to each boy. Well, on this day we barred teacher out, as we called it. We barred the door on the inside and nailed the windows fast. Presently along comes teacher and wanted in. We told him he could stay out. 'I'll see about that,' says he, so he climbed up on the roof and stopped the chimney. We had a big fire on the hearth and presently the room began to fill with smoke. We coughed and sneezed and gasped for breath. The door and windows were nailed so fast that before we could get them open we were almost suffocated."

"What did the teacher do to you," inquired the reporter anxiously. "Nothing. He just said, 'boys, I guess I'll have to treat you anyhow."
[Emmitsburg Chronicle, 6 Mar. 1908]

"Well, when I was about sixteen I came to Emmitsburg to live. The town didn't look very different then from what it does now. It was built up about the square but with an indifferent class of improvements. There were then many log houses in town. They were warmed with big open fireplaces and wood stoves. We knew nothing about coal. We lived well and comfortably, however. Locks on the doors were unknown – we had no thieves. There were no butchers nor bakers. We ate pork more than any other kind of meat. Once in a while a farmer would kill a calf and divide it up amongst the neighbors, each taking his turn at butchering. We wore homespun clothing. Everybody had his own patch of flax. To prepare the flax for spinning the straw was first passed through the breaker which loosened the woody part of the stem; then it was scutched in a hand machine to take out "chive" and waste matter. Next it was hackled by combing to take out the tow. The women spun the flax by the big fireplace in the long winter evenings and then it was taken to the country weavers to be made into linen cloth. That made the fine shirts. I have some of the old country-woven cloth yet. I can't say much for the breeches they made out of the tow. They were mighty uncomfortable but they wore well.

"We raised our own sheep and, of course, had our own wool. There were lots of little woolen mills around the country driven by water power. I remember three that were near Emmitsburg. The cloth was mostly of three shades; gray, brown and black. The town tailor made our clothes for us and if they were not stylish they were, at least, warm and comfortable. As for shoes, the farmers would buy a side of sole leather at the tannery and take it home until the traveling shoemaker, who went around the country with his bench on his back, should arrive. When he came he would make shoes for the farmer and his family. They weren't very comfortable and they didn't keep the water out but we had to get along the best we could.

"We didn't have many games. I only remember two, alleyball and longball. The former was played with a soft ball which was knocked against the side of a house with the bare hand. The fellow who could keep it going the longest without its touching the ground won. The alley alongside of Mr. Motter's house was the favorite place for playing this game. Longball was played in the street with an iron ball about the size of a croquet ball. It was generally played for the drinks – the one who rolled the shortest distance had to buy. It wasn't much of a game and was dying out when I came to town to live.

"Oh, yes. We knew what whiskey was in those days," replied Mr. Rowe to the reporter's inquiry. "It was good whiskey, too. There were lots of distilleries around here. Whiskey only cost twenty cents a gallon and sold in the taverns for a gill for a fip. A fip was a Spanish coin worth six and a quarter cents, a about the size of the old three-cent piece. Most all of our silver coins were Spanish. But about the whiskey. It was usually bought by the barrel for household use and everybody could help himself when he wanted a drink. Ah, those were good old times. There was much less drunkenness than there is now in spite of the fact that whiskey then was almost as cheap as Emmitsburg water is now.

"I was apprenticed to a gunsmith named Armstrong. We used to buy the barrels and make the stocks and other parts. The first barrels were made by welding two bars of iron around a solid core. Later old horseshoe nails were made into gun barrels. Some of the barrels we bought in Lancaster, Pa., and some were made around here. We bored out the barrels ourselves testing the accuracy of the work by squinting through the bore at a bright light; any inequality would cast a shadow on the opposite side of the barrel. When I first apprenticed, the old flintlock muskets were quite common but they were rapidly being replaced by the percussion-cap guns. The choke barrel was unknown in those days. We had lots of game to shoot. Partridges and pheasants were plentiful and the wild pigeons came in clouds. There were deer and wild turkey on the mountains, too.

"I must tell you a story. I used to hear the old folks tell about a preacher in the Lutheran church in the days when the services were held in German. He used to tell his congregation that if they were not careful to mend their ways sermons would someday be preached in English in their church. That was to scare them into good behavior. We were good churchgoers. The farmers mostly came to church on horseback with their wives perched up behind them. The sermons were longer than they are now but I don't know that they were any better.

"Mr. Rowe," said the reporter, "you have drawn a most interesting picture of the old times. Now tell which you like best – the old ways or the new?" Mr. Rowe thought for a moment and then slapping the reporter on the knee said emphatically, "The old ways for me – I like them best. I guess I can't help it. I was raised that way. I don't say that the world hasn't grown better in some respects, but I liked the simpler life we lived. Everybody was independent. We raised our own flax, our own wool and our own food. We made nearly everything we used. There was no stealing and very little drunkenness. Gossip and slander were almost unknown and children were more obedient and respectful to their parents than they are now. But I have no complaints to make of the world as I know it now. I enjoy good health. I can, with the aid of my glasses, read all day, and when the weather is fair I go out for a walk every day. I don't feel good unless I get my exercise in the open air every day. My hearing is a little heavy and I have to be careful about what I eat, but for a man 86 years old I am doing well. I am going to pay *The Chronicle* a visit sometime and then maybe I will tell you more stories of the old days.
[Emmitsburg Chronicle, 13 March 1908]

The first oldest inhabitant *The Chronicle* man met was Mr. Nathaniel Rowe sitting under his vine and fig tree of his house. "Mr. Rowe," said the reporter, "the readers of *The Chronicle* want some more of your reminiscences. "Well, they can have them and welcome," said he. "Since you stirred me up to thinking about the old times much has come back to me that I had forgotten and now I can talk to you to some purpose. How would you like to hear about the old fashioned corn-husking?" "Nothing better, it's all good," said the reporter.

"In the first place," said Mr. Rowe, after he had led the way into the cool, dusky front parlor, "I want to say to you again, that I don't think these latter days are such a wonderful improvement on the past. Take harvesting for example. Many a time I have seen my grandfather reaping grain with a sickle. He would take a great armful, just the right quantity, and cut it off about three inches from the ground with a slashing stroke of the blade; it would fall exactly and neatly as if laid down by a reaper. And he made good speed too. Not as fast as the machine would do it, of course, but we had plenty of time then and we weren't crazy to go through everything we had to do at a breakneck pace. We thought more about doing our tasks well and thoroughly than in getting done in a hurry. I don't say we should go back to the sickle and the cradle, but I do love the days when a man was of more importance than a machine.

"Our methods of thrashing were as primitive as our reaping. Horses trod out the grain as the oxen in the scriptures did. The heads were piled in a big circle on the barn floor and four horses, two and two, walked around and around on them until all the grain was trodden out. We had to keep turning the mass with a fork so as to insure a thorough job. That was cold work. We always thrashed in the winter time and we boys would have to ride the horses to keep them on the grain. Thus we did with wheat and oats. Rye and buckwheat were thrashed with flails, two men striking together. There was a knack about that and if you didn't understand how to do it you were liable to get a crack on the head you would remember. Of course in both methods the grain was passed through a mill to winnow out the chaff.

"But I must get to the corn-huskings. We generally held them them in October when the moon was full. In those days it was the custom to allow the corn to ripen thoroughly on the stalks and it therefore plumped out better than when the stalk is cut and shocked with the ears on. A corn field after it had been topped was a pretty sight. When the corn was full ripe the ears were pulled off and hauled to the barn. The stalks were allowed to stand through the winter and in the spring were pulled up and burnt. We generally seeded a field to oats after it had been in corn.

"The day before the husking the neighbor-women would come and help get ready the harvest supper. What did we have for supper? Good things, let me tell you. Chicken pot-pie, roast pork and apple sauce, cakes and every kind of pie you could think of – plenty of everything. Well, the ears of corn would be laid out near the barn in long rows about three feet high and three feet wide. As many men as could get to a row would fall to with their husking palms. We began about dark and worked until about ten o'clock. If there was no moon great bonfires were made to give us light. Most every farmer had at least one or two slaves and the darkies would bring their banjos and sing the good old songs while we worked. It was a thirsty business and a bucket of water was kept going up and down between the rows. The water bearer would carry the bucket in one hand and the whiskey bottle in the other for some needed a little stimulant to sustain them at their labors.

"Sometimes as many as seventy-five men with their women folk would come to a husking and they could shuck the corn crop of a big farm in one evening. No, the women didn't do any husking. The helped put the finishing touches on the supper and serve it when the men were ready. Between the hard work and the whiskey we had hearty appetites by quitting time and what we could do to a pile of grub would astonish you.

"I believe the young people got more fun out of the apple butter boilings than out of the huskings. The night before the boiling the neighborhood boys and girls would come to core and 'snits' the apples, as they called it, and that was always a great frolic." "Must have been a great opportunity for courting," said the reporter. "Well, I have heard say there was some kissing done on the sly," admitted Mr. Rowe, "but the next night was more interesting for the young people who were inclined that way. In the morning the big copper kettle in the yard would be filled with cider, as much as a barrel, sometimes, and the fire would be lighted. The cider was boiled down one half. That generally took until noon. Then the apple snits were added a little at a time. Then the stirring began and never stopped until the apple butter was done. A paddle fastened at right angles to a pole about six feet long was used for stirring and it was kept going slowly round the kettle until 'way into the night. The young people in pairs would take turns in stirring, one on each side of the pole facing each other. When a boy and girl had hold of the pole you can guess what might happen in the evening when it was dark except for the light of the fire under the kettle," said Mr. Rowe with a sly twinkle in his eye as if his knowledge of what happened was not altogether a matter of hearsay.

"Were politics as interesting when you were a young man as they are now?" inquired the newspaper man. "Yes, more so. You can't imagine the fervor and enthusiasm of the campaign of 1840 when William Henry Harrison, the Whig candidate for the presidency ran against Martin Van Buren, the Democratic candidate, and defeated him. A national election now is a pink tea affair by comparison with the 'Tippecanoe and Tyler Too' campaign. You know Tippecanoe was the nick-name given to General Harrison on account of his defeat of the famous Indian Chief, Tecumseah, in 1811. The battle was fought on Tippecanoe river in what was then the territory of Indiana of which Harrison was governor. He and John Tyler were nominated by the National Whig Convention in December 1839 and during the succeeding year, up to the election, the fight was red hot. It was the most exciting presidential campaign the country had ever experienced and Emmitsburg was not less aroused than the rest of the nation. Political mass meetings and processions were first employed in that campaign to stir up enthusiasm and make votes. Party emblems and watchwords were used as never before. It was also known as the 'log cabin and hard cider campaign.' Harrison lived at a place called North Bend, in Ohio, which was then a wilderness, about sixteen miles from Cincinnati. One end of his house consisted of a log cabin covered with clapboards and it was said that he placed hard cider instead of wine on his table. The Democrats, I believe were really responsible for the log cabin and hard cider becoming issues in the campaign. They ridiculed Harrison for his primitive way of living but the Whigs accepted the challenge and made the log cabin and hard cider emblems of democratic simplicity which, of course, was very effective. In our parades in that campaign we had a log cabin built on a wagon. It was six feet wide, about sixteen feet long and one story high. Coon skins were nailed beside the door and inside on the walls. The door of the cabin had the latch ring hanging out and everybody was welcome to go inside and tap the barrel of hard cider that was kept on hand. The inside walls of the cabin were hung with traps, rifles, powder horns, and buckskin ball pouches. The wagon was driven by a man dressed in hunting shirt made of linen or tow which came to the knees and was hung with a three inch fringe around the bottom and held with a broad belt of buckskin. His hunting breeches of buckskin and a cap of coon skin completed his costume. The horses wore bonnets of coon skin with the heads and tails on. The whole outfit made an impressive appearance.

"When Harrison first settled in Southern Ohio everybody there led the frontier life. They were dependent for food mainly on such game as they could kill. The Democrats said that Harrison had lived on coon meat, hard cider and corn bread. The Whigs added the coon as a political emblem to hard cider and the log cabin. That was the significance of the coon in this campaign.

"We had speech making without end. I think General Harrison made a speech in Emmitsburg during the campaign. He was here at any rate. At the meetings campaign songs were sung by William Webb who is still alive. He lives in Thurmont. I tried once to get from him some of the old campaign songs but he had forgotten them.

"We made a big campaign ball of muslin stretched on a wooden frame. It was twelve feet in diameter. Through the centre of the ball a long pole ran horizontally so that the ends stuck out about five feet on each side. It rolled on the ground on a wooden flange, running around the outside at right angles to the pole. Men would trundle the ball through the streets by taking hold of the ends of the poles on each side and pushing it ahead of them. It was painted with cartoons and political mottoes. I remember one of the cartoons with a picture of a fox getting his paw caught in a trap. The fox's head was the head of Van Buren. James Hickey, professor of drawing and music at Mount St. Mary's did some of the painting and so did my brother-in-law, Joshua Rowe. Once we rolled the ball to Frederick for a big meeting there. We left here in the evening and rolled all night, getting to Frederick after daylight next morning. We took a wagon along with straw on the floor and plenty of provisions. When a crew got tired they would climb into the wagon and go to sleep and another crew would keep the ball rolling on through the night. That showed our enthusiasm. Would any of the young men now do as much for Taft or Bryan? After the election was over a big celebration was held at a tavern kept by a man named Harvey opposite to the toll gate on the Thurmont Pike. You know Harrison gave Van Buren a tremendous licking, the latter getting only 60 electoral votes out of 294. The result was very popular in Emmitsburg. Most of the people were Whigs and everybody joined in the celebration which lasted a day and a night. It was held in the field back of the tavern. We needed plenty of room for our celebrations in those days. People came from all over the county and it was easier to come than to get away for hard cider and other hard stuff flowed like water.

"Of course it is a good thing that such a custom has died out. Political campaigns have improved in another respect. You have no conception of the personal bitterness politics engendered seventy years ago. Outrageous and slanderous attacks on private character were usual, and were often the cause of personal encounters which sometimes resulted seriously. However, there was little, if any, buying of votes. Bribery as we have it now wasn't practiced in those days."
[Emmitsburg Chronicle, 17 July 1908]

Additional information about Nathaniel Rowe is provided by the generous-sized obituary for him published in the Emmitsburg newspaper.

Nathaniel Rowe, a life-long resident of this place and one of Emmitsburg's nonagenarians, died at his home, Saturday afternoon, April 10, 1915. Death was due to the infirmities of age, but despite his advanced years, he had enjoyed unusually good health. Although Mr. Rowe had not taken an active part in the affairs of this vicinity during the past decade, yet his death was a shock to the entire neighborhood, for he seemed in reality to actually belong to this community, because he was a member of it so many years. He was the son of Daniel and Susan Rowe, natives of Frederick County, and would have celebrated his ninety-fourth birthday had he lived until August 8.

In early life Mr. Rowe took up the occupation best suited for his natural qualities as a mechanic, and he became a gunsmith. Later, when his trade gradually began to be displayed by inventions of efficient machinery, and fine workmanship by hand was no longer required, he turned his attention to steam-fitting and plumbing. About this time, the plans for the present system of Emmitsburg's water-works were being developed, and he became the supervising contractor and plumber for the local Water Company. He was also employed in this capacity by Saint Joseph's College and Academy for many years, having been given the contract for heating several of the large buildings of that institution.

Mr. Rowe was born August 8, 1821, near this place, five years before the death of two of America's great names in history — Thomas Jefferson and John Adams. He became a voter the year after the death of General William Henry Harrison, "The Hero of Tippecanoe," and cast his first presidential ballot when James K. Polk was elected in 1845. His right of sufferage was exercised throughout the administrations of nineteen presidents, having voted for Woodrow Wilson, the twenty-eighth President, in November 1913. In 1844 he married his second cousin, Miss Elizabeth Rowe, who died thirty-nine years after their marriage, in 1883. The grandfathers of both Nathaniel and his wife, were officers in the army of the American Colonists during the Revolutionary War, having served as commanders over companies consisting of Marylanders. Mr. Rowe was a remarkable man for his age, perfectly rational up to the time of his death. He was an extensive reader of current literature and took especial interest in events and topics of the day.

The deceased is survived by one sister, Mrs. Eliza Rice, of Washington, D.C., two sons, Albert S., of Columbus, O., and Quincy E., of Emmitsburg, two daughters, Mrs. William H. Sellers, of this place, and Miss Helen J. Rowe, Principal of the Samuel Ready School, Baltimore, Md., and three grandchildren, Robert Rowe Sellers, a member of the faculty of the University of Florida, Gainesville, Fla., Miss Helan E. Sellers, of Emmitsburg, and Charles E. Sellers, a student of Bucknell University, Lewisburg, Pa.

The funeral was held at the home on Tuesday morning, Rev. Dr. Charles E. Reinewald, pastor of the Lutheran Church, assisted by Rev. Dr. U. S. G. Rupp of Frederick city, officiating. Interment in Mountain View Cemetary.

The pall-bearers were nephews of the deceased, Messrs. Charles R. Hoke, Samuel L. Rowe, Edward H. Rowe, Howard M. Rowe, M. Frank Rowe, and J. William Rowe. Forty-eight immediate relatives of the deceased were present at the funeral.
[The Weekly Chronicle, Emmitsburg, Md., 16 Apr. 1915]

RUSSEL, WILLIAM

William Russel and Company Have for Sale a few pairs of 4 lb. cannon and 6, 4, 3, lb. and Grape Shot . . . They will furnish Guns of any size, and of the best kind, at a short notice.
[Maryland Journal & Baltimore Advertiser, 20 August 1782]

S

SAINSBERRY, R. Cockeysville

Sainsberry, R.

1887	Cockeysville	blacksmith & gunsmith

[Maryland Directory & State Gazetteer, 1887, The Baltimore Publishing Co.]

SAVIN, T. S.

This Agreement, made the 21st day of May 1829, between Wm. H. Bell, Lt. of Artillery for and in behalf of the U. States on the one part, and Underhill Dann and [T. S.] Savin of Port Deposit, Maryland, of the other part, Witnesseth . . . such kinds, qualities, quantities and description of the timber and at such time and prices as are expressed for casemate carriages.
[National Archives, Contracts for Ordnance & Ordnance Supplies, 1829-1842, Vol. II, p. 23.]

SCHAEFFER & LONEY Baltimore

Shortly after the Civil War, Rinehart Schaeffer, whose previous occupation was listed as a locksmith, left the firm and it became known as Francis B. Loney and Company.

A large bore, double barrel shotgun marked on the barrel rib "London fine twist" and "made for Schaeffer & Loney Baltimore" on the lock exists. The hammers are dolphin or fish head hammers which make the gun look even more massive.

Schaeffer & Loney

1847-1860	1 & 3 Hanover Street	importers of hardware, cutlery & guns

Loney, F. B. & Co.

1868-72	240 West Baltimore Street	gunsmith

[Baltimore city directories]

SCHARFFE, AUGUSTUS Washington, D. C.

Scharffe, Augustus

1867	319 E north	gunsmith

[Boyd's Business Directory, Washington, D.C., Georgetown & Alexandria]

SCHARFFE, G. Washington, D. C.

Scharffe, G.

1866	375 F north	gunsmith

[Boyd's Business Directory, Washington, D.C., Georgetown & Alexandria]

SCHECKELS, J. W. Washington, D. C.

Scheckels, J. W.

1866	304 D north	gunsmith
1867	319 Pa. Ave.	gunsmith

[Boyd's Business Directory, Washington, D.C., Georgetown & Alexandria]

SCHILLING, A. Washington, D. C.

Schilling, A.

1865	489 11th west	gunsmith
1866	489 11th west	gunsmith

[Boyd's Business Directory, Washington, D.C., Georgetown & Alexandria]

SCHLEY, GEORGE JACOB Frederick

By this deed dated 17 January 1765 William Banks Watts of Fredericktown, Frederick County, Province of Maryland, barber, sold to Jacob Schley of the same place, gunsmith, a tract of land.
[Deed Book J, p. 1025, Frederick CCH]

By this deed dated 28 August 1770 Ludwick Weltner of Frederick County in the province of Maryland sold to Jacob Schley of the same County and province aforesaid gunsmith . . . half that Lott on the portion of ground in Fredericktown, in Frederick County addjoyning to the Lott number 44. . .
[Deed Book N, p. 322, Frederick CCH]

By this deed dated 12 September 1774 Charles Beatty of Frederick County in the province of Maryland and Geroge Hawkins of Prince George County in the province of Maryland sold to Jacob Schley of Frederick County in the province of Maryland a tract of land . . .
[Deed Book W, p. 54, Frederick CCH]

April 20, 1776
Ordered That the said Treasurer pay to Mr. Jacob Schley fifty Pounds on Account of his Contract for Rifles.
[AM, p. 356]

June 27, 1776
Ordered That the Commissioners of the Gun Lock manufactory at Frederick Town, furnish and make for Jacob Schley ten large Gun Locks agreeable to a patent they will receive for that purpose from said Schley.
[AM, p. 526]

July 30, 1776 Council of Safety to Jacob Schley
The Public service requires that you should send to this place, with all the expedition you can, the rifles by you made for the use of the Province; also the ten large rifles contracted by you to be made and delivered on the first day of August next.
[American Archives, 5th Series, Vol. I, p. 667]

October 17, 1776 Council to Schley
We are in great want of the rifles which you were to furnish for the use of this State by the first day of August and desire you will immediately send them down to us.
[AM, p. 356]

Indenture made November 26, 1779 [lease] between Daniel Dulaney of Frederick County province of Maryland of the one part and Jacob Schley of the same aforesaid of the other part . . .
[Frederick CCH, Liber W. R. 2, p. 334]

This Indenture made this 15th day of August 1785 Between Jacob Razor of Frederick Town and Frederick County . . . and George Jacob Schley of Frederick Town . . . Witnesseth that . . . . Jacob Razor for . . . 372 pounds 7 shillings and 1 penny . . .

doth . . . sell . . . unto . . . George Jacob Schley . . . a certain part of a Lott of Ground . . . in Frederick Town adjoining Ground belonging to Stephen Brenner . . . 60 feet [by] . . . 45 feet . . . Jacob Reasor

. . . At the same time came Margaret wife to said Jacob Razor . . .

[Frederick CCH, Land Record Book W. R. 6, p. 53]

In the Name of God Amen I George Jacob Schley of Frederick County . . . being weak in body but of sound . . . mind . . . do . . . make . . . my last Will and Testament . . .
Item it is my will and I do hereby order . . . my Executor . . . to sell . . . my lot of ground situate in Beatty and Hawkin's edition to George Town and also my other lot of Ground situate and fronting on Market Street in Frederick Town, being in front 96 feet and running back across three lots 86 feet more or less . . . and also all such my personal property as I may die possessed of at Public Sale after due notice given thereof in one or more of the news papers of Frederick Town and Baltimore for cash . . . And I do further order . . . that the proceeds of the said sale shall be immediately applied to the payment of all my Just Debts and funeral charges and if any balance remains . . . the same shall go into the General mass of my Estate to be disposed of . . . as hereinafter directed –
Item I do further order and direct . . . my said Executor . . . to sell . . . my other lot of ground situate in Market Street Frederick Town on which my Tavern stand is placed together with all the houses and improvements thereon . . . on the following conditions . . . the purchaser . . . shall pay to my said Executor one fifth part of the purchase money in cash and he or they . . . shall give bonds with Security to my said Executor for the payment of the residue . . . one third on one year and one third in two years and one third in three years . . . with interest from the said time of sale . . . I give . . . unto my Eldest son John Schley . . . one equal fourth part of the fund to be raised by the sales herein before mentioned. To my next son Jacob Schley . . . I give . . . one other equal fourth part of the fund . . . To my Daughter Polly Schroeder . . . I give . . . one other equal fourth part . . . To my aforesaid son Jacob Schley . . . I give . . . the remaining equal fourth part of the fund . . . in Trust for the use of my daughter Peggy Sanderson . . . In Testimony whereof I have set my hand . . . this 8th day of February 1808. George Jacob Schley
Frederick County to wit on the 24th day of July 1811 then came John Schley and made oath . . . that the foregoing Instrument of writing is the true whole will . . . of George Jacob Schley late of Frederick County Deceased . . .
[Frederick CCH, Will Book R. B. 1, p. 203]

Indenture made May 20, 1812 [deed] between Jacob Schley, Henry Schroder and Maryl his wife, and Francis Sanderson and Margaret his wife heirs and deviser of the George Jacob Schley late of Fredericktown, in Frederick County deceased of one part and [to] John Schley the remaining heir and deviser of the said George Jacob Schley . . . Lotts of ground lying and being in the addition to Fredericktown known and distinguished by the numbers 156, 157, 158.
[Frederick CCH, Liber W. R. 42, p. 310]

SCHMIDT, ---------------- Baltimore

A good-quality, large, German-style, singled-edged side knife is known stamped "Schmidt Balto" on the large cast brass handle. This may have been produced by Paul Schmidt who was a brass caster at 46 North Frederick Street during the 1850's.

SCHNEIDER, C. G. Washington, D. C.

Schneider, C. G.

1864	271 Penn. Ave.	gunsmith

[Hutchinsons Business Directory, Washington, D.C., & Georgetown]

Schneider, C.

1865	271 Pa. Ave.	gunsmith

[Boyd's Business Directory, Washington, D.C., Georgetown & Alexandria]

SCHNEIDER, J. Hagerstown

Schneider, J.	1878	Hagerstown	gun & locksmith
Schneider, J.	1880	Hagerstown	gun & locksmith

[The Maryland Directory, 1878, 1880; J. Frank Lewis & Co. Publishers]

SCHROEDER, VALENTINE Baltimore

Schroeder, Valentine

1847	82 Henrietta Street	gunsmith

[Baltimore city directory]

SCHRYER, GEORGE Baltimore

Last name was also spelled Shreyer.

Schryer, George

1810	Franklin Street near Green	gunsmith
1814-15	Hookstown Road	gunsmith

[Baltimore city directories]

SCHUMACHUR, CONRAD Baltimore

Schumachur petitioned the City Committee of Claims in the amount of $630.00 for guns and rifles taken by the mob on 21 April 1861. He arrived at his store on Pratt Street and found it broken into a second time. There he met several persons with guns and was informed that it was all right because they had an order and the city would reimburse him.

Schumachur, Conrad

1858-60	224 West Lee Street	gunsmith
1864-87	58 West Pratt Street	gunsmith

[Baltimore city directories]

SEARS, WILLIAM

October 6, 1777 Council of Safety
That the said Treasure pay to William Sears seven pounds ten shillings for a Musquet and Bayonet.
[AM, p. 392]

SEBBY, JAMES — Washington, D. C.

Sebby, James
1855 — 121 13½ — gunsmith
[Directory, Washington, D.C., & Georgetown]

Probably this listing is a misspelling for James Selby.

SELBY, JAMES D. — Washington, D. C.

1850 Washington, D.C., 7th Ward

Name	age	born		
Selby, James	28	born	Washington, D.C.	gunsmith
" Mary	25		Washington, D.C.	
" Charles W.	5		Washington, D.C.	
" Emma R.	4		Washington, D.C.	
" James L.	2		Washington, D.C.	

[1850 Federal Census]

Selby, James
1858 — h 60 D south — lock & gunsmith
[Business directory, Washington, D.C.]

Selby, James D.
1862 — h 663 Md ave. — gun & locksmith
[Boyd's Washington and Georgetown Directory]

SELF, BRADLEY — Washington, D. C.

Self, Bradley
1855 — 578 7e — gunsmith
[Directory, Washington, D.C., 7 Georgetown]

SHAFFER, BALTZER — Baltimore

I have remaining on hand some old Guns, Pistols, and Gunlocks, which were brought to my Gunsmith shop to be repaired; some of them have been in my care for four or five years past. I therefore give notice, that if the owners do not come and pay the charges on the same, before the last day of June next, I shall put them to public sale, in order to pay my self and get them off my hands as I have given over the business to Charles Shnyder, who carrie on the Gunsmith's business for himself in my shop, Baltzer Shaeffer.
N.B. I have also on hand, some new Rifle Guns, which I will sell as low as they can be bought in any part of Pennsylvania for Cash or produce. B. S.
[Maryland Gazette & Baltimore General Advertiser, 13 March 1787]

SHAW & BURNARD

The Shaws (William) and Notley Burnard at the time of the War of 1812 got out timber for gun-stocks for the Harper's Ferry armory, which they hauled to Westernport and then transported on flat-boats down the Potomac River.
[HISTORY OF WESTERN MARYLAND, p. 1467]

SHAW, JOHN — Annapolis

Was appointed Province Armourer on 3 March 1780. Prior to this he worked with Archibald Chisholm stocking muskets for the Council.

February 23, 1776 Council of Safety
Messrs John Shaw and Archibald Chisholm agreed with the Council of Safety to stock the Gun Barrels made by Mr Isaac Harris for the use of this Province at the price the same are done in Philadelphia.
[AM, p. 180]

April 16, 1776 Council of Safety
Ordered That the said Treasure pay to Messrs John Shaw and Archibald Chisholm twenty two Pounds six Shillings eight Pence for stocking Musquets &c. as Pr. Account.
[AM, p. 333]

August 17, 1776 Council of Safety
Ordered that the Treasurer of the Western Shore pay to Messrs Shaw and Chisholm 72 pounds 12 shillings 11 pence for repairing and stocking Guns.
[American Archives, 5th Series, Vol. I, p. 1354]

February 13, 1777 Council of Safety
That said Treasure pay to Messrs Shaw and Chisholme Forty Eight Pounds and Two pence.
[AM, p. 133]

November 8, 1777 Council of Safety
That the said Treasure pay to Edward Timmons one pound one shilling, thirteen shillings and six pence, four pounds six shillings and three pence due per Accounts passed and the further Sum of four pounds for a Musquet and Bayonet bought of him for the Public by John Shaws receipt.
[AM, p. 413]

February 4, 1778 Council of Safety
That the said Treasure pay John Shaw one hundred and twenty three pounds, seven shillings and two pence per Account passed.
[AM, p. 483]

Annapolis July 7, 1784
WANTED
A JOURNEYMAN GUNSMITH to clean and repair a number of Arms. Any orderly person, qualified for that purpose, may meet with good encouragement and a considerable time of employment by applying to JOHN SHAW.
[Maryland Gazette, 8 July 1784]

SHELLMAN, JOHN — Frederick

Deeds in the Frederick County Court House of 1751, 1759, and 1769 list his occupation as a carpenter and in 1771 as a farmer.

Col. John Shellman, great-great-grandfather of Lewis E. Shellman, came from Germany to America in 1745, with a company of emigrants. He settled in Fredericktown, Md., and carried the chain when the town was marked out. . . . Col. John Shellman was born in 1724, in the Rhineland of Germany. When nineteen years old he came to America just when Fredericktown was coming into existence. He equipped himself and was with the army during General Braddock's defeat, and participated in some of the stirring scenes enacted during that period. He was a master mechanic, and a self-taught one; a worker in wood, brass, iron and silver . . . Here he worked turning out specimens which showed great ingenuity; he constructed a carriage for his use when he was eighty-five, a rifle

at 87, and repaired his own clock when he was ninety years of age. He made his own coffin and often laid down in it . . . He died in the house located where the late Colonel Charles Trail resided, when he reached the age of 92 . . .
[HISTORY OF FREDERICK COUNTY, Vol. II, pp. 1313-1314]

SHERICK, GEORGE C. Baltimore

Sherick, George C.

1884	114 West Pratt Street	gunsmith
1885	114 West Pratt Street & 114½ West Pratt Street	gunsmith
1886	114 West Pratt Street	gunsmith
1887	304 East Pratt Street	gunsmith

[Baltimore city directories]

SHOBEL, J. G. Washington, D. C.

Shobel, J. G.

1863	322 Va. ave	gunsmith

[Hutchinsons Business Directory, Washington, D.C., & Georgetown]

SHOT

The manufacture of shot in Baltimore began in 1822, by a company of which Col. Joseph Jamison was president. The tower was on North Gay Street, and was 160 feet high. This tower was pulled down in 1845. (There was also an old shot tower on South Eutaw Street which was demolished in 1851.) The imposing tower of the Merchant's Shot-Tower Company, which arrests the eye of every visitor to this city was erected in 1828. The corner-stone was laid by Charles Carroll of Carrollton and bears upon it this inscription: June 2, 1828, this stone was laid by Charles Carroll of Carrollton, the only survivor of the signers of the Declaration of Independence of the United States. The diameter of the tower is 50 feet at the foundation and 25 at the top; the height is 250 feet, being by one foot the highest shot-tower in the world. The shot from sizes B to 3T, inclusive, are dropped to the foundation, the smaller sizes from the middle floor. The capacity of this tower is over 100,000 bags in a season, or nearly 500,000 bags yearly.
On Howard and Montgomery Streets the shot-tower of the Baltimore Lead-Works was erected in 1877. It is 110 feet high, 24 feet square at the bottom, and 12 feet square at the top. By the introduction of a current of cold air, the extra ordinary height of the drop is obviated and dispensed with. These towers manufacture the most perfect shot that is offered to the public, and command a preference readily in every market to which they are shipped.
[HISTORY OF BALTIMORE CITY, pp. 420-421]

SHNYDER, CHARLES Baltimore

I have given over the business to Charles Shnyder, who carrie on the Gunsmith's business for himself in my shop, Baltzer Shaeffer.
[Maryland Gazette & Baltimore General Advertiser, 13 March 1787]

SHROYER, MATHIAS Taneytown

His name was also spelled Schroyer or Sroyer.

In the latter part of the eighteenth century there stood on or near what is now the lime-kiln in Taneytown a long, low frame building, in which were manufactured, by a Mr. Sroyer, such implements as fireshovels, tongs, hoes, nails, and guns. The venerable Mrs. Elizabeth Thompson has in her possession a heavy pair of tongs made at this primitive factory, on which is inscribed the date of manufacture, 1796. The establishment was under the supervision of the government, or at least that portion which embraced the manufacture of firearms, and was annually visited and inspected by government officials. The machinery was very crude and simple. Instead of the belts, pulleys, emery-wheels, and ordinary appurtenances of a modern factory, regulated by steam, and by means of which a gun-barrel or other iron implements can be polished in a few moments, the only contrivance then known and used was a huge grindstone turned by an old horse. With these limited facilities, however, many guns were made for the government. The factory burned down early in the present century, and was never rebuilt, the government factory having been subsequently transferred to Harper's Ferry, Va.
[HISTORY OF WESTERN MARYLAND, p. 837]

Contractors with the government under Act of July 5, 1798 for Charleville pattern muskets at $13.40 per stand.

(4) Eli Williams, Williamsport, Md.	2000 muskets
(5) Nicholas White, Thomas Crabb, Jacob Metzger & Christopher Barnhizle, Frederick Town, Md.	1000 muskets
(14) Alexander Clagett, Hagerstown, Md.	1000 muskets
(25) Mathias Shroyer, Taneytown, Md.	1000 muskets
27 contractors	40,200 stands

[U. S. ORDNANCE, Vol. I, p. 20]

Shroyer, Mathias 1798 part of Resurvey on Brothers Agreement 100 acres $140 value Taney Town
[Assessment Book, Frederick County, 1798, Hall of Records #19268]

Indenture made June 29, 1805 [deed] between Mathias Schroyer of Frederick County and state of Maryland of the one part and [to] John McKaleb of the County and state aforesaid of the other part . . . tract or parcel of land lying and being in Frederick County aforesaid being part of a tract called the Resurvey On Brothers Agreement.
[Frederick CCH, Liber W. R. 27, Folio 200]

The following entries are from the "Day Book [of] John M. Kaleb & Co. Taney Town May 9th 1801," which is owned by Edward R. and Helen Flanagan, 25 Flanagan Rd., Thurmont, Maryland. The number preceeding the name is the account number assigned by the keeper of the day book. The book begins early in 1801 and continues until 14 April 1809, but Shroyer's last entry is dated 9 July 1803.

		L	s	d
1802 August 18	88 Mathias Sroyar			
	pr Daughter 1 pr ton cards 3/9		3	9

1802 September 22 void Mathias Sroyar
fiels 15/10
1802 November 6 88 Mathias Shroyer
1 gall whisky 5/4 1 gall Rum 7/ 13 3
1802 November 29 88 Mathias Shroyer
3½ yd R. Coting 22/9 3 yd flanl 11/3
3½ yd Cloth 5/6 2 13 3
1803 January 24 88 Mathias Shroyar
4¾ yds Linen at 3/9 19 9
1803 March 26 88 Mathias Sroyar
per son 1 g. whisley 4 6
1803 July 9 88 Mathias Sroyer
per Daughter ½ Doz. pewter plats wt 5 lb 16/8 16 8

SHULZ, DANIEL Washington, D. C.

Shulz, Daniel
1867 7 A south gunsmith
[Boyd's Business Directory, Washington, D.C., Georgetown & Alexandria]

SHUNK, ISAAC Taneytown

This indenture made the 18th day of February 1801 between Peter Shunk and Philip Cramer both of Frederick County and state of Maryland witnofseth that the said Peter Shunk hath this day bound his son Isaac Shunk unto Philip Cramer Gun Smith for the term of 3 years and 9 months . . .
[Frederick County Indentures, Hall of Records, p. 287]

SILGE, ALBERT Baltimore

Prior to 1887 Silge's occupation was listed as a machinist.

Silge, Albert
1887-89 401 Eastern Avenue gunsmith
[Baltimore city directories]

SIMMONS, ISAAC

August 12, 1776 Council of Safety
Ordered that the Treasurer of the Western Shore pay to Isaac Simmons 5 pounds for two Guns.
[American Archives, 5th Series, Vol. I, p. 1352]

August 13, 1776 Council of Safety
Ordered That Eastern shore Treasure pay to Isaac Simmonds five Pounds for two Guns.
[AM, p. 198]

SIMMONS, JAMES W. L. Cumberland & Baltimore

Simmons, J. W. L. 1878 Cumberland gun & locksmith
Simmons, J. W. L. 1880 Cumberland gun & locksmith
Simmons, J. W. L. 1882 Cumberland gun & locksmith
[The Maryland Directory, 1878, 1880, 1882; J. Frank Lewis & Co. Publishers]

Simmons, James W. L.
1885-89 304 Aliceanna Street gunsmith
[Baltimore city directories]

SISCO BROTHERS Baltimore

Sisco Brothers
1865 95 West Baltimore Street
(ad) Military goods, particular attention given to banners, gilt & plated bullion, frenges, stars, tassels, etc.
1866 95 West Baltimore Street
Sisco Brothers (Charles T. & J. Edward)
1867 14 North Charles Street
1868-70 southwest corner of Charles & Fayette Streets
[Baltimore city directories]

SLEGHEL, GOTTLIEB Washington, D. C.

Sleghel, Gottlieb
1858 h 170 6th west gunsmith
[Business directory, Washington, D.C.]

SMITH, JOHN M. Baltimore

Smith, John M.
1847 Saratoga Street lock & gunsmith
[Baltimore city directory]

SMITH, WILLIAM D. Baltimore

Smith, William D.
1845 16 Pine Street locksmith
1851 16 Pine Street blacksmith
1871-74 Uhler Alley near Charles Street gunsmith
1875-81 East Fayette Street gunsmith
[Baltimore city directories]

SNEIDER, CHARLES EDWARD Baltimore

Experimented with different actions and worked in Europe in 1842. On coming to Baltimore before the Civil War, he was employed by Poultney and Trimble and refined the incomplete and almost forgotten Gilbert Smith carbine. This was patented by Sneider but assigned to Poultney, who reaped great profits. Subsequently several other assignments of his inventions were made jointly to himself and Poultney. By 1864 he was working on his own as a custom gunsmith, later with Duncan Clark and his sons. Sneider's many inventions and patents include pistols, military percussion and metallic cartridge carbines, muzzle to breech loading firearms, magazine firearms, as well as sights, a device for crimping cartridges and other mechanism improvements. His arms bear the stamp "Sneider Baltimore Patenter" or "E. Sneider Baltimore."

Sneider, Edward
1864 34 Barre Street gunsmith
1867 66 South Sharp Street gunsmith
Clark & Sneider (Duncan C. Clark, C. Edward Sneider)
1873-84 214 West Pratt Street gun manufacturers

Sneider, C. Edward

1885-88	155 Paca Street	gunsmith
1889	545 Paca Street	gunsmith

[Baltimore city directories]

SNEIDER, CHARLES LOUIS Baltimore

Sneider, C. Louis

1880-88	155 Paca Street	gunsmith
1889	545 Paca Street	gunsmith

[Baltimore city directories]

SNEIDER, CHARLES W. Baltimore

On 21 December 1875 he and his father, C. Edward, Sneider, patented a locking lever for sporting guns. In the next few years Charles W. was busy applying for other patents for mainly breech loading hammer and hammerless shotguns. Many fine Sneider patents can be illustrated.

Sneider, Charles W.

1875-77	216 Pratt Street	gunmaker
1878-79	153 Paca Street	gunsmith
1880-84	58 South Greene Street	gunsmith
1885-86	214 Pratt Street	gunsmith
1887-89	209 Sharp Street	gunsmith

[Baltimore city directories]

SNYDER, LEONARD Washington County

Snyder is known chiefly for his work as a rifle barrel maker in Franklin County, Pa., immediately north of Washington County, Md., and barrels holding his "L S" stamp are commonly found on the rifles of various gunsmiths of Franklin Co. But he also made rifles. On 29 March 1816 he was a resident of Washington Co., Md., and he signed a document giving freedom to a Negro slave named Richard for the sum of $10. He apparently moved to Franklin Co. the same year and again signed a document insuring Negro Richard of his freedom. He could read and write only German.

SOMES, DANIEL E. Washington, D. C.

August 18, 1863

Be it known that I, Daniel E. Somes of Washington, in the District of Columbia, have invented a new & useful Improvement in Fire-Arms . . . the object of my invention is to produce a gun or cannon of such a form & construction that when it is fired the ball shall be firmly held in the barrel until the maximum explosive force of the charge is obtained.

[U. S. Patent Office, #39,592]

SPARROW, THOMAS Annapolis

A silver-hilted small sword with a touch mark attributed to Thomas Sparrow is in the collection of the Historical Society of Delaware. Considerable research on Sparrow was done many years ago and published in 1930 in the book MARYLAND SILVERSMITHS 1715 -- 1830 by J. H. Pleasants and Hl Sill. From this account comes the following information:

Thomas Sparrow was born about 1746 according to Pleasants, but inasmuch as he advertised himself as a goldsmith and jeweller in 1765, it seems likely that he was born a few years earlier. His father was Thomas Sparrow, Gentleman of Annapolis, who died in 1753. Young Thomas was brought up in the home of Jonas Green, Maryland's notable colonial printer and publisher of the MARYLAND GAZETTE. Apparently he apprenticed in Philadelphia. On 21 March 1765 he placed an extensive advertisement in the MARYLAND GAZETTE:

THOMAS SPARROW, Goldsmith and Jeweller, from PHIL—ADELPHIA, has just open'd SHOP near St. Ann's Church, in South-East Street, Annapolis, Maryland. Makes all sorts of GOLD and SILVER WORK, such as Tureens, Tea Kettles, Bread Baskets, Chasing Dishes, Coffee Pots, Butter Boats, large and small Waiters holding from 1 Glass to 18, fluted Soup Ladles, Tankards of all sizes, Sugar Dishes, Punch Bowls, Tea Pots, Cream Pots, Quart, Pint and Half Pint Cans, Slop Bowls, Porringers, Soup Spoons, Table Spoons of all sorts, Pap Ditto, Tea Ditto, round Salts, Catters, Virgin Candlesticks, Pepper Boxes, Punch Strainers, Tea Ditto, Tea Tongs plain or pierc'd, chas'd, pierc'd, gadroon'd, and plain Sword Hilts, carv'd and plain Stock Buckles, Sleeve Buckles, carv'd, pierc'd, and plain Shoe and Knee Buckles, Heart and Diamond Breast Buckles, Broaches, Rings, Coat and Jacket Buttons of all sizes and Fashions, Tortoiseshell Buttons tipt with Silver, Sleeve Buttons and Studs of all Sorts, Stock Clasps, Shoe Ditto, Pocket Book Ditto, Whistles and Bells chas'd or plain, Whip and Cane Heads, chas'd and plain Snuff Boxes, Shell Ditto tipt with Gold and Silver, Silver Spurs, Scissar and Pincushion Chains, and Hearts, Watch Chains of all Sorts, Seals of all Sorts, Thimbles, Spectacle Rims and Cases, Arm, Hair and Wrist Plates, for *Indians*, Gorgets, and all Sorts of Work for Traders, Garter Buckles, Belt Ditto, Waistband Ditto, Gold Buttons and Studs, Gold Thimbles, Gold Rings, Tortoiseshell Rings lined with gold, Motto Rings, and Stone Motto Rings, with 1, 2, or 3 Doors, to lock, for secret Names, and to contain Hair, (such as was never made before), Gold Lockets, Doctor Instruments of all Sorts, Silver Pipes, Mason Medals and Instruments, Parlour Sconces, Candlesticks, Barometers, Antigulars, Hydrometers, Ivory Cups and Balls, Ivory Rules, Billiard Balls, Dies, Maces, Guns and Pistols mounted with Silver, Engraving Plates, Cyphers, and Letters at a Penny each Letter, and manufactures all Kinds of Stones.

N. B. The said THOMAS SPARROW intends to Sell at the *Philadelphia* Prices, and marks all his Work with the Word STERLING, which will prevent Impositions that are too often practised, to the prejudice of the Buyers, and much to the Discredit of true Professors of this Business.

He gives the best Prices for old Gold, Silver, Chrystal or *Baltimore* Stones, Copper, Brass, Pewter, and Lead; and will

engage to Clean and Keep all his PLATE WORK in Repair for 7 Years *Gratis*.
[Maryland Gazette, 21 March 1765]

The advertisement is notable for the thoroughness with which it covers the broad range of work done by a colonial silversmith. For students of arms making in Maryland it is notable for mentioning silver hilts, plain, chased, pierced, and gadrooned; guns and pistols mounted with silver, gorgets and other silver items for traders, and arm, wrist, and hair plates for Indians. The list, of course, represents the things that he was prepared to make or otherwise supply if he got orders or commissions for them.

Sparrow worked as a confidential agent for the Maryland Council of Safety in 1776. The last record of him is in a grand jury list for Annapolis for February 1784.

SPIKER, BENJAMIN — Allegeny County

Allegeny County
Spiker, Benjamin age 22 born Maryland gunsmith
[1850 Federal Census]

SPILLER, EDWARD N. — Baltimore

Spiller, Edward N. (Meredith, Spencer & Co.)

1858	32 North Fremont	

Spiller, Edward B. (Meredith, Spencer & Co.)

1859-60	168 West Biddle Street	whole sale dry goods merchants

[Baltimore city directories]

STABLER, EDWARD — Sandy Springs

March 14, 1865
Be it known that I, Edward Stabler, of Sand Springs, Montgomery County, and State of Maryland, have invented certain new and useful Improvements in Magazine Fire-Arms . . . Limiting or arresting the movement of the carrier-block of a magazine-gun at any desired point, for the purpose of converting the arm into a single-loader, substantially as described.
[U. S. Patent Office, Patent #46,828]

STEWART, JOHN — Baltimore

He signed "J. Stewart Baltimore" in script on the rib of a small, flintlock, double barrel shotgun and then inlaid his name with gold. A longrifle is also inlaid the same "John Stewart." This was an elegant feature, yet the over all craftsmanship is not above average.

Stewart, John

1810	6 Light Street	gunsmith
1812	dwelling Bank Street	gunsmith
1816-19	28 Light Street	gunsmith

[Baltimore city directories]

STOKES, JOHN — Baltimore

A small, well engraved, double barrel shotgun has survived from his shop and is signed in script on the rib and locks.

Stokes, John

1847-54	48 Aisquith Street	gunsmith
1855-56	154 Dallas Street	gunsmith
1857	134 Mullikin Street	gunsmith
1858-60	Orleans near Central Avenue	gunsmith
1864	Sterling near Monument Street	gunsmith
1875-86	14 West Fayette Street	gunsmith

[Baltimore city directories]

STONESIFER, CHRISTIAN EMORY — Boonsboro

His longrifles are long, very plain, usually large bore and an exact copy of his father, John's. The barrels are signed in script on the top facet "C. Emory Stonesifer."

C. E. Stonesifer, gunsmith, Boonesboro, Maryland
Would take this method of informing the citizens of Boonsboro' and vicinity, that he continues to carry on the White smithing & Gunsmith Business, at the Old Stand, where his father done business, and would invite them to favor him with their calls, feeling assured that from his knowledge of the business, he can render perfect satisfaction to all who may favor him with their work. New Rifles & Shot Guns, of Various Kinds, Always On Hand.
[*Boonsboro Odd Fellow*, 23 August 1860]

C. E. Stonesifer enlisted on Aug. 3, 1861 at Williamsport, Md., in the 1st Virginia union cavalry U. S. A. as a 2nd. Lieutenant under Col. Lemons. Stonesifer transier to the 1st Regiment Co. I Maryland cavalry U. S. A. and resigned Jan. 9, 1862.
[HISTORY & ROSTER OF MARYLAND VOLUNTEERS, WAR OF 1861-65]

STONESIFER, JOHN — Boonsboro

Father of C. Emory Stonesifer. Probably was not a life long resident of Boonsboro because in 1818 the residents advertised that they stood in need of the following mechanics: one whitesmith, one silversmith, one gunsmith, one cabinet-maker and one blue dryer. The author has had three longrifles made by John. All have been plain, large bore, percussion, mountain type rifles. The patch box and toe plates were the limit of his engraving, which was well done. The top of the barrels are signed "J. Stonesifer" in flowing script and one lock plate was also signed this way.

A relief-carved longrifle by Stonesifer has been located, demonstrating that he was a quality craftsman. The patch box has six piercings, finely engraved as are other inlays, showing Shenandoah Valley influence. The barrel is also signed in flowing script as above, except on a silver plate.

Stonesifer, John   Washington County

1 male of 16 years
1 male under 16 years
7 females

[1790 Federal Census]

Washington County, Boonsboro District

John Stonesifer	age 37	sex M	occ. Gunsmith	prop. $350
				birth Maryland
Mary	28	F		Maryland
Emory	12	M		Maryland
Jacob	9	M		Maryland
Robert	4	M		Maryland
Laura	1	F		Maryland

[1850 Federal Census]

STROBEL, JOHN G.   Washington, D. C.

Strobel, John G.

1864	322 Va. Ave.	gunsmith

[Hutchinsons Business Directory, Washington, D.C., & Georgetown]

Strobel, John G.

1866	822 Va. Ave.	gunsmith
1867	322 Va. Ave.	gunsmith

[Boyd's Business Directory, Washington, D.C., Georgetown & Alexandria]

STRONG, SAMUEL   Washington, D. C.

On 19 May 1863 his second patent, #38,643, was a further improvement relating to military gate hinged breech load rifles.

Be it known that I Samuel Strong, of Washington, in the District of Columbia have invented new & useful Improvement in Breech-Loading Fire-Arms . . . consists in mounting the hammer upon & securing it to the hinged gate.
[U. S. Patent Office, #37,208]

T

TAUBER, JOHN — Cumberland

Tauber, John	1878	Cumberland	gun & locksmith
Tauber, John	1880	Cumberland	gun & locksmith
Tauber, John	1882	Cumberland	gun & locksmith

[The Maryland Directory, 1878, 1880, 1882; J. Frank Lewis & Co. Publishers]

TAYLOR, JAMES

May 1, 1776 Council of Safety
Ordered That the Treasure of the Western Shore pay to James Taylor five Pounds Currency in Part Payment for making Gun Carriages.
[AM, p. 395]

June 27, 1776 Council of Safety
Ordered That the Treasure of the Western Shore pay to James Taylor, fifteen pounds, thirteen shillings, and eleven pence, for Gun Carriages.
[AM, p. 524]

August 6, 1776 Council of Safety
Ordered That the Treasure of the Western Shore pay to James Taylor four pounds five Shillings for a Musquets.
[AM, p. 174]

July 4, 1777 Council of Safety
That the said Treasure pay to James Taylor seven pounds seventeen shillings and six pence Curry. due on Acct. passed.
[AM, p. 308]

November 7, 1780 Council Proceedings
That the said Treasure pay to David Poe Esq. Twenty one Pounds, five shillings & seven pence half penny in lieu of Eight hundred & fifty one Pounds, five shillings for the use of James Taylor.
[AM, p. 215]

TERYAR, WILLIAM — Frederick

Teryar, William	1878	Frederick	gunsmith
Teryar, William	1880	Frederick	gunsmith
Teryar, William	1882	Frederick	gunsmith

[The Maryland Directory, 1878, 1880, 1882; J. Frank Lewis & Co. Publishers]

This gunsmith probably is the same person as W. G. Tyeryar of Frederick.

THOMAS, LEWEN W. & PHILIP E. — Baltimore

James H. Merrill established the business in the city in 1845, later incorporating Ferinond C. Latrobe and two Thomas's in the business. The Thomas's were still active with the firm after it became the Merrill Patent Fire Arms Manufacturing Company, but nothing can be found of them before or after the firm was active.

Merrill, Latrober & Thomas

1856-59	6 Hanover Street	gunsmiths

Merrill, Thomas & Company

1860	239 West Baltimore Street	gunsmiths

Merrill Patent Fire Arm Manufacturing Company

1861-67	239 West Baltimore Street	gunsmiths

[Baltimore city directories]

THOMAS, WILLIAM

July 8, 1776 Council of Safety
Ordered That the Treasure of the Western Shore pay to William Thomas six Pounds Currency for two Musquets.
[AM, p. 9]

January 7, 1777 Council of Safety
Ordered That the Western Shore Treasure pay to William Thomas Eighteen Pounds.
[AM, p. 25]

April 12, 1777 Council of Safety
Ordered That the Western Shore Treasure pay to William Thomas eighteen pounds, for Amt. of Acct.
[AM, p. 208]

August 14, 1781 Council Proceedings
That the said Treasure pay to William Thomas Thirty six pounds, fourteen schillings and six pence of the same Emission of the money appropriated as foresaid due him per Account passed by the Depy. Aud.
[AM, p. 559]

THOMIZ, ANDREW — Baltimore

Thomiz, Andrew	1850	age 39	locksmith	birth Germany
Christinia	1850	39		Germany
Hellon	1850	11		Maryland
Margaret	1850	10		Maryland
Caroline	1850	8		Maryland
James	1850	6		Maryland
Andrew	1850	4		Maryland
John	1850	2		Maryland

[1850 Federal Census]

1851-60 Thomiz, Andrew lock & gunsmith 50 Light Street
[Baltimore city directories]

THORNTON, W. — Washington, D. C.

May 21, 1811
Be it known that we W. Thornton & J. Hatkill of Washington city, in the District of Columbia have invented a new & useful Improvement in Fire-Arms.
[U. S. Patent Office]

TILESTON, WILLIAM — Georgetown

Be it known that I William Tileston, of Georgetown, in the county of Washington, in the District of Columbia, have invented a new & useful Improvement in Fire-Arms . . . the object of my invention is to construct a lock for revolving firearms to prevent the cylinder thereof from revolving when not in immediate use.
[U. S. Patent Office, #44,126]

TIMMINS, EDWARD

February 8, 1776 Timmins to Council
I apply for employment in making steel ramrods which I am willing to make as reasonable as Possible which I think cannot be done under five shillings Pr. rod if the gentlemen think proper to employ me I must beg you will assist me in getting some steel as it is very scarce.
[AM, p. 144]

March 21, 1776 Council of Safety
The Council contracted with Edward Timmins for the making of steel Ramrods at five Shillings each.
[AM, p. 272]

May 3, 1776 Council of Safety
Ordered That the said Treasure pay to Edward Timmins seven Pounds Five Shillings and Five Pence for 32 Steel Ramrods.
[AM, p. 402]

July 15, 1776 Council of Safety
That the said Commisary deliver to Edward Timmins 6 Musquets to be repaired.
[AM, p. 47]

August 12, 1776 Council of Safety
Ordered that the Treasurer of the Western Shore pay to Edward Timmons 3 pounds 12 shillings 6 pence for a Musket.
[American Archives, 5th Series, Vol. I, p. 1352]

August 23, 1776 Council of Safety
Ordered that the Treasurer of the Western Shore pay to Edward Timmons 3 pounds for repairing Arms.
[American Archives, 5th Series, Vol. I, p. 1358]

September 27, 1776 Council of Safety
Ordered that Western Shore Treasurer pay to Edward Timmons 10 pounds for three Guns.
[American Archives, 5th Series, Vol. II, p. 639]

October 3, 1776 Council of Safety
Ordered That said Treasure pay to Edward Timmins ten Pounds, for three guns.
[AM, p. 317]

November 9, 1776 Council of Safety
Ordered That said Treasurer pay to Edward Timmins seven Pounds four shillings for two guns.
[AM, p. 433]

February 15, 1777 Council of Safety
That the said treasure pay to Edward Timmonds Three Pounds Ten Shillings for a Gun.
[AM, p. 138]

March 12, 1777 Council of Safety
Ordered That the Western Shore Treasure pay to Edward Timmons Seven pounds ten Shillings for 2 Guns.
[AM, p. 168]

May 15, 1777 Council of Safety
Ordered That the Western Shore Treasure pay to Edward Timmonds twenty six Pounds for seven Musquets.
[AM, p. 254]

November 8, 1777 Council of Safety
That the said Treasure pay to Edward Timmons one pound one shilling, thirteen shillings and six pence, four pounds six shillings and three pence due per Account passed and the further Sum of four pounds for a Musqet and Bayonet bought of him for the Public by John Shaw receipt.
[AM, p. 413]

Ann-Arundel County

Timmins, Edward	3 males over 16 years
	1 male under 16 years
	5 female
	5 slaves

[1790 Federal Census]

TOULSON, ALEXANDER I — St. Mary's

There must have been two Toulsons by the same name because the younger man had indentures in the mid-1700's and was wed to a woman named Mary.

April 9, 1663
Know ye that whereas Alexander Toulson St Marries in the provence of Maryland gunsmith by his petition and for consideration a tract of land.
[Hall of Records, Liber 18A, #15]

TRIMBLE, DAVID B. — Baltimore

The city directories reveal the existence of an earlier firm of Poultney & Trimble as hardware and paint dealers, during a period of about twenty years ending in 1848. It is probable that the same two families were involved in both enterprises. At the conclusion of the re-established firm, David became associated with C. B. Kleibacker. Both were listed as gunsmiths. Shotguns have survived that bear the names "Poultney & Trimble" and "Trimble."

Poultney & Trimble

1860-76	200 West Baltimore Street	gunsmith

Trimble & Kleibacker

1877-80	200 West Baltimore Street	gunsmith

[Baltimore city directories]

TRIMBLE, THOMAS — Baltimore.

Probably a son of David B. Trimble. He carried on the business at the old address for a year.

Trimble, Thomas	1850	age 40	stone factory	birth Ireland
Mary A.	1850	37		New York
Robert M.	1850	15		Maryland
Mary Jane	1850	11		Maryland
Josephine	1850	9		Maryland
Thomas J.	1850	7		Maryland
Hesther	1850	5		Maryland
Matilda Clark	1850	22		Pennsylvania

[1850 Federal Census]

Trimble, Thomas

1881	200 West Baltimore Street	gunsmith

[Baltimore city directory]

TROUTMAN, BEN Frederick

Ben Troutman began his apprenticeship in gunsmithing under George Rizer at Cumberland in 1805 according to William Bowers. According to Kauffman (1960), Benjamin Troutman, gunsmith, is listed on tax records for Southampton Township, Somerset County, Pennsylvania, for the years 1810, 1811, and 1817.

Ben Troutman informs the public, generally, that he carries on the gunsmith business on Market Street at the sign of the gun, where all orders in his line of making or repairing will be punctually attended to and especially rifles.
[Frederick Examiner, 14 September 1810]

TYERYAR, FRANK Frederick

1886 Tyeryar, Frank 41 East 4th St. gunsmith & dealer in sportsmen's goods
[GENERAL DIRECTORY OF FREDERICK CITY & COUNTY]

TYERYAR, FREDERICK F. Frederick

1895 Tyeryar, Frederick F. 41 East 4th St. gunsmith
[GENERAL DIRECTORY OF FREDERICK CITY & COUNTY]

TYERYAR, WILLIAM Frederick

1886 Tyeryar, William 41 East 4th St. gunsmith
[GENERAL DIRECTORY OF FREDERICK CITY & COUNTY]

A halfstock percussion rifle of about .60 caliber with engraved brass furniture and a German silver cap box, signed on the barrel in script "W. G. Tyeryar Frederick Md." was sold at auction near Frederick in September 1976.

William Teryar who was listed as a gunsmith in Frederick in 1878, 1880 and 1882.

U

UNSELL, HENRY — Frederick

His only other deed in the Frederick County Court House was recorded in November of 1768. John Unsell sold the house on lot number twenty in Hagerstown that he had purchased from Abraham Unsell and probably moved to Frederick in Henry's former property. The relationship between the two has not been established.

Indenture made April 24, 1769 between Henry Unsell of Frederick County and province of Maryland locksmith of the one part and Conrad Grosh and John Unsell of the other part . . .
[Frederick CCH, Liber M, p. 189]

UNSELD (UNSELL), JOHN — Frederick

His last name was also spelled Unselt and Unsold. First indenture was recorded 5 September 1766. He lived in Hagerstown until about 1760. Relationship to Henry Unsell has not been established.

Indenture made September 5, 1766 [deed] I Abraham Unsell do hereby sine over all my rights and title unto John Unsell, of house and lot of the number of twenty in Hagers Town.
[Frederick CCH, Liber K, Folio 725]

Indenture made April 24, 1769 [deed] between Henry Unsell of Frederick County and province of Maryland locksmith of the one part and [to] Conrad Grosh and John Unsell of the other part . . . Tract of land lying and being in Frederick County being all that tract of land called Resurvey On Beauty.
[Frederick CCH, Liber M, Folio 189]

Indenture made May 9, 1770 [deed] between Henry Shrock of Frederick County and province of Maryland of the one part and [to] John Unselt of Frederick County and province of Maryland of the other part . . . Lott and portion of ground in Elizabeth Town.
[Frederick CCH, Liber L, Folio 655]

April 4, 1776 Council of Safety
Ordered That Mr Henry Yost and Mr John Unseld deliver to Capt. George Stricker all the Musquets and Bullet Moulds which are now finished by them for use of this Province.
[AM, p. 308]

May 3, 1776 Council to Unseld, Fredk. City
On the 14th of December 1775, you entered into an Engagement to make and deliver on or before the first Day of May 1776 to the Council of Safety or their Order, at George Town or Baltimore Town eighty good substantial proved Muskets 3½ Feet in the Barrell ¾ Inch in the Bore, with good double bridle Locks, black Walnut or Mape Stocks & plain strong Brass mounting, Bayonets with steel Blades 17 inches long, steel Ramrods, double Screws, priming Wires and Brushes fitted thereto with a pair of brass Moulds to cast twelve Bullets on one Side & on the other Side to cast shot of such a size as the Musketts will shamber three of them.
Twenty nine Muskets very roughly made with Pickers and Wipers and one Bullet Mould have been delivered but not one Bayonet. The Residue of the Arms and the bayonetts & are immediately wanted, the Province has been greatly disappointed by your Non Compliance with your Contract, We therefore require you will immediately send them on Pain of answering all the evil Consequences that may ensue.
[AM, p. 401]

June 14, 1776 Council of Safety
Ordered That the Commissary of Stores deliver out of the Magazine at Annapolis to Henry Yost and John Unseld all the Musquets by them respectively made, which were ordered by the late Convention to be returned to the said Yost and Unseld as unfit for service . . .
[AM, p. 489]

June 27, 1777 Council of Safety
Ordered, That the western shore Treasure pay to Dectr. Henry Snebely one hundred & seven pounds Curry. for the use of John Unsold due him on Acct.
[AM, p. 301]

V

VANSANT, J. A. Easton

Vansant, J. A.

1887 Easton gunsmith

[Maryland Directory & State Gazetteer, 1887, Baltimore Publishing Co., 12 E. Baltimore St.]

VAN TROMP, JOHN Baltimore

He produced a 40 caliber, 36 inch, octagon barrel, half stock rifle with a back lock signed in script "J. Van Tromp Balto."

Van Tromp, John

1852 5 Fayette Street gunsmith

[Baltimore city directory]

VON KAPFF, J. HERMAN Baltimore

Walsh, Hoen & Von Kapff

1887-88 27 South Charles Street gun importers

[Baltimore city directories]

W

WAGNER, JACOB — Washington, D. C.

Wagner, J.
1860 478 15th west gunsmith
[Business directory, Washington, D.C., Georgetown & Alexandria]

Wagner, Jacob
1862 381 D north gunsmith
[Business Directory, Washington, D.C., & Georgetown]

Wagner, Jacob
1863 381 D north gunsmith
[Hutchinsons Business Directory, Washington, D.C., & Georgetown]

WALBACH, J. J. B. — Baltimore

November 6, 1866
Be it known that I, John J. B. Walbach, of the city and county of Baltimore, in the State of Maryland, have invented a new and Improved War-Rocket . . . The nature of my invention consists in constructing a war-rocket in such a manner as to secure a long and correct and direct flight, (having wings,) and to insure an explosion of the magazine upon striking.
[U. S. Patent Office, Patent #59,487]

WALRAVEN, JOHN — Baltimore

Walraven, John
1800-1810 137 Baltimore Street merchant
[Baltimore city directories]

WALSH, HOEN & VON KAPFF — Baltimore

The company was composed of T. Yates Walsh, John H. Hoen, and J. Herman Von Kapff.

Walsh, Hoen & Von Kapff
1887-88 27 South Charles Street gun importers
[Baltimore city directories]

WALTERHAVER, CHRIST — Baltimore

Walterhaver, Christ
1851 29 New Church Street gunsmith
1854 101 Pearl Street gunsmith
[Baltimore city directories]

WALTERS, PETER — Frederick County

This is the only known indenture or record of Walters. He may have been a blacksmith, for the tools listed are blacksmith tools and not necessarily gunsmith tools.

Indenture made June 10, 1800 [bill of sale] I Peter Walters of Frederick County state of Maryland for and in consideration of the sum of 75 pounds current money to me in hand paid by Nicholas White, Christopher Barnhizle, and Jacob Metzger of Frederick County state aforesaid the following property – towit 3 iron anvils, 4 vices and all my smith tools of every description, 1 pair large bellows, 2 grindstones and a large wheel that turns . . .
[Frederick CCH, Liber W. R. 20, Folio 19]

WARFIELD, AZEL

September 9, 1777 Council of Safety
Ordered That the western shore Treasure pay to Azel Warfield fifteen pounds for a Gun.
[AM, p. 367]

September 9, 1777 Council of Safety
That the Armourer deliver to Azel Warfield twelve Gun Barrels and twelve Locks to be stocked taking his receite for them.
[AM, p. 367]

September 9, 1777 Council of Safety
Ordered That the Commissary of Stores deliver to Azel Warfield 120 yards Kendal Cotton and 100 yards Cruden for 6 Guns purchased of him.
[AM, p. 367]

December 12, 1777 Council of Safety
That the said Treasure pay to Azel Warfield seventy eight Pounds as per Account passed the Board.
[AM, p. 434]

WARFIELD, A. M. — Washington, D. C.

Warfield, A. M.
1875 902 8th se, h G nr 5th se gunsmith
[Boyd's District of Columbia Directory]

WARNER, ANDREW ELLICOTT — Baltimore

At least two known silver sword hilts were produced in Baltimore by Warner and bear his touch. One is a short-sword with eagle-head pommel and knuckle-bow. The other is a bird-head pommel piece. The essential details of Warner's life have been in print for many years, for he was a prominent silversmith of Baltimore in the 19th century.

Andrew Ellicott Warner, the second son of Cuthbert Warner and Ann Smith, was born in Harford County, Md., 27 Nov. 1786. He was named for Andrew Ellicott, the mathematician. He doubtless learned his trade from his father, a watch maker and silversmith, who had removed from Bucks County, Pa., to Harford County, Maryland, at the close of the Revolution, and from the latter place came to Baltimore about 1798 or 1799. In 1805 Andrew E. Warner formed a partnership with his eldest brother Thomas Warner with their silversmith shop at 5 North Gay Street; this partnership was not dissolved until about 1812. From this date until his death in 1870, "Andy" Warner, as he was called, had his shop at the same place, working continuously there at his forge and anvil until his death. In the War of 1812, he served as captain of the 39th Regiment and took part in the Battle of North Point.

Warner married, by license dated 25 June 1812 Dorothy Litsinger of Baltimore, and had by her several children. He died January 16, 1870, in his eighty-fourth year. His son, Andrew E. Warner, Jr., who had been associated with his father in business, removed the shop the year after the latter's death from Gay Street to 135 West Baltimore Street, where the business was conducted until his retirement in 1893.
[J. H. Pleasants & H. Sill (1930) MARYLAND SILVER-SMITHS 1715 - 1830, p. 193.]

WASSMAN, GEORGE FREDERICK — Georgetown

Wasserman, George F.
1860 — 201 Penn Av. Georgetown — gunsmith
[Business directory, Washington, D.C., Georgetown & Alexandria]

Wassman, Frederick
1862 — 201 Penn. Ave. — gunsmith
[Business Directory, Washington, D.C., & Georgetown]

Wassman, George F.
1863 — 201 Penn. Ave. — gunsmith
[Hutchinsons Business Directory, Washington, D.C., & Georgetown]

Wassman, F.
1864 — 201 Pa. Ave. — gunsmith
[Hutchinsons Business Directory, Washington, D.C., & Georgetown]

Wassman, G. F.
1865 — 201 Pa. Ave. — gunsmith
[Boyd's Business Directory, Washington, D.C., Georgetown & Alexandria]

Wassmann, G. F.
1866 — 201 Pa. Ave. — gunsmith
[Boyd's Business Directory, Washington, D.C., Georgetown & Alexandria]

Wassmann, George F.
1867 — 201 Pa. Ave. — gunsmith
[Boyd's Business Directory, Washington, D.C., Georgetown & Alexandria]

Wassman, George F.
1868 — 201 Pa. Ave. — gunsmith
[Boyd's Business Directory, Washington, D.C., Georgetown & Alexandria]

WATSON, JOHN — Baltimore

Watson, John
1817-18 — Montgomery Street — gun bores
[Baltimore city directories]

WATTS, WILLIAM L. — Baltimore

Watts, William L.
1835-40 — Conway Street adjoining Gist's Factory — shot maker
1847 — Gough Street & East Broadway — gunsmith
1851 — 7 Hillen Street — gunsmith
1855-56 — 107 Gough Street — gunsmith
[Baltimore city directories]

WEANER, JOHN H. b. 1804 — Washington, D. C.

1850 Washington, D.C. real estate value $500

Name	age	born	occupation
Weaner, John H.	46	Germany	gun & Locksmith
" Catharine	36	Germany	
" Sophia	13	Washington, D.C.	
" Charles	10	Washington, D.C.	
" Catharine	9	Washington, D.C.	
Gropley, Edward	21	Germany	gunsmith
Gweeties, Charles	22	Germany	gunsmith
Glarps, George	75	Germany	gunsmith

[1850 Federal Census]

WELNER, HENRY — Washington, D. C.

Welner, Henry
1855 — 75 2 — gunsmith
[Directory, Washington, D.C., & Georgetown]

WERNER, J. H. T. — Washington, D. C.

Werner, J. H. T.
1846 — N side La ave btw 6 & 7 w — gunsmith & turner in metal
[Directory, Washington, D.C.]

WERTSNER, C. S. & COMPANY — Baltimore

Wertsner may have been in business with Morris of Morris and Kaidel who last advertised in 1886.

Wertsner, C. S. & Co.
1883 — 186 West Pratt Street — gunsmith
Wertsner & Morris
1885-88 — 2 Light Street — gunsmith
[Baltimore city directories]

WEST, STEPHEN — Prince Georges County

He had many indentures and his occupation was listed as a merchant. His large estate was called Woodyard and was near the town of Rosaryville. Ownership of this tract had passed from Lord Baltimore's brother-in-law Darnall to Black to West. The large brick mansion burned down during the Civil War. West was a wealthy man and employed his own gunsmiths who repaired and produced muskets for the Council. Wife's name was Haniah.

February 3, 1776 Council to West
The Council of Safety have been informed that you have a Number of Gun Locks to dispose of, if that be the case, you will be pleased to let them know by the first Opportunity what Number you have and the price also whether they are single or double bridled Locks . . .
[AM, p. 137]

February 12, 1776 West to DuVall
. . . As to gun locks, I was desired when I was at Annapolis to keep what I had & what cuttoe knives also. I sent up an account of bothvizt 92 gun locks, price from 7/6 to 16/6 the avera. 13/3, which I believe you'l find among the secred committee papers. the qualities are various some plain, some S. bridle some double, but all pretty good as those prices can afford. We have some locks hereabouts that cost 18/ stg. at London and I have seen some at 37/6 stg. those I have will I believe do pretty well. I could have sold them to better advantage to the Virginia Agents but as I suppose some time or other our people would prepare for defence and my Smiths repairing fire arms & wanting locks I kept them. I am very indifferent about selling them but if the Honourable Council of Safety desire it to have them they are at their service. I believe I may have 100.
[AM, p. 154]

February 14, 1776 Small wood to Jenifer, Woodyard
. . . In my way down called on Mr. West & viewed his Gun Locks, some of which are tolerable good, but not equal to the English Musket locks those at 7/6 are good at the price, & might answer the militia; at this critical time I think it wou'd be well to engage the whole, as they are a very scarce article. Knives will be much wanting in the [?] purchased out of their pay or allowed by council 7c he has 231 cutters wh youll judge of the expediency of taking . . .
[AM, p. 161]

March 19, 1776 Council of Safety
Ordered that the Treasurer of the Western Shore pay to Mr. Stephen West ninety five Pounds eleven Shillings and a penny for Gunlocks & Cutteau Knives.
[AM, p. 265]

May 4, 1776 West to Hall, Woodyard
I received your esteemed favour relative to my Muskett which I apprehend suffered from the rain, it is by no means equal to many that I have & shall have, yet I believe it is much better than what you generally get. I am warranted to say so by the People Officers & Soldiers from your City. I send you one much better finished than what you have got and I desire that it may be compared with those you get. I defy all your People to produce such a Bayonet, Rammer, or a barrel so well turned off, or such a Barrel of their make, light, Plain well poized and Proof, yet these Barrells are by no means so good as the rest that I have which are larger and better, if my opinion was worthy of notice I woud deliver the lightest Arms to the Officers, the largest and heavyest to the Privates for very obvious reasons but you cannot expect good Arms or the Art to be carried to perfection unless you give liberal Encouragement. I have many that will be worth 5 Guinnies at least, and those I send you really stand me 5..7..6 without a great Profit or or allowance for my own time, Expence, and Collecting Materials. Your Rammers and Bayonets of your Countrymade Arms I am told are rough Trash, & the Rammers fly out in Exercise, mine will not, they are kept by a fine spring in the lower pipe, people who are good Judges declare that my Arms on an Average are well worth fitted & finished as I do them 4 Guinnies, my own Carbine from London, not equal any way to the one I now send you cost me 4 Guinnies out of John Baziers shop. We cannot make them so cheap as they do, the price I have asked to indemnifye me is 6 of our Paper not equal to 3½ guineas, here, and I affirm in the face of the sum they are well worth it: I prove all the locks, and if any way defective finish them with the best steel. many Guns that I have seen are fitt only to beat Homminy with, not for men to trust their lives to. pray have you summed up the price your Musketts will come to, from piece work from your favourite Harriss try them fairly and see what they will amount to and then when they are really finished as well as mine which I do not believe all your workmen together can do, please to sum up the amount.

first the Barrell, to Squire Harriss	1..15..
Loops for the Sling Swivels	.. 2..
A screw and nutt for the Butt	.. 1..
A trigger and guard	.. 1..
Fixing the loops and sight to the barrell. qr if the price of 25/	.. 3..
A Sett of brass mounting, if cost only 7/6 might yet before it is put on properly filed & polished it will cost at least 15/	..15..
A Stock.	..15..
A Lock.	..15..
A Rammer	..05..
Screws and pins for the Butts, Guards & loops at least	.. 5..
A Spring for the Rammer	.. 1..
Polishing the Barrell	.. 5..
2 Slings Swivels and screws	5..
A bayonet Scabbard hook & fixing suppose	..4.. 6..
A good well finished Bayonet such as mine 16 or 17 inches in the blade, cannot be made compleat for less than	..12.. 6..
To a workman for putting all the parts together fixing the Lock Brasses, loops, swivels, &c in which a Joiner and Smith must both be employed, fitting in the lock to the stock &c will cost you at least 5/ a gun, try it, when you will.	.. 5..
	6..10..
A woollen case for the Gun, thread & making	2..6
	6..12..6

This estimate my good friend you will find if the work be well done and the workmen paid will be pretty near the value of your Barrells of Mr. Harriss when delivered to you fit for use, there will not be much variation. there may possibly be some saving from your cheap purchase of some of the articles or of their inferior Quality, or the business being ill done, or what is very often the case the undertaker getting nothing by the Job. Harriss asserted to the Committee that arms & accoutrements could be made for 3 or 3..10, but I presently demonstrated to him the impossibility & he would not undertake for 4.. or 4..10, the Pennsylvanians who have many advantages get from Virg 5..6..3 Woodden rammers, and their Guns are not equal any way to those of mine. I hope will allow there is great difference in the value of Arms. I think I can furnish from 70 to 100 of all sorts, they shall be good, and as I went to work with a sincere spirit to serve my Country I think I should have the value of my Labour. if you will appoint any Man of Candour to deal with me I shall treat him as a friend. I want Brasses & Locks. I wish you could spare me some out of your

stock, it would expedite matters . . . The Gunsmith at George Town has 4 for his guns without Bayonets, Rammers or Swivells, & but so, so.
[AM, pp. 406-408]

May 5, 1781 Council Proceedings
That the said Treasure pay to Stephen West Esqr. one hundred Pounds for repairing Arms on Acct.
[AM, p. 426]

August 24, 1781 Council Proceedings
That the said Treasure pay to the said Stephen West Esquire two hundred pounds of the same Emission of the money appropriated as foresaid to be by him expended in purchasing Mounting and repairing one hundred stands of arms belonging to the Public on Account . . . That the said Treasure pay to the said Stephen West one hundred pounds of the same Emission of the money appropriated as foresaid for three hundred and forty Stocks which he is engaged to deliver the State not less than twenty five pairs P Week to be accounted for.
[AM, p. 580]

October 30, 1781 Council Proceedings
That the Armourer deliver to Stephen West Esqr. two hundred Gun Locks and Two hundred Bayonets to be applied in repairing the Public Arms in his hands.
[AM, p. 656]

WHETCROFT, WILLIAM SR. Annapolis

Born in 1735 in Cork, Ireland, where he learned and practiced the goldsmith trade. On coming to America he advertised in 1766 as a goldsmith and jeweller on West Street, Annapolis. He married Frances Knapp in 1769 and later established an iron foundry under the name of Whetcroft and M'Fadon. In 1776 he was made deputy postmaster of Annapolis and in 1777 was operating the Patapsco Slitting Mills near Elk Ridge Landing. He died in 1799.

October 21, 1775 Council of Safety
The Council of Safety promise Mr. William Whetcroft that (in case he will next Spring import) a sufficient number of Workmen, and [will deliver for the] use of this Province for every week for two [years] then next following, fifty good substantial proved Musquets 3½ feet long in the Barrel and of three Quarters of an Inch in the Bore, with good double Bridle Locks, black Walnut or maple Stocks, and plain strong Brass mounting Bayonets with Steel Blades 17 Inches long, steel Ramrods, double Screws, priming Wires and Brushes fitted thereto, with a pair of brass Moulds for every 80 Musquets, to cast 12 Bullets on one Side, and on the other Side to cast Shot of such Size as the Musqet will chamber three of them, wholly made within this Province, the Public will, provided the Differences continue between Great Britain and the Colonies, take the said 50 Stands per Week at the Rate of four pounds common money per stand, or for such part of the said two years as the said Differences be a Reconciliation between Great Britain and the Colonies after the said Workmen shall embark, and before the End of the said two years, that then the public will, if the . . . It is agreed that it shall be at the Option of Mr. Whetcroft to fitt 800 of the Musquets first delivered with imported Locks, provided the prime Cost of the said Locks be not less than five Shillings Sterling each. It is understood that the Importation of the said workmen shall be at the Expence & Risque of Mr. Whetcroft.
[AM, p. 84-85]

April 28, 1777 Council of Safety
That the said Treasure pay to William Whetcroft forty five pounds, one shilling and six Pence for Amt. of Acct.
[AM, p. 231]

December 23, 1779 Journal & Correspondence
. . . One hundred and sixty five Pounds, seven shillings and six pence to be delivered over to William Whetcroft due them . . .
[AM, p. 39]

March 19, 1781 Council Proceedings
That the said Treasure pay to William Whetcroft eighteen pounds, fifteen shillings of the same Emission due him per Acct. by the D. Aud.
[AM, p. 355]

WHETSTONE, DAVID Washington County

David and John (II) Whetstone were sons of John (I) who was a blacksmith and farmer. The family had connections in Bedford County, Pa. David was born 16 Dec. 1780. In January 1796, at age 15, he began his apprenticeship in gunsmithing with John Gonter in Hagerstown. David's older brother, John (II) began an apprenticeship with Gonter at the same time.

David Whetstone migrated to Bedford County, Pa. at some time after completing his apprenticeship, and resided there until the mid-1820's.

Indenture made March 14, 1781 [deed] between Michael Whetstone of Bedford County in the Commonwealth of Pennsylvania farmer and John Whetstone of Washington County in the state of Maryland farmer of the one part and [to] Henry Ash of Washington County in the state of Maryland farmer of the other part . . . Tract or parcel of land called Hazar Otherwise Conquest.
[Washington CCH, Liber B, Folio 441]

Indenture made October 26, 1785 [deed] between John Whetstone of Washington County in the state of Maryland blacksmith of the one part and [to] Henry Ash of Washington County in the state of Maryland farmer of the other part . . .
[Washington CCH, Liber D, Folio 508]

This indenture made the 11 day of January 1796 witnofseth that John Whetstone of Washington County and state of Maryland hath bound, placed and put his son David Whetstone and by these presents doth bind, place and put the said David Whetstone apprentice to John Gonter of the county and state aforesaid Gunsmith . . . David Whetstone shall arrive at or attain the age of 21 years, he being of the age of 15 years the 16 day of December last . . .

his
John X Whetstone
mark

[Washington County Indentures, Hall of Records, p. 28]

WHETSTONE, JOHN (II) Hagerstown

One fine relief-carved flintlock longrifle has survived bearing the script signature, "John Whetstone", on the barrel. The engraved brass patchbox with its eight piercings and the lock screw plate are of the Hagerstown type, and the relief carving shows similarities to the work of John Gonter who was Whetstone's master.

John (II) was born on 21 Aug. 1780, the son of John (I) Whetstone and the brother of David. Both John (II) and David apprenticed under John Gonter in Hagerstown. After completing his apprenticeship in 1801 he moved to Bedford County, Pa., but by 1805 he was back in Hagerstown where he sold his gunsmithing tools in 1805. His wife, Cary Ann was a witness to the sale. In 1806 he advertised that he had removed from Hagerstown to Martinsburg, Virginia and commenced the gunsmithing business there.

This Indenture made the 11 day of January 1796 witnofseth that John Whetstone of Washington County and state of Maryland hath bound, placed and put his son John Whetstone and by these presents doth bind, place and put the said John Whetstone apprentice to John Gonter of the county and state aforesaid Gunsmith . . . John Whetstone shall arrive at or attain to the age of 21 years he now being of the age of 16 years 21 day of August last . . .

John X Whetstone
his mark

[Washington County Indentures, Hall of Records, p. 29]

Indenture made October 23, 1805 [bill of sale] Know all men by these presents that I John Whetstone of Washington County in the state of Maryland convey to Daniel Middlekauff of the aforesaid county and state in consideration of the sum of thirty pounds current money of the state aforesaid goods and chattles following one bellows, one boring wheel, two large vises, one leeter, three rifle barrels, twenty rifle stocks, one sign, one adze, one saw, one large screw plate, six small screw plates, one drawing knive, two beds, one stove, five planes, one small anvil, five hammers, one brace, one [?], three chairs, one table, twenty eight bundles lath.
[Washington CCH, Liber R, Folio 430]

WHIDDON, OLIVER

March 23, 1776 Council of Safety
Ordered That the said Treasurer pay to Oliver Whiddon six Pounds seventy Shillings and six Pence for stocking 11 Musquets.
[AM, p. 280]

April 6, 1776 Council of Safety
Ordered That the said Treasure pay to Oliver Whiddon eight Pounds fifteen Shillings for Stocking 14 Musquets.
[AM, p. 314]

Council of Safety
May 25, 1776 . . . Stocking 20 musketts & 2 Gun locks . . .
[p. 444]

June 8, 1776 . . . Stocking 8 muskets . . .
[p. 472]

July 8, 1776 . . . Stocking 6 muskets
[p. 10]

August 30, 1776 . . . Repair of guns . . .
[p. 248]

September 3, 1776 . . . Mending guns . . .
[p. 255]

September 12, 1776 Council of Safety
Ordered that said Treasurer pay to Oliver Whiddon 3 pounds 10 shillings for a Musket.
[American Archives, 5th Series, Vol. II, p. 631]

September 13, 1776 . . . for a musket . . .
[p. 269]

September 21, 1776 . . . for a musket . . .
[p. 293]

October 2, 1776 . . . for two muskets . . .
[p. 316]

January 2, 1777 . . . for a musquet . . .
[p. 7]

January 25, 1777 . . . for a musquet . . .
[p. 75]

February 14, 1777 . . . for 2 guns . . .
]p. 135]

October 30, 1777 . . . per account passed . . .
[p. 405]
[AM]

WHITE, NICHOLAS Frederick

Nicholas White apparently was at work as a gunsmith in Frederick Town before the Revolutionary War broke out, and he seems to have been active if not prominent there for most of the next 30 years. But in 1788 he sold two pieces of real estate and moved to Berkeley County, Virginia. His name does not appear again in the public records of Frederick County until 1797 when he took Nicholas King as an apprentice. In spite of this he remains largely unknown because signed specimens of his work are very scarce, if they exist at all.

The possibility exists that Nicholas White may have been the father of gunsmith Peter White, but there is no proof of this. The age relationship is appropriate. Two tax assessment records for Bedford Borough, Bedford County, Pennsylvania, for the year 1814 list the name of Nicholas White, and one of them gives his occupation as Hostler. At this time Peter White lived in nearby Colerain Township. It was not uncommon for an aging man to spend his last years living near his son.

Indenture made May 17, 1762 [deed] between Nicholas White of Frederick County of Maryland planter of the first part and [to] John Bail of the same County and province, blacksmith of the other part . . .
[Frederick CCH, Liber G, p. 503]

Indenture made October 2, 1775 [deed] between Michael Hollar of Fredericktown, Frederick County province of Maryland shoemaker of the one part and [to] Nicholas White of Fredericktown, Frederick County province of Maryland aforesaid gunsmith of the other part . . . Ground laid of as an additional lott to Fredericktown known and distinguished by it's number viz 179.
[Frederick CCH, Liber B. D. 2, p. 110]

October 1, 1776 Council of Safety
Ordered That the said Treasure pay to Nichs. White twenty two Pounds, thirteen shillings for Guns.
[AM, p. 313]

December 28, 1776 Johnson to Council
. . . Nicho. White was Ensign. Col. Beatty tells me the Company are desirous he shd. be a Captain, and that he thinks him a proper man, I should therefore be glad you'd send him a commission . . .
[AM, p. 558]

June 17, 1777 Council of Safety
Ordered, That the Supr. of Gunlock Manufactory deliver to Nicholas White one hundred and ten Gunlocks.
[AM, p. 291]

June 18, 1777 Council of Safety
Ordered That the Armourer deliver to Adam Coile for the use of N White of Frederick Town 110 Musquets Gun Barrels, 110 Setts of Brass mounting and 110 Musquet Bayonets taking his Rect. therefore.
[AM, p. 293]

November 20, 1777 Council of Safety
That the said Treasurer pay to Nich's White two hundred & fifty pounds on Acco't advanced him on his Contract to make Arms. Nicholas White agrees with the Governor and Council to make and compleatly finish two hundred Stand of Arms in a Workmanlike manner at the rate of Fifty shillings currency for each Stand the Governor and Council supplying the said Nicholas White with Barrels Locks, Bayonets, Mounting, steel for Rammers and Iron for Britching, at the rate of Forty eight stand per month till the Delivery of the whole number contracted for and have also advanced two hundred and fifty Pounds part of the Price on giving Bond with good Security to comply with this Contract.
[AM, p. 419]

Indenture made March 29, 1784 [deed] between Henry Wehaw of Frederick County in the province of Maryland of the one part and [to] Nicholas White of Frederick County in the province of Maryland of the other part a relative of the said Henry Wehaw . . . Being the one forth part of lott number 126, situated in Fredericktown.
[Frederick CCH, Liber W. R. 4, p. 383]

White Nicholas 1787 juror 5 shillings paid
[Frederick County Levy Book, Hall of Records #19257]

Indenture made February 28, 1788 [deed] between Nicholas White of Fredericktown, Frederick County in the state of Maryland gunsmith, of the one part and [to] Andrew Lowe of Frederick County in the state of Maryland of the second part . . . Beginning for the same part at the end of sixty feet from the southwest corner of said lott on Market Street.
[Frederick CCH, Liber W. R. 7, p. 696]

Indenture made August 2, 1788 [deed] between Nicholas White late of Frederick County and the state of Maryland but now of Berkley County state of Virginia of the one part and [to] Henry Hoffsteller . . .
[Frederick CCH, Liber W. R. 8, p. 146]

This Indenture made the 6th day of May 1797 Witnesseth that Nicholas King an orphan boy aged 16 years the fifth day of March last past hath put himself . . . Apprentice to Nicholas White Gun Smith . . . to serve until he shall arrive at the full age of twenty one . . . And the said Master . . . shall teach . . . the said Apprentice the art or mystery of a Gun Smith and also provide for the said Apprentice sufficient Meat, Drink wearing apparel Lodging and washing fitting for an Apprentice and cause him to be Taught to read write and cast accounts in the English language and send him three months to the German School and at the expiration of said term to give him a freedom suit.
[Frederick County Apprentice Indentures, p. 135, Hall of Records]

White, Nicholas 1797 juror paid 5 shillings
White, Nicholas 1797 juror 2 days paid 1 pound 5 shillings
[Frederick County Levy Book, Hall of Records #19257]

This indenture made 16th day of May 1797 witnofseth that Nicholas King an orphan boy aged 16 years this 5th day of March last past hath put himself and by those presents doth voluntarily and of his own free will and accord put himself apprentice to Nicholas White gun smith with him after manner of an apprentice to serve untill he shall arrive to the full age of 21 years . . .
[Frederick County Indentures, Hall of Records, p. 135]

Contractors with the government under Act of July 5, 1798 for Charleville pattern muskets at $13.40 per stand

(4) Eli Williams, Williamsport, Maryland	2000 muskets
(5) Nicholas White, Thomas Crabb, Jacob Metzger & Christopher Barnhizle, Frederick Town, Maryland	1000 muskets
(14) Alexander Clagett, Hagerstown, Maryland	1000 muskets
(25) Mathias Shroyer, Taneytown, Maryland	1000 muskets
27 contractors	40,200 stands

[U. S. ORDNANCE, Vol. I, p. 20]

White, Nicholas 1798 1/2 lott from Chris^{r} Adams value $200 Frederick Town
[Assessment Book, Frederick County, 1798, Hall of Records #19268]

Indenture made June 10, 1800 [bill of sale] I Peter Walters of Frederick County state of Maryland for and in consideration of the sum of 75 pounds current money to me in hand paid by Nicholas White, Christopher Barnhizle and Jacob Metzger the following property – towit 3 iron anvils, 4 vices and all my smith tools of every description, 1 pair large bellows, 2 grindstones and a large wheel that turns.
[Frederick CCH, Liber W. R. 20, p. 19]

White, Nicholas 1802 juror 1 day paid 11 shillings 3 pence
[Frederick County Levy Book, Hall of Records #19257]

White, Nicholas 1806 juror 1 day paid 11 shillings 3 pence
[Frederick County Levy Book, Hall of Records # 19258]

Indenture made May 18, 1807 [deed] between Joseph Gorney of Frederick County state of Maryland of the one part and [to] Nicholas White of the county and state aforesaid of the other part . . . Land lying in Frederick County state of Maryland in the town of Berlin [Brunswick] on the Potomac River distinguished on the plate of said town by #27 lying on 1st Street Street ∴ .
[Frederick CCH, Liber W. R. 31, p. 68]

WHITE, PETER Emmitsburg

Peter White, the famous gunsmith who worked for part of his life in Bedford Co., Pennsylvania, apparantly spent his early working years in Emmitsburg. Proof positive that he worked there as a gunsmith remains to be found, but there is good evidence that he did.

The tombstone of Peter White can be seen today in Uniontown, Pa., in derelict condition. It holds the following inscription: "PETER WHITE died August 25th 1834 IN 56 year of his Age." Thus he was born in late 1777 or the first eight months of 1778, and came of age 21 in 1798 or 1799. His name first appears in the public records of Bedford Co., Pa., in 1806. It seems likely that he worked in Maryland during the first six or seven years of his career, before moving to Bedford Co.

Where Peter White was born, and who his parents were is not presently known, but it is likely that his beginnings were in Maryland. As a working hypothesis it may be considered that the Frederick gunsmith, Nicholas White, may have been Peter's father. Nicholas White was established as a gunsmith in Frederick as early as 1775 (see below), and worked there until 1788 when he moved to nearby Berkeley Co., Va. By 1797 he was back working as a gunsmith in Frederick and presumably he worked there until at least 1810. It may be significant that the name of Nicholas White appears on two property tax assessment records for Bedford Boro, Bedford Co., Pa., for the year 1814; on one of these he is listed as a hostler. If Nicholas White was the father of Peter, perhaps he moved to Bedford to spend his late years near his son.

The evidence that Peter White worked in Emmitsburg during his early years is good. In the first place, a number of relief-carved rifles bearing his signature which appear to have been made early in his career show strong Emmitsburg-style characteristics. The overall architecture, the profiles of the butt stocks, the design of the patchboxes, the pattern of engraving on the patchboxes, and the design of the relief carving all show a strong resemblance to the work of John Armstrong. Secondly, Peter White's son, John, gunsmith of New Lisbon, Ohio, Uniontown, Pa., and Pittsburgh, Pa., reported in the 1860 federal census that he was 60 years of age and born in Maryland. This essentially proves that Peter White was at work as a gunsmith in Maryland in 1800. In addition, the federal census for 1800 for Emmitsburg lists the name of Peter White as head of a household containing 1 male between 10 and 16 years, 1 male between 16 and 26 years (presumably the gunsmith), and 1 female between 16 and 26 years.

Carrying the speculation about Peter White a bit further, if he worked in Emmitsburg in his early years, it is most likely that he worked in close association with John Armstrong. Armstrong was about five years older than White, and White might have worked with him as a journeyman. Possibly White apprenticed under Armstrong, but the apprentice indenture book for this period does not contain any indenture for White. If White was the son of a gunsmith a formal apprenticeship may have been bypassed. It seems significant that both Armstrong and White preferred to make their own flintlocks rather than to use imported ones, and that the style of White's locks resembles that of Armstrong.

Peter White worked in Bedford Co., Pa., from 1806 until the spring of 1819. In April 1819 he advertised in a newspaper in Uniontown, Pa., that he had begun the business of gunsmithing there. By 1830 White was heavily in debt and was forced to sell his gunsmithing tools and supplies; a document in the Fayette Co. courthouse dated August 1830 lists the entire contents of his gunsmith shop. The new owners of his gunshop allowed him to continue in business with his equipment and tools, which he apparently did until his death in August 1834. John White, Peter's son, carried on the business after Peter's death.

For a more detailed account of Peter White during the time he worked in Bedford Co., Pa., and at Uniontown, Pa., see the book by Calvin Hetrick (1975, Second Edition) THE BEDFORD COUNTY RIFLE AND ITS MAKERS, Geo. Shumway Publisher.

WICKHAM, MARINE TILER

An 18-year old lad named Tiler Wickham became an a apprentice to John Armstrong in 1799. The accidental discovery of a deed made in 1810, but not indexed under Wickham's name, makes it possible to tie this apprentice to Marine Tiler Wickham, a man well-known in the military arms field in Philadelphia from about 1811 until his death in 1833. Reading between the lines of the apprentice indenture and the deed, it appears that after Wickham completed his indenture in 1802 he bought a small lot in Emmitsburg and worked there for a few years, and while there was associated with an otherwise unknown gunsmith named Matthews. He then moved to Jefferson County, Virginia, where Harpers Ferry and Charles Town are located.

Presumably he worked at Harpers Ferry for a year or two at least, and possibly longer, learning about the manufacture of military arms. A rental agreement shows that he left many household possessions in a government house at Harpers Ferry when he moved to Philadelphia. By 1808 he was at work as an arms inspector, presumably at the Schuylkill Arsenal in Philadelphia but possibly with connections at Harpers Ferry. In his book, U. S. MILITARY FIREARMS 1776 -- 1956 (1962, p. 12) Hicks mentions that M. T. Wickham was an Inspector for arms made in New Jersey and Pennsylvania under the contracts of 1808. Hicks also presents correspondance showing that Wickham was the designer of the U. S. Model 1812 Musket and the Model 1813 Pistol (p. 35-47). A Philadelphia directory for 1813 lists him as Inspector of Arms at the U. S. Arsenal, and the 1814 directory lists him as Chief Armorer there.

By 1817 he was in business for himself as a gunsmith, hardware merchant, military and fancy hardware dealer. Beginning in 1829 he was a gun manufacturer. His hardware business was carried on under the name of M. T. Wickham & Co., or Wickham & Co. Marine Tiler Wickham died in 1833 in Philadelphia, leaving a widow named Matilda Wickham. A packet of Administration papers in Philadelphia's City Hall includes an inventory of personal belongings and the contents of his armory.

At least two very beautiful flintlock longrifles have survived that can be attributed to Tiler Wickham, presumably made in Emmitsburg shortly after he finished his apprenticeship. An otherwise unknown gunsmith with the name of Matthews was associated with Wickham in producing these pieces. The design of these rifles closely follows the Emmitsburg school designs and architecture worked out and used by John Armstrong. The engraving is of fine quality amd the relief carving on one of the pieces is magnificent. One rifle barrel is signed in script "Wickham & Matthews" on a brass plate let into the barrel, and the flintlock plate, presumably made by Tiler Wickham, is signed "T. W."

Tiler Wickham, Son of Margaret Wickham Aged Nineteen the 3rd Day of January next being an Orphan Child in this day with the Consent of his said Mother bound an Apprentice unto John Armstrong of Frederick County Gunsmith until he arrives at the age of Twenty One Years the said John Armstrong engaging to find his said Apprentice sufficient Boarding and Cloathes and all other necessaries the said term to cause his said Apprentice to be Taught the said Trade of Gun Smith & Gun Lock making which he now useth as far as shall be in his power During said term Given under the hands & seals of the Subscribers two of the Justices of the Peace of Frederick County the 11th of March 1799.
[Frederick County Apprentice Indentures, p. 211, Hall of Records]

This Indenture made the 18th day of May 1810 Between Marine Tyler Wickham of Jefferson County in the State of Virginia of the one part and George Grover of the City of Baltimore . . . Witnesseth that the said Marine Tyler Wickham for . . . $800 . . . doth . . . sell unto the said George Grover . . . that Lot or portion of ground situate in Emmitsburg in Frederick County . . . distinguished on the plan of said Town by number 10, being one of the short corner Lots in said Town together with all and singular buildings Improvements . . .
M. T. Wickham
[Frederick CCH, Deed Book W. R. 38, p. 226-227]

WILKS, THOMAS — Frederick

Inventory of Work tools Belonging to and now in the Gunlock Factory the 19th August 1778 . . . Thomas Wilks 2 months (let free)
[Hall of Records, Red Book IX, p. 273]

WILL, WILLIAM C. — Baltimore

He was 42 years old, a hardware merchant, born in Hamburg, Germany, according to the 1850 Federal Census.

Will, William C.

1835-40	37 Harrison Street	lock & gunsmith
1845	4 Harrison Street	lock & gunsmith & hardware
1859	12 Gay Street	hardware

[Baltimore city directories]

WILLIAMS, ELIE — Williamsport

First indenture in Frederick County was in 1791 and the last in 1807 with no occupation given.

September 12, 1776 Council of Safety
Ordered That the Western Shore Treasure pay to Elisha Williams six hundred pounds, for to raise the militia Company in Lower District of Frederick County.
[AM, p. 266]

Indenture made October 11, 1791 [deed trust] Conrad Grosh of Frederick County state of Maryland of the one part and [to] Elie Williams of Frederick County state of Maryland of the other part . . .
[Frederick CCH, Liber W. R. 10, p. 317]

Indenture made May 28, 1794 [deed] between Elie Williams and William Beatty both of Frederick County state of Maryland of the one part and [to] Jean De Klegand of the same County and state aforesaid of the other part . . .
[Frederick CCH, Liber W. R. 12, p. 441]

Contractors with the government under Act of July 5, 1798 for Charleville pattern muskets at $13.40 per stand

(4) Eli Williams, Williamsport, Maryland	2000 muskets
(5) Nicholas White, Thomas Crabb, Jacob Metzger & Christopher Barnhizle, Frederick Town, Maryland	1000 muskets
(14) Alexander Clagett, Hagerstown, Maryland	1000 muskets
(25) Mathias Shroyer, Taneytown, Maryland	1000 muskets
27 contractors	40,200 stands

[U. S. ORDNANCE, Vol. I, p. 20]

. . . Mr. L. Jacques who resides at the furnace 10 miles from Williamsport and 15 miles from Hagerstown on the main road leading from those places to Cumberland . . .
August 17, 1805

Elie Williams
Lancelot Jacques Trustees

[The Maryland Herald & Hagers-Town Weekly Advertiser, 30 August 1805]

Death – At Georgetown D. C. in 1823 Col. Elie Williams in the 73 year of his age. Col. Williams was an ardent patriot during the Revolution, and in the latter portion of his life was deeply interested in the Chesapeake and Ohio Canal, being elected to the Legislature from Washington County with the special object of promoting the interests of that great enterprise.
[HISTORY OF WESTERN MARYLAND, p. 1056]

Episcopal Church St. Johns The Vestry of Washington County

Daniel Hughes	Elie Williams	Alexander Claggett
1787	1787	1787
1789	1789	1789
1791	1791	1791
	1793	
	1795	
	1798	1798
	1799	1799
		1802
1807	1807	
1813		

[HISTORY OF WESTERN MARYLAND, p. 1078]

WILLIX, WILLIAM

September 13, 1777 Council of Safety
That the said Treasure pay to William Willix seven pounds for a Gun.
[AM, p. 375]

WILSON, JOHN M. — Washington, D. C.

Wilson, John M.	age 50	born Maryland	gun carriage maker
" Rebecca	37	Washington, D.C.	
" Rutledge	11	Washington, D.C.	
" Charles E.	9	Washington, D.C.	
" Lewis P.	6	Washington, D.C.	

[1850 Federal Census]

WINAN, ROSS — Baltimore

Winan tinkered with steam all his life. In his earlier years he assisted Peter Cooper in the building of the Tom Thumb and by the Civil War era he was known as the greatest manufacturer of locomotive engines and railroad cars in the world. He was a member of the state General Assembly with southern convictions and was arrested with 30 other secession leaders. Charles Dickinson invented a bullet-proof, steam powered cannon which was manufactured at Winan's locomotive works. The cannon was shipped to the Confederacy disguised as farm machinery but was captured on 11 May 1861. Although it was not used during the war it could project anything from the size of a bullet to a hundred pound ball at the rate of 200 a minute a distance of two miles.

At the works of the Messrs. Winans their entire force is engaged in the making of pikes, and in casting balls of every description for cannon, the steam gun, rifles, muskets, etc., which they are turning out very rapidly.
[*Baltimore American,* 23 April 1861]

Winan, Ross

1858	Parkin & Hollins	machinist
1859	McHenry & Poppleton	machinist & Locomotives
1860	Parkin & Hollins	machinist

[Baltimore city directories]

A Baltimorean by the name of Ross Winan, a gray-haired old man of more than three score and ten, a bitter rebel and reputed to be worth $15,000,000.00 had made 5,000 pikes of the John Brown pattern to be used by the rebels of Baltimore to oppose the march of United States troops. I thought that if such a man worth $15,000,000.00 were hanged for treason it would convince the people of Maryland at least that the expedition we were on was no picnic. I knew that he was going to Frederick to make a secession speech and I believed if we captured him he would be a very proper specimen traiter to be hanged.
[AUTOBIOGRAPHY OF BENJAMIN F. BUTLER, Peabody Library]

WINANS, WILLIAM LOUIS — Baltimore

February 21, 1865
Be it known that I, William Louis Winans, of Baltimore, Maryland, United States of America, have invented improvements in the manner of mounting and maneuvering cannon or ordnance on ships or vessels of war and floating batteries . . . to do away with the necessity of port-holes or embrasures in the sides of vessels of war and floating batteries, through which to fire cannon, and more perfectly to protect the men, cannon . . . by raising the cannon from a place of security below the

upper deck or from behind the bulwarks.
[U. S. Patent Office, Patent #46,516]

WINCHESTER, OLIVER J. — Baltimore

Winchester, Oliver J.

1835-37	3 Sharp Street	stock manufacturer
1840	165½ Baltimore Street	stock manufacturer

[Baltimore city directories]

WINTER, WILLIAM — Baltimore

Winter, William

1885	8 Patterson Avenue	gunsmith

[Baltimore city directory]

WINTERS, ELIJAH — Allegany County

His first deed was recorded in Allegany County in 1874. In his last deed in 1881 he sold his mill but reserved the right to use the race and water with the right to repair the same from time to time forever. His wife was Eliza Francis Winter.

Elijah Winters, gunsmith and Miller at Brady's Mill seven miles west of Cumberland.
[HISTORY OF WESTERN MARYLAND, p. 1462]

WINTERS, ELISHA — Chestertown

July 2, 1776 Smyth to Council
The Bearer hereof Mr Winters waites on the Convention with 50 stands of Arms for use of the Province which I think will be highly approved of. Mr Hollingsworth has not furnished him with Gun Barrels agreeable to his contract . . . If Mr Winter could be furnished with Locks from the Frederick Manufactory he could easily compleat 50 stands of Arms every week.
[AM, p. 544]

July 27, 1776 Winters to Council, Chestertown
I gladly embrace this opportunity to inform you I shall have 28 Musquets ready to deliver to your order by Monday 3 August. making up 40 musquets P Month agreeable to my contract.
[AM, p. 129]

March 29, 1777 Council of Safety
Ordered for 200 Arms on Mr Winters and copy of the Resolution of the Gen. Assembly of this Day inclosed.
[AM, p. 194]

July 10, 1777 Council of Safety
That the Armourer deliver to Mr Winters twenty five Gun Barrels, twenty five Locks & Powder to prove them to be Stocked by him.
[AM, p. 316]

July 31, 1777 Council of Safety
That the Armourer deliver to Mr Winters eleven Gun Barrels, fifty six Gun Locks & a sufficient Quanity of Powder & Ball to prove sixty one Barrels.
[AM, p. 323]

September 25, 1777 Council of Safety
Ordered that the Armourer deliver Mr Winters the Musquet Barrels made by Mr Bond, locks sufficient for the Barrels and what mounting he has belonging to the Public.
[AM, p. 385]

WINTKLE, WASHINGTON — Baltimore

Wintkle, Washington

1840	Sharp Street north of Lexington	gunsmith
1847	128 South Canal Street	gunsmith

[Baltimore city directories]

	age	sex	occ	birth
Washington Wintkle	41	M	Gunsmith	Maryland
Ellen	41	F		Kentucky
Susan A.	9	F		Maryland

[1850 Federal Census]

WOLF, ELMER P. — Thurmont

Elmer P. Wolf was born March 31, 1868, a son of Herman and Mary Gorden at Foxville, Frederick County. In his youth he ran a sawmill and did some gunsmithing near his birthplace. He married Eliza Jane Stottlemyer and had two sons, Roger and Paul. In his later years he moved to Eyler Valley on the corners of the intersection of Eyler Valley Road and Black Road, just north of Thurmont. He continued to practice the gunsmithing trade using walnut and curly maple for stocks. The extent of the carving on stocks was limited to checkering and brass inlays which were usually round, oval or diamond-shaped, very seldom using silver. The barrels and locks he did not make but bored them out and rifled them by hand. He had many cherries, ranging from about 32 to 48 caliber.
A few of his longrifles and bullet molds are stamped with his name. He also made small caliber, rolling block, breech loading rifles. These are crude and all examined have a stationary silencer at the end of the barrel. His family takes more pride in a violin that he made than in his rifles.
[Interview with Herbert Wolf, foster son of Elmer Wolf]

WOLF, HERMAN — Foxville

Herman and Mary Gorden Wolf were the parents of Elmer Wolf and Elmer learned the gunsmithing trade from his father. Herman was also a mechanic, working at various trades. One longrifle that he produced is of fair quality and was reworked by Elmer.
[Interview with Herbert Wolf, foster son of Elmer Wolf]

WOOD, LENUEL S. — Washington, D.C.

1865 Wood, Lenuel S. military outfitters 450 Penna. Ave.
[Washington City Directory]

WOODWARD, EDWARD — Baltimore

Woodward, Edward

1855	160 York Avenue	gunmaker
1856	160 Green Street	gunmaker
1857-58	417 West Pratt Street	gunmaker

Woodward, E.
1859 417 West Pratt Street gunsmith
Woodward, Edward
1860 417 West Pratt Street gunmaker
[Baltimore city directories]

WORLEY, THOMAS Washington County

Worley sold all of his real estate in Berkley County, Virginia, on 20 August 1776. One portion including the house in which he and his wife Susannah resided was sold to Philip Sheetz. Indentures recorded in Washington County in 1779 and 1783 give his occupation as a blacksmith and in 1785 as a cutter. All of his land bordered the Potomac River between Williamsport and Sharpsburg. His will was probated in 1806 and the executor was a colleague, Elie Williams. Several longrifles signed on the top facet of the barrel by the Sheetz bear the stamp "T. Worley" on the underside. After the war, Worley lived just across the river from Shepardstown and could apparently continue supplying barrels without too much difficulty. The Sheetz were good rifle makers but obviously over-estimated the demand because when they discontinued their gun trade in the early 1800's they had so many stock blanks left that they were used for a yard fence.

We, the subscribers, gunsmiths of Mecklenburg in the County of Berkley, do certify that we can make and deliver, at our shop in this town, to Mr. Thomas Rutherford, or any other person legally appointed by the Honble the Committee of Safety to Secure and Pay for them, twenty-four good and well fixt Rifle guns per month, at the rate of Four Pounds and Ten Shillings Virginia Currency each, or in lieu thereof twenty-four good and well fixt muskets with sufficient bayonets at the rate of Four Pounds . . . Like Currency each.
Certified under our hands this 28th Day of May, 1776

Thomas Worley
Philip Sheetz
Henry Sheetz

[Martinsburg, West Virginia, CH]

WHEAT, THOMAS Washington, D.C.

Thomas Wheat delivered to the Office of Indian Trade 66 large squaw axes at 75 cents each, 67 middle squaw axes at 62½ cents each, and 62 small squaw axes at 50 cents each.
[1817 Treasury Records, National Archives]

Y

YANTIS, JOHN — Washington, D. C.

Yantis, John

1862 bds 707 7th st gunsmith

[Boyd's Washington and Georgetown Directory]

YOST, HENRY — Frederick

On 14 December 1775 Henry Yost and John Unsell agreed to produce muskets together for the Council of Safety. In June of 1776 29 rough muskets with pickers, wipers and one bullet mould were returned to them, classed unfit for service.

Ordered That Mr Henry Yost and Mr John Unseld deliver to Capt. George Stricker all the Musquets and Bullet Moulds which are now finished by them for the use of this Province.

April 4, 1776 Council of Safety
Ordered That Mr Henry Yost and Mr John Unseld deliver to Capt. George Stricker all the Musquets and Bullet Moulds which are now finished by them for the use of this Province.
[AM, p. 308]

YOST, JOHN — Georgetown

January 17, 1776 Youst to Council, Georgetown
Having applied to The Committee of observation for a sufficient quantity of Powder for proving the guns, which I have engaged to furnish The Council of Safety, which I have refused without your approbation should be glad you would please to nominate a man in town to see the Guns proved as it is inconvenient to me, at present to leave the Business. Mr Deakins has left town, which prevent his writing, to you according to agreement, I hope gentleman youl Please to let me know what I am to expect as the guns cannot be delivered or finished otherwise than by your granting a Licence for Powder.
[AM, p. 100]

January 20, 1776 Council of Safety
Ordered That Mr Stephen West deliver to John Youst or Ordered the Quantity of ten Pounds weight of Gunpower (out of that purchased by the Convention from him) to prove the Musquets made by the said Youst for the use of this Province.
[AM, p. 99]

March 8, 1776 Council of Safety
Ordered, That the Treasure of the Western Shore to John Youst two Pounds eleven Shillings, and seven Pence Currency for repairing Guns.
[AM, p. 214]

March 27, 1776 Council to Youst
The Council of Safety desire you will be as expeditious as you possibly can in supplying the Muskets &c you engaged to make for the Province, and inform us whether you have any now completed as we are in very great Want of them and will send for them as soon as you have a Number ready.
[AM, p. 293]

July 7, 1776 Council of Safety
The Council contracted with John Yost for the making of 300 Muskets at 4 pounds 5 shillings each and 100 Rifles at 4 pound pounds 15 shillings each.
[American Archives, 5th Series, Vol. I, p. 1331]

August 1, 1776 Yost to Council
I have received your letter dated the 23 of July and finding you desirous to know what Compliment of Arms I had ready for the Publick. I have them all ready of the first contract including the Banyonets which I expect this day with an Express that I have sent for them. I have also been much detained in the last contract by repairing old arms for the militia the Cort. finding it very necessary . . .
[AM, p. 159] [American Archives, 5th Series, Vol. I, p. 707]

September 13, 1776 Yost to Council, George Town
In Consequence of your Favour of the 6th. Instant, I have to inform you that I have erected a Horse Mill for boring Gun Barrels; that I am now employed with all my Workman I have in making Locks, screws, Mounting and forging Barrels ready for boring, but cannot proceed to that part of the Work before I receive the Materials (Steel in particular) which I purchased at Philadelphia sometime ago; . . . I was told by the manager of the Gun Lock Manufactory at Frederick Town, that they forge Gun Locks much faster then they can finish them off; as that is the Case, I should be glad to furnish myself from thence with 300 ready forged Locks, provided the Terms are admissible. If this Proposal is agreeable, I hope I may shortly be furnished with proper Authority through you, or some other Department, to receive that Quantity, as it will greatly further my work. I have nine Hands employed at present and have engaged two more, who I expect will be at work for me in less than a fortnight.
[AM, p. 271]

March 27, 1777 Council of Safety
That the said Treasure pay to John Yost two hundred pounds in advance on his Contract for arms.
[AM, p. 191]

July 16, 1781 Murdoct to Council
. . . John Yost, who has already repaired several public Arms, and is now employed about repairing those you last sent to this County, has made frequent Applications to me for money and now he tells me that he cannot go on with the work, nor longer Maintain his Family without regular Payments, so that he must stop unless he can be paid for it . . .
[AM, p. 351]

Yost, John Frederick County
3 males of more than 16 years
3 males under 16 years
5 females
[1790 Federal Census]

YOUNG, C. E. — Cumberland

Young, C. E.

1887 Cumberland gun & locksmith

[Maryland Directory & State Gazetteer, 1887, The Baltimore Publishing Co.]

Figure 100b. The Merchants' Shot Tower was built on Eutaw Street in Baltimore in 1828 and is still standing today. It was 50 feet in diameter at the base and 250 feet high. This photograph of it was made in the 1870's or 1880's. Courtesy of Erick Davis.

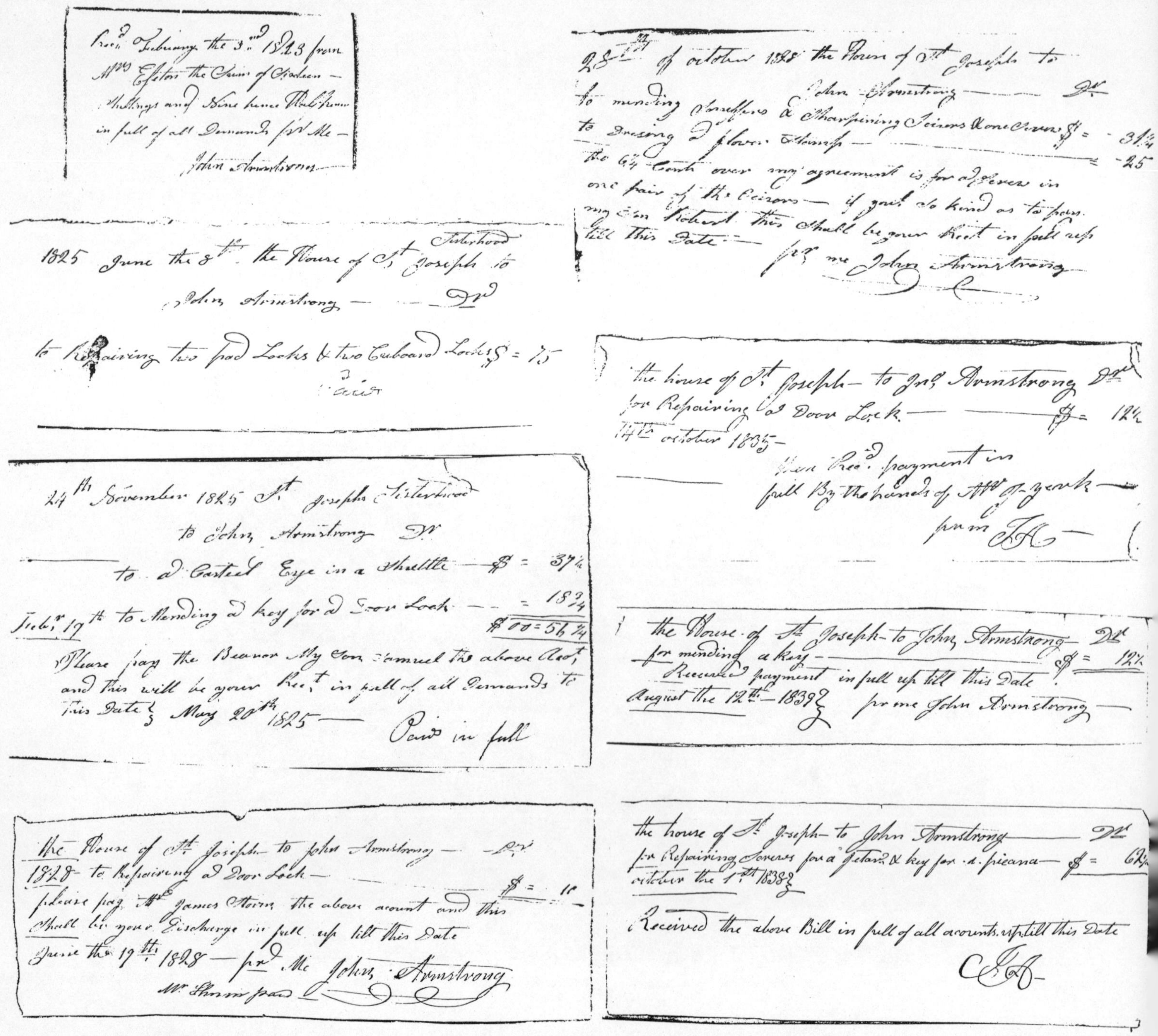

Recd February the 3rd 1813 from Mrs Seton the Sum of Sixteen Shillings and Nine pence Half penny in full of all Demands pr Me — John Armstrong

28th of October 1828 the House of St Joseph to John Armstrong Dr
to mending Scissors & Sharpening Scissors ... $ = 31¼
to Dressing a flower Stamp — = 25
the 6¼ Cents over my agreement is for a Screw in one pair of the Scissors — if you'l be so kind as to pay my Son Robert this shall be your Rect in full up till this Date — pr me John Armstrong

1825 June the 8th the House of St Joseph Sisterhood to John Armstrong — Dr
to Repairing two pad Locks & two Cuboard Locks $ = 75
Paid

the house of St Joseph — to Jno Armstrong Dr
for Repairing a Door Lock — $ = 12½
14th October 1835 —
then Recd payment in full By the hands of Mr ... — pr me JA —

24th November 1825 St Josephs Sisterhood to John Armstrong Dr
to a Casted Eye in a Shuttle — $ = 37½
Feby 17th to Mending a key for a Door Lock — = 18¾
$00 = 56¼
Please pay the Bearer My Son Samuel the above Acct and this will be your Rect in full of all Demands to this Date May 20th 1825 — Paid in full

the House of St Joseph — to John Armstrong Dr
for mending a key — $ = 12½
Received payment in full up till this Date
August the 12th — 1839 pr me John Armstrong —

the House of St Joseph — to John Armstrong — Dr
1828 to Repairing a Door Lock — $ = 1[illegible]
please pay Mr James Storm the above acount and this shall be your Discharge in full up till this Date
June the 19th 1828 — pr Me John Armstrong
Mr Storm paid

the house of St Joseph — to John Armstrong Dr
for Repairing Screws for a Gitar & key for a pianna — $ = 62½
October the 1st 1838
Received the above Bill in full of all acounts up till this Date
CJH —

Figure 100 c. Receipts in the handwriting of Emmitsburg's famous gunsmith, John Armstrong. In addition to his work as a gunsmith, Armstrong served as a locksmith and mechanic. These receipts are for locksmithing work and for some other handyman work – repairing and sharpening scissors, putting a screw in a guitar, and repairing a piano key. The invoice for 24 Nov. 1825 mentions John Armstrong's son, Samuel, and the invoice for 28 Oct. 1828 mentions his son Robert. The earliest invoice is dated 3 Feb. 1813, and the latest shown here is dated 1 Oct. 1838. One additional invoice exists, for mending a key for a chest lock, and it is dated 13 March 1841. These receipts are present among the old records of St. Joseph's Provincial House in Emmitsburg, and were provided through the courtesy of Sister John Mary Crumlish, D.C., Archivist. See pages 124 & 125 for additional details.

Appendix

Figure 101. Engraved label of James Haslett of Baltimore, considerably enlarged in this illustration. This label is present inside the lid of the case for the saw-handled flintlock pistols illustrated in Figures 63 through 70.

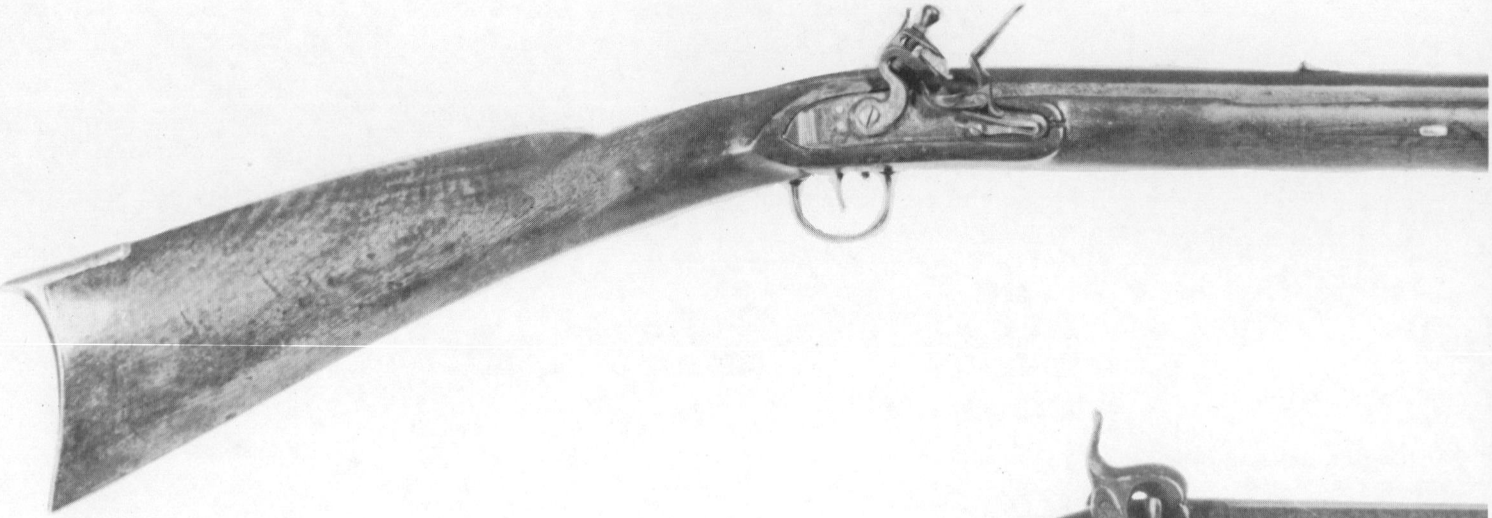

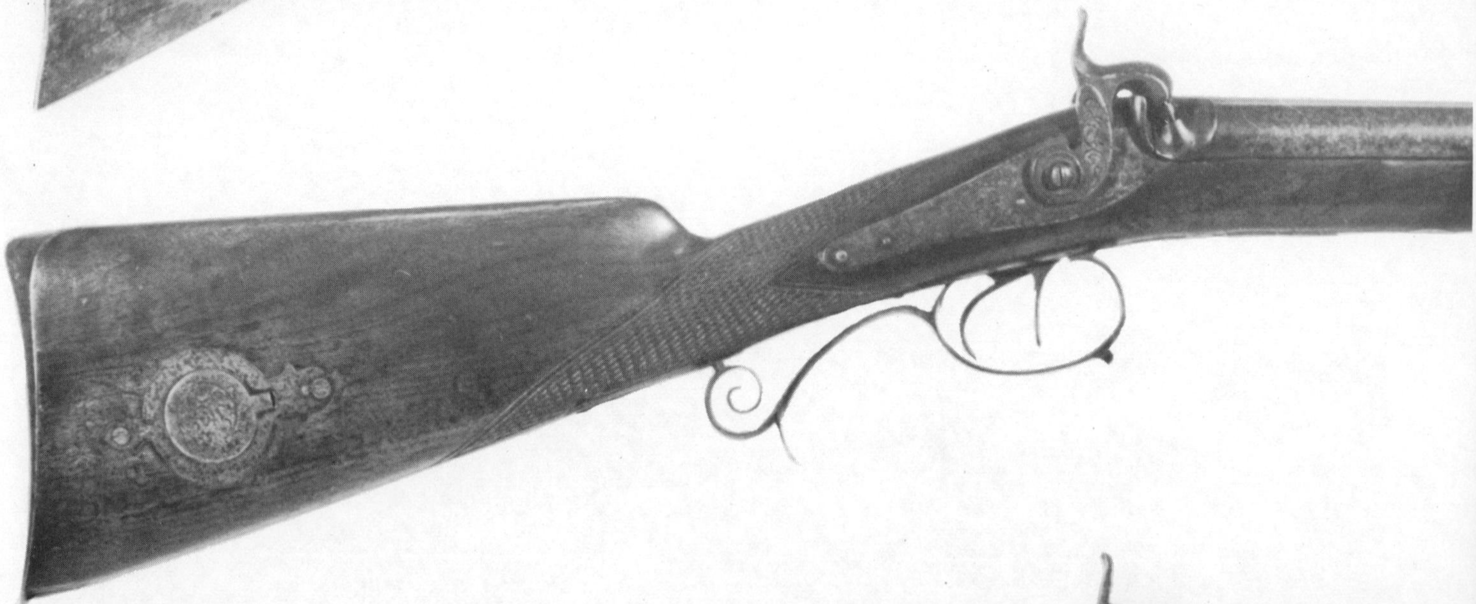

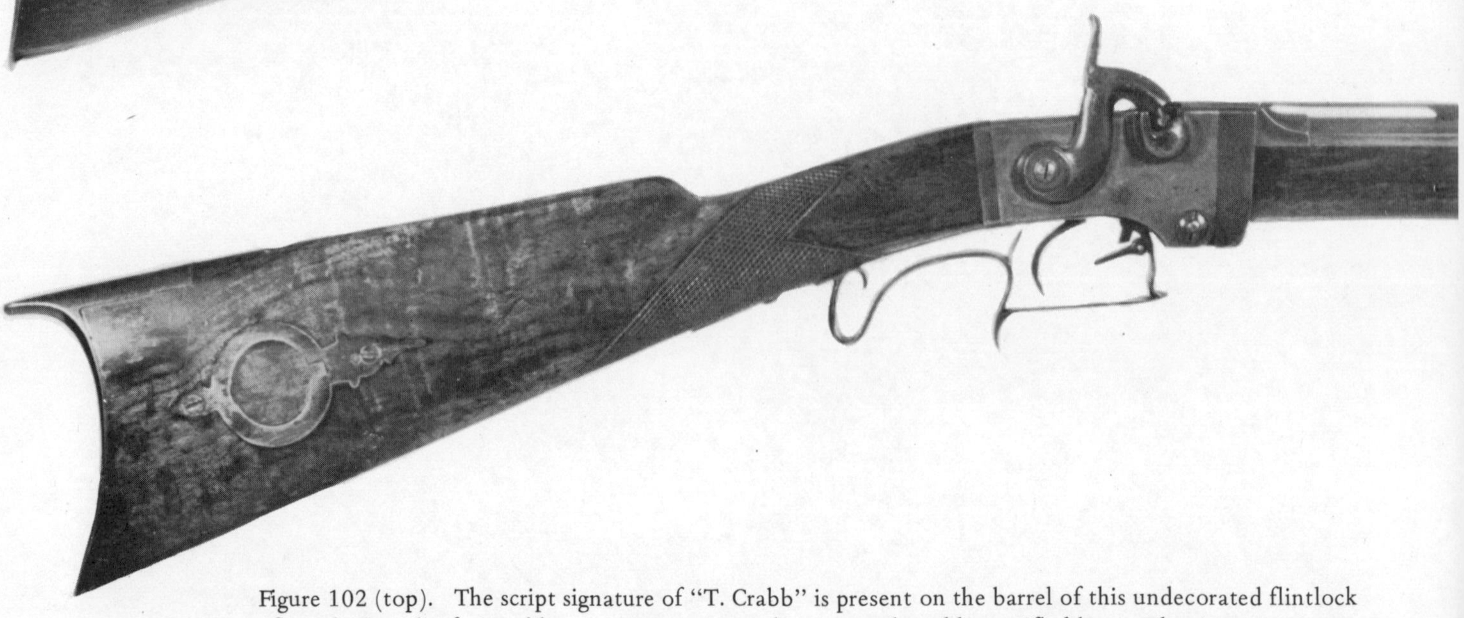

Figure 102 (top). The script signature of "T. Crabb" is present on the barrel of this undecorated flintlock rifle. The barrel, of .48 caliber, is part octagon and part round, and has a rifled bore. The signature resembles Thomas Crabb's signature on a contract for muskets of 1798 (Fig. 12). The arm is 63½ inches in overall length, has a silver front sight, and a tapered-and-flared barrel. Daniel D. Hartzler collection.

Figure 103 (second from top). Double-barreled percussion gun marked "A. Escherich Baltimore". Both barrels are 47 inches in length and of .50 caliber, but only the right barrel is rifled. The arm carries extensive engraving on the metal parts, and is fitted with sling swivels. Harry D. Berry, Jr., collection.

Figure 104 (third from top). "Poultney and Trimble Baltimore Md." is stamped on the barrel of this rare .36 caliber breechloading Smith rifle. The arm incorporates the early unlocking breech lever described in Patent No. 27600 of 1860 by C. Edward Sneider.

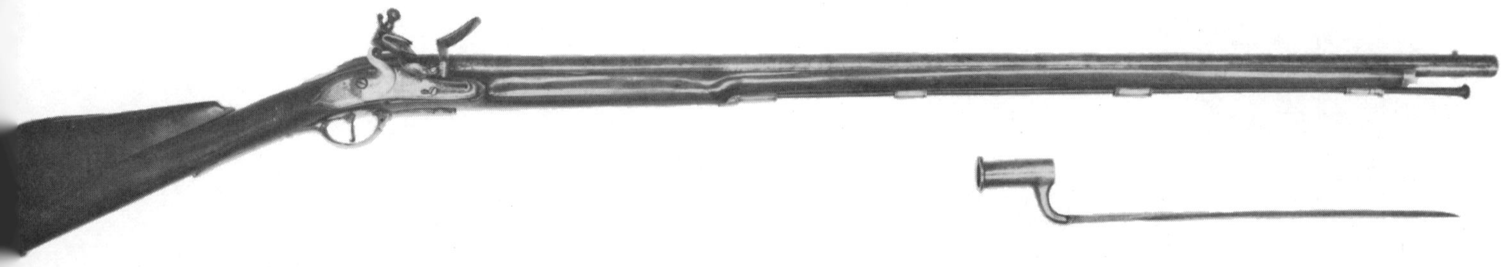

Figure 105 (top). This flintlock military musket of the Revolutionary War period probably was a piece contracted for by the Maryland Council of Safety in 1775 or 1776. It was styled after the British Brown Bess musket, is of .76 caliber, has an overall length of 58 3/4 inches, and a barrel length of 42 7/8 inches. On the left side of the fore-stock in front of the side plate is branded the word "MARYLAND". The flintlock is marked with the name "I. I. BEHR", which in 18th century usage stood for "J. J. BEHR". The top of the barrel is marked "I. BEHR BALTIMORE TOWN". The numeral "10" is marked on the barrel tang, and the butt plate is stamped "R. B. N. 2". James A. Knowles collection. Photos courtesy of the National Rifle Association of America.

Figure 106 (second from top). Overall view of flintlock musket marked "I. BEHR BALTIMORE TOWN".

Figure 107. The script signature of Ben Trautman appears in this form on a rifle barrel, engraved on a silver plate let into the barrel. Unfortunately the remainder of the rifle has been destroyed.

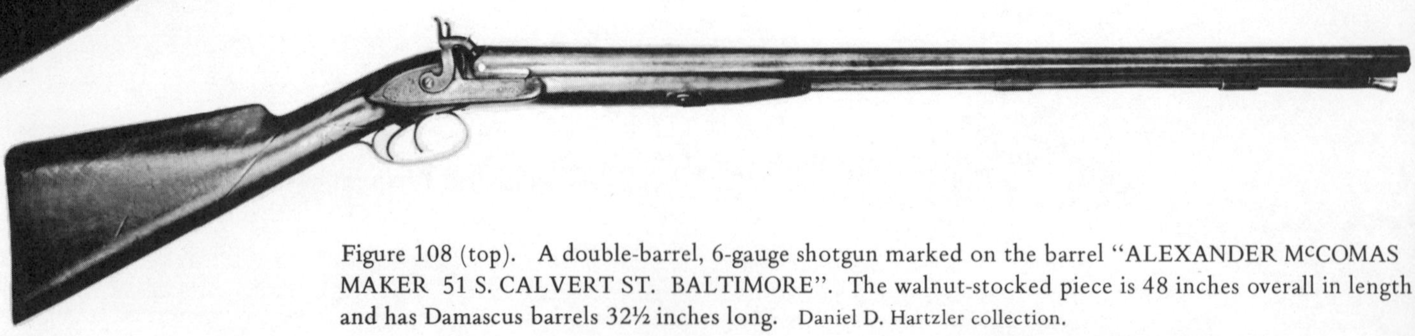

Figure 108 (top). A double-barrel, 6-gauge shotgun marked on the barrel "ALEXANDER McCOMAS MAKER 51 S. CALVERT ST. BALTIMORE". The walnut-stocked piece is 48 inches overall in length and has Damascus barrels 32½ inches long. Daniel D. Hartzler collection.

Figure 109 (second from top). Overall view of 6-gauge double-barrel percussion shotgun by Alexander McComas of Baltimore.

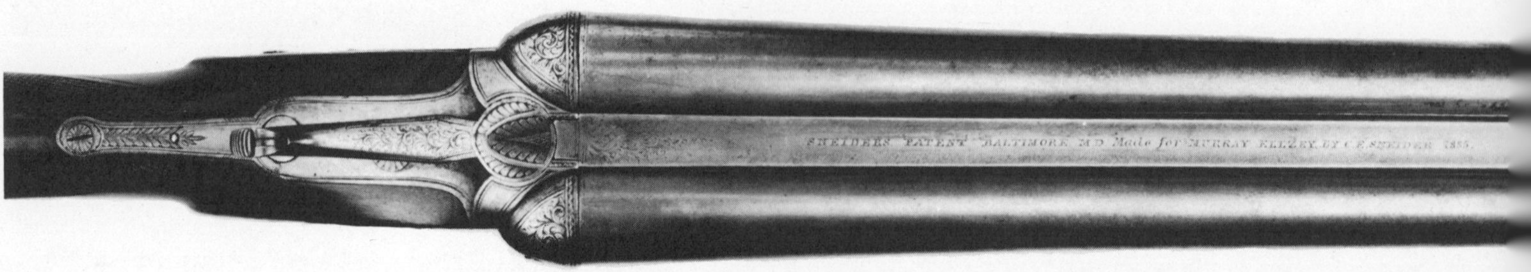

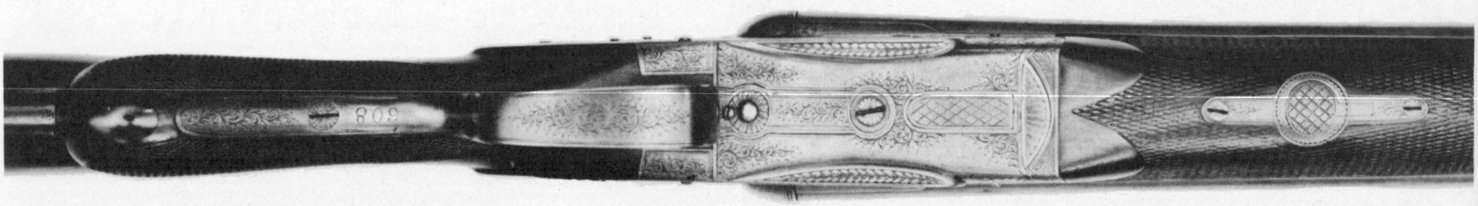

Figure 110 (third from top). Charles Edward Sneider made this 12-gauge, hammerless, double-barrel shotgun, serial number 608. The barrels are engraved on the rib "SNEIDER PATENT BALTIMORE MD. 1885". This beautifully-engraved shotgun has the unique thumb safety on the break lever which Sneider patented. Daniel D. Hartzler Collection.

Figure 111 (fourth from top). Bottom view of 12-gauge shotgun by Charles E. Sneider, showing an abundance of engraving.

Figure 112 (top). A. T. Baxter of Baltimore made this massive 6-gauge, 15-pound, water-fowl shotgun. The double-barrels are beautifully browned and are 37½ inches in length. The iron furniture is finely engraved, and the locks are marked "A. T. BAXTER". The hammers are shaped to be dolphin heads. Daniel D. Hartzler collection.

Figure 113 (second from top). Overall view of 6-gauge shotgun by A. T. Baxter.

Figure 114 (third from top). Top view of the barrels of the percussion shotgun, signed with the maker's name, "A. T. BAXTER".

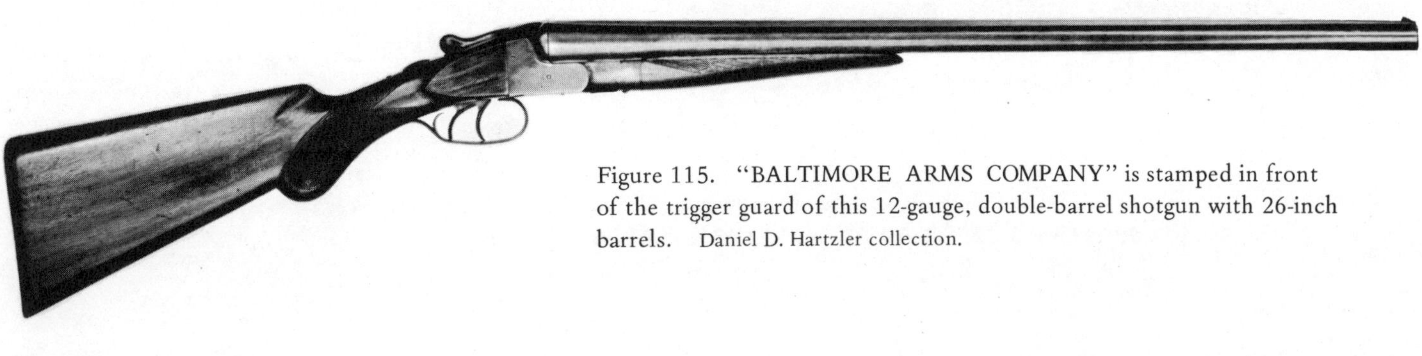

Figure 115. "BALTIMORE ARMS COMPANY" is stamped in front of the trigger guard of this 12-gauge, double-barrel shotgun with 26-inch barrels. Daniel D. Hartzler collection.

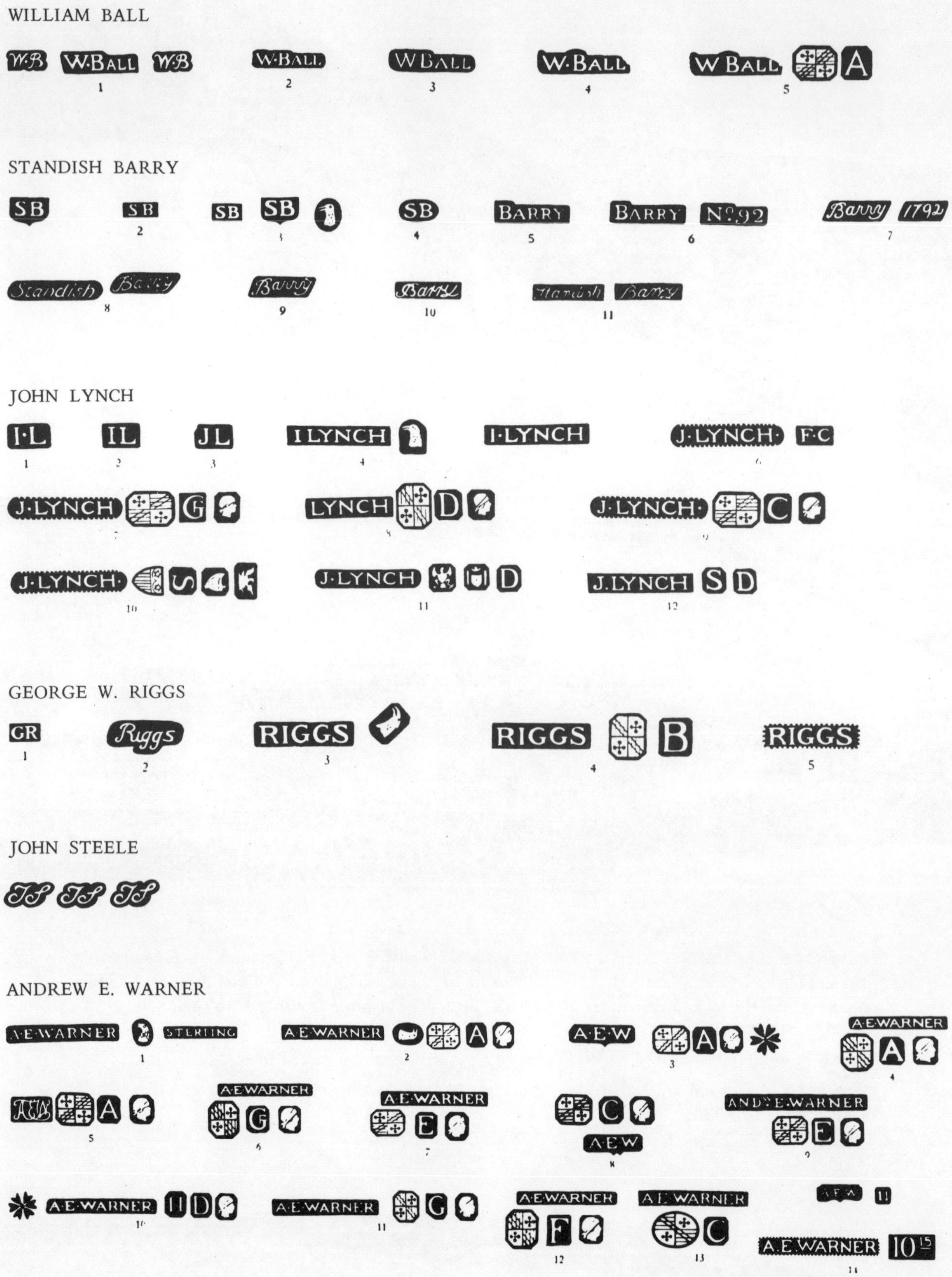

Figure 116. Touch-marks of some Baltimore silversmiths known to have made silver sword hilts. From Pleasants & Sill (1930) MARYLAND SILVERSMITHS 1715 - 1830.

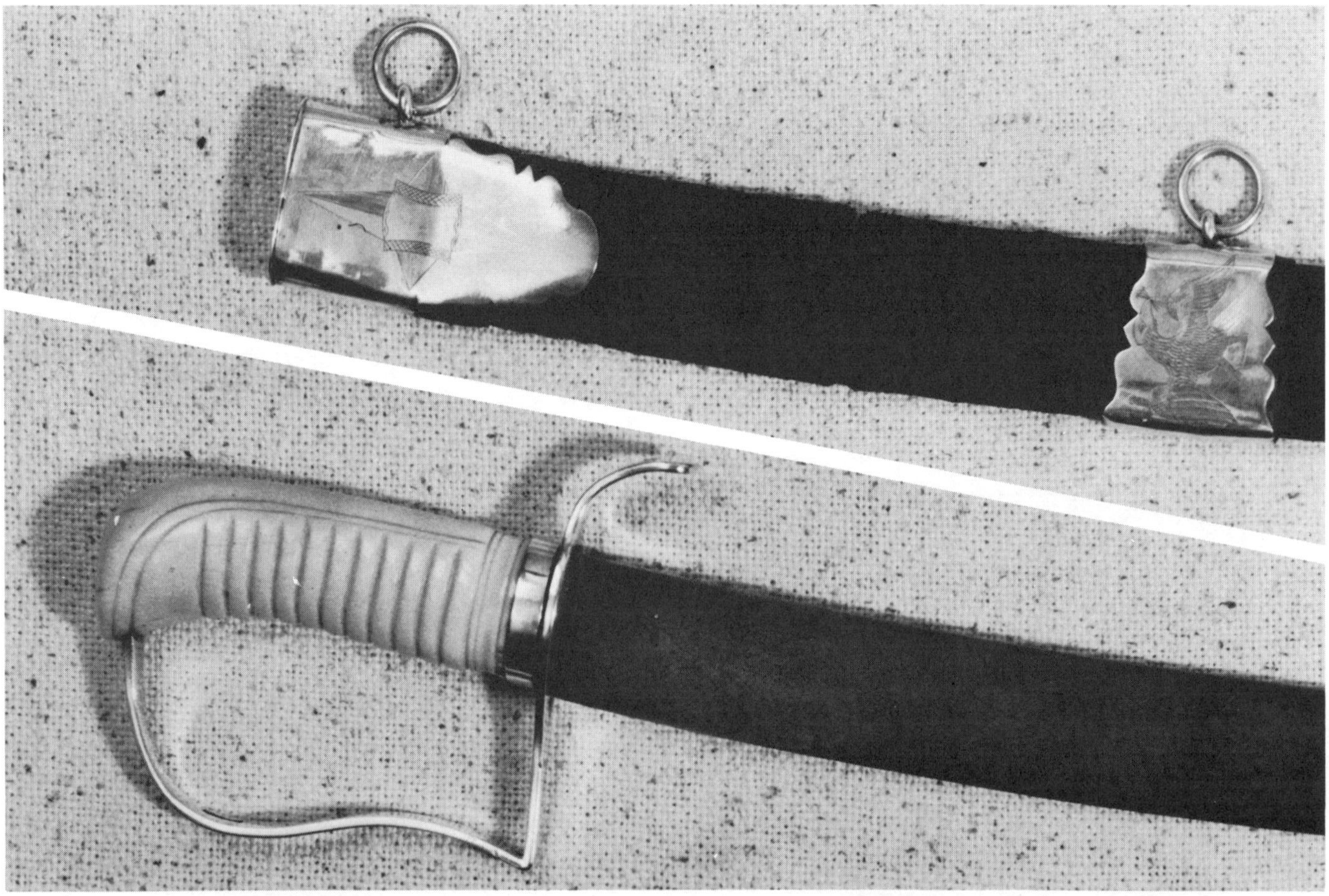

Figure 117. A silver-hilted sword and silver-furnished scabbard made in Baltimore and marked inside the knuckle-bow with the touch "RIGGS". George W. Riggs worked from about 1805 to 1810 in Georgetown, and from 1810 to 1840 in Baltimore. Robert H. McCauley, Jr., collection.

Figure 118. Charles Halbach of Baltimore was an instrument maker in 1835; by 1850 he was listed as a hardware merchant, age 40. A number of flintlock pistols such as the one illustrated here are known, bearing the name "Halbach & Son" on the lock. The styling of these pistols suggests that they were made abroad. Daniel D. Hartzler collection.

Figure 119. Standish Barry of Baltimore (1763–1844) was one of the town's outstanding silversmiths. Silver sword hilts with his touch are known. He had a long military career in the Maryland Militia and rose to be commander of his regiment, with the rank of colonel. This portrait, attributed to Rembrant Peale, shows Col. Barry holding a sword inscribed "Presented by the late Members of the Baltimore Independent Blues to their former Captain –– Standish Barry, Esq. –– now Lt. Col. of the 1st Rifle Regiment, M. M., as an evidence of their respect of his character –– Baltimore, Sept. 13, 1824." Reproduced from Pleasants & Sill (1930) MARYLAND SILVERSMITHS 1715–1830.

Sadsbury Feb. 27th 1810

Dear Sir.

I am extreemly sorry it has not been in my power to fulfill my contract respecting the Gunstalks, but circumstances have been such, that it has been ~~such~~ impossible for me to do it. I had a partner in the business in the first place, altho' I was the only contracter, and know myself to be only responsible to you) I had relied on his promptly attending to the business, but was disappointed. I have been also very much disappointed with the Sawyer respecting the plank, as his Mill broke and other things got out of repair that it was impossible to procure the plank to make them of. I am now Sir, willing to use every exertion in my power and bring on the quantity contracted for as soon as possible, I should hope unless other unforeseen accidents would occur (which I think will not) to have them all deliver'd against the first of July next. If you can consent to wait till that time, I think you will not be disappointed, if you cannot consent to wait, I am willing to come forward immediately and make any Satisfaction required. You will please to do me the favour to inform me by a line your determination respecting the business. Mr Cottrell the bearer of this will call at any time you name, for a line and convey it to me.

I am Sir, very Respectfully Your Obt Hbt Servt William McKim

Tench Cox Esquire

Figure 120. Letter of 1810 from William McKim of Baltimore to Tench Cox, Purveyor of Public Supplies, concerning a contract for gunstocks. Correspondance of Ordnance, General R.G. 92, Box 359, National Archives.

Figure 121 (above). Elmer Wolf, rifle-smith of Thurmont, was in his 70's when this photo was taken in the late 1930's. It shows him standing in front of his workshop with his rifling bench which utilized an old rifle barrel to guide the cutter. This photo and two others, together with an account of a visit to Elmer Wolf at his shop, appeared in Ned Roberts' book THE MUZZLE–LOADING CAP LOCK RIFLE which was first published in 1940.

Figure 122 (left). Nathaniel Rowe, gunsmith of Emmitsburg, as he appeared in a photo reproduced in the *Emmitsburg Chronicle* for 6 March 1908 when he was interviewed concerning his knowledge of Emmitsburg in the days of his youth.

DEATH OF EMMITSBURG'S OLDEST CITIZEN.

Nathaniel Rowe, a life long resident exer is d throughout the administra-
of this place and one of Emmitsburg's tions of nineteen presidents, having

Figure 123. Nathaniel Rowe, gunsmith of Emmitsburg, died there on 10 April 1915. This photograph of him was reproduced in the *Emmitsburg Chronicle* for 16 April 1915 along with an informative obituary. Rowe was born 8 August 1821 and learned the gunsmithing trade as an apprentice to John Armstrong.

Figure 124. Abel Browning, gunsmith of Garret County, was born in 1854, and died in the early 1940's. This photograph of him was obtained by Maurice H. Miller when he interviewed Browning at 84 years of age, on 19 July 1939.

Figure 125 (below). This photo of Abel Browning also was taken on 19 July 1939 by Maurice H. Miller.

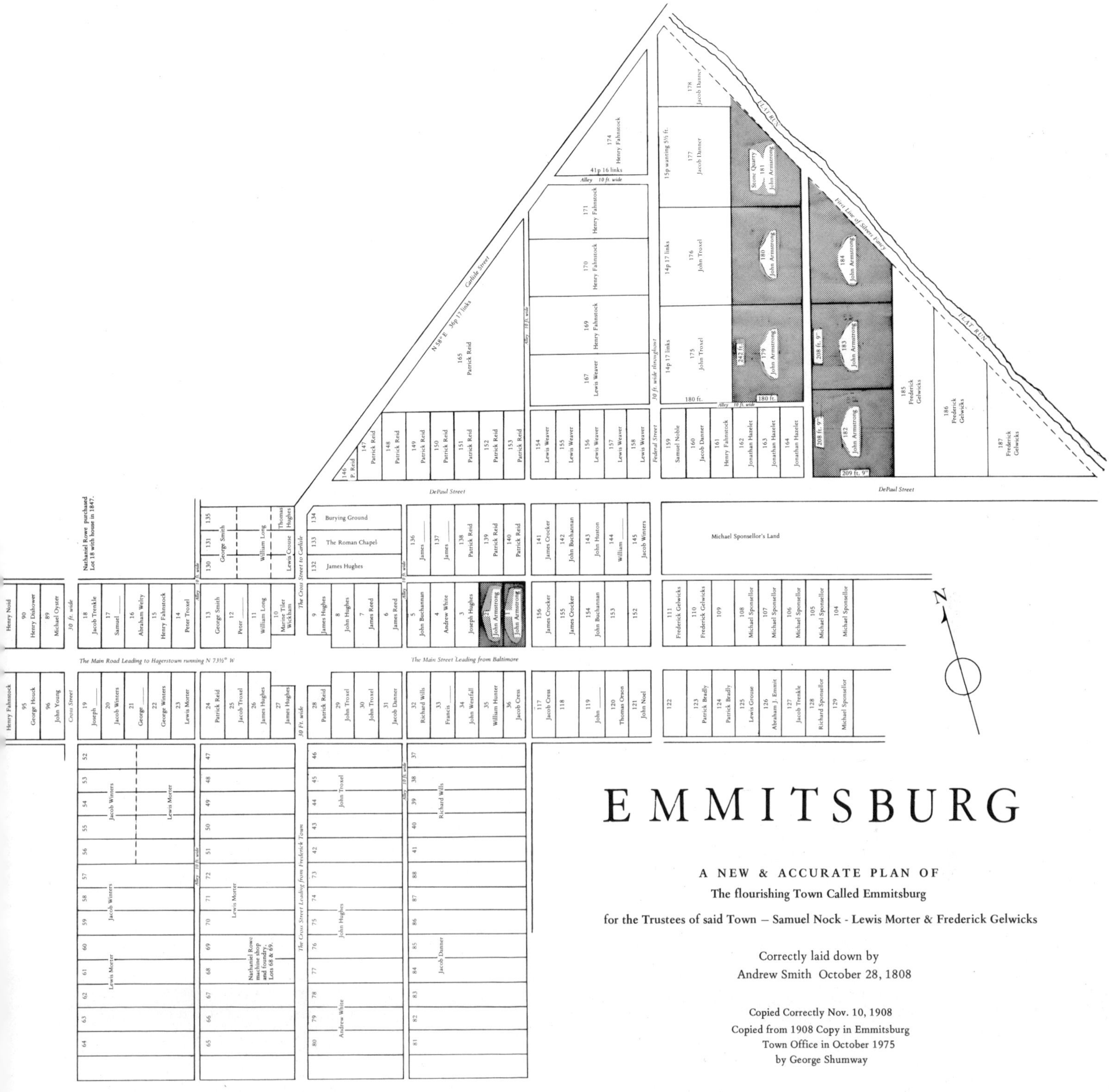

Figure 126. A plan of the town of Emmitsburg was drawn up in 1808 showing how the lots were laid out, giving their numbers, and the names of the owners. The famous Emmitsburg gunsmith, John Armstrong, owned lots 1 and 2 close to the center of town, and lots 179, 180, 181, 182, 183, and 184 near the edge of town. His residence and gunsmith shop was located on lot 2. Lot 10 at the northwest corner of the square was owned for a few years by the gunsmith Marine Tyler Wickham, but this was not shown on the 1808 map. In later years the gunsmith Nathaniel Rowe purchased lot 18 to the west of the square, and he had his machine shop and foundry on lots 68 and 69.

Patents

After the Industrial Revolution got well underway in the 1830's and 1840's, many attempts were made to devise new forms of firearms and artillery pieces. Maryland had her share of inventors. The desirability of having these inventions patented had been proven early, by such men as John Hall with his breechloading flintlock and Samuel Colt with his revolver.

Arms patents granted to men residing in Maryland and Washington, D.C., are listed below. This list represents more than 150 pages of patent descriptions and drawings, and it is not practical to reproduce all of them in this present book. But on the following pages a number of selected patent drawings are reproduced.

Patent Number	*Patent Date*	*Patent Issued to*	*Subject of Patent*
	21 Jan. 1811	W. Thornton & J. Hatkill of Washington, D.C.	Improvement in Fire-Arms
	21 May 1811	John H. Hall of Portland, Maine, and William Thornton of Washington, D.C.	Improvement in Fire-Arms [Hall's breechloader]
	9 Nov. 1832	A. Myers of Boonsbourough, Washington Co.	Gunstock Sawing Machine
	13 Jan. 1835	R. Beale of Washington, D.C.	Percussion Lock for Cannon
	29 Feb. 1835	R. Beale of Washington, D.C.	Percussion Lock to Ignite Cannon
	14 May 1836	William H. Bell of Washington, D.C.	Cannon Traverse Board
6,612	31 July 1849	Benjamin Chambers of Washington, D.C.	Moveable Breeches for Fire-Arms [cannon]
8,126	27 May 1851	Edward Maynard of Washington, D.C.	Tape Primer for Fire-Arms
8,690	27 Jan. 1852	Charles V. Nickerson of Baltimore	Improvements in Fire-Arms
14,077	8 Jan. 1856	J. H. Merrill of Baltimore	Improvements in Fire-Arms [breechloading gun]
14,319	26 Feb. 1856	J. H. B. Latrobe of Baltimore	Percussion-Lock for Fire-Arms
20,954	20 July 1858	J. H. Merrill of Baltimore	Improvement in Breech-Loading Fire-Arms
22,618	11 Jan. 1859	William H. Bell of Washington, D.C.	Improvement in Self-Priming Fire-Arms
23,545	12 Apr. 1859	Francis H. Bell of Washington, D.C.	Self-Priming Gun Lock
	9 Aug. 1859	Charles S. Dickinson of Baltimore	Steam Cannon
26,076	15 Nov. 1859	Wm. H. Arnold of Washington, D.C.	Improvement in Breech-Loading Fire-Arm
26,361	6 Dec. 1859	Edward Maynard of Washington, D.C.	Improvement in Breech-Loading Fire-Arms
26,362	6 Dec. 1859	Samuel W. Marsh of Washington, D.C.	Improvement in Breech-Loading Fire-Arms
26,364	6 Dec. 1859	Edward Maynard of Washington, D.C.	Tape Primer for Fire-Arms
27,518	20 Mar. 1860	Wm. H. Bell of Washington, D.C.	Improvement in Revolving Fire-Arms
27,600	20 Mar. 1860	C. Edward Sneider of Baltimore, Assignor to Thomas Poultney of Baltimore	Improvement in Breech-Loading Fire-Arms
30,537	30 Oct. 1860	Edward Maynard of Washington, D.C.	Improvement of a Tape Primer for Fire-Arms
32,032	9Apr. 1861	James H. Merrill of Baltimore, Assignor to Merrill Patent Fire Arms Manufacturing Co.	Improvement in Breech-Loading Fire-Arms
32,033	9 Apr. 1861	James H. Merrill of Baltimore, Assignor to Merrill Patent Fire Arms Manufacturing Co.	Improvement in Breech-Loading Fire-Arms
32,450	28 May 1861	James H. Merrill of Baltimore, Assignor to Merrill Patent Fire Arms Manufacturing Co.	Improvement in Fire-Arms
32,451	28 May 1861	James H. Merrill of Baltimore, Assignor to Merrill Patent Fire Arms Manufacturing Co.	Improvement in Fire-Arms
33,536	22 Oct. 1861	James H. Merrill of Baltimore	Improvement in Breech-Loading Fire-Arms
33,655	5 Nov. 1861	Samuel W. Marsh of Washington, D.C.	Improvement in Breech-Loading Fire-Arms
34,703	18 Mar. 1862	C. Edward Sneider of Baltimore	Improvement in Revolving Fire-Arms
36,721	21 Oct. 1862	Josiah V. Meigs of Washington, D.C.	Improvement in Breech-Loading Fire-Arms

37,208	16 Dec. 1862	Samuel Strong of Washington, D.C.	Improvement in Breech-Loading Fire-Arms.
38,643	19 May 1863	Samuel Strong of Washington, D.C.	Improvement in Breech-Loading Fire-Arms
39,592	18 Aug. 1863	Daniel E. Somes of Washington, D.C.	Improvement in Fire-Arms
39,707	25 Aug. 1863	C. E. Sneider of Baltimore, Assignor to himself and Thomas Poultney of Baltimore	Improvement in Breech-Loading Fire-Arms
40,884	8 Dec. 1863	James H. Merrill of Baltimore, Assignor to Merrill Patent Fire Arms Manufacturing Co.	Improvement in Breech-Loading Fire-Arms
44,126	16 Sept. 1864	Wm. Tileston of Georgetown, D.C.	Improvement in Safety-Stops for Revolving Fire-Arms
44,290	20 Sept. 1864	Wm. C. Dodge of Washington, D.C.	Improvement in Breech-Loading Fire-Arms
44,586	11 Oct. 1864	John J. Aliscs of Washington, D.C.	Improvement in Ball-Screws for Fire-Arms
46,054	24 Jan. 1865	Charles E. Sneider of Baltimore, Assignor to himself & Thomas Poultney of Baltimore	Improved Method of Converting Muzzle- into Breech-Loading Fire-Arms
46,220	7 Feb. 1865	Arthur DeWitzleben of Washington, D.C.	Improvement in Ball-Screws for Fire-Arms
46,516	21 Feb. 1865	William Lewis Winans of Baltimore	Improvement in Operating Ordnance on Gun-Boats
46,612	28 Feb. 1865	Charles Edward Sneider of Baltimore, Assignor to himself & Thomas Poultney	Improvement in Revolving Fire-Arms
47,755	16 May 1865	C. Edw. Sneider of Baltimore	Improvement in Breech-Loading Fire-Arms
48,423	27 June 1865	Edward Maynard of Washington, D.C.	Improvement of a Tape Primer for Firearms
49,130	1 Aug. 1865	Edward Maynard of Washington, D.C.	Improvement in Fire-Arms
52,547	13 Feb. 1866	William C. Dodge of Washington, D.C.	Improvement in Fire-Arms [shotgun]
54,934	22 May 1866	Josiah V. Meigs of Washington, D.C.	Improvement in Breech-Loading Fire-Arms
56,846	31 July 1866	A. M. Bacon of Washington, D.C., Assignor to himself and Geo. E. H. Day	Improvement in Breech-Loading Fire-Arms
58,790	16 Oct. 1866	William C. Dodge of Washington, D.C.	Improvement in Breech-Loading Fire-Arms [shotgun]
59,487	6 Nov. 1866	J. J. B. Walbach of Baltimore	Improvement in War-Rockets
64,701	14 May 1867	Thomas Poultney of Baltimore & Silas Crispen of New York, Assignors to Thomas Poultney	Improvement in Breech-Loading Fire-Arms
66,596	9 July 1867	E. C. Kirk & E. Sneider of Baltimore	Improvement in Magazine Fire-Arms
69,919	15 Oct. 1867	William F. Kussmaul of Baltimore	Improvement in Safety Gun-Locks
74,478	6 Feb. 1868	Frederick W. Alexander of Baltimore	Improvement in Muzzle-Loading Ordnance
85,252	22 Dec. 1868	C. E. Sneider of Baltimore	Improvements in Breech-Loading Fire-Arms
86,566	2 Feb. 1869	Edward Maynard of Washington, D.C.	Improvement in Fire-Arms
86,971	16 Feb. 1869	Isaiah B. Conklin of Baltimore	Improvement in Breech-Loading Fire-Arms
97,717	7 Dec. 1869	Charles Edward Sneider of Baltimore	Improvement in Sights for Fire-Arms
105,799	26 July 1870	L. Hailer of Washington, D.C.	Improvement in Gun-Locks
112,763	14 Mar. 1871	William C. Dodge of Washington, D.C.	Improvement in Breech-Loading Fire-Arms
113,408	4 Apr. 1871	William C. Dodge of Washington, D.C.	Improvement in Breech-Loading Fire-Arms
114,653	9 May 1871	William C. Dodge of Washington, D.C.	Improvement in Breech-Loading Fire-Arms
114,794	14 May 1871	William C. Dodge of Washington, D.C. & Philip Tell Dodge	Improvement in Breech-Loading Fire-Arms
118,350	22 Aug. 1871	William C. Dodge & Philip Tell Dodge of Washington, D.C.; said P. T. Dodge Assignor to Wm. C. Dodge	Improvement in Breech-Loading Fire-Arms
127,683	11 June 1872	William C. Dodge & Philip Tell Dodge of Wash. D.C.	Improvement in Breech-Loading Fire-Arms
131,188	10 Sept. 1872	Charles E. Sneider of Baltimore, Assignor to himself & Duncan C. Clark	Improvement in Devices for Crimping Cartridges
171,442	21 Dec. 1875	Charles E. Sneider & Charles W. Sneider of Baltimore	Improvement in Breech-Loading Fire-Arms
202,126	9 Apr. 1878	Charles W. Sneider of Baltimore	Improvement in Breech-Loading Fire-Arms
227,135	4 May 1880	Charles W. Sneider of Baltimore	Breech-Loading Fire-Arms
257,097	25 Apr. 1882	Charles W. Sneider of Baltimore	Breech-Loading Fire-Arm
417,594	17 Dec. 1889	Charles E. Sneider of Baltimore, Assignor of one-half to Martin Gillet Gill of Baltimore	Breech-Loading Magazine Fire-Arm

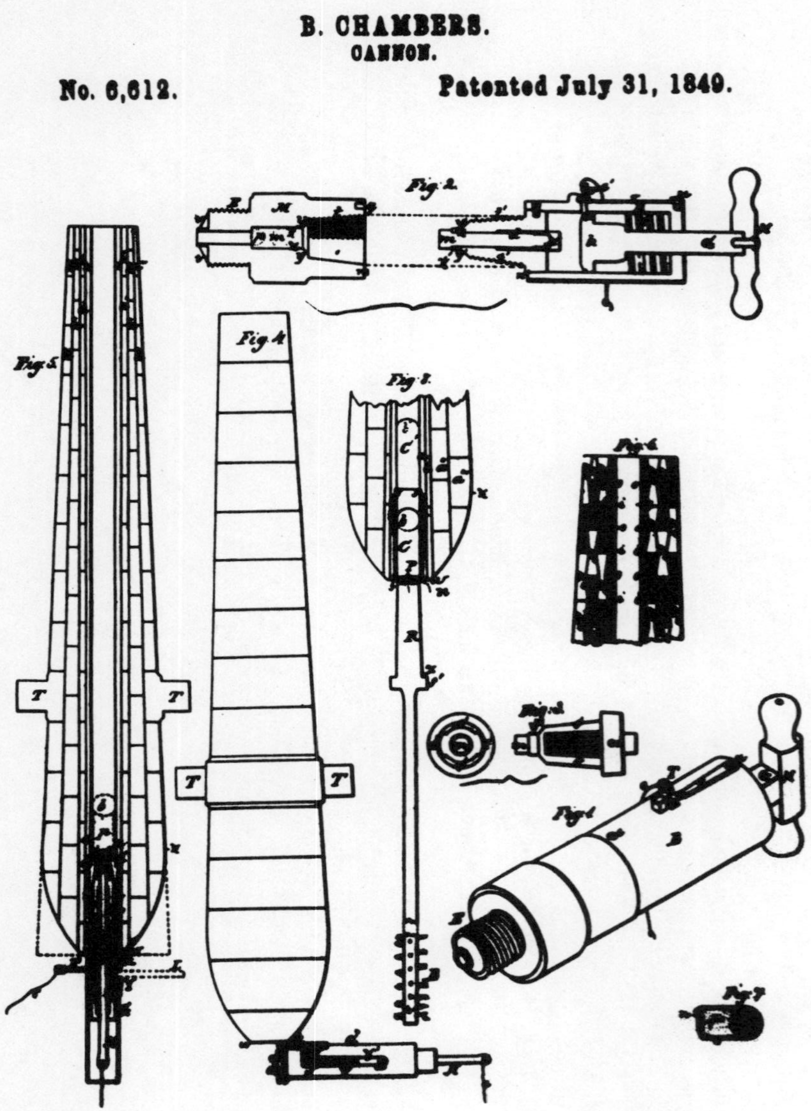

Figure 127.

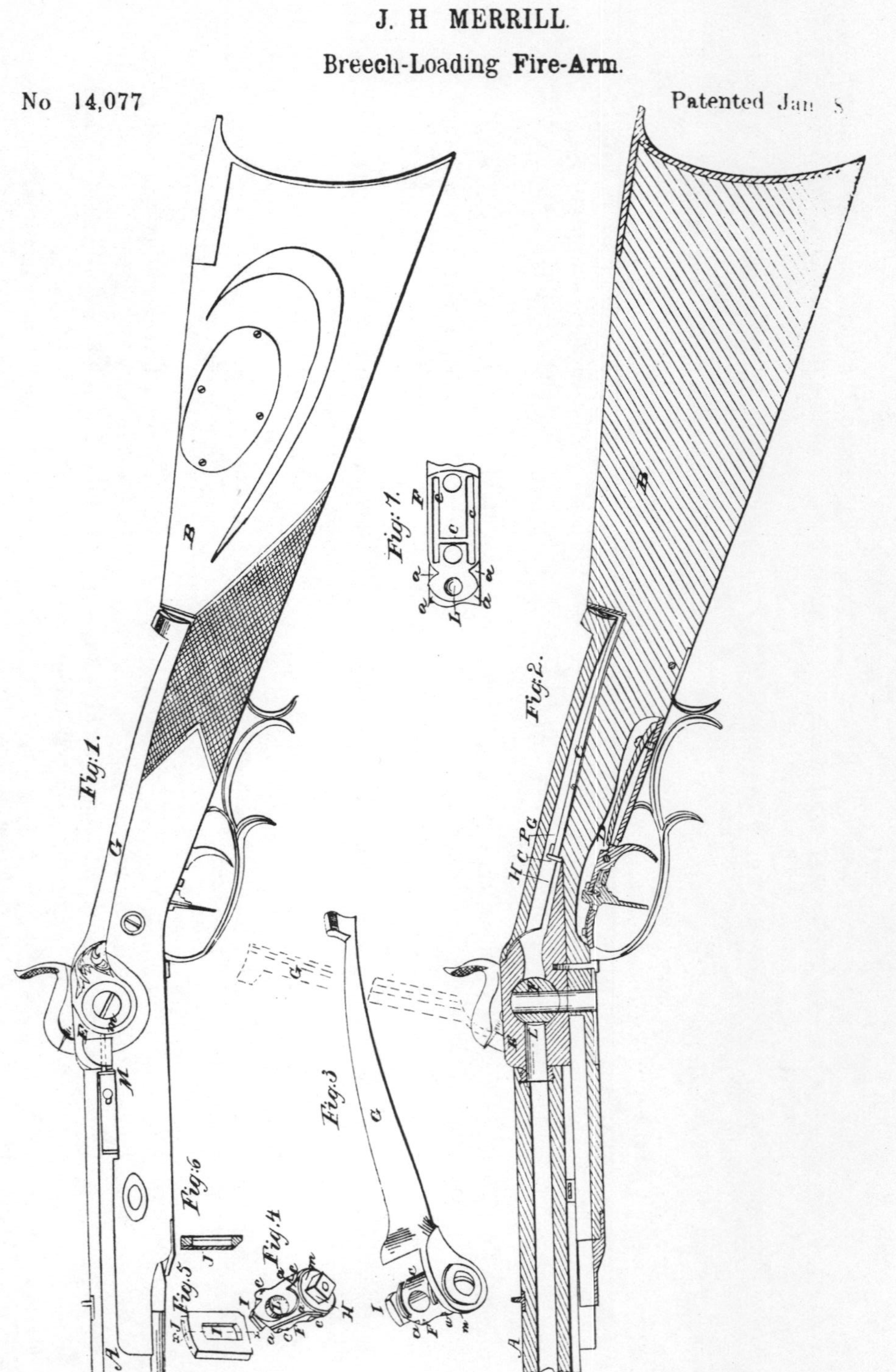

Figure 128.

J. H. B. LATROBE.

Priming Cock.

No. 14,319. Patented Feb. 26, 1856

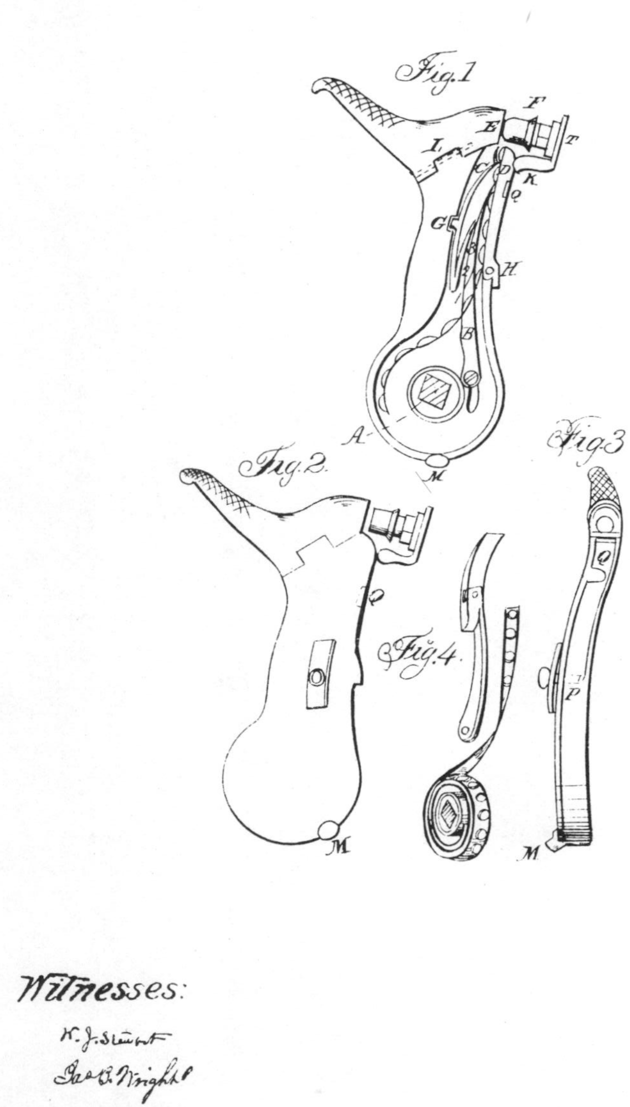

Figure 129.

J. H. MERRILL.

Breech-Loading Fire-Arm.

No. { 2,532, 33,536. } Patented Oct. 22, 1861.

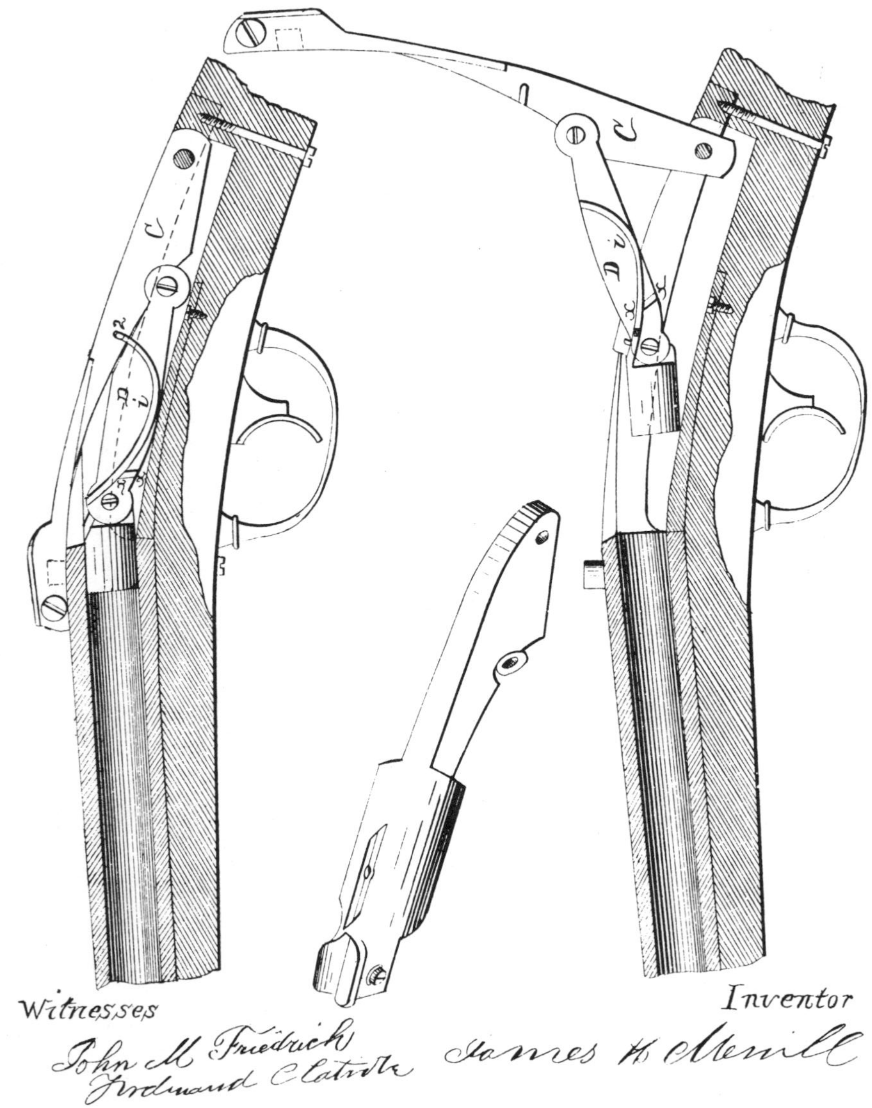

Figure 130.

J. H. MERRILL.

Breech-Loading Fire-Arm.

No. { 1,028, 32,032. } Patented Apr. 9, 1861

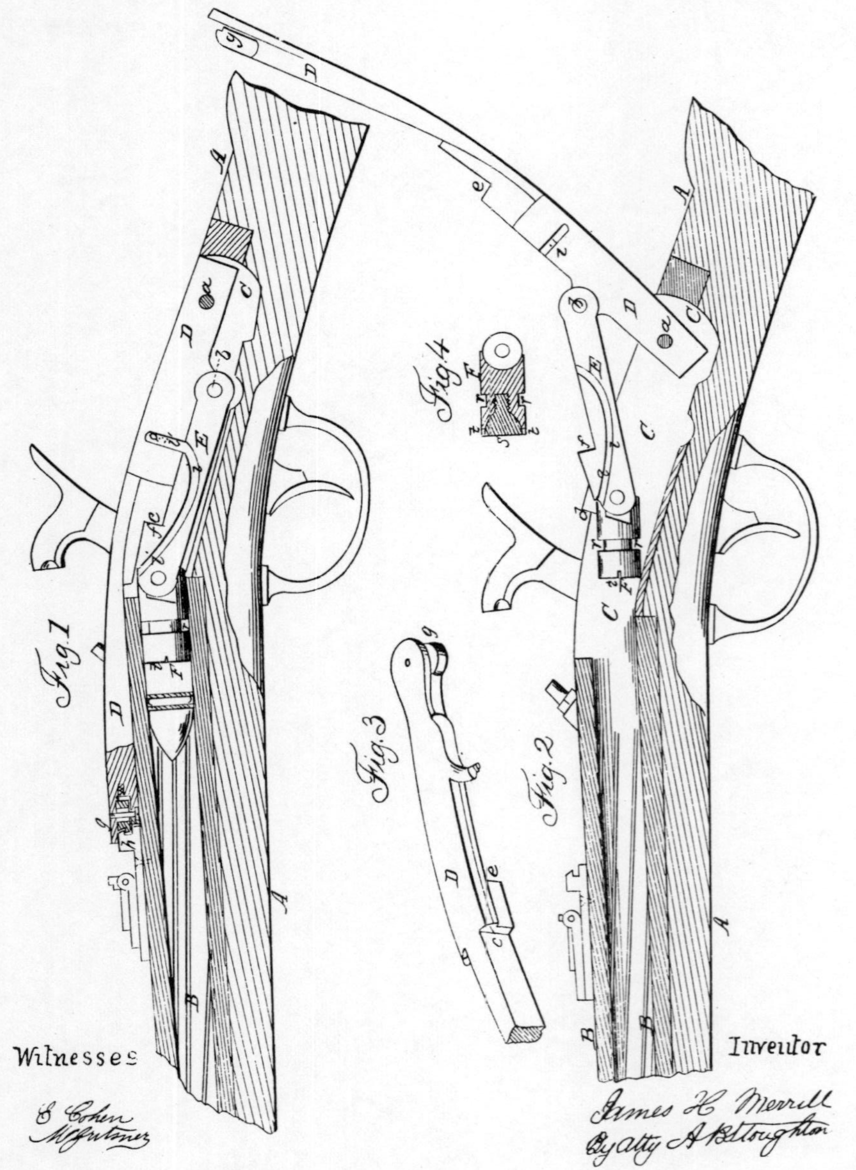

Figure 131.

C. E. SNEIDER.

Revolver.

No. 34,703. Patented Mar. 18, 1862

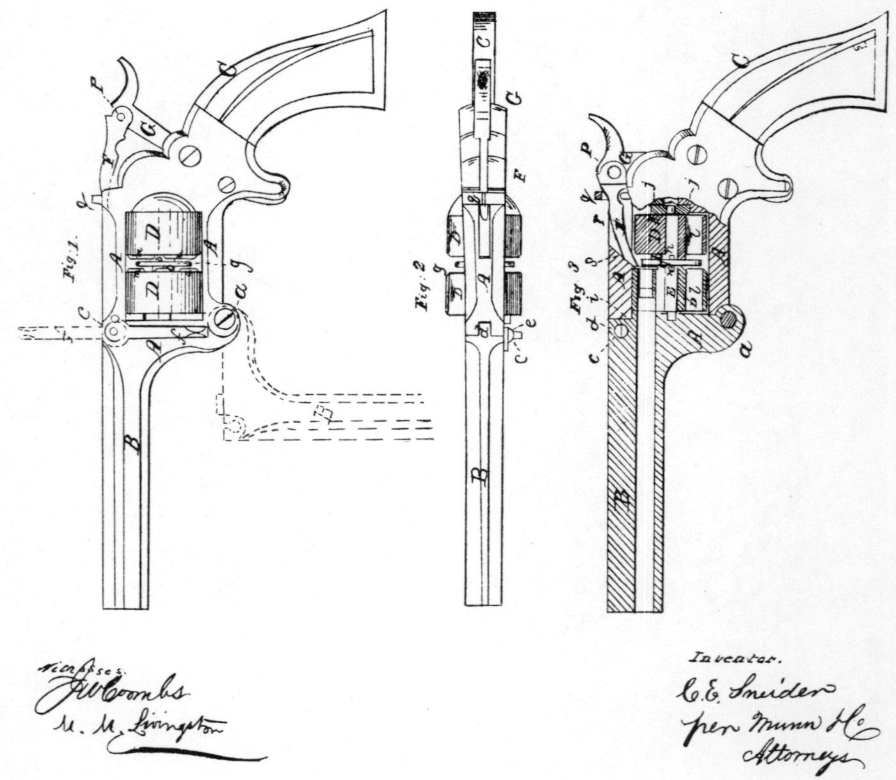

Figure 132.

J. H. MERRILL.

Breech-Loading Fire-Arm.

No. 40.884 Patented Dec. 8, 1863.

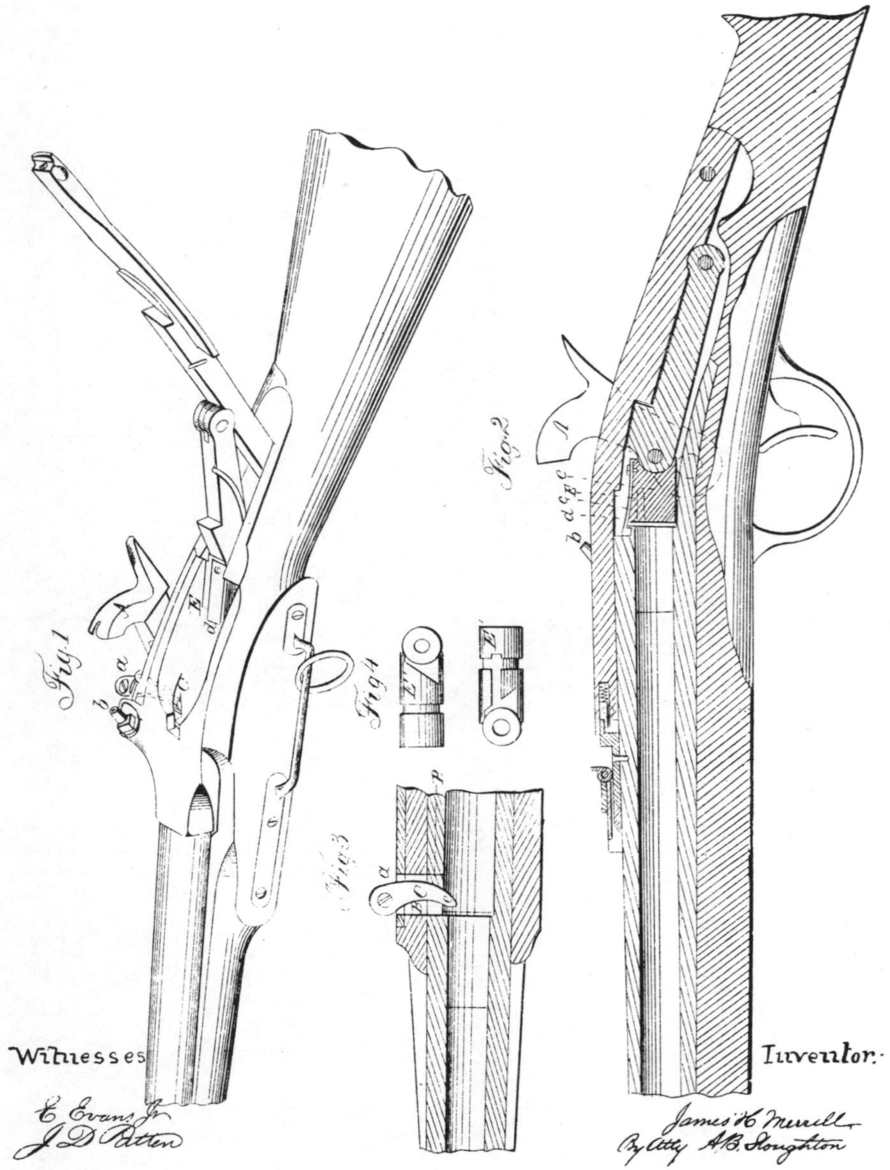

Figure 133.

C. E. SNEIDER.

Breech-Loading Fire-Arm.

No. 46,054. Patented Jan. 24, 1865.

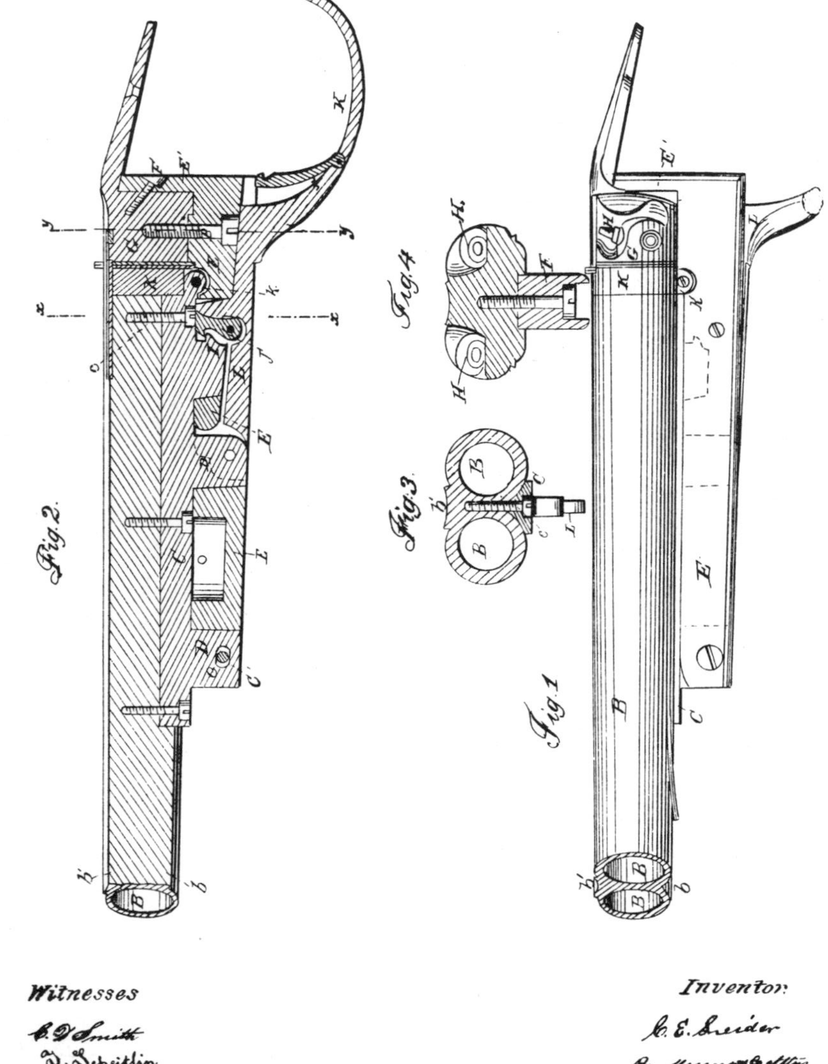

Figure 134.

W. L. WINANS. 3 Sheets—Sheet 1.

Gun-Carriage

No. 46,516. Patented Feb. 21, 1865.

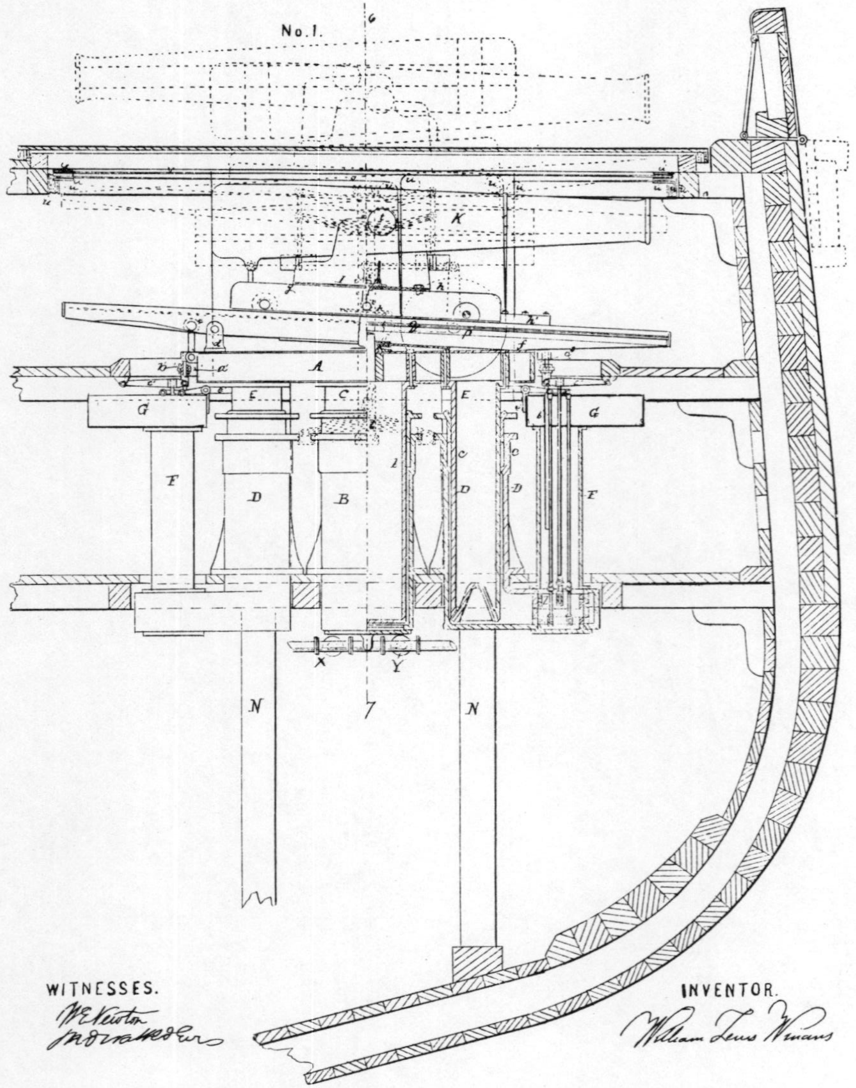

Figure 135.

C. E. Sneider.

Revolver.

No. 46612 Patented Feb. 28. 1865

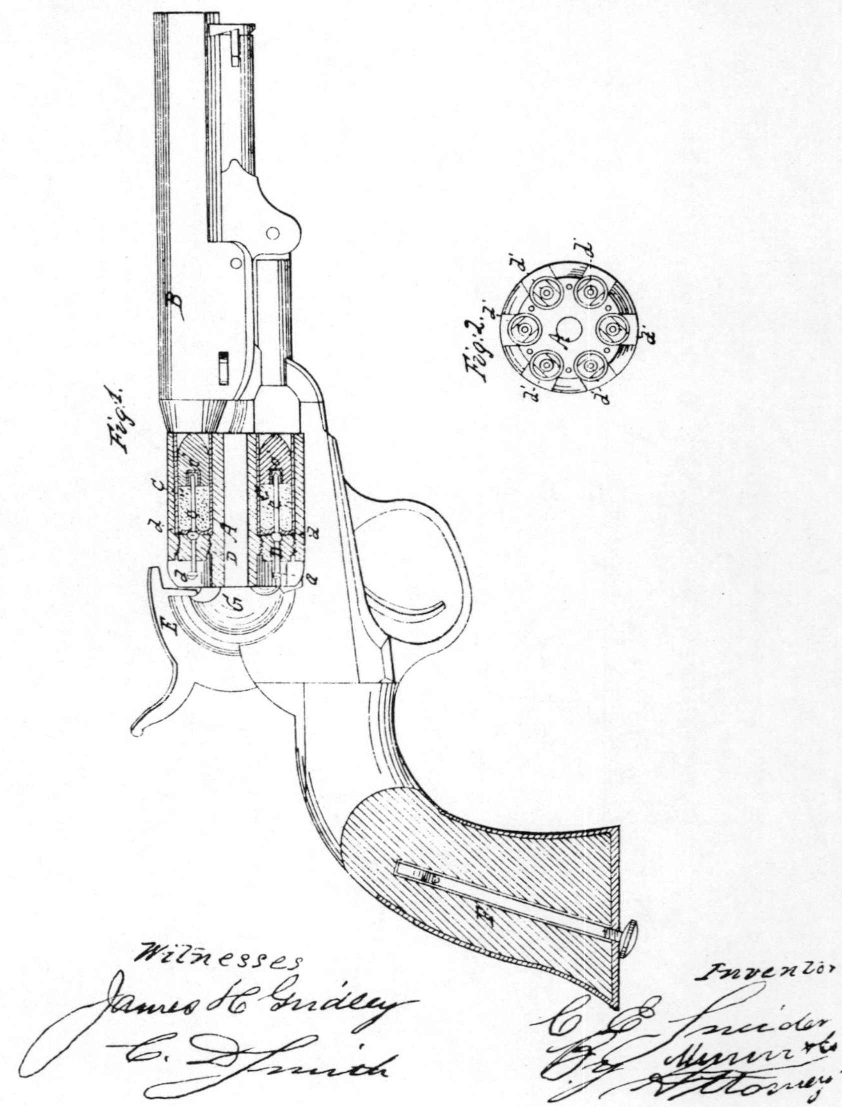

Figure 136.

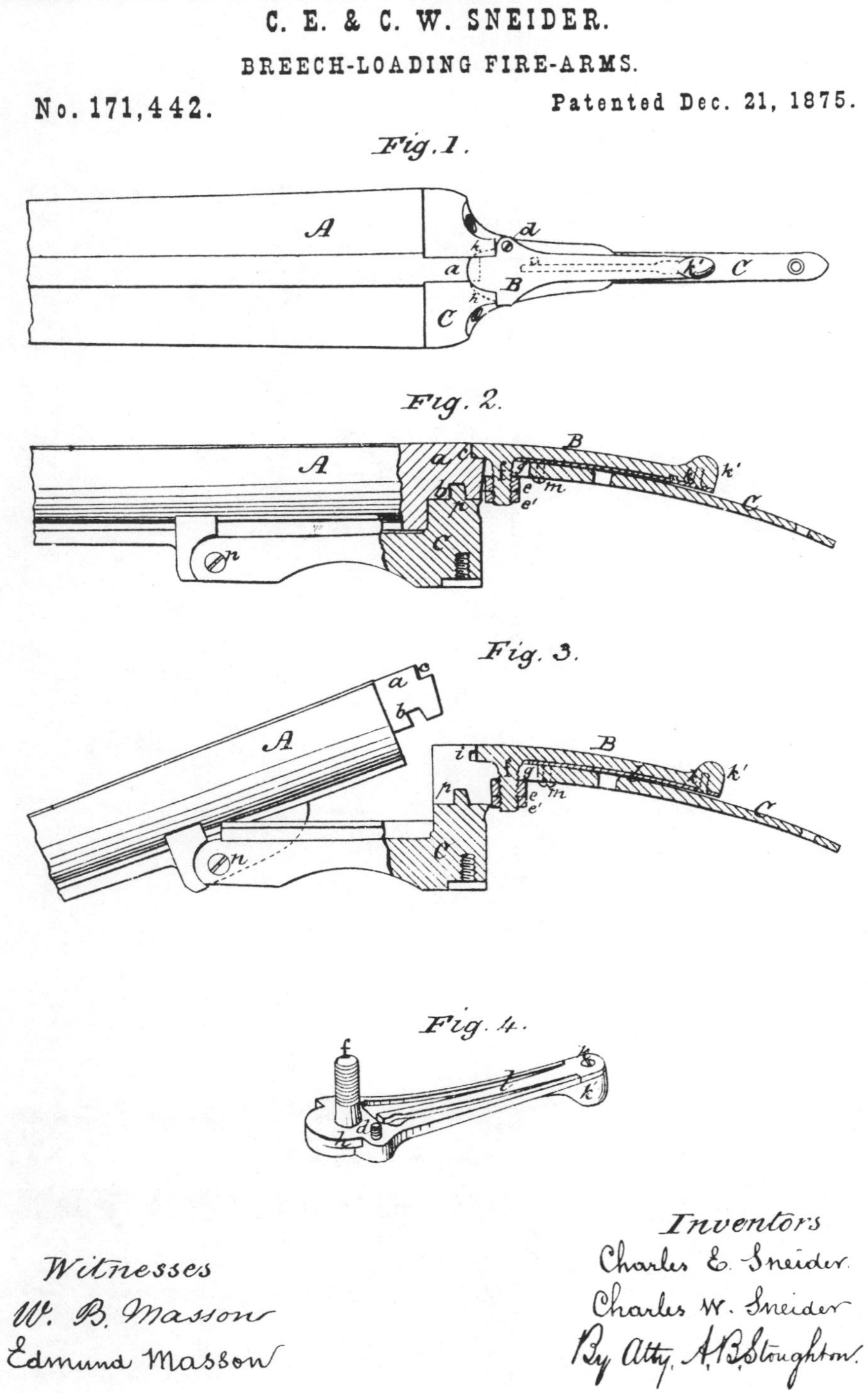

Figure 137.

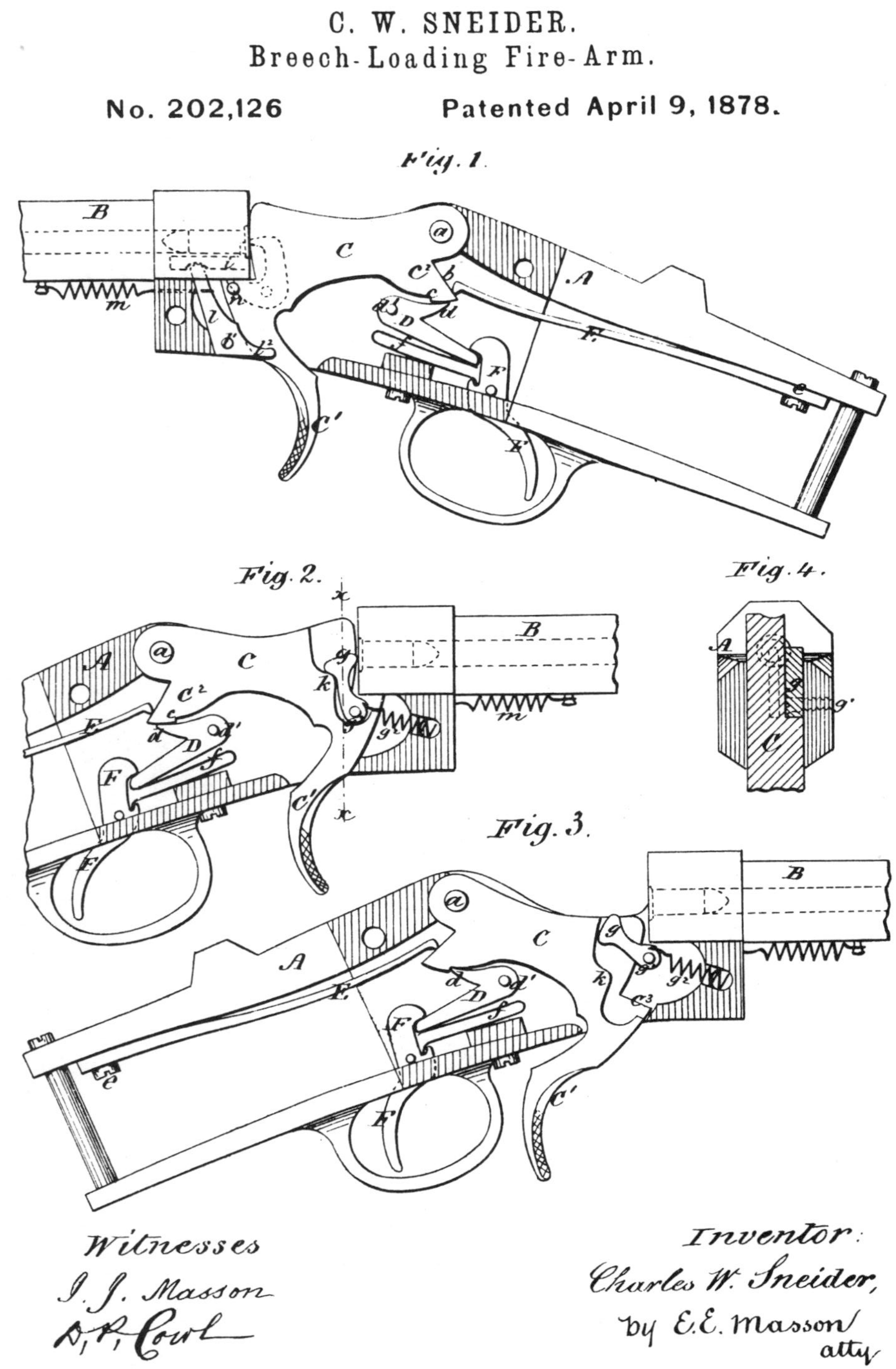

Figure 138.

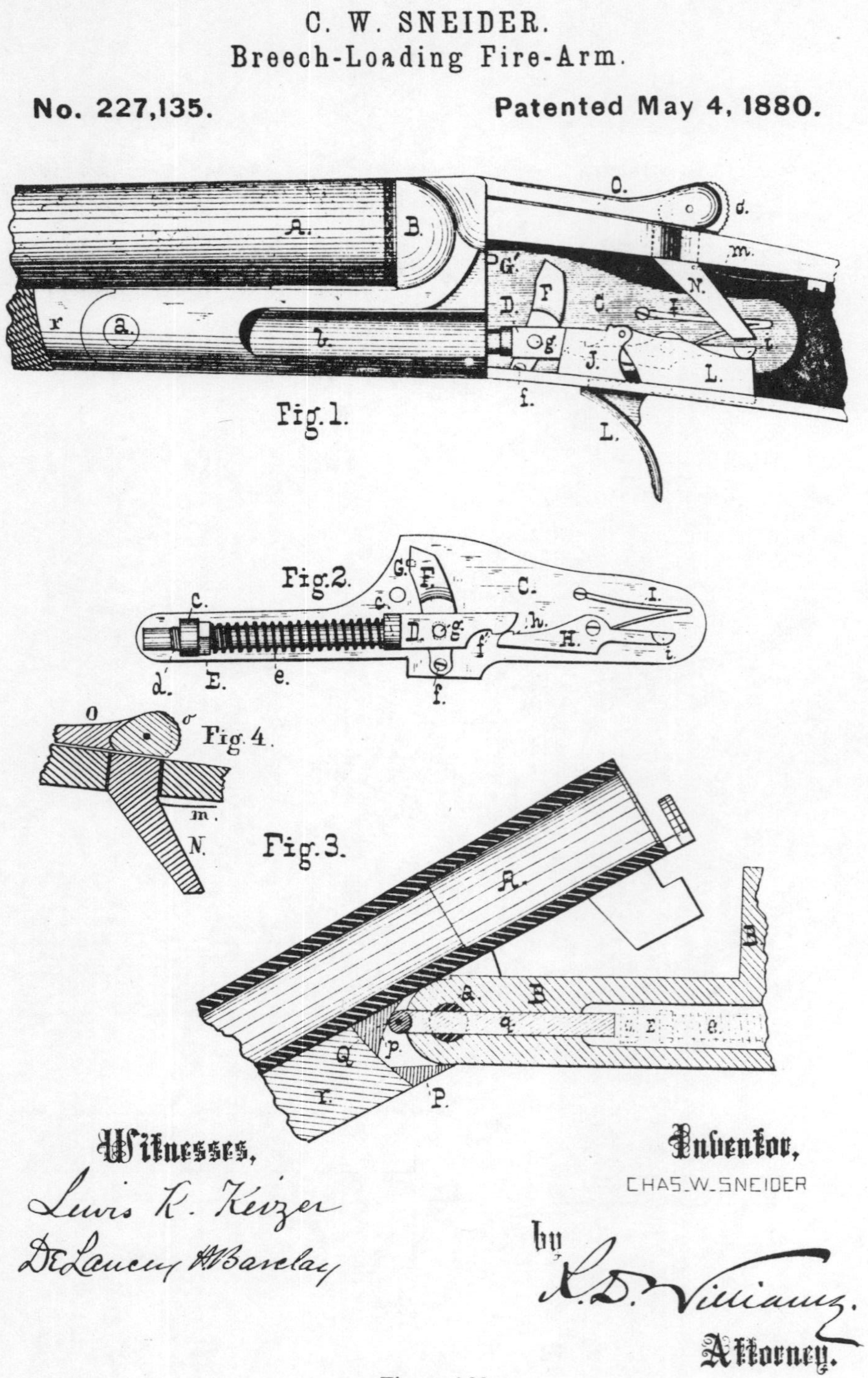

Figure 139.

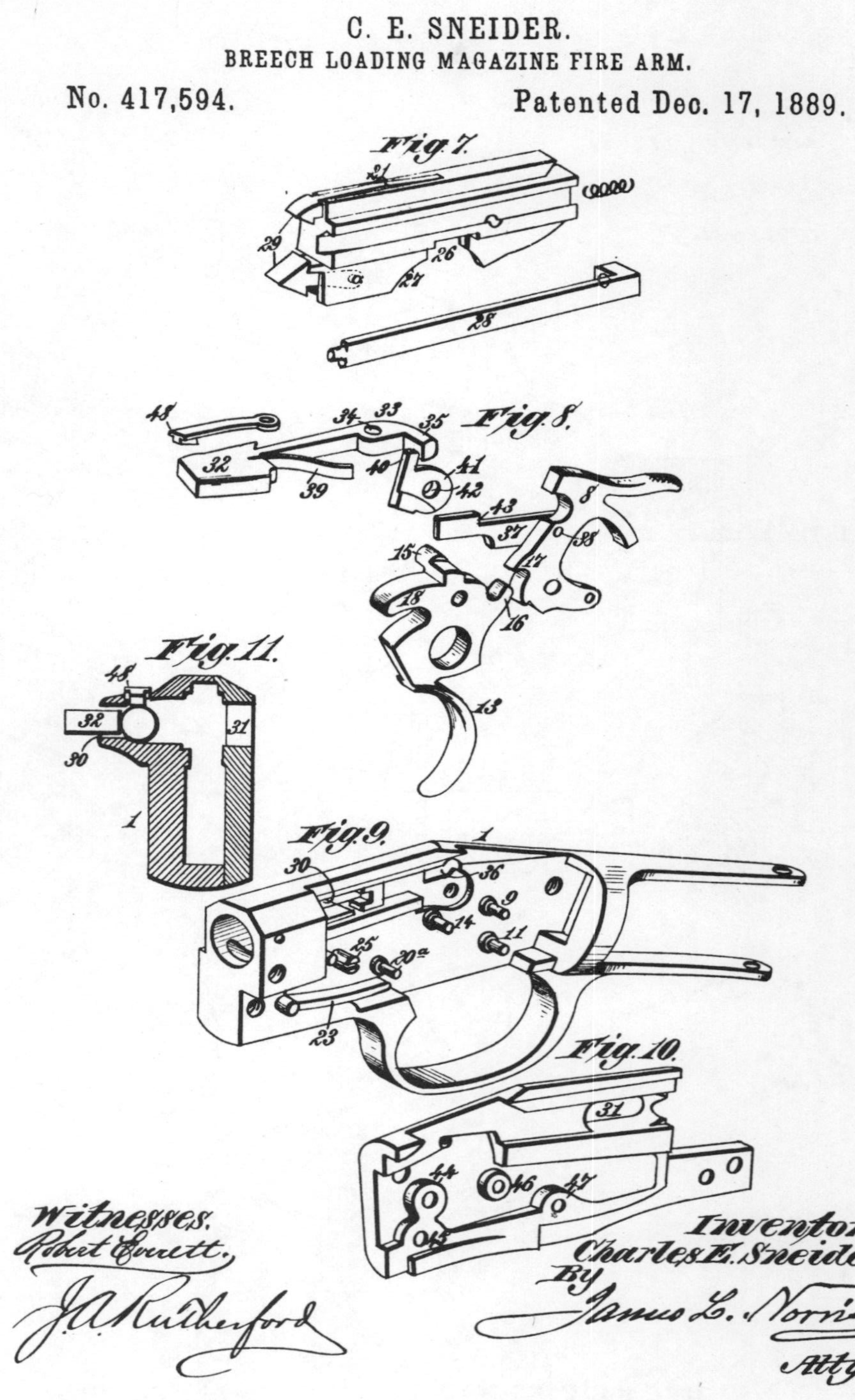

Figure 140.

The Author

Daniel David Hartzler was born in 1941 and raised in the small town of New Windsor in north-central Maryland. His parents being avid sportspersons, at an early age he was introduced to the pleasures of fishing and hunting in the beautiful rolling hills of the Free State.

Dan's six uncles on his mother's side returned from World War II with a heterogenous assortment of souvenirs which the young lad proudly displayed in a closet. In 1953 his dad bought a Civil War musket at a country sale and Dan soon began using it to shoot with, under the supervision of his parents. By the time he graduated from high school Dan had disposed of his World War II memorabilia in favor of antique guns, and his collection graduated from the closet to the attic.

Having completed his studies at Eckels College of Mortuary Science, a branch of Temple University, and a two-year apprenticeship, Dan passed his state board examinations. Joining both of his parents, who are licensed morticians, he became a fourth-generation member of the firm of D. D. Hartzler & Sons. Dan, his wife Kit, his son, Jeb, and his daughter, Sandra, live today in the town of New Windsor.

In 1962 Dan purchased a Maryland presentation sword for his collection, and was so thrilled after researching the veteran who had carried the blade, that his assorted antique arms began to be traded for pieces made in Maryland. It occurred to him that the Old Line State had been sadly neglected as far as research on the arms and the men who made them was concerned, and so he began to dig into this almost virgin field. His Marylandia arms collection now contains about one hundred pieces, and still is growing.

Dan has been an avid hunter with muzzle-loading arms for years. For squirrels, rabbits, pheasants, quail and grouse in the rolling hills, and geese and ducks on the shore, his front-end stuffers have cracked. The flintlock longrifle which he is holding in the illustration above has dropped a buck, but not yet a gobbler. Dan hopes that the future will, among other things, bring more big game to his sights, and also that it will provide further opportunity to study and learn about the arms makers and arms of early Maryland.

Index

Aaron,George W., 117
Adams, Christopher, 270
Adams, Joseph, 218
Adelsberger, Daniel G., 247
Adlum, John, 141
Albright, Christian, 242
Alexander, ———, 200
Alexander, ———(Capt.), 194
Alexander, Frederick W., 117, 292
Alexander, George (Col.), 161
Alexander, John B. (Capt.), 183
Alexander, Walter (Capt.), 97
Alford, A.G. & Co., 117
Alford, Nelson E., 117
Aliscs, John, J., 117, 292
Allen, William, 232
Allen, Zachariah, 117
Allison, Catherine, 174, 186
Allsworth, Samuel, 117
American Arms Co., 71
American Machine Works, 71
Annan, Robert L. 121, 122
Annapolis, John Fulford (Capt.), 100
Archer, Dr. John, 146
Armendt, George, 117
Armendt, John E., 117
Armstrong, ———, 249
Armstrong, Agnes, 119, 124
Armstrong, Anna, 45, 118
Armstrong, Elizabeth, 45, 118
Armstrong, James, 45, 117, 118
Armstrong, Jane, 45, 118
Armstrong, John, 117, 232, 271, 272, 246, 247
Armstrong, John (Jr.), 36, 44-47, 49, 57, 58, 118-121, 123, 124, 133
Armstrong, John (Sr.), 45, 118, 120
Armstrong, Quintain, 123
Armstrong, Robert, 45, 118, 124
Armstrong, Samuel, 45, 118, 122, 124
Armstrong, William, 45, 118
Arnold, Wm. H., 73, 291
Arnote, Thomas, 183
Ash, Henry, 268
Assoiction of Freemen, 23
Aster, Rosa, 187
Auman, J. S., 248

Bacon, A.M., 292
Baden, N. (Capt.), 139
Bahr, Rudolph P., 55
Baird, John D., 169, 180
Baker, ———, 196
Ball, William, 78
Ball, William Jr., 80
Baltimore Council, 194
Baltimore Gazette & Daily Advertiser, 164
Barach, ———(Col.), 33
Barbour, James, 220
Barley, William, 242
Barnard & McAinn, 80
Barney, Joshua, 87
Barnhassel, Christopher, 129
Barnhizle, Christian, 29
Barnhizle, Christoph, 27-29,49, 51, 139
Barnhizle, Christopher, 129, 140, 224, 255, 265, 270, 273
Barry, Francis G., 129
Barry, Lavallin, 129
Barry, Standish (Col.), 129
Barsotti, John, 169
Barstow, ———, 129
Barstow, Frederick S., 129,217
Bartlett, A., 129
Bates, Frederick, 186
Baxter, Arthur A., 130
Bayly, Eleanor, 140
Bayly, Mountjoy, 140
Bazier, John, 24, 267
Beal, Thomas (Capt.), 154
Beale, R., 130, 291
Beals, Issac, 243
Bean, Lucy, 118
Bear, George, 134, 154
Beatty, ———, 199
Beatty, ———(Col.), 270
Beatty, Charles, 23, 41, 154-155, 161, 252
Beatty, Charles (Capt.), 41, 130
Beatty, Cornelius E., 131
Beatty, William, 273
Beckenbaugh, Christina, 212, 213
Behr, J.J., 131
Behrens, Christian F., 33, 131
Bell, Charles M., 131
Bell, Francis H., 73, 131, 291
Bell, Jackson W., 131
Bell, James E., 131
Bell, Lloyd C., 131
Bell, Mary E., 132
Bell, N.H. & Co., 196
Bell, Sarah B., 131
Bell, Sephi (M), 131
Bell, Thomas, 132
Bell, Thomas H., 131
Bell, Walter, 131
Bell, William H., 73, 131-132, 146, 252, 291
Bell, Willie, 132
Belleson, George W., 132
Belleson, William, 132
Bellona Powder Mills, 91
Belschner, John L., 132
Beltzhoover, Elizabeth, 178
Beltzhoover, Melcher, 178
Benedict & Burnham, 112
Benedict & Burnham Manu. Co., 112
Benedict & Coe, 112
Benedict, A., 112
Benet, Hugh Jr., 162
Bennett & Packson, 21, 132
Bennett, F.W., 132
Bentz, George, 141
Bentz, Mrs. William, 224
Biggs, Benjamin, 33
Bingham, Martin, 157
Bird, Henry, 132
Bird, Nightingale & Neff, 132, 228-229
Bishop, John E., 145
Bitner, Anne E., 132
Bitner, Elizabeth, 132
Bitner, William G., 132
Black, John, 122
Blackburn, Alexander, 120, 133
Blackburn, John, 45, 47, 120, 132-133
Blakslee, B.E., 133
Blocker, George, 243
Boerstler, Daniel, 133
Bohrer, Julius L., 133
Bonaparte, Jerome, 65
Bond, Richard, 133
Boon, Samuel, 24, 130-131
Boone, Samuel, 32, 133, 134, 154-155
Boone, William, 134
Borden, Daniel,134, 142
Borden, William. 134
Border, Daniel, 134
Bosley & Howe, 219
Bosley, Arthur, 135
Bossell, V.E., 135
Boteler, John D., 135
Bower, Catherine, 243-244
Bower, David, 182
Bowers, William, 262
Bowers, William S., 246, 259
Bowford, George (Lt.Col.), 220
Bowford, John (Lt.Col.), 220
Bowlus, David, 214
Bowman, George, 224
Boyd, Andrew G., 191
Boyd, James, 135
Boyer, Henry, 142
Brada, Conrad, 135
Braddock's Road, 19
Bradley, Patrick,121, 142
Brady, Thomas, 121
Brenner, Stephen, 253
Brice, Jacob (Capt.), 98
Briggs, E.G. & Co., 135
Bright, Jacob, 135
Bringman, Martin, 204
Bringmas, M., 160

Brotherhouse, William, 135
Brown, John, 199
Browne, David (Col.), 218
Browning, Abel, 55, 135
Browning, John, 135
Browning, Meshach, 75, 135
Bruce, ———(Col.), 203
Bruff, Charles Oliver, 80
Bruner, David, 216
Bryan, Francis, 136
Bryant, C.W., 196
Buchanan, ———, 195
Buchanan, John, 204, 206
Budenbaugh, John, 136
Budenbaugh, Martin, 53, 136, 204
Burges, Edward (Capt.), 218
Burgess, ———(Capt.), 195
Burkhardt,Philip, 136
Burnard, Notley, 33, 254
Burnett, Garland, 136
Bursking, William, 182
Burt, Ed & Co., 136
Butler, ———(Gen.), 104-105

Calderwood, Thomas A., 137
Caldwell, Samuel, 121
Calhoun & Cowton, 137
Calvert & Parkhurst, 137
Calvert, ———(Gov.), 217
Campbell, Daniel, 137
Campbell, William (Capt.), 137
Canfield & Brother, 137
Canfield Bros., 106-107
Canfield Bros. of Baltimore, 80
Canfield, Ira C., 137
Canfield, William B., 137
Carroll, Charles, 255
Carter, Thomas, 137
Cashell, R., 138
Catoctin Furnace, 195
Cattey, Edward, 138
Chambers, Benjamin, 138, 291
Chambers, William, 138
Charrier, James, 138
Charrier, John, 138
Chanier, Pete, 138
Chanier, Peter, 138 , 140
Chase, A.C., 138
Chisholm (Chisholme), Archibald, 25, 138, 254
Christopher, A., 138
Christopher, E., 138
Christopher, J., 138
Clagett, Alexander, 27, 129, 139-140, 255, 270, 273
Clagett, William, 132
Clagett, Wiseman, 139
Clapsaddle, Daniel (Capt.), 97
Clark & Sneider, 139
Clark, Duncan, 256
Clark, Duncan C., 139, 256
Clark, Matilda, 261
Clarke, John, 156, 163
Cleaveland & Charrier, 138, 140
Cleaveland, George, 140
Clendenin, Daniel, 139
Clenney, John, 140, 199, 227
Cline, George Philip, 213
Coile, Adam, 270
Cole, Richard, 217
Colt, Samuel, 231
Compton, Alexander, 140, 157
Compton, John, 140, 157
Conklin, Isaiah B., 140, 292
Cook, Charles, 140
Cookman, William, 140, 190, 191, 201
Cooper, Peter, 273
Copley, Lionel (Col.), 218
Cox, Jonathan, 243
Coxe, Tench, 243
Crabb, Thomas, 27-29, 33, 49, 51, 139-141, 255, 270, 273
Cramer, Christina, 141-142
Cramer, Leah, 188
Cramer, Philip, 141-144, 256
Cramer, Margaret, 145
Crandell, John, 141
Crawford, ———, 156
Crawmer, Philip, 141, 142
Creager, Christian, 150
Creamer, Casper, 144
Creamer, Philip, 40-42, 45, 134, 141-145, 256
Cresap, Michael (Captain) 39, 41
Cresap, Thomas (Colonel), 39, 41
Criner, Deloro, 205
Criner, William, 53, 158, 205
Crispin, Silas, 233, 292
Crist, Jacob, 189
Criswell, Elizabeth Ellen Kiger, 145
Criswell, Joseph, 145
Cromwell, Levi, 36, 145
Cromwell, Levi, Son & Co., 145
Cromwell, Oliver, 145
Cromwell, T. Elvin, 145
Crouse, Margaret, 144
Crouse, Sarah, 144, 145
Crouse, Valentine, 145
Crumbaugh, John, 182
Cruzen, Jacob, 145
Curley, James M., 145
Currey, John, 189
Custard, John, 212

Dalhouse, Richard, 146
Dallam (Dillam), Richard, 146, 220
Dann, Underhill, 146, 252
Darnell, Henry (Col.), 161, 218
Davidson, Patrick, 121
Davis, William, 190
Daw, Annie, 146-147
Daw, Charles, 146-147
Daw, Edward, 146-147
Daw, E.J., 146-147
Daw, Ruben, 146-147
Dawson, Tom, 169, 175
Day, George E.H., 292
Dayly, Sam, 181
Deam, Frederick (Capt.), 100
Dean, John (Capt.), 99
Deakins, ———, 276
Deaver, John (Lt.), 98
Dehoff, Jacob, 119
DeLancey, ———, 218
Delawter, George, 214
Demuth, John, 51, 146, 161, 202
Denmead, A. & W. & Son, 105, 147
Denson & Buck, 147
Dern, Susie Koons, 203
Derringer, ———, 232
Deshler, David, 186
Devlin & Co., 107,
Devlin, Daniel, 107
Devlin, Jeremiah, 107
DeWitzleben, Arthur, 147, 292
Dick (Dicked), George, 242
Dickinson, Charles, 273
Dickinson, Charles S., 147-148
Diehl, Anna Belle Hawkins, 170
Digges, William (Col.), 21, 161, 218
Dilman, Henry, 181
Distelbart, George, 33, 148
Dinwiddie, ———, 218
Dittihunt, William, 181
Dixon, Jeremiah, 23
Dobson, Henry (Capt.), 99
Dodge, Philip Tell, 148, 292
Dodge, William C., 148, 292
Doermer, J., 148
Doirsy, Doctor Fred, 182
Dollar, Cath, 148
Dollar, Jos, 148
Dollar, Lawrence, 148
Dorsey, Allen, 148
Dorsey, Daniel (Capt.), 98
Dorsey, John E., 148
Dorsey, Richard (Capt.), 100
Dorsey, Samuel, 162
Dowler, Nancy Ann, 212
Downey, Samuel, 171
Doyle, George, 177
Drake, Matthew, 36, 149
Druck, Conrad, 199
Drum, Peter, 183
Dulaney, Daniel, 252
Dunan, Adolphus (Dr.), 167
Dunbar, George Towers, 108
Dyke, ———, 37
Dyke, Samuel E., 204

Ealer, Catharine, 150
Ealer, Elizabeth, 150
Ealer, Franklin, 150
Ealer, Lewis W., 150
Ealer, Sarah, 150
Ealer, Susan, 150
Ealer, Susanna, 150
Easterley, Christopher, 150
Eastman, Joseph, 150
Eberhard, ———, 150
Ebert, Benjamin, 150
Ebert, Benjamin & Sons, 150
Ebert, John, 150
Eckenrode, Jacob, 123
Edon, John, 150
Edward, C., 139
Edwards, George, 217
Edwin, Marshall, 151
Egg, Durs, 163
Eisenbrandt, Christopher H., 103
Elbert, John, 224
Ellicott, ———, 150
Elliot, David, 244
Elliot, Elizabeth A., 244
Emmitt,William, 119-120, 142
Engle, Barney, 150
Engle, Christian, 150
Engle, Ezra, 150
Engle, Peter, 51

Engle, Peter Jr., 150
Engle(s), Peter Sr., 51, 150
Engle, Susanna, 150
Engles, Anna Marie, 150
Engles, John, 150
Engles, Peter, 150
Ernst, Jacob, 150
Eschbaugh, ————, 213
Escherich & Son, 74
Escherich, A., 69, 76
Escherich, Anton & Son, 151
Escherich, Anton, 36, 151
Escherich, Charles, 151
Esgenbach, Henry, 151
Esherich, Ferdinand, 151
Esherich, Francis, 151
Esterday, Michael, 207
Evans, David, 129
Evatt & Brothers, 151
Evatt, Agnes, 152
Evatt, Columbus, 217
Evatt,Edward I, 151
Evatt, Edward II, 33, 151
Evatt, John, 151
Evatt, Mary, 151
Ewell, Thomas, 91
Ewing, ————, 200
Ewing, Nathaniel (Capt.), 97
Ewing, Thomas (Capt.), 152
Ewing, Thomas (Col.), 97
Eyster, John, 152

Favier, Peter, 153
Feckty, Christian, 181
Feinour, Alfred, 153
Fenner, Albert, 153
Ferguson, Alfred B., 153
Ferguson, Catharine, 153
Ferguson, Elizabeth, 153
Ferguson, Emily, 153
Figely, Peter, 181
Findlay, Robert & Co., 153
Fischach, George, 205
Fischer & Shaefer, 153
Fisher, G. Frederick, 153
Fisher, James, 153 F
Fisher, John, 178
Fitzhugh, William (Col.), 195
Flanagan, Edward R., 124, 142, 246, 255
Flanagan, Helen, 142
Flegel, George, 205
Fleming, Robert, 119
Fletcher, Catharine, 153
Fletcher, H. Clay, 153
Fletcher, Hestor, 153
Fletcher, John L., 153
Fletcher, Stephen G., 153
Fletcher, William F., 153
Fletcher, William G., 153
Flohr, Leonard, 214
Floming, Robert, 120
Flower, Daniel, 153
Fornshell, John, 153
Fort Cumberland, 22
Fort Frederick, 22
Fort McHenry, 33
Foy, Harriet, 154
Foy, James, 36, 154
Foy, James H., 154
Foy, James & Son, 154
Fredericktown Gun Lock Factory, 23-24, 133-134, 154-155, 252
Friederick, Henry, 155
Fritz, Catharine, 170, 176
Friend, Charles, 49, 155
Friend, Gabriel, 49, 155
Friend, Jacob, 49, 155
Fugith, Edward, 155
Fugitt, Charles, 155
Fugitt, James P., 155
Fugitt, Josephine, 155
Fugitt, Mary E., 155
Fugitt, Samuel, 155
Fugitt, Thomas M., 155
Fugitt, William H., 155
Fugitt, Ugene, 155
Fulford, ————(Capt.), 156
Funderlinn, Michael, 144
Furnival, Alexander (Capt.), 100

Gafford, J., 58
Gafford, Joseph, 58, 156
Galloway, Benjamin, 155, 199
Gansevoort, H.S. (Col.), 196
Garrett, Amos, 156
Gary, Elizabeth, 243
Gary, Margaret, 243
Gary, Peter, 243
Gaver, John, 215
Gaver, George, 214
Gaylord, Amos, 181, 182
Geesey, Frederick, 214
Gephart, Peter, 243
Gephart, Rosina, 243
Getter, ————, 213
Gibson, John, 156
Gibson, Thomas, 156, 163
Gibson, Woolman, 156
Gibson, Woolmand, 156
Gilbert, Arlan K., 91
Gill, Martin Gillet, 292
Gillingham, George, 167
Gist, ————(Major), 22, 25, 152
Gist, Mordecai (Col.), 97
Glarps, George, 156, 159, 266
Glatt, Nicholas, 156
Glaze, George W., 54
Gminder, Jacob, 103, 106, 107, 156
Goodrick, Sir Henry, 218
Goodwin, Vaughn K. (Lt. Col.), 169, 180
Gonter, John, 53, 140, 156-157, 161, 227, 268-269
Goppert, George, 158
Goppert, Margaret, 158
Goppert, P., 158
Gordon, John, 134
Gordon, George, 158
Gordon, Herman, 274
Gordon, Mary, 274
Gorney, Joseph, 271
Gough, Hanna E., 227
Gough, Harry D., 227
Gouter, John, 156
Grable, Samuel, 160
Grant, John, 158
Great Wagon Road, 19, 49, 177, 185
Green, Jonas, 257
Greenberry, Nicholas (Col.), 218
Greener, W. W., 158, 210
Greenwalt, Benjamin, 106
Greiner, William, 158-159, 204-205
Grice, John, 216
Griffeth, J.A. & Co., 109
Griffie, Ronnie, 55
Griffith & Griffith, 109
Griffith, Henry, 159
Griffith, John A., 109, 159
Griffith, John A. & Co., 109
Griffith, Kinzey, 159
Griffith, Samuel (Capt.), 98
Grillot, Nellie Marker, 212, 214-215
Gropley, Edward, 156, 159, 266
Grosh, ————(Capt.), 41
Grosh, Conrad, 263, 272
Grove, Catharine, 159
Grove, J., 159
Grover, George, 272
Grumbaugh, John, 181
Gunby, John (Col.), 100
Gunter, William, 142
Gusler, Wallace, 185
Gweeties, Charles, 156, 159, 266
Gwynn, William, 101

Hachen, Elizabeth, 170, 175
Hachen, George, 170, 175
Hachen, John, 170
Hachen, Juliana, 170, 175
Hachen, Nicholas, 170, 175
Haffner, Jacob, 160
Hager, Jonathan, 157, 160, 176-177 205, 236
Hager, Mary, 157
Hagerstown Fire Company, 178
Hahn, George A., 161
Hahn, J. H. F., 109
Hahn, Mary A., 227
Hailer, L., 161, 292
Haimes, William, 21, 161
Halbach & Sons, 61, 161
Halbach, ————, 24
Halbach, Charles, 161
Hall, ————, 24
Hall, John H., 291
Hall, Josias C. (Col.), 97-98
Hammell, Samuel, 53, 157, 161
Hamilton, Hugh, 161
Hamilton, James, 161
Hamilton, Levin (Capt.), 99
Hanson, ————, 199
Hanson, John, 23, 91, 130, 154-155
Hanson, John, Jr., 23-24, 161
Hanson, Walter, 91
Hapwood, David T., 161
Harig, Frederick, 162
Harbourt, Peter, 242
Hariss, ————, 24-25
Hariss, Isaac, 162
Harner, Andrew, 142
Harner, Jacob, 47, 122
Harpers Ferry, 174-175
Harris, Isaac, 25, 38, 162, 254
Harris, James, 146
Harris, Robert (Capt.), 99
Harris, William, 162, 221
Harrison, William Henry (Gen.), 51, 250
Harry, John, 181
Harsh, Conrad, 119
Hartman, John, 181

Hartogensis, Henry S., 162
Hartzler, Daniel D., 180, 188
Harwood, Thomas (Capt.), 97
Haslett, Eliza, 165, 167
Haslett, Emily, 165, 167
Haslett, James, 33, 55, 58, 61, 65-68, 162 -168
Haslett, James Jr., 164, 167
Haslett, Mary, 163, 164, 167
Haslett, Sarah, 165, 167
Haslett, William S., 162
Hatkill, J., 169, 260, 291
Hawken, Ana Long, 185
Hawken, Anna Maria, 188
Hawken, Catherine, 171
Hawken, Catherine Ann, 190-191
Hawken, Christian, 52-53, 57, 176-181, 189, 191, 205
Hawken, Christian Jr., 53, 171, 175, 177, 183, 185, 191
Hawken, Christian Sr., 169-171, 174-176, 180 -181, 183-187, 189, 191-192
Hawken, Christopher M., 175, 186
Hawken, Daniel T., 53, 175, 183
Hawken, David, 177, 180, 185
Hawken, Elizabeth, 171, 176-177
Hawken, Elizabeth Ann, 187
Hawken, George, 53, 174-175, 180-183, 187, 191-192, 198, 252
Hawken, Hazel, 190
Hawken, Jacob, 52-53, 171, 174-177, 180 -182, 185-188, 191
Hawken, Jacob Jr., 186
Hawken, Jake, 169, 186
Hawken, John, 52-53, 174-177, 180, 183, 185, 187, 190-192, 198
Hawken, John George, 170, 176-177, 183
Hawken, John Turney, 175
Hawken, Julia Ann, 186
Hawken, Julian, 180
Hawken, Juliana, 170, 177, 179, 182-183
Hawken, Juliann, 171, 174, 176-177, 180-181, 183-189
Hawken, July Ann, 187
Hawken, Henry, 53, 174-175, 182-183
Hawken, Henry Jr., 175
Hawken, Henry Clay, 170, 175
Hawken, Leah, 174, 189-190
Hawken, Lewis, 187-188
Hawken, Margaret, 171, 180, 189
Hawken, Margareth, 176-177
Hawken, Mary, 190-191
Hawken, Nancy, 191
Hawken, Nancy Ann, 174-175, 177, 180-181
Hawken, Nicholas, 53, 170, 174-176, 182
Hawken, Rosanna, 186-188
Hawken, Rosannah, 171
Hawken, Samuel, 53, 169-171, 180-181, 186, 191
Hawken, Samuel T., 174-175, 177-178, 183, 185-188, 191
Hawken, Sophia, 184
Hawken, Susannah, 174, 183
Hawken, William, 52–53, 140, 170, 174-175, 178, 180-183, 187-191
Hawken, William L., 171, 189
Hawkey, Henry, 182
Hawkey, Nicholas, 176
Hawkin, Christian, 177
Hawkins, Catherine, 184
Hawkins, George, 184
Hawkins, John, 185
Hawkins, Katie, 184
Hawkins, Paul E., 169
Hawkins, Sophia, 184
Hawkins, Thomas, 33
Hawkins, Thomas Paine, 170, 182
Hayes, Elizabeth, 124
Hayes, Thomas, 118, 124
Hays, Thomas,122
Hayward, Bartlett & Co., 105, 192
Hawky, Jno., 198
Heard, Jane, 124
Heard, Samuel, 118, 124
Hedrick, John, 216
Hedrick, Margaret, 216
Heffleigh, Peter, 205
Heffner, Michael, 215
Heiberger, Francis J. & Co., 192
Heiberger, Washington, 109
Heider, Henry, 192
Heinz, Richard, 192
Henry, J.J., 30, 232
Henry, James (Capt.), 99
Herald, Frederick, 33
Herr, John Peter, 179
Herring, John, 214
Hetrick, ————, 37
Heyser, William, 177-178, 205
Hibner, Abraham, 192
Hibner, Barbara, 192
Hibner, John, 192
Hibner, William, 192
Hickey, James, 251
Hill, Edmund, 192
Hill, Jacob, 224
Hindman, Edward (Capt.), 98
Hineline, Thomas, 192
Hiss, George R.A., 192, 201
Hitzelberger, Stephen V., 192
Hobbs, Joseph, 192
Hoblitzell, Jacob, 244
Hochen, Christian, 177
Hockey, Christian, 177-178
Hocky, Nicholas, 176
Hodgson & Co., 193
Hodgson & Thompson, 192
Hodgson, John W., 192
Hoen, John H., 193, 265
Hoepner, Benjamin, 193
Hoepner, Casper B., 113
Hoff, David T., 193
Hoff, John, 157
Hoffenbridel, Lewis, 193
Hoffman, George, 244
Hoffman, Margaret, 244
Hoffsteller, Henry, 270
Hogan, John, 193
Hogg, Thomas M., 155
Hoggins, John, 198
Hoglan, Issac, 193
Hoke, Charles R., 251
Hoke, Frances A., 247
Hoke, Peter, 247
Holiday, Henry, 106
Holland, Francis (Capt.), 97
Hollar, Michael, 270
Holliday, Thomas, 193
Hollingsworth, ————(Col.), 87
Hollingsworth, Henry, 24, 193-196
Hollow, William, 160
Hopkins, Gerrard, 201, 211
Hopkins, Thomas Montgomery, 186
Horst, George, 195
Horstman(n) Brothers & Co., 109-110
Horstman(n), Sigmund H., 109
Horstman(n), William H. & Sons, 109-110
Horstmann, William J., 109
Hough & Clark, 139
House, William C., 195
Howard, Henry, 83, 195
Howard, McHenry, 106
Howard, William, 195
Huges, ————, 195
Hughes, Barnabas, 195
Hughes (Hughs), Daniel (Col.), 83, 87, 195-196, 273
Hughes, James, 121
Hughes, John, 121
Hughes, Samuel, 83, 87, 195-196
Hughey, William, 186
Humrichouse, William W., 190
Humrickhouse, Charles W., 181, 190-191
Hunot, Gabriel, 186
Hunt & Goodwin, 196
Hunt ,William, 218
Hunter, Adam, 219
Hupman, Nicholas A., 196, 201
Hutchins (Hutchings), E. & Co., 196
Hutton, Jonathan, 232
Hyder, Margaret Koons, 203
Hyne, Andrew (Capt.), 99

Jackson, Andrew, 45, 141, 167
Jackson, Samuel, 103
Jacob, David C., 199
Jacobs, A. & Sons, 110
Jacques, L., 273
James, ————(Miss), 118
James, Agnes Armstrong, 45, 119, 124
Jamison, Joseph (Col.), 255
Jenifer, ————, 146
Johnson Iron Furnace, 199
Johnson & Reat, 199
Johnson, ————, 193, 199
Johnson, Baker, 199
Johnson, Bradley T., 106
Johnson, Hannah, 187
Johnson, James, 23, 154-155, 199
Johnson, Joseph, 187
Johnson, Josiah (Capt.), 130
Johnson, Oscar, 140, 199, 227
Johnson, Richard, 199
Johnson, Rodger, 199
Johnson, Thomas, 195, 199
Johnson, Thomas Jr., 130
Jones, ————, 195
Jones, Sarah Jane, 241
Joseph & Andrew, 199
Jotul, Carl, 199

Kaidel, ————, 36
Kaidel, Henry, 200
Kaleb, John M., & Co., 142, 144, 255
Kane, George (Police Marshall), 104, 106, 148
Katz, Jacob, 157
Kauffman, ————, 37
Kauffman, Henry J., 204

Kausler, John, 183
Kayser (Kayson), John C., 200
Kealhofer, Henry, 157
Keener, John I, 200
Keener, John II, 200
Keener, George P., 200
Keener, Margaret, 200
Keener, Peter, 200-201, 219
Keener, Samuel, 25, 152, 200
Keller, Joseph W., 140, 190-191, 201
Kennedy, Thomas, 207
Kenner, Alexander, 201
Kenner, George, 201
Kenner, James, 201
Kenner, Jane Robison, 201
Kenner, Maria, 201
Kenster, John, 201
Kettner, Francis G., 201
Key, John Ross (Major), 203
Key, Francis Scott, 203
Kienborts, Daniel, 183
Kindell, ————, 196
Kindig, Joe, 37, 203, 229
King & Hiss, 192, 201
King & Hupman, 196, 201
King, Andrew L., 201
King, Henry, 201
King, Henry S., 201
King, John, 181, 201-202
King, Nicholas, 49, 202, 270
Kirk, E. Clarence, 202, 292
Kirk, Samuel & Sons, 202
Kirselner, Elizabeth, 160
Kirshner, Jacob, 181
Kiry, John, 182
Kleibacker, Christian B., 202, 261
Kleinhenn & Dudley, 202
Kleinhenn, August, 202
Kloke, Anna, 202
Kloke, Anton, 202
Kloke, Catharine, 202
Kloke, Eliza, 202
Kloke, John, 202
Kloke, Mary, 202
Knapp, Frances, 268
Knel, Philip, 202
Knight, J.W., 202
Knowles, James A., 131
Koch, George, 147, 202
Koch, Henry, 51, 147, 202
Koons, Abraham, 203
Koons, Catharine, 203
Koons, Elizabeth, 203
Koons, George, 203
Koons, George, 203
Koons, Henry, 202-203
Koons, Henry Jr., 203
Koons, Issac, 203
Koons, Jacob, 203
Koons (Koontz), John, 203
Koons, Joseph, 203
Koons, Margaret, 203
Koons, Mary, 203
Koons, Peter, 203
Koons, Susie, 203
Koons, William, 203
Kraft, Catherine, 204
Kraft, Jacob, 204, 236
Krauspe, Julius, 204
Krebs, George, 205
Krell, John, 204
Kreps, George, 52-53, 136, 158-159, 178, 206-207
Kreps, George Jr., 53, 204, 207
Kreps, George Sr., 204-205
Kreps, John, 207
Kreps, Martin, 205
Kreps, William, 157, 207
Kuhn, Henry, 33
Kuntz (Kunz), Jacob, 203
Kussmaul, William F., 292

Lakenan, ————, 186
Latrobe, Ferdinond C., 71, 209, 221, 260
Latrobe, John H.B., 209, 291
Lauderman, William H., 33, 209
Laurwing, Dermenico, 209
Lavin, M.E. & Co., 209
Lawrence, Upton, 179
Laypole, Ann C., 209
Laypole, Eliza B., 209
Laypole, John H., 209
Laypole, Julia A., 209
Laypole, Martha, 209
Laypole, Mary, 209
Laypole, Susan A., 209
Laypole, Theopholes, 209
Leatherman, Henry, 215
Leavenworth & Scoville, 111
Leavenworth, Hayden & Scoville, 111
Lee, David, 209
Lee, Robert E. (Gen.), 106
Lemmer, Frederick, 209
Lenz, Michael, 165
Lepper, Henry G., 209
Lettigg, Peter, 219
Levy, Charles J.D., 247
Levy, Charles V.S., 247
Lewis,————, 218
Lewis, Washington, 209
Lewis, William, 181, 217-218
Libig, William, 209
Liddle, Robert, 209
Lilley, M.C., 107
Lilley, M.C. & Co., 110
Lincoln, Abraham (Pres.), 148
Lindell, John, 183
Lindenberger, George, 91
Lindenburg, Henry, 110
Linebaugh, Polly, 245
Linthicum, William H., 158, 210
Lippy, David, 181
Litsinger, Dorothy, 266
Littig, Peter, 211
Little, John, 210
Lloyd, Edward (Gov.), 61, 164-165, 167
Lock, Nathaniel, 103
Locke, William M., 168
Lockwood, C.L. & Co., 210
Loenyi, Albert, 151
Lolliger, John, 210
London, Francis, 181-182
Loney, Anne E., 210
Loney, Francis B., 210, 252
Long, Ana, 185
Long, Daniel, 181
Lord Baltimore III, 21, 217
Lords of Trace, 22
Louden, Samuel, 117, 123-124
Loudonn, Samuel, 121
Low, Andrew, 224
Lowdon (Lowden), Samuel, 121-122
Lowe, Andrew, 270
Lowe, Nicholas, 217
Lowry, ————(Major), 101
Lowry, Harrison, 216
Lowrides, Christopher, 134
Ludke, Annie, 210
Ludke, Elizabeth, 210
Ludke, Francis Joseph, 210
Ludke, Joseph, 210
Ludke, Mary, 210
Lunerick, J.A., 110
Lux, ————, 195
Lydick, Philip, 211
Lydick, Peter, 25, 152, 210-211
Lynch, John, 80, 211
Lynch, John (Capt.), 99

Maedel, Adolf, 212
Maedel, Charles B., 212
Manahan, Timothy, 205
Manley, David, 212
Manson, Anna, 124
Manson, James, 118, 124
Manton, John, 163, 167
Mantz, Francis, 138, 141
Marbury, Joseph (Capt.), 98
March, Morris, 212
Marker, Amelia, 215-216
Marker, Barbara, 213
Marker, Catharine, 213
Marker, Christina Beckenbaugh, 215
Marker, Daniel, 214-216
Marker, Daniel Jr., 212-213
Marker, Daniel Sr., 50-51, 212-213
Marker, Elizabeth, 213, 215-216
Marker, Ezra, 213, 215-216
Marker, Franklin, 213, 215-216
Marker, George, 212, 216
Marker, George (I), 214
Marker, George (II), 214
Marker, George Jr., 51, 213-215
Marker, Jacob, 50, 213, 215-216
Marker, James, 36, 50-51, 77, 212-213, 215-217
Marker, John, 213
Marker, Joshua, 213
Marker, Julia, 213, 215
Marker, Louisa, 213, 215
Marker, Mary Ann, 213
Marker, Michael, 213
Marker, Paul, 50-51, 212-214, 216-217
Marker, Peter, 213, 215
Marker, Sarah Ann, 213, 215-216
Marker, Susan, 213
Marker, William, 213, 215-216
Marks, Andrew, 217
Marks, Jacob, 217
Marriorr & Barstow, 217
Marriott, J.H. Wilson, 129, 217
Marsh, Samuel W., 217, 291
Marshall, Edwin, 217
Martin, John, 21, 217
Martin, Susanna, 179
Martin, Thomas, 179
Maryland Arms Purchases, 217

Maryland Council of Safety, 17, 21, 23-26, 36, 49, 83, 87, 91, 130, 133-135, 137, 138-139, 141, 146, 150, 152-156, 158, 161-162, 193-196, 199-201, 210-211, 217-220, 229, 231, 236, 252-254, 256, 258, 260-261, 263, 265-270, 273-274, 276
Maryland Line, 95, 104
Maryland Sportsman Supply Co., 219
Maslin, Michael M., 61, 220, 228
Mason, Charles, 23
Mason, John, 141, 207, 220, 243
Massachusetts Arms Co., 71
Matel, Adelfenia, 212
Matel, Adeline, 212
Matel, Atolf, 212
Matel, Carolina, 212
Matel, Charles, 212
Matel, Rosater, 212
Matthews, ———, 195, 272
Matthews, George, 83, 220
Mattocks, George, 101
Maxwell, Frances A., 247
Maxwell, James (Capt.), 97
May, James, 146, 220
Mayfield, John, 220-221
Mayfield, M., 221
Mayfield, S., 221
Mayfield, Thomas, 221
Maynard, Edward, 73, 221, 291-292
McAfee, Robert, 183
McAinn (Barnard & McAinn), 227
McAkron, Elizabeth, 245
McClellan, John, 91
McComas, Alexander, 55, 60, 70, 140, 199
McComas, Harry Gough, 227
McComas, Jno.Wm., 102
McComas, Mary, 140, 199
McComas, Nicholas A., 227
McComas, Preston, 227
McCormick, Robert, 156, 163, 165
McCreary, James, 121
McIntosh, William, 53
McIntoch, William, 157
McIntoch, William Jr., 227
McKaleb, John, 255
McKean, ———(Gov.), 164-165
McKelly, John, 203
McKennsy, James H., 111
McKim & Brother, 61 , 228
McKim & Maslin, 228
McKim, William D., 61, 228
McLaughlin, Mathew, 228
McNair, Samuel, 121-122
Meigs, Josiah V., 73, 221, 291-292
Melcher, Edward O., 221
Melcher, Harman, 221
Melchoir, Edward, 221
Melchoir, Henriette, 221
Melchoir, Nathaniel, 221
Merchant, Nicholas Lowe, 21
Meredith, Benjamin, 33, 221
Meredith, Joseph H., 107, 137
Merrill, James H., 71, 103, 209, 221-222, 260, 291-292
Merrill, Jeremiah L., 222
Merrill, Latrobe & Thomas, 71, 209, 222
Merrill, William Y., 222
Merrill Patent Fire Arms Manufacturing Co., 71, 260, 291-292
Messersmith, ———, 200
Messersmith, Sam, 25
Metz, Philipipina, 224
Metzger, Christina, 51, 224
Metzger, Jacob, 27-29, 49, 51, 139-140, 255, 259, 265, 270, 273
Metzger, Jacob, Jr., 51, 129, 224
Mey, James, 219
Meyer, Andrew, 167
Meyer, John C., J., 74, 224
Michael, Bertha, 224
Michael, Henry, 224
Michael, Josephine, 224
Michael, Virginia, 224
Middlekauff, Daniel, 269
Mier, Jacob, 51
Mier, Samuel, 51
Miles, ———, 163
Military History Department of the U. S. Army, 15
Miller, Charles W., 83
Miller, Daniel, 178
Miller, David, 178
Miller, Jacob, 181
Miller, John, 178, 242
Miller, Julian, 180
Miller, Margaret, 180, 191
Miller, Robert, 180
Missersmith, Sam, 25, 152
Missouri Historical Society, 145, 171
Mohler, Mary, 212-213
Monocacy Road, 17, 41
Moore, Thomas, 166
Moran, Edmn, 155
Morgan, Daniel (Capt.), 19, 41
Moritz & Kaidel, 200
Moritz, ———, 36
Moritz, J. D., 200
Mortimer, H. W., 163
Mosby, John Singleton (Lt. Col.), 196
Motter Maxwell & Co., 247
Motter, ———, 249
Motter, Joshua, 123, 247
Motter, J. Taylor, 247
Mount Pleasant Furnace, 195
Moyer, Jacob, 181
Mt. Etna Furnace, 83, 195
Murphy, John D., 133
Murray, Alexander (Capt.), 97
Myers, A., 291
Myers, Jacob, 181
Myers, John H., 214

Naeff, Amelia, 212-213
Naval Historical Foundation, 15
Navigation Acts, 23
Neal, W. Keith, 168
Neff, P., 132, 228
Neill, Alexander, 180-181, 189, 191
Nelson, ———(Gen.), 33
Nelson, R., 243
Netz, John, 228
New York Clothing House, 110, 228
Nestle, Frederick, 74, 77, 228
Nestle, Josephine, 228
Nichol & George, 228
Nicholson(s), ———(Col.), 79, 150
Nickerson, Charles V., 71, 228, 291
Nightingale, Samuel, 132, 229
Nippes & Co., 26, 31
Nock, Joseph, 229
Norris & Beatty, 131, 229
Norris & Brother, 229
Norris, Richard & Sons, 229
Norris, Richard Sr., 229
Northampton Furnace, 83, 195, 229
Norwood, Edward (Capt.), 97
Nottingham Company Furnace, 229
Noxes, Richard (Mrs.), 224
Null, Rachel, 134, 142
Nunemaker, Jane, 247
Nunemaker, George, 229
Nunemaker, Philip, 247
Nusz, F.C. (Mrs.), 224

O'Bannon, Presly, 199
Oberteuffer, Adila, 230
Oberteuffer, Charles A., 230
Oberteuffer, Mary, 230
Oberteuffer, Ulrich, 230
O'Brien, Charles C.C., 55, 230
Oehm & Co., 80
Oehm, Charles N., 110, 230
Oehm, Franz F., 110
Oehm's Acme Hall, 230
Oehms Acme Hall Clothing House, 110
Oehms, Charles N. & Bro., 110
Office of Indian Trade, 53, 243, 275
Ogborn, Mary Koons, 203
Oldham, Edward (Capt.), 98
Oldwine, Barney, 181
Oliver, John, 230
Ordnance Department, 71
Orms, Archibald, 134
Orr, Robert C., 55
Oster, Daniel, 181
Oster, Rosanna, 174, 187
Ott, Adam, 178
Ott, Mary Ann, 203
Owings, Aleysius, 211

Page, George & Company, 105, 231
Pagenhardt, C.A., 132
Paine, H.E. (Maj.Gen.), 80
Park, John, 231
Parker, John, 231
Parkhurst, ———, 231
Parsons, Hiram, 231
Paschall, Stephen, 79
Patterson, Betsy, 65
Patterson, C. Meade, 167
Patterson, William P., 231
Payne, E. Scott, 231
Peabody, John J., 231
Peale, Rembrandt, 129
Perry, Roger, 243
Petry, Philip, 171
Pettibone,Manufacturing Co., 110-111
Pettibone, James, 110
Phipps, John, 232
Pierson, ———, 232
Pilkerton, Thomas, 26
Pikesville Armory, 232
Piper, George, 45, 47, 119-120, 232
Piper, Peter, 232
Pipino, Jacob, 74, 192, 232-233
Pipino, John L., 233

Pipino, Lewis, 233
Pippert, Carl, 55
Pirtern, Elenor C., 161
Poe, David, 260
Poffenberger, Andrew, 206
Polk, James K., 251
Polk, William S., 71
Popplen, G*N Jr., 233
Porter, Abel & Co., 111
Porter, Peter B., 220
Porter, William, 120
Poultney & Trimble, 104, 202, 233, 256, 261
Poultney, Thomas, 71, 233, 291-292
Powles, Daniel, 206
Prather, Thomas (Col.), 155
Pratt, William, 71-72, 233
President of the United States of Congress Assembled, 23
Price, Thomas (Capt.), 39, 41
Price, ———(Major), 154
Price, Thomas (Col.), 97
Purucker, John W., 233
Purvis, N.Y., 111
Pyes, ———(Col.), 161

Quinnebaug Rifle Co., 196

Ragan, ———(Col.), 166, 179, 187
Randell, Richard H., 162
Reasor, Jacob, 219
Recontre, Antoine, 186
Reddick, ———, 200
Reed, Robert, 24
Reid, Patrick, 121
Reid, Robert, 33, 194
Reinewald, Charles E., 251
Reinhardt, Charles, 103
Reisner, William Jr., 180
Rentch, Peter, 160
Reser, Peter, 236
Resor, Catherine, 204, 236
Resor, Elizabeth, 236
Resor, Jacob, 204, 236
Resor, Johan, 236
Resor, Peter, 204, 236, 238
Ressel, Arthur J., 169
Retner & Thomas, 236
Rettberg, Jacob, 236
Rettberg, Lewis, 236
Rettberg, Louisa, 236
Rettberg, Rosa, 236
Reuter, Andrew, 236
Reynall, Richard, 238
Rezor, Peter, 236
Rice, Eliza, 251
Rice, Joseph, 129
Richards, A.E. (Lt.), 196
Richards, Jacob, 238
Richardson, W.H., 91
Richardson, William (Col.), 97, 99
Ridabock & Co., 111
Ridabock, Henry C., 111
Ridabock, James H., 111
Riddick, ———, 200
Ridenour, Adam, 179
Ridenour, Cepi Henry, 179
Ridenour, Issac, 181
Ridenour, Nicholas, 174
Ridgely, Charles, 229
Ridgely, Henry, 238
Ridgely, Henry (Capt.), 98
Ridgen, Thomas, 201
Riechter, Peter, 236, 238
Riehl, Jacob, 224
Riggs & Griffith, 238
Riggs, Amelia Dorsey, 238
Riggs, George W., 238
Riley, John W., 238
Rinehart, Thomas, 157
Ritchie, Hamilton, 171, 188
Ritchey, Martha, 174, 188
Rizer Family, 238
Rizer, Albert A., 244
Rizer, Anthony, 238
Rizer, Catharine, 241, 244
Rizer, Christina, 241
Rizer, Eliza, 241
Rizer, Florence, 244
Rizer, George, 241, 245, 262
Rizer, George (Sr.), 55, 238
Rizer, George Hoffman, 244
Rizer, George Huth, 238
Rizer, George Washington, 238
Rizer, Jacob, 55, 244-245
Rizer, Martin L., 54, 238, 241, 243-244
Rizer (Riser), Martin Jr., 241-243
Rizer, Martin Sr., 238, 241-242
Rizer, Martin Jr. (III), 244
Rizer, Mary Jane, 244
Rizer, Mary Magdalena, 241-243
Rizer, Mathias, 241
Rizer, Matthias (I), 238, 245
Rizer, Oliver Mortimer, 238
Rizer, Philip, 241
Rizer, Phillip, 55, 245
Rizer, Polly, 241
Rizer, Rosanna, 244
Rizer, William L., 55, 245
Robertson, Charles, 121
Roesser, Mathias, 236
Roft, William, 232
Rohrer, Jacob, 236
Rohrer, Samuel, 179
Rose, Adam L., 245
Rose, Ann E., 245
Rose, Marah C., 245
Rose, Mark, 245
Rose, Mary, 245
Rosenfeld Brothers, 111, 245
Rosenfeld & Co., 111, 245
Rosenfeld, Goody, 245
Rosenfeld, Michael, 245
Ross, Charles W., 247
Roth, Christian, 246
Roth, Frederick, 246
Roth, Heinrich, 246
Roth, Henry (Capt.), 246
Roth, Henry Sr., 51, 245
Roth, John, 246
Roth, Peter, 246
Roth, Ulrich, 246
Rothanheafer, Lawrence J., 246
Rothanheofer, L., 246
Rowe, Albert S., 246, 248, 251
Rowe, Ann Eliza, 247
Rowe, Arthur, 248
Rowe, Catherine A., 247-248
Rowe, Charles C., 246, 248
Rowe, Charles F., 247
Rowe, Daniel, 246-247, 251
Rowe, E. Kate, 246
Rowe, Edward H., 251
Rowe, Elias S., 247
Rowe, Elizabeth, 246-248, 251
Rowe, Emma K., 248
Rowe, Emma R., 247
Rowe, George, 247-248
Rowe, Helen J., 246-248, 251
Rowe, Howard M., 251
Rowe, Jainey, 247
Rowe, James A., 247
Rowe, Jemey E., 247
Rowe, Joshua, 251
Rowe, M. Frank, 251
Rowe, Michael, 248
Rowe, Nathaniel, 45-47, 57, 118, 123, 246-250
Rowe, Quincey Edgar, 246, 248, 251
Rowe, Samuel L., 251
Rowe, Sarah, 247
Rowe, Susan, 246, 251
Rowe,William, 251
Ruckles, Thomas C., 90, 94, 101
Ruggles, A., 196
Rumsey, ———, 156, 193
Rupp, U.S.G., 251
Russel, William, 251
Rutherford, Thomas, 275

Sanderson, Peggy, 253
Savin, T. S., 252
Saw, John, 254
Saylor, Mary, 243-244
Schaeffer & Loney, 210, 252
Scharffe, Augustus, 252
Scharffe, G., 252
Scheckels, J.W., 252
Schilling, A., 252
Schley, George Jacob, 252-253
Schley, Jacob, 154, 219, 252-253
Schley, John, 253
Schmidt, ———, 103, 253
Schnebly, John, 181
Schneider, C.G., 253
Schroder, Henry, 253
Schroder, Maryl, 253
Schroeder, Polly, 253
Schroeder, Valentine, 253
Schroter, George, 253
Schroyer (Schreyer,Schreyor), George, 40-41, 45
Schryer (Shreyer), George, 45, 55, 253
Schumachur, Conrad, 253
Schuylkill Arsenal, 49
Schwartz, Rachael S., 169
Scott, Glenn R., 169, 185
Scott, Mary Juanita Weiker, 169
Scoville & Co., 111
Scoville & Buckingham, 111
Scoville, J.M.L. & W.H., 111
Scoville Manufacturing Co., 107, 111
Scritchfield, Joshua, 242
Sears, William, 253
Seawright, John, 244
Sebby, James, 254
Seeger, Jacob, 101
Selby, Emma R., 254
Selby, Charles W., 254
Selby, James D., 254

Selby, Mary, 254
Self, Bradley, 254
Self, Rezin, 141
Sell, Jacob, 232
Sell, Jacob (Jr.), 47
Sellers, Charles E., 251
Sellers, Helan E., 251
Sellers, Robert Rowe, 251
Sellers, William H. (Mrs.), 251
Sellman, Jonathan (Capt.), 98
Sex, John, 203
Shafer, Carl W., 83
Shaffer, Baltzer, 254
Shank, Susannah, 203
Sharpe, Horatio (Gov.), 22, 218-219
Shaver, George, 176
Shaw & Burnard, 254
Shaw, John, 25, 30, 137-138, 261
Shaw, William, 33, 254
Shecter, Wintell, 176
Shee, ———(Gen.), 163
Sheetz, Frederick, 185
Sheetz, Philip, 275
Shellman, John, 254
Shellman, Lewis E., 254
Shellman, William, 224
Sherick, George C., 255
Shields, Barnet, 214
Shinn, Albert, 244
Shinn, Howard, 244
Shinn, George, 244
Shinn, Martha, 244
Shinn, Mary Jane, 244
Shnyder, Charles, 255
Shorb, John, 203
Shobel, J.G., 255
Shot, 255
Shriner, Sarah Ann, 213
Shrock, Henry, 263
Shroyer (Schroyer, Sroyer), Mathias, 27, 45, 139, 255
Shroyer, Susannah, 203
Shryock, George (Capt.), 187
Shulz, Daniel, 256
Shumway, George, 169, 246
Shunk, Issac, 134, 142, 256
Shunk, Peter, 142, 256
Siebert, John, 110
Silge, Albert, 256
Simmonds (Simmons), Issac, 256
Simmons, James W.L., 256
Sinsberry, R., 252
Sisco Brothers, 106, 256
Sisco, Charles T., 256
Sisco, J. Edward, 256
Skinner, John, 229
Sleghel, Gottlieb, 256
Slicer, Henry, 227
Small, Jacob, 166
Smallwood, ———(Col.), 95
Smith, Gilbert, 71
Smith, John, 189
Smith John Donnell (Capt.), 106
Smith, John M., 256
Smith, Nathaniel (Capt.), 100
Smith, Peter, 181
Smith, Samuel E., 168
Smith, Sophia, 171
Smith, William D., 256
Smithsonian Institution, 15
Smyth, Thomas (Capt.), 97
Snebely, Henry, 263
Sneider, ———, 60
Sneider, Charles Edward, 70-71, 233, 256, 291-292
Sneider, Charles Louis, 139, 257
Sneider, Charles W., 77, 257
Sneider, Edward, 202, 256
Snyder, Leonard, 257
Somes, Daniel E., 257, 292
Sparrow, Thomas, 78-79, 257-258
Spencer, Joseph, 202
Spiker, Benjamin, 258
Spiller, Edward B., 258
Spong, Ervin, 216
Spong, Matthias, 216
Spycher, Peter, 134
Sroyer, ———, 27
Sroyer, Mathias, 255-256
St. Clair, John, 22
Stabler, Edward, 258
Stamp Act, 23
Stansbury, ———(Gen.), 166
Stanford, George, 55
Steam cannon, 84-85, 291
Steel, David, 206
Steele & Johnson, 107, 112
Steele & Johnson Button Co., 112
Steele & Johnson Manufacturing Co., 112
Steiner, Christian, 154
Stephen, Adam, 242
Steuart, George H. (Gen.), 106
Steuart, Richard D., 168
Stewart, John, 55, 165, 167, 258
Stimmell, Catharine Koons, 203
Stinson, ———(Capt.), 41
Stoddert, Benjamin, 87
Stodest, David, 86-87
Stokes, James, 212
Stokes, John, 258
Stone, John H. (Col.), 97
Stonesifer, Christian Emory, 36, 51, 258
Stonesifer, Emory, 259
Stonesifer, Jacob, 259
Stonesifer, John, 51, 54, 258-259
Stonesifer, Laura, 259
Stonesifer, Mary, 259
Stonesifer, Robert, 259
Stottlemire, Jacob, 215
Stover, Frederick, 259
Stowell, A. Jr., 106
Stricker, George (Capt.), 154, 276
Strobel, John G., 259
Strohfeldt, Thomas A., 244
Strong, Samuel, 73, 259, 291-292
Strum, Margaret, 213-214
Stull, Daniel, 157
Stultz, Conrad, 203
Sullivan, Albert M.,Jr., 47, 123
Sung, Phil, 79
Sutten (Sutton), Catherine, 212-213
Swope, Jacob, 171

Tauber, John, 260
Taylor, James, 247, 260
Tecumseah, 250
Ten Stars, 112
Teryar, William, 260
Thomas, Lewen W., 209, 221, 260
Thomas, Philip E., 209, 260
Thomas, Phillip E., 71, 221
Thomas, William, 260
Thomiz, Andrew, 260
Thomiz, Caroline, 260
Thomiz, Christinia, 260
Thomiz, Hellon, 260
Thomiz, James, 260
Thomiz, John, 260
Thomiz, Margaret, 260
Thompson, Elizabeth, 27, 255
Thornton, W., 169, 260, 291
Thralls, Arthur A., 241
Thralls, George R., 241
Thralls, Hetty Ann, 241
Thralls, Maria Jane, 241
Tickehoof, Fred, 182
Tileston, William, 260, 292
Tilgman, Mathew, 202
Tilson, William, 183
Timmonds (Timmins), Edward, 261
Toulson, Alexander, 21, 261
Toulson, Mary, 261
Trade Acts, 23
Trail, Charles (Col.), 255
Trimble & Kleibacker, 233
Trimble, David B., 60, 71, 202, 233, 261
Trimble, Hester, 261
Trimble, Issac R., 104
Trimble, Josephine, 261
Trimble, Mary A., 261
Trimble, Mary Jane, 261
Trimble, Robert M., 261
Trimble, Thomas J., 261
Tripp, ———(Capt.), 161
Troutman, Ben, 48, 51, 262
Troxal (Troxel), John, 120-121
Troxel, Sarah J., 247
Trucks, John (Capt.), 203
Turner, Margaret J., 167
Turner, Zephaniah, 133
Turney, Daniel, 182-183
Turney, Susannah, 182
Tyeryar, Frank, 262
Tyeryar, Frederick F., 262
Tyeryar, Mary E., 262
Tyeryar, William, 262
Tyler, John, 250
Tyler, John (Dr.), 141

Ulrich, Catherine, 246
Ulrich, Elizabeth, 246
Unseld (Unsell), John, 25, 263, 276
Unsell, Abraham, 263
Unsell, Henry, 263
Unselt, John, 263
U.S. Navy Department, 87
U.S.S. Constellation, 87

Vanbibber, ———, 25, 152
Vansant, J.A., 264
Vrebs, George, 160
Van Buren, Martin, 250-251
Van Swearinger, ———Jr., 155
Van Tromp, John, 264
Von Kapff, J. Herman, 264-265

Wachler, Michael, 215
Wachter, George, 215

Wachter, Michael, 214
Wagner, Jacob, 265
Walbach, John J.B., 265, 292
Walraven, John, 165, 265
Walsh, Hoen & Von Kapff, 193, 264-265
Walsh, T. Yates, 265
Walters, Peter, 129, 224, 265, 270
Walterhaver, Christ, 265
Warfield, Azel M., 265
Warner, Andrew Ellicott, 265
Warnsfelts, Mary Polly, 213
Warrenfeltz, Elizabeth, 213, 216
Washington, George, 22, 27, 161, 194
Wassman, George Frederick, 266
Wasserman, George F., 266
Waterbury Button Co., 107, 112
Waterbury Jewelry Co., 112
Waters, Elizabeth, 195
Waters, Jonathan, 232
Waters, Richard (Col.), 30
Watson, Issac, 179
Watson, James, 183
Watson, John, 266
Watson, William, 183
Watts, William Banks, 252
Watts, William L., 266
Weaner, Catharine, 266
Weaner, Charles, 266
Weaner, John H., 156, 159, 266
Weaner, Sophia, 266
Webb, William, 250
Wehaw, Henry, 270
Wells, Daniel, 227
Welner, Henry, 266
Weltner, Ludwick, 252
Welty, Joseph, 189
Werner, J.H.T., 266
Wertebecker, Elizabeth, 213
Wertsner, C.S. & Co., 266
West, Haniah, 266
West, Stephen, 24, 266-268, 276
Westen, John, 229
Wheat, Thomas, 275
Whetcroft, William, 25, 27
Whetcroft, William Sr., 268
Whetstone, Cary Ann, 268-269
Whetstone, David, 53, 268-269
Whetstone, John, 53, 157
Whetstone, John (I), 268-269
Whetstone, John (II), 268-269
Whiddon, Oliver, 158, 269
White, John, 47, 271
White, Nicholas, 27-29, 49, 129, 139-140, 201-202, 211, 219, 224, 255, 265, 269-271, 273
White, Peter, 47, 49, 269, 271
Wibb, Daniel, 176
Wickham, M.T. & Co., 49
Wickham, Margaret, 119, 272
Wickham, Marine Tiler (Tyler), 45-46, 49, 119, 271-272
Wickham, Matilda, 272
Wilks, Thomas, 272
Will, William C., 272
Willey, Naomi, 211
Williams, Eli, 27, 139-140, 224, 255, 270, 273, 275
Williams, Elie, 272-273
Williams, Elisha, 272
Williams, Elisha O. (Capt.), 100
Williams, Henry, 119
Williams, Otho H. (Col.), 99, 179
Williams, T.J.C., 33
Williamson, James, 232
Willick, ————, 103
Willis, Levin C., 171
Willix, William, 273
Wilson, Charles E., 273
Wilson, John M., 273
Wilson, Lewis P., 273
Wilson, Rebecca, 273
Wilson, Rutledge, 273
Wilson, Thomas, 167
Wilson, Woodrow, 251
Winan, Ross, 105, 148, 273, 292
Winans, William Louis, 273
Winchester, Oliver J., 274
Winder, Levin (Capt.), 97
Winter & Reid, 193
Winter, Eliza Francis, 274
Winter, William, 274
Winters, ————, 154
Winters, Elijah, 274
Winters, Elisha, 154, 194, 274
Wintkle, Ellen, 274
Wintkle, Susan A., 274
Wintkle, Washington, 274
Wolf, Elmer P., 55, 274
Wolf, Herbert, 274
Wolf, Mary Gordon, 274
Wood, Ellen, 123
Wood, James, 123
Wood, Lenuel S., 274
Woodward, Edward, 274
Worley, Susannah, 275
Worley, Thomas, 275
Wren, Tabitha Ann,
Wright, Elizabeth, 244
Wright, Robert, 61, 165, 167

Yantis, John, 276
Yate, Thomas (Capt.), 99
Yost, David G., 180
Yost, Henry, 263, 276
Yost, John, 23-24, 83, 130, 154, 201, 219
Young, C.E., 276
Young, Frederick, 152, 277
Young, William, 202
Yourtee, Annie, 199
Youst, John, 276

Zeek, Dietrick, 247
Zeek, Mary, 247
Zimmerman, Ezra A., 247
Zimmerman, John, 123-124

The Longrifle Series

Among the many antique objects which in their totality hold America's heritage in decorative art, none surpass the longrifle in originality, sophistication of design, and skill in execution. For this reason the American longrifle deserves to be studied in great detail and the results of these studies need to be available in a convenient and attractive form to all who care to learn about them.

This series has been created as a means of integrating the important studies in this field and of publishing them with uniform format. Research on the longrifle is far from complete, and thus it will be that over the years this series will grow as new data is brought to light and as new studies are completed.

Within the broad field of the rifle, the art work upon it, the gunsmiths who made it, the history and evolution of it, the mechanics of making it, the mechanics of using it, and the sociological significance of it both as a tool and as an object of artistic endeavor and appreciation, there are many areas of study still available for the attention of serious students. The publisher urges those who might care to pursue such work to contact him.